PLANT SERVICES AND OPERATIONS HANDBOOK

Other Books and Handbooks of Interest from McGraw-Hill

Kohan and Spring • BOILER OPERATOR'S GUIDE

Woodruff, Lammers, and Lammers • STEAM-PLANT OPERATION

Elliot • STANDARD HANDBOOK OF POWERPLANT ENGINEERING

Cedarleaf • PLANT LAYOUT AND FLOW IMPROVEMENT

Rosaler • STANDARD HANDBOOK OF PLANT ENGINEERING

Hodson • MAYNARD'S INDUSTRIAL ENGINEERING HANDBOOK

Mulcahy • WAREHOUSE DISTRIBUTION AND OPERATIONS HANDBOOK

Wrennall and Lee • HANDBOOK OF COMMERCIAL AND INDUSTRIAL FACILITIES MANAGEMENT

Nayyar • PIPING HANDBOOK

PLANT SERVICES AND OPERATIONS HANDBOOK

Anthony Lawrence Kohan, BME, P.E., CSP

Member, American Society of Mechanical Engineers;
Certified Safety Professional; Member, National Society
of Professional Engineers; Member, American Welding
Society; Comissioned National Board Inspector

McGraw-Hill, Inc.

New York   San Francisco   Washington, D.C.   Auckland   Bogotá
Caracas   Lisbon   London   Madrid   Mexico City   Milan
Montreal   New Delhi   San Juan   Singapore
Sydney   Tokyo   Toronto

Library of Congress Cataloging-in-Publication Data

Kohan, Anthony Lawrence.
    Plant services and operations handbook / Anthony Lawrence Kohan.
        p.   cm.
    Includes index.
    ISBN 0-07-035940-7
    1. Plant engineering—Handbooks, manuals, etc.   2. Plant maintenance—
Handbooks, manuals, etc.   I. Title.
TS184.K64   1995
658.2—dc20   94-48612
                                                                      CIP

1 2 3 4 5 6 7 8 9 0   DOC/DOC   9 0 9 8 7 6 5

ISBN 0-07-035940-7

*The sponsoring editor for this book was Robert W. Hauserman, the
editing supervisor was Nancy Young, and the production supervisor
was Donald F. Schmidt. This book was set in Times Roman by
Renee Lipton of McGraw-Hill's Professional Book Group
composition unit.*

Printed and bound by R. R. Donnelley & Sons Company.

This book is printed on acid-free paper.

McGraw-Hill books are available at special quantity discounts to
use as premiums and sales promotions, or for use in corporate training programs. For more information, please write to the Director of
Special Sales, McGraw-Hill, Inc., 11 West 19th Street, New York,
NY 10011. Or contact your local bookstore.

For the many persons who operate, test, inspect, maintain, and repair the vital facility services and associated equipment that are so essential for keeping the diverse commercial, institutional, and industrial plants operating safely, efficiently, and continuously.

CONTENTS

PREFACE

Plant services that have been traditionally provided for commercial, institutional, and industrial properties require plant service operators to have a broad knowledge of heating systems; water supply and disposal of waste; compressed air systems; high-pressure steam systems for process use or power generation; prime movers, such as steam engines and turbines and internal combustion engines and turbines; electrical circuits and power supply; motors, generators, pumps, fans and blowers, transformers, rectifiers, and lighting systems; refrigeration and air conditioning; and maintenance of the physical plant. This book addresses these traditional functions of facility service operation, which may be applicable in diverse occupancies.

The service function for facility operation has been expanded to include recently enacted governmental regulations that can have an impact on how the services are operated and maintained, such as the Clean Air Act, Clean Water Act, Hazardous Wastes, Spills and Disposal, PCB regulation, and regulation of toxic substances in soil and water. Chapter 1, which is about these environmental regulations, has been included primarily to familiarize the reader with them.

OSHA safety regulations must be applied to the workplace, and facility service and operational procedures must maintain a safe facility by these standards; Chap. 2 discusses the impact of workplace safety regulations.

Fire protection and life safety from smoke and fire are now part of most facility services operators' responsibility; this most important subject is covered in Chap. 3.

Plant service operators must have a technical interdisciplinary range of knowledge that crosses many fields of engineering in order to operate, inspect, and safely and efficiently maintain the services required in a facility. In a single volume on a broad field of technical subjects it is necessary to concentrate on the practical application of theory and problem solving. Fundamentals are introduced in general terms with simple equations to stress the basic theory and its application to problem solving. The text is not for design professionals but rather for the people responsible for running the services in a facility. Information in many cases is of an introductory nature for the beginner who is entering this field, but it is also expanded enough to provide management information for people who are *responsible* for providing *all the services* in a diverse range of occupancies.

Each chapter has a Question and Answer section that reviews or expands on some of the basics. It includes problems that require simple mathematics to solve. This follows the pattern of operator license examinations that are required by many jurisdictions on boilers, steam-using machinery, refrigeration, and air conditioning. Appendix D lists the jurisdictions that have such requirements. The Questions and Answers will also assist those people who are entering the facility service field or seeking promotions in this field as well as facility managers, inspectors, repairers, and others in outside services catering to the facility service function.

The author gratefully acknowledges the many sources from which useful information was obtained. This included the following technical magazines: *Power, Mechanical Engineering, Chemical Engineering, Occupational Health and Safety,* and *Welding Engineering.* Much useful information was derived from the ASME Boiler and Pressure Vessel Code, National Fire Protection Association Codes,

American Welding Society, Factory Mutual Engineering, Industrial Risk Insureres, Hartford Steam Boiler Inspection and Insurance Co., The Royal Insurance Co., and Nalco Chemical Co., as well as others not mentioned whose assistance is gratefully acknowledged.

While due diligence and care were used in gathering the information in this book, no legal responsibility is assumed for the information and the accuracy contained herein or for the possible consequence of the use thereof. The author, however, will appreciate being advised of any errors or omissions that the reader may note in the text or illustrations so that necessary changes can be made in the next printing.

Anthony Lawrence Kohan

PLANT SERVICES AND OPERATIONS HANDBOOK

CHAPTER 1
FACILITY SERVICE FUNCTIONS AND ENVIRONMENTAL REGULATIONS

Plant services have an interdisciplinary nature since they provide the following essential services for the location: heat; process steam if the facility is a process plant; water and waste disposal; electric power, which could include power generation in some industries; compressed air; refrigeration; air conditioning; and waste disposal. The chief function of facility service operators is to operate the equipment involved with these services and maintain them so that safe, reliable service is provided for the location. At times, plant service operators may become involved with the design of additions to the facility as the location expands.

Keeping the many services in a facility functional requires a broad knowledge of the equipment involved as well as an understanding of how vital this service may be to the commercial building, institution, manufacturing operation, or other activity for which the facility may have been designed.

Many jurisdictions in the United States and Canada require licensed operators for boilers, refrigeration, air conditioning, and prime movers such as turbo and diesel generators. To secure a license, appropriate practical experience under a licensed person must be obtained prior to taking an examination. Appendix 4 lists some of the jurisdictions that have operator license laws.

Environmental regulations and restrictions have expanded the functions of facility service operators from the normal one of just providing services to the facility. For example, in the area of air pollution regulations, federal, state, and local regulations now impose limits on SO_2 and NO_x emissions which can affect boiler operation and controls. Facility management expects facility service operators to operate and maintain the equipment so that government regulations are complied with.

Fire loss prevention is another function that facility service operators become involved in through maintenance of standpipes, fire pumps, fire extinguishers, alarm systems, etc. A knowledge of fire codes will assist the well-trained facility service operator. Life safety for the occupants of the facility is an expanding field in which facility service operators can play an important part.

EXTENT OF ENVIRONMENTAL REGULATIONS

The following regulations all can have an impact on a facility:

Clean Air Act

Clean Water Act

Resource Conservation and Recovery Act (hazardous waste)

Spills and release reports

PCB regulations

Underground storage tanks

Superfund on toxics in soil and water

Community Right to Know Act. This involves hazardous substance notification and annual reporting on toxic chemicals at the facility.

In addition to environmental regulations, service operators should be familiar with *safety rules and regulations* as they affect their own working habits and procedures and the occupants of the building who are using the facilities they are responsible for maintaining.

EXTENT OF OSHA REGULATIONS

The following Occupational Safety and Health Administration (OSHA) 29 CFR 1910 regulations affect facility safety practices:

General Safety and Health

Lockout/Tagout

Walking/Working Surfaces

Powered Platform, Manlifts, Vehicle-Mounted Work Platforms

Occupational Health and Environmental Control

Hazardous Materials

Personal Protective Equipment

General Environmental Controls

Medical and First Aid

Fire Protection

Compressed Gas and Compressed-Air Equipment

Materials Handling and Storage

Machinery and Machine Guarding

Hand and Portable Powered Tools/Other Hand-Held Equipment

Welding, Cutting, and Brazing

Special Industries

Electrical

Commercial Driving Operations

Toxic and Hazardous Substances, including OSHA's Permissible Exposure Limits (PELs) and Confined Space Entry

The costs of injury and illness in the workplace have climbed, as have liability costs. There is new emphasis on maintaining a facility without hazards. Potential citations by governmental bodies such as OSHA have further emphasized the duty of facility service operators to maintain a facility that is safe for occupants, including facility service staff people.

POLLUTANTS

A pollutant is defined as a substance or condition which if present above an established threshold limit may have a detrimental effect on the human body. Pollutants can be solid or in particulate form, gases such as SO_2 and NO_x, and in the liquid state, such as PCBs. Measurement of source emissions, whether confined or fugitive, is a part of facility service equipment operation in order to demonstrate compliance with regulations and also to make adjustments in the controls to meet performance specifications or regulations. *Impact studies* are now required on any expansion of facilities to show the environmental effect, if any, on the surroundings. Environmental *site selection* or expansion includes such factors as air inversion causing smog, drainage, waste treatment, and waste disposal.

Air Pollution

The air can be polluted from burning fuels with *high sulfur content,* which produces SO_2. This gas forms sulfurous acid when combined with water, aptly called acid rain because the combination occurs during any rainy period. Acid rain erodes building surfaces and also affects vegetation and fish life in lakes.

If high-sulfur fuels, such as coal, are burned, regulations now require the gases from the combustion process to be treated before they are emitted to the atmosphere. Fluidized bed burning is being used in coal-burning plants to combine the SO_2 with pulverized limestone, which produces calcium sulfate in the furnace. This can be disposed of in landfill areas. Other treatments include *scrubbers* or *flue gas desulfurization.*

Grade 6 fuel oil is the heaviest residue of the crude oil refining process and may have sulfur impurities that will require flue gas treatment as required by local jurisdictions.

NO_x *emissions* in burning fuels affect not only boiler operation but also any fired heater, such as is used in refineries and chemical plants and gas turbine emissions. For example, the Los Angeles area requires that furnaces with capacities of less than 40 million Btu/h must release less than 40 ppm of NO_x, while for furnaces with capacities over 40 million Btu/h, the limit is 25 ppm of NO_x. Smog formation in urban areas is a result of NO_x emissions. NO_x combines with reactive organic gases to form ozone (O_3). While ozone is necessary in the upper atmosphere, it is considered carcinogenic and thus presents a health threat when mixed with ambient air.

NO_x is formed by burning fuel that contains nitrogen, but most of it is formed by air combining with the combustibles in the fuel since air contains approximately 79 percent nitrogen. This NO_x is called *thermal NO_x.*

NO_x *reduction* in the combustion process takes many forms, such as:

Combustion staging by lowering the peak flame temperature.

Flue gas recirculation.

Water or steam injection.

Fuel with no nitrogen content.

Air/fuel ratio control. Low excess air burners work on the principle that low levels of excess air suppress NO_x formation.

Postcombustion techniques that include selective catalytic and noncatalytic reduction and chemical scrubbing of the flue gas.

In the *selective catalytic* reduction process, ammonia is injected into the flue gas stream upstream of a catalyst bed. A chemical reaction converts the NO_x and ammonia to an ammonium salt intermediary, and this in turn decomposes to form elemental nitrogen and water. Catalysts used are titanium, vanadium, platinum, zeolites, and ceramics, which are usually shaped in honeycomb plates, rings, or pellets.

In the *selective noncatalytic* reduction process, postcombustion control is used that reduces NO_x to N_2 and H_2O. Ammonia is injected into the upper part of the combustion chamber or into a thermally favorable location downstream. Urea-based reagents are also being introduced in place of ammonia because it is easier and safer to handle.

Smoke, which consists of soot, ash, grit, and gritty particles, is an old air pollutant. The black smoke that sometimes comes from stacks is a sign of incomplete combustion due to not having the correct air-fuel mixture in burning. Black smoke is a sign of too rich a fuel mixture or of incorrect fuel viscosity for the burner, improper air register position, a dirty atomizer, an atomizer in wrong position causing flame flutter, carbon formation on the throat tile, and wrong fuel pressure, to name some burner-related problems. These are all within the operator's control and adjustment responsibility.

Ringelman charts are an old method to measure the density of smoke and are used quite often by regulatory inspectors who check for compliance with clean air regulations (see Fig. 1.1). The chart is a white card with black cross-hatching. The hatching increases by divisions of 20 percent for each Ringelman number. When placed at a distance from the observer's eye, the crosshatch lines merge and appear as a uniform shade of gray. Instruments have been developed, such as a telesmoke, through which the smoke can be viewed. A shaded filter is superimposed on part of the image, and then the Ringelman number is determined. Check with your local ordinance for acceptable Ringelman numbers.

Other remote sensing methods have been developed to detect volatile organic compounds, or *VOCs,* as a result of governmental restrictions on the release of such volatiles as benzene, vinyl chloride, toluene, mercury, and beryllium. The basic principle is to measure the interaction of light with matter. Some of the remote sensing techniques being applied for VOC measurement are shown below (see also Fig. 1.2).

Method name	Operating principle
Infrared (IR) spectroscopy	A standard IR spectrum is used to compare with field readings to determine constituents and concentration.
Ultraviolet spectroscopy	Ultraviolet spectra are collected over an absorption spectral region with the differential absorptions of the compounds from the air determining the identity and concentrations of the contaminants.
Gas-filter correlation	A sample of the gas to be detected is used as a reference, and this is then compared to the gas measured in the air to determine concentration.
Laser absorption	One or more lasers are used to measure absorption at different wavelengths.

The above list is only a sample of air pollutant measuring devices; this field is expanding since the public and governmental bodies demand cleaner air for the environment.

Infrared *flue gas analyzers* measure SO_2, CO, CO_2, and CH_4, or methane. Gases absorb radiation at specific wavelengths, and this is used by detectors to determine if a gas is present and what its concentration is.

NO_x analyzers are available that rely on the chemiluminescence method to convert the NO_x to NO, which is then combined with ozone. The resulting light in this reaction is measured by a photomultiplier tube, with sensitivity down to 10 ppm possible.

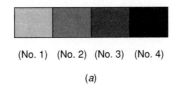

(No. 1)   (No. 2)   (No. 3)   (No. 4)

(a)

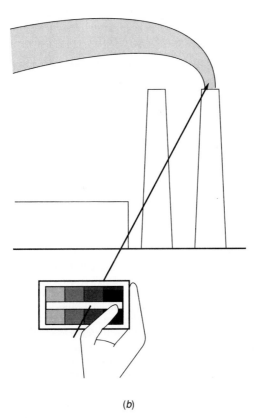

(b)

FIGURE 1.1   Ringelman charts are used to grade smoke emissions from 0 to 5, which are all black, by comparing readings to standard. (a) Black crosshatches on white cards vary in black color density for grading smoke emissions from 0 to 5, which are all black; (b) instrument read-out grades smoke to Ringelman standard.

Opacity measurements are required by the Environmental Protection Agency (EPA) and consist of a single beam of light being transmitted across a stack. The reflections received from a mirror, as shown in Fig. 1.2, are then compared to standards of opacity.

Continuous emissions monitoring of stack emissions is now required by the EPA for large flue gas emitters such as utilities. The following are required to be monitored on a continuous basis:

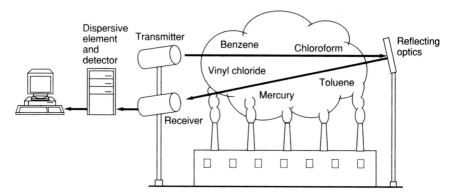

FIGURE 1.2   Remote sensing and measurement of gas pollutants. Ultraviolet, infrared, or laser light transmitters send these signals across the measurement path, and the signals are reflected back by a mirror to a receiver-detector. The interaction between light and any pollutants in the path affects wavelengths, and this is measured electronically in the dispersive element or detector. These types of pollutant measuring devices are replacing the Ringelman chart method, with one such device being the opacity meter.

Opacity

SO_2 and NO_x

O_2 and CO_2

CO

Total reduced sulfur

Portable gas analyzers are now available for fine-tuning combustion in boiler furnaces and fired heaters to boost burning efficiency and to reduce environmentally harmful stack gases to acceptable levels. The hand-held analyzer has sensors that measure CO and O_2, with hydrogen compensation for improved accuracy. With this data the analyzer can calculate the CO_2 and combustion efficiency. Other measurements that may be included on the portable analyzer are excess air to determine heat loss up the stack and gas inlet temperature to the stack. This instrument can be used on smaller boilers to fine-tune air-fuel ratios and thus improve burner performance.

Dusts and Fumes (Particulates)

Dusts are fine powders which, depending on source, can be a fire hazard or can attack human tissue to cause cancer; poisonous dust can be absorbed into the blood stream.

Fumes are smaller particles than dust and result from the chemical reaction or condensation of vapor; they are as potentially harmful as dust. For example, welding fumes are solid particles of the base metal, electrode, or filler metal. When metals are molten, a vapor is formed. As the vapor cools, it condenses at room temperature to become a fine metal solid. Fumes are always spherical in shape and are tiny, about 1/25,000 of an inch.

Dust particles are measured in microns (μm), or one thousand of a millimeter. The smallest particle visible to the eye is between 50 and 100 μm, and the most dangerous sizes are between 0.2 and 5 μm. Concentration of dust is important. Substantial concentrations are classified as those of 10 mg/m^3 for 8-h weighted average of total inhalable dust. However, this varies by the type of dust or fumes per OSHA regulations.

TIME PERIOD EXPOSURE LIMITS

OSHA has regulations for the exposure of plant personnel to hazardous chemicals, and this can apply to dust and fume exposure if any of 500 listed chemicals are in the particulate. OSHA has developed *PELs* for dosage exposure to dangerous chemicals, noise, radiation, and workplace temperatures. Two time periods are used in establishing threshold limits: short-term exposure expressed as a given concentration over a 15-min span and daily time-weighted average over an 8-h period.

Dust and Fume Control

Three methods have been established to control the exposure of personnel to harmful dust and fumes:

1. Administrative controls require a facility to establish a watch system so that the amount of time a worker is exposed to the pollutant is limited and so that the overall 8 h of exposure is below the established PEL. Many devices are now on the market for a worker to wear that display the accumulated exposure during the working day.
2. Personal controls include the use of respirators, gloves, safety clothing, and shoes to limit the amount of contact a person working in the affected area has with the pollutant.
3. Engineering controls include the design, installation, and maintenance of hoods, ducts, fans, and collection bins for collecting the dust and fumes from the workplace. OSHA regulations prefer to have in-place engineering controls, with administrative and personal controls permitted only where engineering controls may not be possible. The two basic methods of engineering controls are source capture and collection and general ventilation to dilute the concentration. Source capture can be hoods or extension arms or ducts where more worker movement may be required (see Fig. 1.3). General ventilation consists of fans or blowers to keep fumes out of the worker's breathing zone by pulling out the contaminated air and exhausting to the outside or, if required by EPA regulations, into a collection system, and bringing in adequate makeup air.

Filtering contaminants out of the polluted airstream from the workplace can take many forms and includes cleaning flue gas from fired heaters such as boilers. Dry-type collectors include *electrostatic precipitators and cyclone collectors* (see Fig. 1.3), which depend on centrifugal force separating out particles, *bag filters, cartridge filters,* and, for wet separation, *scrubbers.*
   See Chap. 3 for the fire and dust explosion hazard.

Water Pollution

Water testing (see Chap. 4) can become an important function for facility service operators. There are a wide variety of water sources that require testing for the presence of pollutants. Conventional testing is by sampling and laboratory analysis. Continuous monitoring can include measuring flow rates, temperature, pH, conductivity, dissolved oxygen, turbidity, cyanide, and total suspended solids. Conventional testing by sampling and lab analysis must be performed for biological and chemical oxygen demand, oil and grease content, total suspended solids, heavy metals, ammonia, nitrogen, total organic carbon, and quality control sampling of treated water and water that is treated

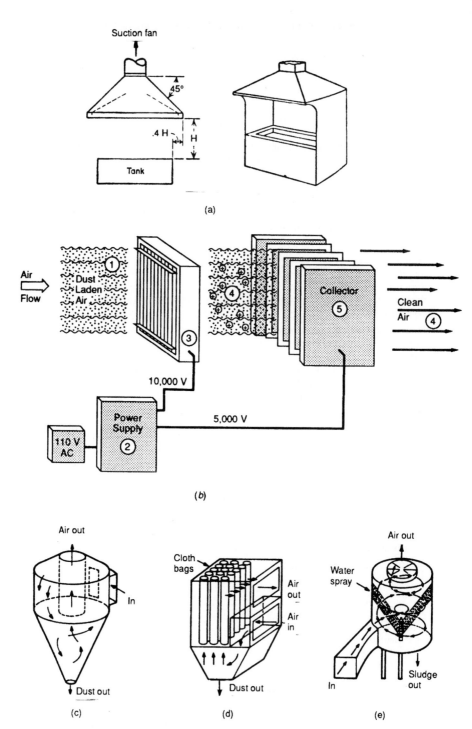

FIGURE 1.3 Fume and dust collectors. (*a*) Hoods over work area; (*b*) electrostatic precipitator; (*c*) cyclone collector or separator; (*d*) baghouse filter; (*e*) wet scrubber.

before disposal. Storm water must be sampled at depths that correspond to high and low runoff rates. The samples are to be collected the first 30 min of a storm.

Hazardous Waste

Hazardous waste management is closely regulated by such federal agencies as EPA, OSHA, and DOT. The regulations include packaging and shipment requirements, minimizing the risk of spills, explosions, reactions, corrosion, and storage requirements. Labeling and documentation of hazardous waste locations, content, etc., are also required. Also included is the use of proper personal protective equipment when handling hazardous wastes. Approved disposal methods are part of the regulations.

Most labeling falls into the following categories: poison, flammable liquid, flammable solid, corrosive, toxic, and oxidizer.

EMERGENCY RESPONSE ACTION TRAINING

Facility service operators should be aware of the requirements established in Code 29 of Federal Regulations (CFR) 1910.120 and the OSHA May 1989 promulgated regulations that define training requirements for personnel engaged in emergency response actions and hazardous waste operations.

To date, OSHA has not developed an accreditation program for the emergency response or hazardous waste training programs. It is the responsibility of the employer to evaluate the quality and applicability of offered training programs per OSHA recommendations and to determine if they are suitable for the facilities exposure and if they also satisfy the requirements of the OSHA regulations. Most organizations providing this training offer certificates of competency after the attendee passes written and skill examinations. The field of hazardous waste control is expanding the opportunities for facility service operators in plants with such exposures.

Hazardous material storage may be under local fire and building code requirements. A few known rules are:

1. Explosive or fire potential hazardous substances require fireproof storage rooms, and, generally, these should have sprinklers. See Chap. 3 for the class of extinguishers that may be needed.
2. Make plans for spill disposal.
3. There must be a secondary exit for areas greater than 200 ft^2.
4. Mechanical ventilation is recommended at an exchange rate of 1 m^3/ft^2 or greater of area in storage.
5. Electrical equipment may have to be explosionproof depending on the properties of the hazardous material stored.
6. Nearby showers and eye wash stations are required in most codes.
7. When the materials stored include explosives, poisons, or alcohols, security against unauthorized access must be provided.

Asbestos

Asbestos was extensively used in the past as an insulator around steam pipes, boiler casings, heat engine cylinders, and steam turbine casing covers for conserving heat and to protect the room or building from excessive heat release. It was also used around structural steel as a fire protection shield to prevent the steel columns and beams from buckling from the heat of a fire. Asbestos powders are now linked to health problems such as respiratory problems, asbestosis, lung cancer, and other respiratory diseases. OSHA and EPA have established regulations for the use, handling, and disposal of asbestos-containing material; contact the nearest EPA or OSHA office for a copy of these regulations on asbestos. "Asbestos-only" surveys are conducted on bulk insulation and other such coverings to determine the percentage of asbestos in the bulk. Bulk samples are now required to have 1 percent or less of asbestos to be considered asbestos free.

When removing asbestos, the area of removal must be isolated from the rest of the property to prevent the spread of asbestos dust. Workers must wear proper clothing and use proper respiratory masks. The asbestos has to be disposed of in approved containers and at disposal sites.

Cancer-causing fibers, besides asbestos, are also receiving similar attention as an environmental-type pollutant; EPA and OSHA rules on asbestos also apply to these fibers.

Underground Storage Tanks

In 1988, the EPA established new requirements for underground storage tanks that contain bulk liquid chemicals and petroleum products to minimize the risks to the environment, which include harmful vapor releases and leaks that could affect ground water. The standard is 40 CFR 280. Among the requirements is retrofitting underground storage tanks with secondary containment or providing acceptable leak-detection instrumentation. Underground storage tanks are required to have overfill prevention devices. For aboveground storage tanks, this could be double-wall construction, concrete dikes, or another approved mounding structure that is required to catch leaks. Leak detection on underground tanks may include vacuum and pressure monitoring, level monitoring, and the use of gas/liquid sniffers.

Facility service operators who are responsible for underground storage tanks that contain bulk chemicals or petroleum products should obtain a copy of standard 40 CF 280 from the EPA.

Odors

Odors are being increasingly recognized as a source of nuisance, and if they affect personal comfort and well-being, they violate the Federal Clean Air Act as an environmental hazard. Odors come from various sources, such as chemical and industrial plants, water and waste treatment plants, municipal landfills, and spills during transportation of various odorous compounds. Odors are emitted from *point sources* such as leaks from compressor seals, pump seals, valves, flanges and connections, and pressure-relief devices. *Fugitive* or *area sources* are cumulative leaks from many sources that are difficult to identify. For example, a worker blowing out a tank and then closing the valve may spread a fugitive odorous gas or liquid into the environment.

Odor releases in plants are minimized by good maintenance practices such as correcting leaks from point sources and also perhaps tracing the source of leaks with portable instruments. These include soap bubble tests, where a soapy solution is applied to a suspect area, such as a pipe joint, to see if soap bubbles form around the joint. More sophisticated field screening instruments include the flame ionization detector; the photoionization detector for nonmethane hydrocarbons; the odormeter, which gives a comparative odor reading for different gases; and various sulfur analyzers.

The best odor abatement program is to implement a leak prevention maintenance program, including replacement of worn components when leaks are found and the establishment of a periodic monitoring program for source leak possibilities.

Building Syndromes

Unacceptable indoor air quality in many office buildings is also an environmental problem. The symptoms of building syndromes include skin irritations, headaches as the day progresses, tight chest, fatigue, and irritation of the eyes, ear, and throat. Industrial hygienists are sometimes employed to identify the cause of some of these effects. The American Society of Heating, Refrigeration and Air Conditioning Engineers (ASHRAE) stresses air flow measurements and adequate *clean* air ventilation. ASHRAE has established a limit of 1000 ppm for carbon dioxide in indoor air and a recommendation of 15 cfm/person of fresh outside air into the building as a means of keeping the buildup of carbon dioxide to acceptable levels. Growth of microorganisms in humidifier and dehumidifier pans in the air distribution system requires these pans to be cleaned or treated periodically so that the accumulated water does not spawn microorganism growth.

Maintenance procedures to avoid building syndromes include:

1. Checking the proper operation and cleanliness of humidification and dehumidification systems to avoid microorganism growth
2. Filter replacements to avoid contamination buildup
3. Fan and louver operation checks
4. Duct cleanliness maintenance
5. CO_2-level measurement to see if they comply with ASHRAE standards
6. Fresh intake operation to see if makeup air is adequate

Absolute standards for the avoidance of building syndromes have not been established but may occur in the future.

FACILITY MANAGEMENT

Physical conditions of a property that also need attention are:

Grounds and yard condition

Building foundations

Exterior doors and windows

Roof conditions

Interior walls, exits, and fire walls and doors

Ceilings

Plumbing and repairs or alterations

Heating systems

Cooling and ventilation

Electrical service and distribution

Lighting

Motors, generators, pumps, boilers, turbines, compressors, switchboards, transformers, and supply cables

A facility inspection program will assist in keeping the equipment and building services operating in a safe and reliable manner. It is essential, however, to establish an effective maintenance program. This should state the objectives of the program and whether corrective actions are to be taken by in-house staff or outside service organizations. An inventory control should also be set up on the equipment that will require periodic maintenance work; in large organizations, a work order system should be established to control maintenance costs; and history cards should be maintained on the maintenance and repair that has been performed on vital facility equipment.

Condition reports need to be reviewed, and corrective work that may be recommended should be classified as *urgent,* requiring immediate correction, *secondary* for next planned maintenance outage, and *future* to extend life expectancy of the property or equipment.

Urgent work is that which can affect the safety of the property or employees. It also includes equipment where there is no standby available, and thus would affect the business immediately if it failed.

FACILITY SECURITY

Security is required to guard a premises against damage to assets from fire, explosion, and water damage. It is also important to prevent criminal actions on the premises, such as industrial espionage, vandalism, fraud, theft, and malicious mischief.

Security systems may have layers of protection. These could include:

1. *Physical barriers.* Wire frames on windows, gates at entrances, break-glass detectors, vibration sensors and alarms, pressure pads placed under carpets in critical rooms, magnetic reed contacts, which are glass-encapsulated magnetically operated reed switches that are placed in doors and windows.

2. *Motion detection devices.* An infrared light beam across an opening, ultrasonic and microwave movement detectors, special closed-circuit TV cameras. These can set off alarms when picking up motion, alerting a staffed security station.

3. *Security patrols and guards.*

4. *Mechanical key systems* to prevent entrance to designated vital areas.

5. *Magnetic card systems* to gain entrance to a key area.

All security systems with alarms require periodic checking and maintenance to make sure they are always functional; otherwise they will fail from neglect or tampering. It is in this check of functional security and maintenance that facility service operators can become involved as part of their duties.

QUESTIONS AND ANSWERS

1. What does the term *particulate matter* mean in air pollution?

    Answer.   Any solid or liquid that is airborne, but it does not include uncombined water vapor.

2. What is the difference between soot, fly ash, and odor?

Answer. Soot is unburnt fuel particles that result from incomplete combustion in a furnace, Fly ash is the fine particles that are emitted from the furnace toward the stack when pulverized coal is burned. Odors are obnoxious smells that irritate the nose and human respiratory system.

3. What fuel oil is relatively sulfur free?

Answer. Grade 6 is the heaviest residue oil from refineries and, therefore, has the most impurities such as sulfur. As the grade decreases, more refinement has taken place, and generally grades 3, 2, and 1 are relatively sulfur free.

4. What are the primary pollutants found in stack emission from furnaces burning coal?

Answer. SO_2 and various forms of nitrogen oxides, named NO_x.

5. What is thermal NO_x?

Answer. Thermal NO_x is the combining of nitrogen in the air, which is 79 percent nitrogen, with the oxygen in the air at high temperatures. Below about 1900°F, this chemical combination does not occur. Thermal NO_x thus differs from the chemical combination of *nitrogen in fuel* with oxygen in the combustion process.

6. On what principle do low-excess air burners work?

Answer. By burning with the lowest air possible to obtain proper fuel burning, less air is introduced into the burning process, and this suppresses NO_x formation.

7. What is flue gas recirculation as a means of lowering NO_x formation?

Answer. In flue gas recirculation systems to control NO_x formation, a portion of the flue gas is extracted from the stack and is returned to the burner with combustion air. This lowers the peak flame temperature and reduces thermal NO_x formation. In addition, oxygen content is reduced in the combustion air, further reducing NO_x formation since less oxygen is available. Recirculating flue gas temperature should not exceed 600°F, and the recirculating rate is limited 15 to 25 percent in order to sustain combustion.

8. What does the term *SCR* mean in NO_x reduction?

Answer. SCR is selective catalytic reduction, which involves injecting ammonia into the flue gas upstream of a catalyst bed. The NO_x and NH_3 combine to eventually release elemental nitrogen and water vapor into the stack. The optimum range for SCR is 600 to 700°F.

9. What is a PEL and how is it used?

Answer. OSHA has regulations on the exposure of plant personnel, especially in industrial facilities, to hazardous chemicals, noise levels, radiation, and extreme temperature. PELs have been established for these occupational hazards. A PEL is further defined as the exposure that a worker can endure for an 8-h day or a 40-h week for a working lifetime without suffering serious ill effects or sickness. Time-averaged concentrations are used for chemical, fumes, and dust exposures. These are short-term exposure limits expressed as a given concentration over a 15-min period and a daily, or 8-h time-weighted average. For example, for benzene the short-term PEL is 5 ppm, while the daily PEL is 1 ppm on a volume basis.

10. What is breathing monitoring?

Answer. Breathing monitoring is used to check on the exposure from chemicals, fumes, or vapor in the workplace to make sure it does not exceed the OSHA PEL guidelines. Monitoring is performed via a small sampling medium that is fastened to the worker's collar or lapel and worn for the varying periods of time depending on short- or long-term exposure classifications. To test for dusts, vapors, and gases, personal filters, sorbents, and liquid-filled impingers can be used singly or in combination. Air can be sampled with small battery-powered pumps attached to a worker's belt and then analyzed by a lab for the type of contaminant and concentration.

11. Do NO_x emission regulations apply to small boilers or fired heaters?

Answer. This depends on the locality; however, the EPA law requires lower governmental bodies to adopt and enforce federal rules. At times, because of geographic location, local laws are more severe than the federal law. For example, for Southern California, the following limits have been established:

Boiler, hp	Pounds of steam/h	Btu input, $\times 10^6$/h	NO_x and CO limits
To 120	4,000	2–5	30 ppm NO_x 400 ppm CO
To 240	8,000	5–10	40 ppm NO_x 400 ppm CO
To 960	32,000	10–40	40 ppm NO_x 400 ppm CO
Over 1,000	Over 32,000	Over 40	30 ppm NO_x 400 ppm CO

12. What is the ASHRAE recommendation for fresh air makeup in nonindustrial occupancies?

Answer. At least 15 cfm per person in the building of fresh, clean, or filtered outside air to counteract the carbon dioxide generated by people in the building.

13. What are some of the benefits in having an environmental testing program?

Answer. The testing program develops data that can be used for obtaining permits and passing compliance audits that may be made at the facility. The program can also be used to improve pollution control equipment performance so that waste minimization and pollution prevention are maintained at the facility. It is also used to determine health risks in the workplace or surroundings and assists in planning for emergencies if an accidental release of toxics, fumes, etc., does occur.

14. What must be continuously monitored on stack emissions for large Btu inputs?

Answer. This monitoring applies to large boilers and fired heaters as used in utilities and petrochemical plants. Opacity, SO_2 and NO_x, O_2 and CO_2, CO, and total reduced sulfur, should be monitored on a continuous basis.

15. What are five methods to filter particulates or contaminants?

Answer. Electrostatic precipitator, cyclone collector, baghouse filters, cartridge filters, and wet scrubbers.

16. What are the five layers of security protection systems?

Answer.   Physical barriers such as fences, motion detection devices such as infrared light beams, patrols and guards, mechanical keys on doors and gates, and magnetic cards for selected areas of the premises.

CHAPTER 2
SAFETY IN THE WORKPLACE

Workplace safety is now a national goal with the adoption of OSHA by Congress in 1970. States always had labor laws, but they varied from one section of the country to another. OSHA standards are designed to promote uniform rules and regulations for safety practices throughout the country. A system of adopting recognized existing standards, based on consensus, was adopted by OSHA initially. From this base, additions have been and are continuing to be made as experience indicates a need for such a standard. For example, as a result of several severe explosions with death and injury, OSHA adopted a Process Safety standard recently.

Traditionally, trained plant service operators are aware of the potential risks in the equipment under their control and thus develop a "safety" awareness as they perform their duties. As the previous chapter indicated, their responsibilities could include other facility safety work as their job function is expanded into fire protection, spill cleanup, noise reduction, accident investigations, and electrical safety, to name a few. It is also essential for the staff involved in facility services to follow good safety practices and for the supervisors to "police" the practices of their coworkers so this is properly controlled. This includes making sure that *outside contractors* also follow good safety practices.

COMMON INJURIES AND THEIR PREVENTION

There are many ways that injuries can occur to the human body such as back problems from improper lifting, eye injuries from foreign object damage, feet injuries from a heavy object falling on them, falls from ladders, respiratory injury in a dust-laden or hostile chemical environment, burns from touching hot objects, electrical burns from touching live or energized equipment, asphyxiation from entering a space with insufficient oxygen, and chemical burns to skin, eyes, nose from the release of irritant-type chemicals (see Fig. 2.1). The list of possible sources for injury is huge.

Safety Factors

Hazard identification is an important function of all facility service operators as is maintaining controls over the hazards, which should be a facilitywide practice. Understanding the nature of the hazard exposure will determine the methods to be used to minimize the exposure.

Safety organizations and committees are used today in many industries to implement safety practices and to review existing health or safety regulations. This includes

FIGURE 2.1   Workplace injuries occur for many reasons, and the medical attention required varies. (*Courtesy, Krames Communications, Inc.*)

chemical hazard identification and other exposure analysis. Worker participation is essential in the safety committee meetings since they may know, from their own experience in having to do the work, the hazards and procedures that are practical to adopt in the workplace. Another function of safety committees is to implement *training programs* that have as their main objectives hazard or risk awareness and the safety measures that should be applied.

Another function of safety committees is to implement *safety audit inspections* by in-house people. This could be as simple as a foreman checking to see that safety practices are being followed by employees under his or her control, or it could mean central office personnel who are even equipped with portable instruments that monitor the surroundings for noise, fumes and dust, hazardous chemical concentrations, and similar "hidden" hazards.

Accident review committees are essential in any safety practices program since they assist in determining the reasons for the accident and what additional measures may be needed to prevent future occurrences.

Back Injuries

Back injuries are a common occupational hazard nationwide. The back is made up of the spinal column and the ligaments and muscles that support it. The bones of the spine, the vertebrae, support the entire body. There are spongy discs or cushions between the bones in the spine. Lifting excessive weight, or doing it improperly, can put great stress on the back to the extent ligaments may tear or a disc may rupture. Good safety practices in lifting include limiting the weights that are too heavy for one person, flexing the knees for lifting with the leg muscles instead of the back, keeping the weight close to the body, and lifting in a manner that allows forward movement so as to avoid twisting the spine. Recommended average person maximum weights in lifting are based on age and sex as follows:

| | Maximum lifting weight | |
Age	Men	Women
16–18	44	26
18–20	51	30
20–35	55	33
35–50	46	28
Over 50	35	22

Strengthening both the back and the stomach muscles by exercise will help to avoid back injuries, but the amount and manner of weight lifting still requires good safety practices. OSHA has regulations for safe lifting in their general duty clauses, as does the National Institute for Occupational Safety and Health (NIOSH). Repeated back lifting injuries require corrective actions by the facility. These include:

1. Investigate the occurrences and take corrective procedures.
2. Review stressful lifting tasks with workers and supervisors and use this information to evaluate job changes or to monitor proposed procedural changes.
3. Institute training programs on safe lifting practices.

4. Use occupational therapists to assist in worker rehabilitation from back injuries and encourage them to teach proper exercise, stamina buildup, and training in safe lifting habits.

5. Encourage storing heavier loads near the ground so that they can be lifted by pallet load lifting devices. If over 70 lb, storage should be within 17 in of ground level.

Eye Injuries

Eye injury prevention begins with assessing the nature of possible eye hazards and instituting a safety eyewear program in order to prevent damage to the eye from foreign object impact damage, for example. More than half of eye injuries result from foreign objects getting into the eyes because workers did not wear the proper goggles or face shields. OSHA has a broad mandate on eye protection in 29 CFR 1910.132. It is necessary to consider five basic eye hazard categories: *impact protection, chemical irritation, dust and fumes, heat,* and *optical radiation.* From the analysis of the hazard that may exist in the workplace, proper protective equipment can be specified in the facility's safety program. The extent or scope of where eye protection must be worn is also an important consideration. Should it be plantwide or only in certain areas of the facility? Eyewash stations may be needed and first-aid instructions may need to be posted for the eyewear area involved. Worker acceptance includes individual selection of vision correction glasses and adjustable frames so the glasses can be fitted to each person's face.

Eye strain and fatigue should also be considered in some occupancies. Examples are glare from lights, poor lighting in the work area, small scripts or prints, reflecting glares from glass, paper, and mirrors. Age can be a factor in eye fatigue.

Head and Foot Injuries

Hard hats are now a recognized method to prevent head injuries in any area where falling stray objects may exist at the facility or where pipes and valves are located low enough that accidental striking of the head on these protruding objects is possible. Worker acceptance of wearing hard hats in designated areas of a facility is materially improved if hats are selected that fit and are comfortable.

Safety shoes protect workers' feet from injury by providing stiff shoe surfaces. Injuries arise from heavy objects falling on the feet, such as in construction or maintenance activity or while lifting.

Hearing Injuries

Sound, noise, and vibration are somewhat interrelated; however, some sounds and vibrations are considered nuisances and distractions and affect people's nerves, while noise can harm their ears. High-intensity noise can cause physical injury to workers, especially to their hearing. Excessive vibrations on machinery will eventually cause fatigue failures. Vibration tracking is an excellent way of monitoring performance and to note developing conditions on a machine, which assists in determining when maintenance is required. Some sounds are generated by liquid and gases flowing in pipes and ductwork, and many times these are very annoying to people who have to work nearby, as is vibration of floors, walls, and workstations.

The *human ear* can be damaged by excessive sound energy, which can cause loss of hearing. This loss of hearing can come from short-term high sound levels of exposure or from long-term lower sound levels. To understand noise limits for the ear, it is necessary to review sound energy and how this is expressed. Sound is a sensation caused by pressure variations in the air. It must occur more rapidly than barometric pressure to be heard. *Audible* sound has a 20- to 20,000-cycle frequency range. Sound *intensity* is a measure of energy and is expressed as watts per meter. Intensity is proportional to the square of pressure, or

$$\frac{I_1}{I_0} = \frac{P_1^2}{P_0^2}$$

A bel is expressed as the logarithm to base 10 of the ratio of the perceived pressure squared to the reference pressure square. A decibel is one-tenth of this, or

$$dB = \frac{1}{10}\left(\frac{\log I_1}{\log I_0}\right)$$

As can be noted, the decibel scale is not linear but logarithmic (i.e., 20 dB is 10 times as intense as 10 dB).

Noise is measured in dBAs. The A factor stems from the fact that noise meters are equipped with filters designed to mimic the human ear's reception of sound frequencies. Noise levels measured with these filters are designated dBA levels. Standards refer to the A-scale reading. Noise levels between 85 and 120 dBA can affect the hearing of workers. Pain to the ear can start from short-term exposures of 140 dBA.

OSHA has established 85 dBA as the limit for noise exposure for an 8-h day. OSHA also requires engineering controls in the workplace to limit the sound level by mechanically keeping the noise away from the workplace. Secondary protection such as earplugs and earmuffs worn by employees to deaden the noise can only be used where engineering controls are not possible. Some *noise reduction* methods that are used to keep noise levels low are:

1. Using sound-absorbing material between the origin of the sound and the workplace.

2. Using porous material to absorb sound as sound energy is retarded by friction within the material's barriers of solid and void paths for the sound.

3. For sound in ducts, resonant absorbers, known as a Helmholz resonators, are used. This consists of a properly designed chamber connected by a narrow neck to the duct that is carrying the sound. The volume of air in the chamber will resonate at a frequency determined by the volume of the chamber and thus attenuate the noise in the main duct. Also, the noise generated in a duct system depends on the velocity of flow; therefore, a larger cross-sectional area of ductwork will reduce the velocity of flow and its noise level. Absorber materials can be installed inside the duct to reduce the outside noise level.

Annoying vibration can be reduced by mounting the machinery on rubber or in other ways, such as enclosing the machinery with decorative sound-absorbing walls. Electrical machinery vibration is often traced to loose or improperly braced coils. Some sliding objects develop frictional forces that create a vibration nuisance problem. Better lubrication may cure this type of vibration. Distancing people from the noise source is another method of reducing the noise hazard.

Other Hazards

Respirators are required in many workplace environments in order to protect people from potentially debilitating or deadly damage to their breathing systems. For example, in welding, damaging fumes and toxic gases such as carbon monoxide, nitrogen dioxide, ozone, and phosgene may be released. The latter gas can be formed when welding is conducted adjacent to vapor-degreasing tanks or near parts that are wet with chlorinated solvents.

The proper selection and use of a respirator is important because if the wrong respirator is chosen for the hazard faced, the wearer may run the risk of injury. NIOSH has published criteria for selecting the proper respirator, which is a facility management responsibility since they know the chemicals and substances that may be involved in their operation. Per NIOSH, the individuals selecting the respirators should be familiar with all pertinent regulations and be able to assess and evaluate the hazardous environment that may be or may develop in the facility and should be able to select the proper respirator by knowing the limitations of each class of respirator. The environmental conditions that determine the classification of the type of respirator to be used have been divided into Levels A, B, and C, with Level A being the most severe. The Level A environment contains dangerous and toxic gases that can affect the skin and respiratory system.

Protective clothing for certain environments must consider the nature of the hazardous substances, the type of work that will be performed, and the duration of the work. The level classification is almost the same as for respiratory protection—from Level A for protection against severe hazards to skin to Level D for nuisance contamination. The selection of barrier fabric such as nitrile, neoprene rubber, or natural rubber for gloves will depend on the hazard to be encountered on the job, the severity of the exposure, whether it is liquid or gas, and similar concerns. OSHA 29 CFR 1910.120 also requires a written personal protective equipment program for all employers whose personnel handle or are exposed to hazardous materials in the performance of their jobs. This may include written procedures for emergency conditions, such as accidental spills of hazardous liquids or gases.

Radiation

There are two types of radiation classifications:

1. *Ionizing radiation* includes x-rays, fluoroscopy, and nuclear scans. This type of radiation has cumulative and long-term effects that may damage tissue; therefore, in some occupancies, workers must wear monitors to measure the dosage they may be exposed to. They are also required to be protected from scatter and direct exposure. Facility service operators may become involved with x-ray inspections at their site, such as checking welds or old riveted joints for plate cracking.

2. *Nonionizing radiation* includes ultraviolet light, laser beams, magnetic fields, and radio frequencies. These are considered less hazardous than ionizing radiation. High-intensity ultraviolet light as emitted in welding arcs can have some portions of the electromagnetic spectrum that causes sunburn on the skin. Therefore, it is essential to have protection for the eyes, neck, face, and arms against this potential hazardous burn if there is exposure to this type of radiation.

Confined Space Entry and Work

As a result of asphyxiation, fire, and explosions that caused injury and death incidents, OSHA has adopted a national confined space rule, effective April 1993. The rules include requirements to have training programs for workers who are to work in confined spaces with the following hazards:

1. Insufficient oxygen
2. Engulfing a worker inside due to fire, smoke, vapors, or fumes
3. A tight configuration so that a worker cannot easily exit from the space
4. Any other recognized serious hazard, such as electric shock

The regulations require atmospheric testing of the inside of the confined space for sufficient oxygen for human occupancy and, of course, for the presence of toxic, volatile, and flammable substances.

Employers must establish a permit system for workers who enter such confined spaces. Employers are required to identify and label all confined spaces that could endanger workers, evaluate the hazards that may exist in the space, and then develop and implement procedures and practices necessary to make the entry safe for the workers involved. A special "hot work" permit is required to perform any operations that could be a source for ignition, such as welding. The rules also require that an "entry supervisor" sign written permits for entry and cancel the permits when the work is completed. The rules require the presence of outside attendants who can respond if a worker inside the confined space needs help. On-site rescue teams must be equipped properly and must have specialized rescue training based on the site conditions or hazards. Refresher training is specified for workers entering confined spaces and for attendants and rescue teams.

In general, there are four hazardous atmospheres: *flammable* and *explosive, toxic, irritant* and/or *corrosive,* and *asphyxiating* from lack of sufficient oxygen. Remember also that even though a confined space is gas free, some work procedures can generate flammable mixtures, such as spray painting, coatings, and the use of solvents for cleaning. There have also been instances of flammable gases being released during sludge and scale removal inside the vessel or space. Toxic atmospheres are those that exceed the permissible exposure limit for any substance in the confined space that may be considered toxic per national standard references such as OSHA, NIOSH, or material safety data sheets (MSDSs).

PROCESS SAFETY STANDARD

Some severe explosions and fires in petrochemical plants that injured and killed workers are listed below:

City and country	Year	Fatalities
Flixborough, United Kingdom	1974	23
Mexico City, Mexico	1984	Over 650
Bhopal, India	1984	Over 2000
Romeoville, Illinois	1984	8
Pasadena, Texas	1989	23
Deer Park, Texas	1990	17

They caused OSHA to publish new rules for process safety, known formally as 29 CFR 1910.119; "Process Safety Management of Highly Hazardous Chemicals,

Explosives and Blasting," published February 1992, which became effective in May 1992. The standard applies to:

1. Process involving listed hazardous chemicals in amounts that exceed the listed threshold limits for a possible violent reaction.
2. Processes involving over 10,000 lb of *flammable* liquids or gases in one location, except for hydrocarbon fuels to be used at the workplace for services such as heat, steam generation, and refrigeration.

OSHA rules in 29 CFR 1910.119 should be secured for a review if your facility has a hazardous process.

EMERGENCY ACTION PLANS

Employers with more than 10 employees must have written action plans per OSHA 29 CFR 1910.38 "Employee Emergency Plans and Fire Prevention Plans." Part of this planning should include: (1) Potential nature of emergency, such as fire, dangerous spills, toxic gas or vapor release, and confined space entrapment, (2) potential damage to people, property, business interruption, and perhaps the community, (3) what measures will be taken to control the incident, such as handling it by in-house trained personnel or calling designated outside assistance such as the local fire department, ambulance service, industrial firms specializing in spill cleanup, and (4) designating the people responsible for taking the actions required to get the emergency actions started and all others who will be assigned responsibilities in preventing injuries to employees and protecting the safety of the facility.

ILLNESS AND INJURY RECORD KEEPING

OSHA rules require a facility to keep records of injuries and illnesses that are work related. This applies to employers with 10 or more workers, but clerical-type industries, such as banking, insurance, and similar "white collar" occupancy are excluded. The regulations for affected industries require a log and summary to be maintained for occupational injuries or illnesses that cause loss of a workday, require medical treatment, or cause a fatality. Notification to the nearest OSHA office is required within 48 h if an occurrence at work results in a fatality or the hospitalization of five or more employees.

This OSHA record-keeping requirement imposes a reporting procedure duty to foremen and supervisors to immediately complete a report of an accident which results in lost workdays or in an injury in the workplace. This accident report should be forwarded to the responsible person in the organization whose duty it is to maintain the accident log of the facility. Otherwise OSHA may levy fines for not maintaining injury records.

GENERAL POWER EQUIPMENT SAFETY

All plant service operators will be involved with boilers, pumps, fans, compressors, perhaps steam turbo-generators, or diesel generators, as well as electrical machinery and equipment and therefore are required to follow good safety practices when working with this type of equipment. A well-managed plant will make sure that good safety practices are followed by all employees who operate, inspect, maintain, or repair this type of power equipment in addition to any OSHA requirements. Most manufacturers supply precautionary instructions for their equipment, and these should also be followed. The following guidelines are meant to be reminders for the beginner of power equipment operation and are general in nature:

Hazard identification is the first responsibility of a well-managed plant that has power equipment. Examples are the precautions to be followed in blowing down a water column, testing a low-water fuel cutout, and welding a crack inside a pressure vessel, to name a few.

Good housekeeping is essential, and that means cleaning up oil and water spills so that no slipping accidents occur and removing things such as oily rags.

Protective clothing such as hard hats, protective eye glasses, and hard-toed leather and rubber footwear should be worn, depending on what the job requirement may be. Loose-fitting or torn clothing are easily caught in moving machinery.

Power and hand tools should be in good condition. Examples are hammers with tight heads, wrenches without worn corners, dry electrical equipment without frayed or worn insulation, and having ground connections.

Lift weights that will not cause back injuries. Place a limit on your ability, such as maximum of 45 lb.

Avoid touching hot objects above 120°F and use the back of the hand for this so that the palm will be able to do work.

Cool equipment before working on it, preferably with the surrounding air.

Reduce pressures to atmospheric before removing any covers such as manholes and handholes on pressurized equipment.

Keep all guards, rails, fences, and shields in place on operating machinery.

Isolate, immobilize, or switch off any machine to be worked on so that somebody else does not start it. Tag-out and lock-out procedures must be followed.

Keep covers on all electrical equipment. This will prevent dirt and moisture from getting into the equipment and will avoid a possible fire originating from an electrical failure.

Store all oxygen, acetylene, propane, or any other gas cylinder in a chained, upright position in a separate room from power equipment.

Electrical testing, maintenance, and repair should be performed only by qualified people trained in electrical safety.

Electrical equipment must be kept clean and dry and connections must be tight.

A permit system should be established when any work is to be performed on boilers, pressure vessels, electrical equipment, and prime movers such as turbines and diesels.

Fire fighting procedures should be posted, and all personnel operating power plant equipment should be trained on what procedures to follow in the event such an emergency arises. Also see Chap. 3.

QUESTIONS AND ANSWERS

1. What are four eye injury sources?

   Answer. Foreign object impact damage, chemical irritation, dust and fume irritation, excessive heat, and radiation.

2. What is the OSHA decibel limit for noise exposure for an 8-h day?

   Answer. 85 dBA.

3. What is a Level A environment per EPA regulations?

   Answer. Level A environment has dangerous and toxic gases that can affect the skin and respiratory system, thus requiring protective clothing and breathing apparatus of the approved type.

4. What type of radiation are x-rays?

   Answer. X-rays are classified as ionizing radiation, which has cumulative and long-term effects that may damage human tissue.

5. What are four possible hazards in working in a confined, unapproved space?

   Answer. Insufficient oxygen, dangerous or toxic fumes in the space, tight enclosure restricting exiting from the space, and electric shock from defective electric equipment used in the space.

6. What is the first thing to do if electrical equipment catches on fire?

   Answer. Deenergize the equipment so that no more electrical energy feeds the electrical failure or causes more short circuits to occur, causing the fire to spread.

7. What is the purpose of a permit system?

   Answer. Its main purpose is to make sure that proper safety procedures are to be followed on the piece of equipment where work on it is to be done. This means it will be safe to enter, is properly tagged and locked out, its internal atmosphere has been checked, a fire watch may be present, and an authorized person will do the necessary repairs, testing, or alterations with the full knowledge of plant management.

8. What is the danger in directing the stream from CO_2 extinguisher on skin or clothing?

   Answer. CO_2 forms dry ice and will freeze the skin and clothing, leading to injury to the person so struck by the CO_2.

9. When are oxygen-breathing apparatus used?

   Answer. The face piece for breathing is supplied oxygen by an oxygen cylinder. This separate source of oxygen is necessary if there is insufficient oxygen in the surrounding air or if the air has dangerous concentrations of fumes or other toxic-type gases, where an ordinary gas mask would not be suitable.

10. What are some options available to a site in emergency planning for a fire?

    Answer. This is covered in OSHA Standard 1910.156. The options permitted are (*a*) No action is taken by the employees to suppress the fire. An evacuation is ordered per the site's Emergency Action Plan, and response to the fire is by the local fire department. (*b*) All employees are trained to fight incipient stage fires using small hose streams or portable fire extinguishers. (*c*) Only *designated employees* in the workplace will fight the incipient fires with small hose streams or fire extinguishers. All other employees must evacuate per the emergency plan. (*d*) Use a fire brigade for incipient fires, using small hose streams or fire extinguishers. This group does not use personal protective equipment. The remaining employees must evacuate. (*e*) Organize and maintain a structural fire brigade that is trained to fight incipient and interior structural fires until the local fire department responds. The brigade has and uses personal protective equipment.

11. What are important considerations for personnel who may have to rescue a person in a confined space?

Answer. Many people have been killed rushing in to rescue a person in a confined space. Before a rescuer enters the confined space, make sure there is sufficient oxygen, and if not, use an oxygen-breathing apparatus. Depending on the previous contents in the confined space, it may be necessary to first check for dangerous fumes or toxic gases by the use of gas-detecting instruments. It may be possible to blow fresh air into the enclosure. If not, make sure to wear the proper protective equipment. Finally, determine before entering the best method for getting the injured person out and what additional equipment or help may be needed to do this safely.

12. What does rusting do to a confined space that has been sitting idle such as a boiler or tank?

Answer. Moist steel surfaces consume oxygen by rusting, and this can lower the oxygen content well below the 19 percent specified by OSHA as minimum oxygen content for a confined space.

13. What is the danger of inhaling ammonia that may be used in a refrigeration plant?

Answer. Ammonia can become dangerous if it mixes with moisture to form aqua ammonia, or ammonium hydroxide. This substance is an irritant to skin and eyes and in severe concentrations can cause skin and lung burns and even may cause blindness. Inhaled ammonia mixes with the moisture in mouth, throat, and lungs to start forming aqua ammonia, and this is the reason any facility handling ammonia is required to have proper gas masks that should be worn if a leak develops in the system. Protective clothing should be worn under these conditions so no skin burns occur. This also applies to the hands, where recommended gloves should be worn.

14. Can ammonia leaks cause explosions?

Answer. Ammonia burns at high temperature and decomposes into nitrogen and hydrogen at about 1800°F. Hydrogen has a broad range of explosive mixtures with air, and it is the released hydrogen that usually causes the explosion. Fired objects should be kept out of any room with ammonia, as should any other high-temperature heat source.

15. How may a welder be exposed to lead fumes?

Answer. Lead fumes may be formed when welding or burning surfaces that were previously painted with lead-based paint. Lead adversely affects the brain, central nervous system, circulatory system, kidneys, and muscles; therefore, proper respiratory and protective clothing should be worn when working on any parts that contain lead-based paints.

16. What are the dangers of carbon monoxide and carbon dioxide gases as produced in the combustion of fossil fuels?

    Answer. Carbon monoxide usually results from the incomplete combustion of fossil fuels. It is a colorless odor-free gas. The OSHA-permissible exposure limit is 35 ppm. Carbon monoxide is extremely dangerous and toxic. Low levels of exposure result in pounding of the heart, ringing in the ears, and throbbing headaches. High concentrations can be fatal since the gas interferes with the body's ability to transport oxygen to the bloodstream. Note that a separate supply of oxygen is needed when entering any space that contains carbon monoxide above the OSHA threshold limit. Carbon dioxide poses the danger of suffocation from lack of sufficient oxygen in an atmosphere that may have excessive carbon dioxide. For example, at 0.50 percent carbon dioxide, headaches and drowsiness appear. Larger doses can cause a person to black out, and when the concentration reaches about 8 percent carbon dioxide, a person may die from suffocation. Again, because the surrounding air with carbon dioxide in it can be dangerous, and not suitable for breathing, a separate supply of oxygen is needed for suitable face masks in a carbon dioxide filled room or enclosure.

17. What are the steps required in making an appropriate chemical protective clothing choice?

    Answer. It is necessary to identify (*a*) the chemicals and their concentrations, (*b*) the hazards involved with the chemicals, such as effects to skin, eyes, lungs, toxicity, threshold limits, combustibility, explosion hazard, (*c*) the physical environment of the exposure-confined space, spill in liquid form, open space vapor-cloud, pin-hole leak from a weld, and similar considerations, (*d*) the potential activity that the worker may be exposed to in combating or cleaning up the hazardous substance, and (*e*) the length of the exposure, which will depend on the extent of the release of the hazardous material.

18. What is the difference between a *vapor cloud* fire and explosion and a *fireball?*

    Answer. Both involve the leakage of flammable liquids from containers. *Vapor cloud fires* result when a volatile liquid is released from a container and is not ignited immediately. The volatile liquid will start to vaporize and form a cloud or plume and drift with the prevailing wind currents until a source of ignition is found. If the air-fuel mixture is in the right proportions, rapid and instantaneous combustion takes place that can reach the speed of an explosion and injure and kill people in its path, as well as cause huge property damage. *Fireballs* result when a boiling liquid inside a container is heated by internal or external sources so that the pressure in the container causes the tank to rupture. This releases a volatile liquid that rapidly vaporizes and forms a combustible mixture with air because it has already been superheated inside the tank prior to rupture. When ignition occurs, the resultant severe fire can be over 1000 ft in diameter and is called a fireball. It can injure or kill people in the vicinity.

19. What are first-, second-, and third-degree burns?

    Answer. The most severe are second and third degrees, which require medical treatment by a doctor or hospital. *First-degree burns* turn the outer layer of skin pink or red similar to a mild sunburn. Usually washing and applying a first-aid ointment to the affected area will help relieve the pain. *Second-degree burns* affect both the outer and inner layer of skin and usually blister. If nerve endings are exposed to air, or are affected by swelling, this type of burn can be painful; it requires medical treatment. *Third-degree burns* destroy all skin layers in the burned area, producing a charred or white appearance. The person so affected can go into shock and requires fast medical attention. In all cases, general recommendations are to cool the skin to give some pain relief until professional medical help is provided.

20. How is radiation measured?

Answer. *Ionizing radiation* such as x-, gamma, and cosmic rays are measured in rems, with two threshold limits, rems/year and millirems (mr) per hour. The maximum allowed by the Federal Drug Administration (FDA) is 0.5 mr/h. *Nonionizing radiation* is measured in watts (W) per cm^2 with two limits per FDA regulations: For ultraviolet light, it is 1×10^{-7} W/cm^2; for visible infrared, it is 1×10^{-2} W/cm^2.

21. Are there any threshold limits for ozone exposure?

Answer. Per OSHA the threshold limit is 0.1-ppm average exposure over a given day. Industrial hygienists recommend 0.3 ppm of *maximum* exposure at any time.

22. What is the hazard in PCB release from transformers or similar equipment?

Answer. PCBs can cause liver damage and birth defects. The heat of a flame causes PCBs to break down and produce dioxins and dibenzo furans—toxic substances. Toxic substances so released can get into the bloodstream through the lungs and then affect other parts of the body. OSHA's threshold limit for PCBs is $1 \ \mu g/m^3$.

CHAPTER 3
FIRE HAZARDS AND PREVENTION

Operators of plant services learn ways of avoiding plant or facility fires in their duties of protecting the property. This includes the obvious one of keeping trash and rubbish from piling up so that it does not become a source of combustion. Operators also can become involved in fighting fires if they do occur. It is their responsibility to take action to prevent the spread of fire, and they must institute life preservation measures for the occupants of the facility.

FIRE CODES

Fire detection and suppression are also governed by local, state, and federal OSHA rules and regulations, which facility operators should be familiar with, and which have as their main priority the preservation of life. Municipal codes are legal requirements and therefore must be complied with. This also applies to state and federal statutes. Municipal codes may start with building codes that govern the construction and occupancy of a building. Fire-resistant construction is often specified for the type of building and occupancy by the local building department that administers the code. Sprinkler protection may also be specified on new construction.

Fire prevention codes are involved with local fire departments, which have the responsibility of regulating potential fire hazards, such as gases and flammable liquid storage, and the methods to be used in fighting fires.

Building codes are patterned after nationally recognized model codes, such as the Uniform Building Code or National Building Code. Fire prevention guidelines are detailed in the National Fire Protection Association (NFPA) standards on many facility equipment; they include electrical guidelines and ones for burners, stoves, and similar premises facilities.

Insurance companies are interested in *life support systems,* and with respect to property coverage, place additional emphasis on the preservation of the building and its contents. Thus, operators must become familiar with insurance companies' property conservation recommendations. Insurance companies' engineering representatives can provide information on property conservation, based on their company or national standard requirements or on incidents they have experienced with other similar properties. They generally specialize in identifying fire sources and what can be done to reduce the exposure. Many times these are based on NFPA standards and may have been incorporated into local legal requirements because these standards have received national recognition as good standards to follow. For copies of NFPA standards write to NFPA, Batterymarch Park, Quincy, Mass. 02269.

OSHA FIRE PREVENTION PLAN

After several fires killed workers in an industrial plant, OSHA Labor Fact Sheet No. 91-41 was issued. This requires employers/owners of a facility to implement a *written fire prevention* plan with the following features:

1. The employer must provide proper and workable exits and fire-fighting equipment, have an emergency plan for contingencies that may arise based on the occupancy, and have an employee training program to prevent fire casualties in the workplace.
2. There must be housekeeping procedures for the storage and cleanup of flammable materials and waste. There is a recommendation to recycle flammable waste such as paper, but safe packaging and handling must be followed.
3. Workplace ignition sources such as smoking, welding, and burning-type operations must have written procedures to prevent igniting any flammable substances on the premises. Flammable material must be kept away from all sources of heat. This includes leaks and spills. All heat-producing equipment such as burners, boilers, heat exchangers, ovens, stoves, fryers, and similar equipment must be properly maintained so they function safely, and they must be kept clean and free of leaks and flammable residue.
4. The written plan must be available to employees for their review and input. The employees must be appraised of all potential fire hazards as they relate to their jobs and of the procedures to follow if a fire erupts, as specified in the employer's fire prevention plan. All new employees are to be similarly instructed. If the occupancy or conditions in the plant change, all employees must be so instructed.

FIRE TERMS

Fire terms are briefly reviewed here to give plant service operators an understanding of the quite complex subject of fire prevention engineering. *Fire* is generally defined as a chemical reaction that involves "oxidizing" a substance, solid, liquid, or gas; the visible evidence of this is a fire or flame and is accompanied by the release of heat. Combustion takes place in the presence of oxygen, a flammable substance, and a source of ignition. All three must be present to sustain it.

Fire protection engineers have classified fires as follows:

1. *Class A.*   The burning of ordinary combustibles such as wood, paper, and cloth where water can easily lower the temperature of ignition and thus put out the fire.
2. *Class B.*   Flammable liquid fires such as oil, gasoline, and grease where blanketing or smothering is needed to put out the fire (removing the oxygen).
3. *Class C.*   Electrical equipment fires where an agent that does not conduct electricity is needed to fight the fire.

Fires in metals include the burning of magnesium, powdered aluminum, zinc, sodium, potassium, and similar hazardous metals in which ordinary extinguishing agents do not work to put out the fire. Special powdered or granular substances must be used to cover the burning metal and exclude oxygen from the fire.

A *flash point* is the temperature at which a substance (solid, liquid, or gas) gives off enough vapor to make an inflammable mixture with air. The *ignition, or burning, point* is that temperature at which a substance ignites or burns and continues to burn as long as flammable material and oxygen are present.

Flame luminosity is an indication of the intensity of combustion by the amount of heat liberated. It is related to flame temperatures as follows:

Flame color, heat	Flame temperature, °F
Red	977
Dark red	1292
Cherry red	1562
Pale red	1742
Yellow	2012
Gray-white	2372
Full-white	2732

The *explosive range* is a range or percentage of volume mixture in air that limits the combustibility of the substance and rate of burning when mixed with air (see Fig. 3.1). In a lean mixture the particles of the substance are so widely separated that those set on fire by ignition will not set fire to the other particles nearby. When the particles are so close together that they exclude the proper mixing of air or oxygen with the particles, the mixture is called rich. The lean mixture's point of combustibility is called the lower explosion limit (lel), while the corresponding rich mixture's threshold limit is called the upper explosion limit (uel). Thus, in Fig. 3.1, hydrogen has a lel of 4.1 percent and a uel of 74 percent. The *ignition points* of some common gases are shown in Fig. 3.2.

CAUSES OF FIRE

It may be well worth reviewing the many causes of fire because by knowing the potential sources, facility operators can identify the symptoms and take corrective action. The causes are:

1. Direct contact of combustibles with open flame or glowing materials.

2. Long-time exposure to low heat of combustibles until the released gas ignites and burns the surroundings. Hot steam pipes without sufficient insulation can ignite wood, dust, etc., that touch the steam pipes; this is an example of long-time heating effects.

3. Spontaneous heating or combustion. This is a slow oxidation at ordinary temperatures of material that is a poor conductor of heat. The heat of oxidation is not carried away, and as a result the temperature rises, with the reaction speeding up, until a temperature is reached at which a self-sustaining combustion, or burning, begins. Some heat is caused by fermentation such as hay or grass cuttings in a poorly ventilated pile.

Gas or vapor	Vapor density (air = 1)	Rate of diffusion (air = 1)	Explosive limits, % in air by volume	
			Lower	Upper
Acetic acid...	2.1	0.69	4.0	
Acetone...	2.0	0.71	2.0	9.0
Acetylene...	0.90	1.05	2.6	82.
Alcohol (ethyl alcohol).............................	1.6	0.79	3.5	19.0
Ammonia..	0.59	1.30	16.0	25.
Amyl acetate..	4.5	0.47	1.1	4.0
Amyl alcohol..	3.0	0.58	1.2	5.0
Benzene...	3.0	0.58	1.4	5.9
Benzol (benzene)....................................	2.7	0.61	1.4	8.
Butane..	2.0	0.71	1.6	6.5
Butyl acetate..	4.0	0.50	1.7	
Butyl alcohol..	2.6	0.62	1.7	18.
Carbon bisulfide....................................	2.6	0.62	1.0	50.
Carbon dioxide......................................	1.5	0.82	Nonflam.	
Carbon monoxide....................................	0.97	1.02	12.5	74.
Dichlorethylene.....................................	3.35	0.55	5.6	11.4
Dichlorodifluoromethane (F-12)......................	4.2	0.49	Nonflam.	
Dichlorotetrafluoroethane (F-114)...................	5.9	0.41	Nonflam.	
Ethane..	1.03	0.98	3.3	10.6
Ether (ethyl ether)..................................	2.6	0.62	1.9	22.
Ethyl acetate..	3.0	0.58	2.5	11.5
Ethyl bromide.......................................	3.76	0.51	6.0	11.0
Ethyl chloride.......................................	2.2	0.67	3.7	12.0
Ethylene..	0.97	1.02	3.0	34.0
Ethylene oxide......................................	1.5	0.82	3.0	80.
Gasoline..	3.5	0.53	1.4	6.
Hydrocyanic acid....................................	0.93	1.04	5.6	40.
Hydrogen..	0.069	3.81	4.1	74.
Hydrogen sulfide....................................	1.17	0.92	4.3	46.
Illuminating gas....................................	0.65	1.24	5.0	31.
Kerosene..	4.5	0.47	1.1	6.
Methane...	0.55	1.35	5.6	13.5
Methyl acetate......................................	2.6	0.62	4.1	14.
Methyl (wood) alcohol...............................	1.1	0.95	6.	36.
Methyl bromide.....................................	3.3	0.55	13.5	14.5
Methyl chloride.....................................	1.7	0.76	8.1	17.2
Methylene chloride (dichloromethane)...............	2.9	0.58	†	†
Methyl formate.....................................	2.1	0.69	4.5	20.
Monofluorotrichloromethane (F-11)..................	4.7	0.46	Nonflam.	
Pentane...	2.5	0.63	1.45	7.5
Propane...	1.5	0.82	2.3	7.3
Propylene...	1.45	0.83	2.2	9.7
Sulfur dioxide.......................................	2.2	0.67	Nonflam.	
Toluol (toluene)....................................	3.14	0.56	1.4	7.
Turpentine..	4.7	0.46	0.8	
Xylol (xylene)......................................	3.7	0.52	1.0	5.3

FIGURE 3.1   Upper and lower explosive limits, flammable vapors and gases. + Indicates practically nonflammable at ordinary temperature. (*Courtesy, National Board of Fire Underwriters Bulletin.*)

4. Explosions or rapid propagations of flame. These result from rapid decomposition of a substance with large release of heat (exothermic reaction) and also rapid flame travel because of released flammable vapors or by rapid release of pressure.

5. Lightning igniting combustible material.

6. Dust fires and explosions. Dust can accumulate from such substances as lamp-black, grains, wood, flour, starch, sugar, wool, fibers and fluff from shearing, metal powders, resins, celluloids, and plastics.

Gas	Ignition, in oxygen, °F	Ignition, in air, °F	Autoignition, °F
Hydrogen	580–590	580–590	1076
Carbon monoxide	637–658	644–658	1204
Ethylene	500–519	542–547	842
Acetylene	416–440	406–440	635
Hydrogen sulfide	220–235	346–379	500
Methane	556–700	650–750	999
Ethane	520–630	520–630	959
Propane	490–570		871
Ammonia	700–860		1204

FIGURE 3.2   Ignition temperatures of some gases.

7. Electric sparks. Electricity can be a source of intense heat, and if sparks develop from poor connections, failure of insulation, overload, or similar causes of electrical breakdowns, these sparks can ignite combustible vapors, liquids, or solids.

8. Chemical reactions. These can cause the release of flammable gases which in severe cases can form vapor clouds that when finally ignited cause very destructive explosions and fires.

9. Friction causes heat to develop, and this heat raises temperatures of the surroundings until combustion of any nearby flammable material may result.

10. Static electricity. This is produced by friction, or the rubbing together of two surfaces, which can produce a buildup of electric charge. The charge can reach a point where a spark-over occurs that can ignite combustible materials. Of great concern are flammable liquids flowing in pipes that generate static electricity and hold it if there is no ground drainage for it. It is necessary to ground any conducting-type discharge nozzle or tip so that the static is drawn off.

Autoignition temperature, as shown in Fig. 3.2, is the lowest temperature at which a solid, liquid, or gaseous substance will initiate a self-sustained combustion in the absence of a spark or flame. This temperature can vary considerably since it is influenced by the nature, size, and shape of the igniting surfaces.

Other Fire Causes

Fires are also set by humans either deliberately (arson) or by accidental actions or behavior. A careless disposal of matches or cigarettes has started many fires. The spread of fires has also been accelerated because the water supply had been shut off to repair a leak, no one reopened it, and a fire erupted. Lack of emergency plans, faulty alarms, and similar shortcomings also help to spread fires.

Another fire protection engineering term is *fire loading,* usually expressed as the weight of combustible materials within the structure and the contents in the building, commonly measured as Btu's per square foot of floor area. For example, wood and paper are calculated as having 7000 to 8000 Btu/lb, while flammable liquids are assigned 15,000 Btu/lb. Fire loading is used to measure the severity of a fire and also to design the sprinkler system that may be required.

Fire-resistance ratings of various building materials and structures are determined by tests standardized by national testing organizations, such as ASTM, Underwriters' Laboratories (UL), and Factory Mutual (FM). These tests measure heat transmission through the material and/or construction, extent of damage when exposed to flame, and even the effects of hose streams. From the tests, the time required to spread a fire without sprinkler protection is determined.

Fire ratings are based upon the time element of exposure to fire and heat and can range from $\frac{1}{2}$ to 8 h depending on the material or construction. Thus, to prevent steel columns and beams from buckling in a fire, they are encased in concrete or sprayed with a fire-resistant mineral fiber. Many fireproof constructions are specified by local codes, and facility operators should become familiar with them as they affect their facility so that the original design integrity is maintained during any alterations or repairs.

CONTENTS PROTECTION

Besides sprinkler systems, most insurance company fire protection engineers also stress compartmentalization of the internal space so that barriers to the spread of fire exist. *Fire walls* are used as barriers between buildings or large floor areas. They must be of fireproof construction since their value depends on insulating quality, tightness, stability, and strength. The combustibility and value of the contents determine the location of fire walls, whose integrity should be maintained. Fire walls should extend above the ceiling to the roof and even beyond in certain occupancies in order to avoid the fire spreading in the attic spaces.

Fire doors are used in openings of fire walls that separate important premises areas or that may enclose a hazardous process area. They should have the same fire rating as the fire wall, and where heavy process traffic exists, automatic closing of the door is usually specified and should be enforced by plant service operators.

Automatic fire doors or *dampers* are also required in ventilating or air conditioning ducts so that fire or heat cannot be spread from its origin to other parts of the facility through the ducts. These should be kept in good condition by periodic inspection, testing, and maintenance.

In general, the hazard presented by the occupancy in a facility will determine how much compartmentalization is needed against the spread of fire. Once this is determined, it is important to keep the barriers, and sprinklers, if so equipped, in workable condition at all times.

FLAMMABLE LIQUIDS AND GASES

Flammable liquids and gases present serious fire and explosion hazards because they ignite easily, are difficult to extinguish, and burn fast with intense heat. Many local, state, and federal rules apply to their storage, shipment, pumping, and processing, and these are beyond the scope of this book. Insurance companies have also drawn up some guidelines. For example, for hazardous liquids such as acetates, acetone, alcohols, benzol, carbon disulfide, ether and ethyl, gasoline, toluene, and xylol, the guidelines include:

1. Isolate the liquid hazard by storing it away from buildings or by use of fire cutoffs.

2. Confine the liquid in closed pipes or containers in order to avoid a source of easily ignitable fuel.

3. Ventilate the areas in which flammable liquids are used in order to reduce the vapor concentration to a safe level.

4. As backup where flammable liquids are handled, explosion vents in the room will reduce the potential pressure buildup in case an accidental ignition does occur and thus will limit the damage to the building.

5. Eliminate as much as possible ignition sources such as welding and electric sparking by isolating the electrical equipment from the flammable material or using explosionproof construction, banning smoking, removing oily rags, etc., from the process area.

6. Employees must be trained to follow safe practices and be advised of the dangers inherent in the flammable liquid or gas process being used. This should include an awareness of defective equipment, controls, and safety devices that should be reported immediately to plant management. OSHA now requires a process safety program for employees that includes written instructions for proper operating procedures for emergencies and actions to be taken in the event a fire or explosion occurs. Plant service operators should be involved with these procedures.

7. Good, workable fire protection is needed in the areas involved with flammable substances. Sprinkler systems are required to back up fireproof construction, fire doors, and fire walls.

HAZARDOUS CHEMICALS

There are many chemicals that have fire or explosion hazards as a result of reacting with other substances. These are chemicals that oxidize or react with organic combustibles that can cause ignition to occur. Consult with your fire insurance company representative on the chemicals so classified. These types of chemicals should be kept away from combustibles, open flames, heat, or potential ignition sources.

DUST FIRES AND EXPLOSIONS

Dusts can range from particle to flour size, which can pass through a screen of 400 mesh/in^2. The finer the dust, the more rapidly an explosion can occur and the more violent it can be. Dust becomes flammable because of its ability to adsorb air and other gases or vapors to make an explosive mixture. Dust explosions occur in two stages: (1) an ignition stage that forms an atmosphere of hydrocarbons and carbon monoxide, and (2) this explosive mixture then is rapidly ignited by the initial flame of burning dust. The two actions can be very rapid and almost indistinguishable. Organic and metallic dusts that liberate oxygen in decomposition while burning can cause the most violent of dust explosions.

The lel is the minimum concentration of combustibles in the air that will propagate a flame, and experiments indicate this averages 0.065 oz of combustible dust per cubic feet of air. Dust explosion intensity depends upon the rate of pressure rise and the maximum pressure resulting from the explosion. Dust explosion pressures range from 3 to 49 psi for powdered metals to 34 to 51 psi for fertilizer dusts.

The prevention of dust explosions has received great attention, especially in the grain industry, which has had some destructive explosions with loss of human life. Generally, the following is recommended to prevent dust explosions:

1. Hazardous processes with a dust explosion potential should be compartmentalized in separate buildings or separated by fire walls and doors.

2. Process equipment should be kept dust tight, and frequent cleaning is recommended to prevent the accumulation of any dust cloud and to prevent dust from starting a local fire.

3. All possible sources of ignition should be eliminated by (*a*) using explosionproof electrical equipment or compartmentalizing the electrical equipment in separate fireproof rooms as much as possible, (*b*) using indirect heating to prevent dust from collecting on heated surfaces, and (*c*) grounding all equipment to drain off any static electricity.

4. Any conveying, collecting, or processing equipment must have features that will not produce metallic sparks or frictional heat that could ignite the powder being handled.

5. Vents such as explosion doors, light roofs, and large hinged windows sized for the anticipated pressure rise can help to release explosion pressures and thus prevent structural damage.

6. Fire fighting emergency plans should be in place to fight the type or class of fire that can be expected for the installation.

7. Inert gas, such as nitrogen, can be used in storage and piping systems that convey the combustible dust in order to prevent ignition.

A *hazardous location* is defined as one in which a highly ignitable material is present; the material can be a liquid, gas, or dust. While liquids having a flash point lower than 200°F are considered flammable, it is apparent that if any combustible liquid is heated, it will produce flammable vapors above the 200°F flash point. For example, heavy fuel oil under atmospheric temperature may have a flash point of 0°F but when heated above 300°F may release vapors as flammable as gasoline. It is generally good practice to treat all combustible liquids as potentially hazardous.

Explosionproof electrical equipment are generally motors and switches that are so constructed that if a vapor or gas is ignited within the enclosure, such as may occur from an electrical failure, the resulting flame is prevented from propagating to the outside surrounding atmosphere. This prevents a fire or explosion from spreading because of electrical sparking.

All plant service operators should have a copy of the National Electrical Code, available in paperback form from the NFPA, Batterymarch Park, Quincy, Mass. 02269. This code details the features that are required of all equipment that is installed in the Classes and Division locations. This includes necessary nameplate information from the manufacturer of the equipment so that it can be matched to the conditions existing at the installation location.

FIRE PREVENTION

Even though a building may have various degrees of fireproof construction and may also be fully or partially sprinklered, plant service operators should always try to prevent fires by inspection, testing, and good maintenance practices. Areas requiring attention are discussed below.

Electrical Fires

Electrical failures are caused by dirt and grease on or inside equipment that can lower insulation resistance and cause a failure to ground or between phases—a dead short. The same effect is produced if electrical equipment becomes wet. Thus equipment must be kept *clean and dry*. Insulation resistance tests to note trends are a good way to check to see if deterioration is occurring.

Arcing at loose joints or connections creates intense heat that can ignite nearby combustibles, or fuse controller contacts, causing in some cases continuous unsafe operation or single phasing of three-phase equipment. All joints, including contacts on controllers or relays, should be checked at least annually for arcing or the closely related *sparking* on the joints, connection, or contact points. Thermographic surveys are used to detect "hot spots."

Overheating may be caused by overload, single phasing of three-phase equipment, and the blocking of cooling passages by leaves, lint, powders, grease, and similar substances. The equipment will fail from "roasting" and thus could cause a fire to start.

Electrical fires are Class C fires, so operators should know how to use carbon dioxide extinguisher, which is normally used.

Transformer failures may be due to loose internal or external connections causing single phasing, lightning strikes, or deterioration of insulating oil, to name a few. Transformers with critical loads should receive yearly insulation tests on the windings and on the oil as well as inspection and maintenance on controlling relays and switches.

Generators should receive the inspection, testing, and maintenance usually specified by the manufacturer. This includes checking tightness of windings, cleanliness, and the condition of bearings, oil, hydrogen if so cooled, and conducting insulation resistance tests.

Motor maintenance closely follows the generator items mentioned above. This will also be influenced by the kind of motor. For example, motors with brushes, such as dc, synchronous, or wound rotors need more attention.

Wiring and cables need to be checked for damage due to moisture, corrosion from acids and vapors, heat, loose connections, chafing at entrance or exits from conduits, and rodent damage.

Moisture, loose connections, and worn contact points, causing arcing across the points, are the chief hazards for *switches*. Many failures of incoming switches have been caused by rodents entering a switch to keep warm. Switches should be kept closed to prevent any arcing or flame from damaging the surrounding area.

Lightning arrestors should be checked for tightness of connections and continuity of grounding; the resistance to ground should not exceed 5 Ω per code requirements.

Electrical equipment needs careful attention due to the inherent energy in the form of heat that can become uncontrolled, and thus start fires.

Spontaneous Ignition

Heating and ignition can occur from an oxidizing chemical reaction, not from a flame, spark, source of heat, or friction. If ignition occurs after reaching that temperature, it is called *spontaneous combustion.* Substances that can cause spontaneous combustion are oily rags, wool waste, wood sawdust, sugar and gunny sacks, lampblack, and printed matter with linseed oil or print in piles, to name a few. Hay and grain ferment, giving off heat that can cause spontaneous combustion. There are also many chemicals that pose this risk, including barium oxide, sodium peroxide, hydrides of phosphorus, and silicon in air. Charcoal and coal piles need attention as well as vegetable oils and fats. Some chemicals such as quicklime, potassium, sodium, and acetylene start giving off heat with the adsorption of moisture. Plant service operators should practice good housekeeping on some obvious sources of spontaneous combustion, such as oily rags and similar waste material. Chemicals require special attention which should be available from the supplier or manufacturer.

Smoking

Because of the health hazard, the trend is to limit smoking to designated plant areas, which should be reasonably fireproof. Signs should be posted reminding people to extinguish all cigarettes in such areas.

Open Lights or Flames

Open lights or *flames* should come from equipment that is noncombustible and should not be left unattended. Fire fighting equipment should be nearby.

Ovens and Driers

The installation of *ovens* and *dryers* should follow NFPA guidelines. Leaking fuel lines should be repaired immediately. Flame failure safeguard systems are usually recommended where there is a chance for fuel buildup in the furnace. Late ignitions have caused furnace explosions. Proper clearances from combustibles are needed. Cleanliness around the furnace must be maintained.

Welding and Cutting

Welding and *cutting* operations require all surrounding combustibles to be protected by steel plates or fireproof blankets. Fire extinguishers should be readily available. A "fire watch" should be in attendance during any welding. Cylinders of oxygen should not be stored in the same room with cylinders such as acetylene and should be kept away from heat sources or combustibles.

Friction

Mechanical friction creates heat and should be avoided. This mostly involves bearings and belts in most facilities. Proper inspection and maintenance should avoid this hazard.

Leaks

Leaking lubrication or fuel lines are especially dangerous near any hot oven, steam lines, or similar hot spot. Steam turbogenerators have had huge oil fires as a result of lubricating oil from a cracked pipe hitting hot steam pipes. Diesel fires have occurred from oil accumulation around the engine. The source of any oil leak should be corrected immediately.

Rubbish and Litter

Any undue accumulation of *rubbish* and *litter* can start a fire due to spontaneous combustion from oily waste or papers or can be a source of combustible material ignited by a carelessly thrown match or cigarette. Rubbish should be placed in fire-resistant receptacles and disposed of before it decomposes and becomes a health and fire threat.

Static Electricity

Static electricity is created by friction or bringing together and separating two unlike substances. Electric charge can build up to the point that sparking occurs. Some industries, such as textile manufacture, use humidity control to drain the charge, or they prevent its buildup on dry goods in the process of manufacturing. Others prevent the accumulation of charge by grounding all moving parts and containers. This is especially applicable to flammable liquids in motion through pipes and nozzles, moving dust, finishing and coating operations, and on belts and conveyors in motion. It is thus essential to periodically check the grounding to avoid sparkovers.

Molten Metal

The areas where metal is molten require noncombustible floors and walls. Incidents involving electric melting furnaces include molten metal igniting electric cables that were supplying power to the furnaces. Furnaces require periodic inspection for cracks or thinning. The pouring out of any hot substance from a container requires disciplined, trained operators to prevent the molten material from coming in contact with combustibles.

FIRE PROTECTION

Fire protection starts with the building site selection and construction and with local, state, and federal construction and installation requirements. Its true purpose is to preserve lives, property, manufacturing operations, and, in most cases, finished goods in storage waiting to be shipped or sold. The type of fire protection needed depends on management's concern for its workers and the community, legal requirements, and the financial risk that a fire can impose on an enterprise. There is one more important factor that is often overlooked—the loss of customers due to curtailment of operations from a serious fire. Insurance cannot compensate for some of the fire losses—loss of customers, skilled workers, records and process parameters, or one-of-a-kind process machinery. There are many choices available to management for fire protection—fireproof construction, fire walls and doors, trained fire fighting workers, strategically placed portable fire extinguishers, and permanent sprinkler systems. Insurance can be bought to cover direct property losses and indirect losses, such as loss of profits, damage to finished stock, and spoilage. It is not the intent of this section to name all the options available in preventing fire losses but rather to point out the considerations that make up fire protection methods and systems and to help plant service operators become familiar with the systems on their premises and what is available for any upgrading or additions.

Fire protection engineers always state, "assume a fire occurs. What protection is available to prevent the spread of fire and to put out the existing fire?"

Water Supply and Fire Fighting Plans

A good public water system is usually the most economical source of fire protection. This may have to be supplemented, depending on the occupancy or where the public supply is primarily for domestic water. However, all plant service operators should be familiar with the *fire fighting plan* for the property. The following should receive attention:

1. Familiarity with the property's ground plan and the type of buildings on the property.
2. If one is not available, make a sketch of public water connections to the property including pipe and valve sizes and where they are located.
3. Also sketch check valves, hydrants, fire pumps, and hose connections.
4. Know where fire extinguishers and first aid equipment are located.
5. If there are sprinkler systems, know how they are activated and what valves must be open for them to be effective. This also applies to foam and fog systems.
6. Have readily available fire department alarms and phone numbers so fire brigades, police, and executives can be called in case of a fire or fire fighting impairment.
7. Know all exits and the procedure for occupants to follow to leave the building.
8. Become thoroughly familiar with the plant's emergency plan for disconnecting electrical equipment, shutting down process equipment, and staffing fire fighting equipment.
9. Determine what kind or class of fire exists so that proper extinguishing agents will be used in fighting the fire.

Fire protection engineers classify a *primary water supply* as one that will supply the gallons per minute (gpm) needed by sprinklers within the first few minutes following the start of a fire or prior to a secondary water supply being made available. Primary demand can be 100 percent of *total demand* in occupancies involving flammable liquids or gases, hazardous chemicals, or explosive dusts. Primary demand has been shown by experience to involve about 25 or more sprinklers—roughly 500 gpm. Total demand is the gallons per minute required to supply *all* sprinklers.

Primary water supplies can be (1) public water systems, (2) elevated gravity tanks and reservoirs, where public water is not available or there is insufficient pressure or flow, and (3) pressure tanks used with gravity tanks in multistory buildings to supply high pressure to sprinklers in the upper stories.

Secondary water supplies are always required where the primary supply is insufficient. Sources for secondary supplies are (1) gravity tanks, (2) fire pumps, (3) booster pumps that take suction from a low-pressure public water supply, and (4) fire department pumper connections.

Standpipes are used to obtain fire streams at the upper stories of high-rise buildings. Standpipes are of two types: (1) small water hose streams to be used by the building employees and (2) large water hose streams to be used by the public fire department or by a fire brigade.

The standpipe sizes and numbers are determined by the floor area and occupancy to be covered and also by the distance from the outlets to the source of water supply. The number of standpipes is recommended to be so arranged so that all parts of every floor can be reached within 30 ft by a 100-ft hose connection to a standpipe.

Water supply to standpipes from public water systems generally is recommended to be 250 gpm for one riser and 500 gpm for two or more risers. The supply must be capable of delivering 40- to 50-psi water pressure for at least 1 h through a $2\frac{1}{2}$-in outlet.

Where fire pumps are used, for a 150-ft-high building, delivery must be 500 gpm for 4-in standpipe, 750 gpm for a 6-in standpipe or for two 4-in pipes, and 1000 gpm for two or more 6-in pipes.

Water pressure should be available at all times in standpipes. It is also generally required that at least one fire department connection be provided per standpipe. Check this with your local fire department.

Flow tests of public water supply are performed by local fire departments and fire insurance company representatives to determine if sufficient flow is available at the property site 24 h a day. It is a good way to determine if there are any obstructions in the water pipes and whether there is sufficient water available to fight a fire.

Testing procedures are based on equations of water flow in pipes by converting pressure drops to velocity of flow and then multiplying this by the cross-sectional area of the nozzle opening and also correcting this for friction, pipe characteristics, and Reynolds number. Capacity flow tests are made by opening one, two, or more hydrants. Two pressures appear: "static" pressure at the hydrant with no flow and "residual" (flow) pressures with test streams flowing. Flow tests compare the test results with recognized national standards.

Most fire protection engineers use hydraulic-flow curves. Some approximations can be made. For example, capacity tests of public water systems are quite often converted to the minimum discharge flow available at 20 psi residual pressure. Let us assume the static pressure is 56 psi, and it is known that previous tests developed 31-psi residual with a calculated flow of 2500 gpm through the nozzle, the flow for residual pressure of 20 psi can be found by comparing the square root relationships of the two pressure drops. Remember the rate of flow is approximately proportional to the square root of the pressure drop. For example,

$$56 - 20 = 36 \text{ pressure drop at 20 psi residual}$$

$$56 - 31 = 25 \text{ pressure drop at 31 psi residual}$$

$$Q = \text{flow at 20 psi residual}$$

$$\frac{Q}{2500} = \frac{\sqrt{36}}{\sqrt{25}} \quad \text{or } Q = \frac{2500 \times 6}{5} = 3000 \text{ gpm at 20 psi residual}$$

Fire insurance loss-prevention representatives can assist plant service representatives in checking the adequacy of water flow to their facility.

Fire Pumps

Fire pumps are generally secondary water supply sources; however, they should not be neglected by plant service operators. Their construction and installations are governed by fire insurance requirements or by local fire department rules and regulations. Some requirements on pump rooms are:

Maintain pump room temperature above 40°F.

Do not use the pump room for storage.

Provide lighting and emergency lighting at all times. Battery-operated emergency lights are acceptable.

Access to the pump room must also include access from outside the building.

Pump rooms must be located and constructed so as to protect the pump and its controller from falling machinery or floors and from a fire that may drive the operator away or damage the pump when it is needed the most.

Maintain drainage and ventilation of the pump room. For internal combustion engines, ventilation should be sufficient to provide combustion air and for the removal of exhaust vapors.

Engine-driven pumps should be installed above grade to avoid flooding the engine. Electric motor-driven units and their controllers require dry cooling air so that moisture or humidity does not affect the electrical insulation. This is especially applicable in winter weather and in rainy seasons. Make sure the room is heated to control the humidity. It is preferred that suction be from a source that is above the tank and not through suction lift. Stored water sources should be able to supply the pump at 150 percent of pump-rated capacity long enough to protect the property. Residual pressure cannot go below 20 psi.

For *motor-operated pumps,* requirements are:

The motor should be suitable for across-the-line starting.

An ammeter should be provided on the controller so that load can be checked.

A pilot light should show that electric power is available.

Some locations require an audible alarm or visual signal to indicate when no electric power is available to the pump.

For remote pump locations, an alarm should sound if the fire pump starts to operate.

For *engine-driven pumps,* requirements are:

Make sure the fuel storage capacity is such that the engine can operate for at least 8 h.

Exhaust from the engine should be piped outside.

Engines are required to have workable *tachometers* for engine speed indication, *oil pressure* gages, *temperature gages* on water cooling-type engines, *ammeters* to indicate battery charging rate, and *pilot lamps* to show that the ignition is on.

It is important for everyone in a plant who may have to start the fire pump to be trained to do so. This includes guards. The pump should always be maintained in good condition for immediate operation if needed. It is important to maintain water in the suction side of the pump at all times. If priming is required, do not start the pump until there is a steady flow of water at the umbrella-vent cocks. Follow the pump manufacturer's instructions in starting, operation, and maintenance to avoid problems.

Pump Tests.   Pumps should be tested weekly and run at full speed, discharging water at a convenient outlet per most insurer's recommendations. Similar operational checks should be made on the pump driver. On steam turbine drives it is essential to drain condensate out of casing and steam supply lines to prevent water from "slugging" the turbine blades. Make sure the speed and pressure governors are working properly. The same applies to overspeed trips.

Capacity Tests.   These tests are usually made with a fire insurance company representative, usually on an annual basis. Their purpose is to make sure that the pump can still deliver the gallons per minute that the system requires. Impairment of rated delivery is due to obstructions or defective valves on the suction side. Plugged impellers are another reason. Low water supply pressure from the public service can be another. Closed incoming supply valves due to street repairs is another common finding in capacity tests.

Insurance company representatives compare the percentage of rated flow to poor, fair, good, and excellent pump charts. A poor rating requires immediate attention for correction. Fair ratings also need attention by making arrangements for repairs. Some causes of poor ratings are air-bound pump, impeller damaged or seized, low net head, pump not primed, plugged pump strainer, cavitation damage, obstructed suction line, excess bearing friction from dry bearings, pump and driver misaligned causing vibration, worn or eaten up turbine blades from entrained moisture during idle periods, and faulty steam turbine governor.

Fire Extinguishers

The term *fire extinguisher* is applied to portable fire extinguishers strategically located near any combustible or fire source operation, such as welding, in order to immediately fight a fire at inception, before it becomes too large. Extinguishers must be selected and placed near where one may expect a Class 1, 2, or 3 fire (see Fig. 3.3). The portable extinguishers help in putting out smaller fires by cooling the burning substance usually with water below the ignition point, by excluding the air supply, and by preventing arcs in electrical equipment with halon or similar extinguishers in computer installations or CO_2 on electrical motors and generators in power plants.

The spacing of fire extinguishers depends on occupancy per the following classes:

Class 1 occupancy. For light hazard, such as schools, offices, and public buildings, the requirement is one $2\frac{1}{2}$-gal tank within 100 ft of a person and placed on the floor and at least one unit for each 5000 ft^2 of floor area.

Class 2 occupancy. For ordinary combustible occupancy, such as department stores, warehouses, and light manufacturing, the requirement is a 50-ft distance from any point for a person to reach the unit and one unit per 2500 ft^2.

Class 3 occupancy. This is an extra hazardous occupancy, such as wood-working, paint spray, dipping, etc., and requires a 50-ft maximum distance for a person to reach the extinguisher and a minimum of one unit per 2500 ft^2 of floor area, with a strong recommendation to have additional units if the severity of the hazard requires it. For example, one unit per oven or spray booth.

Wheeled extinguishers for larger hazards should be placed within 200 ft of all ordinary hazards and 75 ft of extra hazards.

Fire pails are also used, and 5- to 12-qt pails are the equivalent of a $2\frac{1}{2}$-gal soda-acid extinguisher.

Extinguisher Types. Extinguishers have been developed to fight the different classes of fires (see Figs. 3.3 and 3.4). A brief description of each type and its use follows:

1. *Soda-acid extinguisher* for Class A fires. Contains bicarbonate of soda to be dissolved in water and liquid sulfuric acid. Reaction of the acid with the soda produces pressure which expels the extinguisher up to 40 ft.
2. *Plain water extinguisher* for Class A fires. Contains plain water and a CO_2 cartridge to provide the pressure. Discharge streams to 40 ft.
3. *Calcium chloride extinguisher,* also called antifreeze. Used for Class A fires. Contains plain water or antifreezing solution for use in unheated or outdoor winter fires. The solution is made of granulated or flake calcium chloride and water. Pressure to expel may be created by a CO_2 cartridge or by chemical reaction. Stream can reach 40 ft.
4. *Pump tank extinguisher* for Class A fires. The contents are the same as for calcium chloride extinguisher, but pressure to expel liquid is created by a hand pump that is installed in the tank.

Type of portable extinguisher	Suitable for following fires		
	Class A (wood, textile paper, rubbish)	Class B (oil, gasoline, grease, paint)	Class C (electrical equipment)
Water pail	Yes	No	No
Soda-acid	Yes	No	No
Plain water	Yes	No	No
Calcium chloride (antifreeze)	Yes	No	No
Pump tank	Yes	No	No
Vaporizing liquid	No (Can be used for small fires)	No	Yes
Carbon dioxide	No (Can be used for small fires)	Yes	Yes
Dry chemical	No	Yes	Yes
Foam	No	Yes	No

FIGURE 3.3   Portable fire extinguishers and the type of fire to which they can be applied.

5. *Vaporizing liquid* extinguisher for Class B and C fires. The contents are specially treated nonconducting liquids of bromochloromethane base. Pressure to expel may be created by built-in pump or by using compressed air or CO_2 gas. The liquid is heavier than air and will not support combustion. Its freezing point with components added can be as low as $-50°F$. The stream can reach 30 ft.

6. *Carbon dioxide* extinguisher suitable for Class B and C fires. It contains CO_2 in liquid form. At ordinary temperatures and pressure CO_2 is a gas, which is inert, and will not support combustion. It is used primarily to fight flammable liquid fires and on electrical equipment. The stream of liquid expelled under pressure forms an inert gas that suffocates a fire from lack of oxygen. Precautions are needed because it may also suffocate human beings if the concentration is high.

7. *Dry chemical* extinguishers are used to fight Class B and C fires. They contain bicarbonate of soda which has been chemically processed to make it waterproof and free flowing. Pressure to expel comes from a CO_2 cartridge. Streams up to 14 ft occur. If a wheeled extinguisher is used, streams can reach 35 ft.

8. *Foam extinguishers* are used to fight Class A and B fires (see Fig. 3.4*f*). Contents are dry bicarbonate of soda, a foam stabilizing agent such as soap bark dissolved in water in the outer compartment, and aluminum sulfate dissolved in water for the inner cylinder. Reaction between the aluminum sulfate and the bicarbonate of soda produces CO_2 gas in the form of bubbles and also some aluminum hydroxide in the bubble film, which expels the foam up to 40 ft.

Fire extinguishers need to be maintained so that contents are present in the correct amount. Except for carbon dioxide, dry chemical, and vaporizing liquid extinguishers, they should be hydrostatically tested at least once in 5 years or where corrosion or

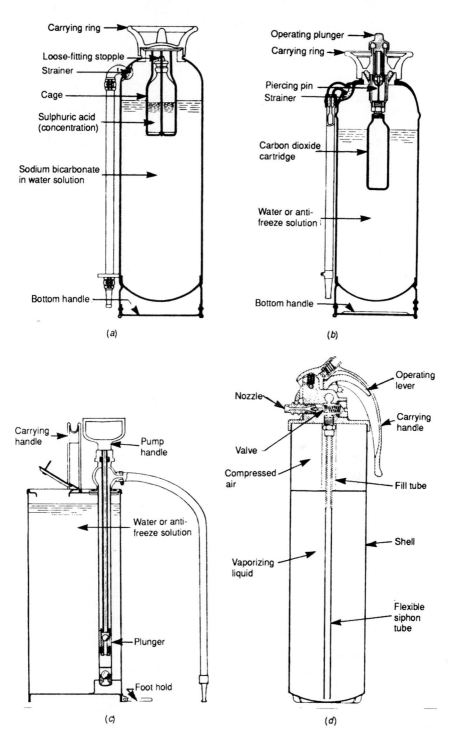

FIGURE 3.4 Illustrations of portable fire extinguishers. (*a*) Soda-acid; (*b*) water-filled gas cartridge; (*c*) pump tank; (*d*) vaporizing-liquid. (*Courtesy, Underwriter's Laboratories.*)

3.17

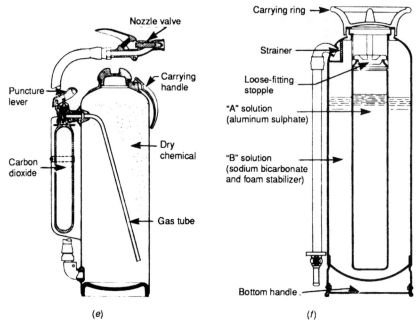

FIGURE 3.4 *(Continued)* (*e*) dry chemical; (*f*) foam. (*Courtesy, Underwriter's Laboratories.*)

damage is noted. This should be done by the manufacturer or firms specializing in this type of work. Plant service maintenance people should check at least annually to see if such service is needed. Key employees on each shift should be trained to operate the extinguishers by actually using them on an artificially created fire in a safe place such as a yard.

Sprinkler Systems

Automatic sprinkler systems permit the water supply, whether public or private, to be conducted to part or all of the building. They use a network of interconnected piping, properly sized, and with automatic spray valves, also strategically sized and placed, that are activated by sensors such as smoke or heat detectors so that water is sprayed on the affected areas to extinguish the fire by water. The value of sprinkler systems has long been recognized by property insurers, who can cite statistics showing how the average fire loss was drastically reduced when there were automatic sprinklers installed. Depending on occupancy, history, and location, most sprinklered properties merit lower insurance premiums from the fire carriers.

Building codes are also requiring sprinklers to be installed in such hazardous areas as hotel kitchens, underground garages, nursing homes, shopping malls, and high-hazard industrial occupancies. In any addition or alteration, plant service personnel should become familiar with local building and fire department codes. Part of their responsibility is to inspect and maintain this equipment, sometimes accompanied by fire department or fire insurance loss-prevention personnel.

Sprinkler designs vary depending on the following factors: nature of the occupancy, possible fire intensity and concentration based on fire loading, the area exposed to potential fire, and considerations for internal and external fire attack.

Types of Systems. The present classifications are wet pipe, dry pipe, deluge, preaction, and fire cycle system. The systems can be briefly described as follows:

1. *Wet pipe system.* This is the most common system (see Fig. 3.5). The sprinkler piping is filled with water under pressure, and when the sprinkler head on the pipes is activated by heat or another sensor, water is discharged and continues to flow until shut off.

2. *Dry pipe system.* This system is used in unheated buildings where the water in the pipes would freeze. The pipes are filled with air under pressure instead of water. When a fire develops, a sprinkler opens up and lowers the air pressure, which in turn automatically trips a dry-pipe valve to let water into the piping system and to the sprinkler head (see Fig. 3.6b).

3. *Deluge system.* This system is designed to supply water to all sprinkler heads for a designated area in order to inundate a large area with water at once. The sprinkler heads are open, and the pipes are empty. When a fire erupts, a signal from a heat-responsive sensor opens a deluge valve, and this permits water to flow to all the open sprinkler heads.

4. *Preaction system.* This system can be wet or dry. The sprinkler heads are closed, heat sensors are strategically placed in the sprinklered area, and water is admitted to the piping by the operation of an automatic water-control valve if a sensor is activated by a rise in temperature. Water discharges from a sprinkler head, but only after the temperature rises high enough so that the fusible element operates to open the sprinkler head, hence the term *preaction.*

5. *Fire cycle system.* This system was developed to minimize water damage as a result of sprinklers operating. It uses heat detectors and electrical controls that continuously go on and off while controlling a fire and shut off the water automatically when the fire is extinguished. The system can be installed in zoned areas where water damage is to be minimized.

Alarms are required on sprinkler systems in order to notify those responsible for their care and also authorities such as fire departments or those that have supervisory responsibility for the sprinklers—American District Telegraph (ADT), etc. *Supervisory systems* are those which have outside agencies responding to alarms, such as ADT. *Proprietary systems* are those within a plant that automatically notify responsible employees and fire brigades.

Temperature ratings of sprinkler heads per FM guides are as follows:

Sprinkler head rating	Rated temperature, °F	Color	Maximum temperature at sprinkler level, °F
Ordinary	160–175	Half black	to 100
Intermediate	175–212	White	150
Hard	250–286	Blue	225
Extra hard	325–360	Red	300
Special quartzoid	400	Green	375
Special Saveall	415	Green	380
Special quartz	500	Orange	465

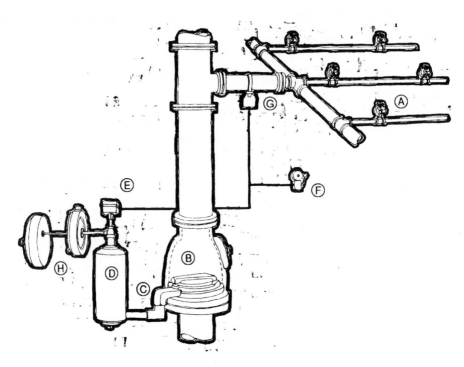

When sprinkler (A) opens in Wet Pipe System, water discharging from system lifts Clapper (B) from seat in Main Alarm Valve and opens Auxiliary Valve (C). Water flows to Retarding Chamber (D), building up under Pressure Switch (E) and sounding Electrical Alarm (F). Waterflow Indicator (G) also activates alarm. Water flows to Water Motor Alarm (H), sounding mechanical signal. During surges or pressure fluctuations, clapper opens momentarily, trapping excess pressure in system, allowing small amounts of water into Auxiliary Valve and Retarding Chamber, preventing false alarms from either water pressure surges or water flow.

FIGURE 3.5   Viking *wet pipe* sprinkler system. (*Courtesy, The Viking Corp.*)

Sprinkler heads on the highest pipeline require a minimum water pressure of 15 psi and a flow rate per head of not less than 100 gpm for ordinary occupancies.

Most sprinkler heads (Fig. 3.6*a*) are of the solder type, using a metal strut and fusible solder that melts at design temperature to release the water. The operating parts of a sprinkler head should never be painted.

Maintenance of sprinklers is essential. Some problems caused by lack of inspection and maintenance are: water is shut off, water supply insufficient or defective, frozen pipes, dry system does not operate with loss of air pressure, wrong temperature rating on sprinkler heads, alarms not working, gravity tank has low water, fire pump fails to deliver required water flow, and outside sprinkler heads have been plugged.

Inspections and Tests.   Most insurance companies prefer a periodic inspection and testing program be arranged by the owner of the property with the manufacturer of the sprinkler system or their local representative, supplemented by frequent in-between inspections by responsible plant employees. Plant employees should be trained to

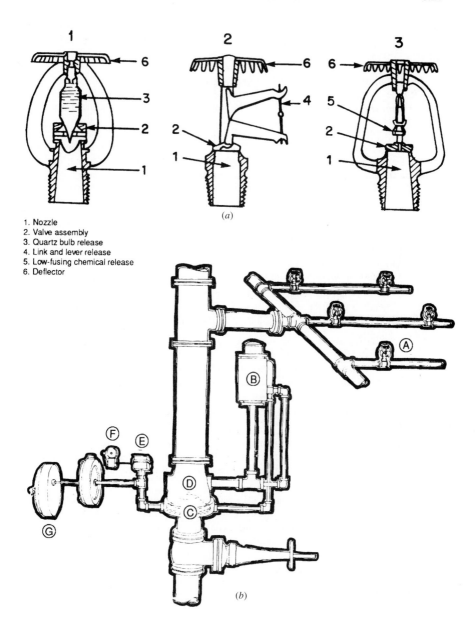

1. Nozzle
2. Valve assembly
3. Quartz bulb release
4. Link and lever release
5. Low-fusing chemical release
6. Deflector

(a)

(b)

When Sprinkler (A) fuses, system air pressure falls, actuating Accelerator (B). Accelerator forces air into Intermediate Chamber (C), eliminating pressure differential and opening Clapper (D) immediately. Build-up of water under Pressure Switch (E) actuates Electrical Alarm (F). Flow of water from Intermediate Chamber operates Mechanical Alarm (G). In open position, Clapper is latched out of water-way, thereby preventing water columning and friction loss

FIGURE 3.6 (*a*) Sprinkler heads with part names; (*b*) Viking *dry-pipe* sprinkler system with operating instructions. (*Courtesy, The Viking Corp.*)

understand the system that has been installed and recognize conditions that will prevent the system from operating as designed or will have a future effect on the system, such as corrosion, and they should have the authority to correct any adverse conditions found by instituting repair procedures.

Most manufacturers of sprinkler systems provide instructions for the plant to follow. Some fire insurance companies provide the plant with inspection forms. They also recommend that these inspections be performed on a weekly basis. Features of these forms include the following:

Sprinkler valve conditions.   Note if valves are open and closed and if impairment has been reported to the authorities, insurance company, and management. Full flow drain tests are required after a valve is reopened. On dry pipe systems, air pressure is of concern, including provisions for moisture removal in unheated areas.

Sprinkler alarms.   These should be tested during any inspection tour.

Water supply.   Pressure is to be noted where public water supply enters plant and on the plant side of any shutoff valve. On secondary systems, operate the fire pump to note suction and discharge pressures and general conditions in the pump room. Sprinkler water tank levels should be recorded, freeze protection should be checked, and circulation should be checked for obstructions.

Sprinklers.   Missing, plugged, or leaking heads should receive immediate attention. Check to make sure all valves to sprinkler sections open. Freeze protection should receive attention. High-piled stock that could interfere with the water spray needs attention.

Fire doors.   They require attention to make sure they are not blocked open.

Fire extinguishers.   These require a review to note if they are in assigned areas and if contents are up to proper level.

Flammable substance areas.   Pay attention to cleanliness, safety operating practices, and whether sources of ignition are being properly controlled by operating personnel.

Electrical equipment.   Equipment such as switchboards should be checked to note if they are operating with normal temperatures and are clean and dry and if connections in the back of the switchboard are tight. Infrared inspections can also be made for tightness and temperature checks.

Housekeeping.   This is a catch-all obvious item, but it needs special attention where dust, oily rags, and flammable or explosive chemicals may be handled. Exhaust systems will require attention as will any overhead sprinkler systems. Cleanliness practices are important as are storage and disposal of all wastes.

Hose equipment.   Hoses need to be checked to make sure they are in place on standpipe systems and not frayed or worn.

Waterflow alarms.   For a wet pipe system, follow the instructions for testing the inspector's test connection at the top of the system. This connection normally matches the flow of a single sprinkler. Make sure the cocks and small valves to pressure switches are in the open position. Electric alarms can be tested by the normally installed test switch.

Special Extinguishing Systems

Special automatic types of extinguishing methods have been developed for the special hazard where water cannot be used.

Foam Systems. These systems depend on the ability of foam to flow over and upon many flammable liquids, while also adhering to the liquid's surface. This quality of foam cuts off the oxygen from flammable liquids and also prevents the formation of burnable gases. Foams vary in composition, depending on their expected application. For example, two common types are (1) for use with flammable liquids not miscible with water applied mostly to petroleum products and (2) for use on flammable liquids that are miscible with water, such as alcohols, esters, and acetones.

Chemical foams produce carbon dioxide gas by the chemical reaction of aluminum sulfate and sodium bicarbonate in the presence of water and foaming agents. There are other foam systems that are used. The main function of a foam system is to act as an immediate extinguishing agent, and because of limited installed capacity, they are not designed to stop long-term fires from spreading or to insulate tanks or equipment against exposure to fire.

Carbon Dioxide Systems. These systems are used to produce a smothering effect on flammable liquids in open and enclosed containers, rooms containing flammable liquids, electric equipment such as hydrogen-cooled generators, and sensitive electrical equipment such as controllers, vaults, libraries, and similar occupancies where fires may be extinguished by this gas in order to avoid water damage or where water extinguishing may spread the fire due to shorts, for example, on electrical equipment. The high-pressure system uses cylinders with the gas at a pressure of about 850 psi at 70°F. The cylinders should be weighed annually to make sure enough gas is present.

The low-pressure system has carbon dioxide stored in liquid form, thus requiring the tanks to be kept refrigerated so that the normally designed 325-psi pressure rating of the tank is not exceeded.

Water-Spray Fog Systems. These systems develop a fine spray with fog nozzles. They cool the fire as the fog absorbs heat and evaporates. There is also some smothering effect from blocking out oxygen for combustion. Hose lines are used to apply the fog as are fixed systems of piping that are similar to sprinklers.

Halon Systems. Halon systems are used in computer installations to prevent the spread of electrical fires by the injection of a nonconducting gas into the computer equipment room. Halon, which belongs to the freon family of gases, also blocks out oxygen.

Plant Fire Brigades

In a broad sense, a fire brigade consists of workers, technicians, office people, and other personnel who are trained, trusted, and reliable; their purpose is to prevent and extinguish fires. The fire brigade should be completely familiar with water supplies, the location of valves, extinguishers, sprinklers, alarms, and notification procedures to follow in case of fire or fire fighting equipment impairment.

Fire attack requires a knowledge of the plant layout and process flow so that critical areas receive the highest attention first, thus preventing serious injury or production impairment as a result of the fire. This also applies to raw and finished goods storage. The flow of the process, concentration of people, value of stock, hazards inherent in the process or storage area, and susceptibility of people, stock, and equipment to smoke or water injury or damage are a few of the items that need to be addressed in a preplanned fire attack for the location.

Impairment

Impairment is a term used by insurance fire protection engineers when fire protection sprinklers, fire pumps, detection devices, valves, alarms, water supply, gravity tanks, and similar equipment that is vital for fighting a fire is out of service either deliberately for maintenance or accidentally due to a mechanical or electrical failure or to building alterations or additions. During the impairment there is always the risk of a fire erupting. Whole properties have been destroyed, for example, because the water supply was shut off in the street to make pipe or valve repairs. *Concealed impairments* can be dangerous because they occur from some unknown action by an unauthorized person either on the property or on the water supply entering the plant. Regular inspections and tests of the fire protection system by plant personnel have the main goal of finding concealed impairments of the fire protection system so that corrective actions can be taken.

Planned impairment for repairs or alterations should be placed under the supervision of the plant's fire protection chief so that emergency plans are put in place for the actions to take during the impairment. Some actions to be taken if an impairment occurs are:

Notify the local fire department and key people involved with the plant's fire fighting plan of the impairment and its expected duration. Many fire insurance companies also want to be notified.

Establish a watch system in areas that contain combustibles or a flammable process.

Provide extra portable fire extinguishers in the area where sprinklers are out of service.

Alert all employees of the areas to take precautions when disposing of trash and combustibles.

Limit any source of ignition activity, such as cutting or welding, in the affected areas.

It is preferred that work on restoration be on an "around the clock" basis because of the risk involved during the impairment.

It is also preferred that hazardous process areas not be operated during the impairment.

It is essential to test the system for proper operation after the repairs are completed.

Notify plant management, fire departments, and fire brigades as well as the fire insurance company when the system is back to normal. This includes outside alarm agencies, such as ADT.

Nitrogen Blanketing

Nitrogen blanketing is used in many hazardous operations because nitrogen does not support combustion. It is also used to purge lines that handle flammable liquids or gases. It is essential to have properly installed check valves and isolation valves to prevent the nitrogen line from being filled with water or flammable liquids or gases that may be at a higher pressure or when the nitrogen pressure drops due to leaks or maintenance activities.

Life Safety Code

Fire protection life safety has received increased attention for such occupancies as office buildings, hospitals, health care facilities, schools and colleges, hotels, theatres, rooming houses, and similar facilities where a large number of people may be immediately affected by smoke and heat from a fire *before* they can exit from the building. The NFPA has established a separate Life Safety Code for such occupancies, and many jurisdictions have included these NFPA requirements in local ordinances or laws. The objective is to minimize the effect of heat and smoke to the occupants of such facilities so that they can be safely removed from the premises without serious injury.

All plant service operators should become familiar with life safety systems. Among the items stressed in these codes are the following:

1. Have sufficient usable exits per floor, based on occupancy load, that can direct the occupants outside of the building safely. Stairways must be of fireproof construction.
2. Floors and buildings should be compartmentalized by installing approved fire walls and doors to limit the spread of fire, smoke, and heat.
3. All such facilities should have an emergency communication system that will make it possible to give instructions to the occupants on what to do in an emergency.
4. Automatic sprinkler protection is generally recommended in high human occupancy areas.
5. For theatres and auditorium-type occupancy, correct sizing guidelines are provided for aisle accessways.
6. For high-rise buildings, some of the items stressed are:
   (*a*) Stairways must have battery backup for lighting.
   (*b*) Above a certain height or square footage, emergency generators are required, sized for the critical emergency load of the building. Other occupancies such as hospitals are also required to have emergency generators.
   (*c*) Elevators must be automatically directed to the first or ground floor in case of fire and be available only to fire fighting personnel.
   (*d*) In some cities, high-rise buildings are required to have a designated and approved fire prevention engineer on site, who is responsible for all fire safety for the facility.

Plant service operators may become involved in life fire safety code equipment maintenance and inspection and should become familiar with the NFPA Life Safety Code as well as local laws that govern human occupancy requirements for the facility in which they work.

QUESTIONS AND ANSWERS

1. How are fires classified, and what substances are involved?

   Answer. There are three classes; Class A involve ordinary combustibles such as wood, paper, cloth, which normally can be extinguished by water. Class B applies to oils, grease, gasoline, paint, and similar substances that cannot be extinguished by water. Class C applies to electrical equipment. Some jurisdictions classify burnable metals such as magnesium and soda as Class D.

2. What are the portable fire extinguishers that are only suitable for Class A fires?

   Answer. Water pail, soda-acid, plain water with CO_2 cartridge, calcium chloride (antifreeze), and pump tank.

3. For what type of fire is the foam extinguisher used?

   Answer. Class B for oil, grease, gasoline, and paint fires.

4. What extinguishers are suitable for electrical fires?

   Answer. Vaporizing liquid, carbon dioxide, and dry chemical.

5. What six chemicals become hazardous when in contact with water?

   Answer. These are Group E chemicals per insurance company listings: Calcium nitrate, sodium potassium, potassium hydroxide, sodium hydroxide, sodium, and sodium peroxide. Check with your fire insurance company representative.

6. What six chemicals may explode in a fire?

   Answer. These are Group H. Ammonium perchlorate, barium chlorate, potassium chlorate, potassium perchlorate, sodium chlorate, and chromium trioxide.

7. Will hydrogen be explosive if its concentration by volume in air is 3.5 percent? How about 20 and 85 percent?

   Answer. The mixture is too lean at 3.5 percent so it will not explode. At 85 percent the mixture is too rich; therefore, it will not explode. It will explode at 20 percent.

8. How would you define autoignition?

   Answer. It is the lowest temperature that is needed to start combustion of a solid, liquid, or gas without requiring a spark or flame to get the combustion started.

9. How is fire loading measured?

   Answer. This is a measure of the combustible contents on a floor and is expressed as Btus per square feet of floor area.

10. What is the purpose of a standpipe?

    Answer. Standpipes are used to get water to upper floors of tall buildings in order to fight a fire. Plant people use hoses that are located on each floor and connected to the standpipe. Standpipes also help the fire department get water quickly to the higher floors of the building.

11. Is there any classifications for standpipes?

    Answer. This may depend on the jurisdiction. A common classification system is based on whether it is a wet pipe or dry system, classified as follows: Wet systems have an open supply valve and water under pressure at all times. This can create a water leakage hazard. A second system uses approved devices to admit water automatically by opening a hose valve. Another system admits water through the manual operation of approved, remote-control devices located at each hose station. Dry standpipes get water through the fire department pumper connection and are usually located outside the building for street connection to the pumper.

12. Where should hose outlets on standpipes be located?

    Answer. Hose outlets should be near or in stairway enclosures or even on fire escapes on older buildings so that they are readily accessible to the local fire department.

13. What types of sprinkler systems are used to protect a property against the spread of fire?

    Answer. Wet pipe, dry pipe, deluge, preaction, and firecycle.

14. Why are deluge systems used?

Answer. They are primarily used to inundate a large area at once, such as in hazardous areas. This will prevent the spread of fire due to fumes and gases that may be generated by the fire. Instead of only one sprinkler head operating, all sprinkler heads in the designated area are open at all times, and when a fire erupts, a deluge valve is opened by strategically placed sensors in the designated area that are activated by temperature or other criteria such as smoke, and this permits water to flow to all the open sprinkler heads in the designated area.

15. What type of sprinklers are used in unheated buildings?

Answer. Dry pipe systems are generally used. Wet pipe systems require antifreeze solutions and must be properly separated from the public water supply to prevent contamination. This would require an antifreeze solution storage tank that is properly sized for the exposure. Since deluge systems have open sprinkler heads and are activated by sensors that admit water if a fire is sensed, deluge systems that were properly designed on the water supply side to avoid freezing of the supply could also be used.

16. Do dry pipes need to be checked for possible freeze damage?

Answer. Before any freezing weather, it is wise to check all low-point drains for water to prevent freeze damage of the dry pipe.

17. How does a dry-chemical extinguisher work?

Answer. These are used for Class B and C fires. The extinguisher should be directed to a corner of the fire, and then the stream should be swept across the flames. The chemical releases a smothering gas on the fire, while a fog or dry chemical shields the operator from the heat. The spray range is 8 to 12 ft.

18. What does the term *fire wall* mean?

Answer. A fire wall is a wall of brick or reinforced concrete which subdivides or separates a building to restrict the spread of fire. It starts at the foundation and extends continuously through all stories to and above the roof, except when the roof is fireproof or semifireproof and the wall fits tightly against the underside of the roof slab.

19. What is a fire partition?

Answer. A fire partition subdivides a building to restrict the spread of fire or to provide an area of refuge. It may not be continuous through all stories to and above the roof like a fire wall. It has a fire resistance rating of at least 2 h.

20. What vertical openings require noncombustible construction?

Answer. Elevator shafts, stairways, chutes, ventilating shafts, and ramps in multistory buildings require noncombustible construction to prevent the fire from spreading upward in the building. Local fire codes should be consulted for other details of required fire prevention for these openings.

21. What does the term *hazardous location* mean?

Answer. This is a location which has highly ignitable material—solid, liquid, gas, or dusts. Hazardous locations require special treatment for process machinery, electrical equipment, and fire protection. This also includes the grounding requirements on electrical equipment, ventilation to be provided, and explosion venting.

22. What are flame arresters?

Answer.   Flame arresters are required on tank vent pipes to prevent flashbacks into tanks that contain explosive materials. Wire screens of 40 mesh psi are used. Arresters constructed of banks of parallel metal plates or tubes are also used, especially if the metal screens have large areas to dissipate the heat. The openings on these are larger than screen deflectors and thus are less liable to clog.

23. Why are gas masks important to fire fighters?

Answer.   During a fire, especially in low points such as basements, irritant and poisonous gases can be liberated, with carbon monoxide being especially danger-ous since it is formed from incomplete oxidation in the burning process and is or can be lethal if inhaled. Canister-type gas masks are used where there is sufficient oxygen in the atmosphere. The mask consists of a rubber mask with a window attached by hose to a canister, arranged so that all the air breathed comes through the canister, which contains absorbing substances for removing poisonous gas. Make sure the canister is a type N as approved by the U.S. Bureau of Mines. This type of mask is used by most fire departments and is called an "all-service" mask. Modern canisters have pointers to show how long a canister has been used. One revolution is about 2 h, and normally this indicates that the chemicals in the canis-ter have been used up, and the canister should be replaced. Canisters should not be used where the oxygen level is below 19 percent. For those cases where there is insufficient oxygen, *oxygen-breathing apparatus* is used with the face mask connected to portable oxygen cylinders.

CHAPTER 4
WATER SUPPLY, TREATMENT, AND DISPOSAL

Water is so essential that ancient civilizations called it one of the four elements that sustain life, the others being earth, air, and fire. The only pure water is carefully distilled water. Natural waters contain dissolved solids and gases, and it is necessary to apply water chemistry in order to determine what the impurities are, and depending on process requirements, what treatment will be required to remove the impurities to a threshold limit deemed satisfactory.

Impurities occur in water mainly because of the action of water dissolving materials which it encounters in its flow to and on the earth. The most pronounced are salts of sodium, potassium, magnesium, calcium, and iron. There is also organic matter from decaying plant and animal debris that water comes in contact with. Water may also have suspended matter in it such as clay, sand, and fragments of organic matter of plant and animal origin, and also may have living microorganisms.

Potable water is water which is fit to drink, and the treatment required differs from that of water to be used for fire fighting purposes and process use. The main emphasis in potable water treatment is to make sure it is free from suspended impurities and has no pathogenic (disease-producing) bacteria. Town and city supplies of water are a matter of public health and, therefore, are under health regulations that in many cases involve city, county, state, and federal rules and regulations. The treatment method for potable water is heavily influenced by the local source of the water and requires licensed chemists for analysis and treatment. In general, suspended impurities are removed by filtration, such as filtering beds of sand and gravel. By adding alum and lime, or ferrous sulfate and lime, a gelatinous, slimy coating is made on the sand grains. This helps to entangle and retain the very fine particles and bacteria that would normally pass through the sand filter. This treatment method is called the *coagulation treatment.*

After the water supply is free from suspended matter, *bacteria* are killed by the addition of small amounts of chemicals such as chlorine, bleaching powder, ozone, and similar chemicals as determined by the bacteria present. The dosage applied is closely regulated; it has a minor effect on higher forms of life but destroys the lower forms. In addition to bacteria, some waters contain microscopic plants called *algae,* which give the water an odor or color. Algae problems are usually treated by injecting controlled amounts of cupric sulfate into the water.

WATER TREATMENT

Industrial plants, and those plants with high-pressure boilers, have adopted many of the practices of the municipal plants that supply potable water, namely, know your source of water, test it, determine treatment, establish testing controls to maintain the quality of the water, and make adjustments in the treatment as conditions change. Undesirables or impurities in the water, if not properly treated or controlled, can cause corrosion, scaling, or fouling of heat-transfer surfaces and microbial fouling that can affect process applications of the water such as in the plastics industry and electronics chip manufacturing processes.

It is essential for industrial plants to determine all the various constituents in the water that may be objectionable to the particular process in their facility so that the best method of treatment may be determined. The following table details some of the impurities and treatment methods used for it:

Impurity	Chemical formula	Treatment for removal from the water
Suspended solid		Settling, coagulation, sedimentation, filtration
Calcium (hardness)	Ca	Chemical or zeolite (ion-exchange) softening, distillation
Magnesium (hardness)	Mg	Same as above
Sodium	Na	Hydrogen ion-exchange, distillation
Iron, manganese	Fe, Mn	Aeration, coagulation and sedimentation, filtration, ion-exchange, chemical softening
Carbonate, Bicarbonate	CO_3, HCO_3	Hydrogen ion-exchange, demineralization, dealkalization, distillation
Sulfate, chloride	SO_4, C_1	Demineralization, distillation
Free acids	H_2SO_4 HC_1, etc.	Neutralization, ion-exchange
Carbon dioxide	CO_2	Aeration, deaeration, neutralization
Oxygen	O_2	Deaeration, scavenging
Oil, grease		Separators, filtration, coagulation
Organic matter		Filtration, coagulation

Suspended solids, such as slime, algae, and dust, can restrict water flow in cooling towers by settling on heat-transfer surfaces. Minerals, salts such as calcium and magnesium, and alkalinity can cause scale in cooling towers and especially in boilers, which can also impede heat transfer and can cause overheating damage to tubes and other surfaces exposed to the heat of combustion. Serious ruptures may result.

Raw water preparation consists of (1) sedimentation, (2) filtration, (3) softening, (4) removal of dissolved solids, (5) removal of dissolved gases, and (6) internal and external chemical treatment of the process or apparatus, such as a boiler.

Sedimentation

Sedimentation allows the solids to settle out of the water by dropping to the bottom of a basin or impounding reservoir. It can be assisted by the use of coagulants such as alum or aluminum sulfate, ferrous sulfate, ferric chloride, sodium aluminate, and magnesium oxide. The raw water that is available will determine the coagulant to use. In addition, with many waters it is necessary to add an alkali, such as lime or soda ash, to bring the water to the best or required pH value.

Natural sedimentation as shown in Fig. 4.1 combines (1) the mixing of chemicals to aid in the suspended solids clinging to the coagulants, and (2) the small particles are then brought together, *called flocculation,* by gentle mixing to form larger particles that settle out faster. This is performed by baffles as shown in Fig. 4.1*a* or by mechanical mixers as shown in Fig. 4.1*b. Pressure filters* are also used to remove smaller amounts of suspended solids, such as makeup water for boilers, as shown in Fig. 4.2. The advantage of mechanical and pressure settling or filtering is that the raw water needs to be retained for less time to remove the suspended solids.

Color in some waters is removed chemically by aluminum sulfate and chlorinated copperas. These compounds react with color in water to form a precipitate that settles with the sludge in the sedimentation process.

Filtration

Filtration differs from sedimentation in that smaller and lighter particles of suspended and coagulated matter remain after sedimentation and now must be removed by filtration (see Fig. 4.2). Most common filters use suitably graded beds of sand or anthracite coal. When the fine particles enter a filter, they settle out in the top few inches of the bed and, with time, build up on the surface. If this is not corrected, it starts to restrict flow.

Backwash is used to remove the particles from the filter bed. Water is passed upward through the bed at a rate of 4 to 7 times the filter rate, and the accumulated suspended particles are washed out of the bed and sent to waste.

Activated carbon filters are used to *remove odors* and improve water taste. Their construction is similar to that of the sediment filter and usually is under pressure. However, the carbon is not a strainer, but an *absorber* of the odorous substance, and must be periodically replaced.

Water Softening

Water containing appreciable amounts of calcium and magnesium compounds in solution is called *hard water*; the name was derived from the fact that when using soap with hard water, it is difficult to obtain a lather. Hard water is especially objectionable for use in boilers because the calcium and magnesium salts deposit on the tubes, forming a stonelike layer on the inner walls of the tubes, commonly called scale. This scale acts as an insulator, preventing proper heat transfer between the fire and water sides of the tubes. This adds to fuel consumption. Severe scaling can cause tubes to become overheated and rupture, which can be dangerous in a boiler plant operation.

There are two types of hardness. One is called *temporary hardness,* or temporary hard water, which contains large quantities of calcium bicarbonate, $Ca(HCO_3)_2$. This water may be softened by boiling it, with the calcium carbonate, $CaCO_3$, precipitating out of solution, and carbon dioxide gas being released. In industrial plants, water is softened by the addition of a sufficient amount of lime to precipitate out the calcium carbonate. This soft sludge must be "blown out" of the boiler to prevent buildup as a precipitate in drums or headers.

Water that contains calcium and magnesium sulfates is not softened by boiling and is referred to as having *permanent hardness.* It can be softened by the addition of sodium carbonate (Na_2CO_3). Calcium and magnesium carbonate precipitate out.

The *cold-lime and soda softening* process treats raw water with lime, calcium hydroxide and soda, or sodium carbonate to partially reduce hardness. The water usually then requires further internal treatment in such water-using equipment as boilers (see Fig. 4.3*a*).

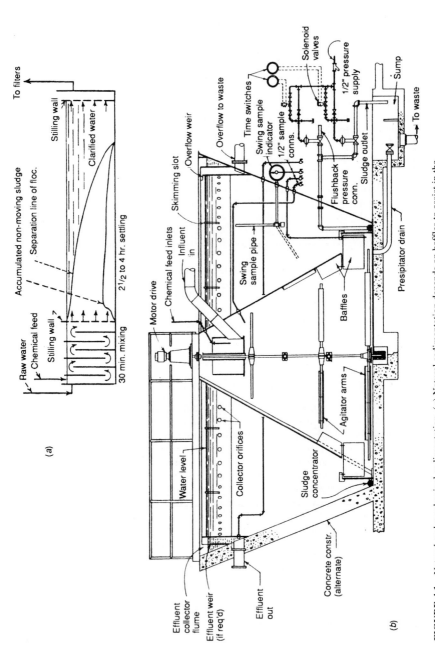

FIGURE 4.1 Natural and mechanical sedimentation. (*a*) Natural sedimentation depends on baffles to assist in the mixing of coagulants with raw water; (*b*) mechanical mixing of coagulant and raw water speeds up the sedimentation process. (*Courtesy, The Permutit Co.*)

4.4

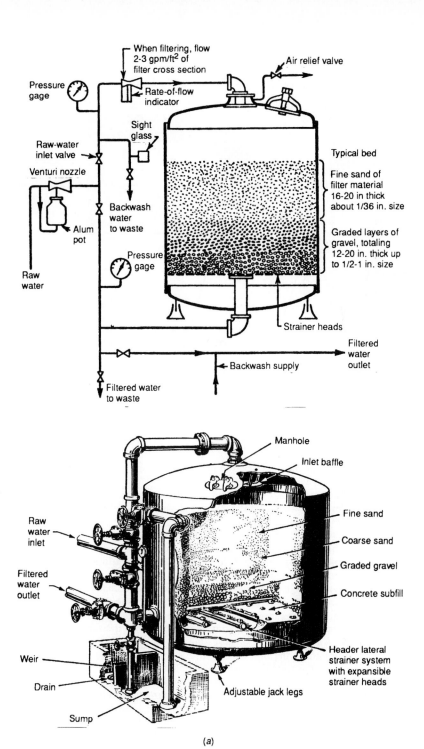

FIGURE 4.2 Pressure filters may be of the vertical or horizontal tank type. (*a*) Vertical pressure filters use an alum pot for finer filtering of suspended solids in water. (*Courtesy, The Permutit Co.*)

4.5

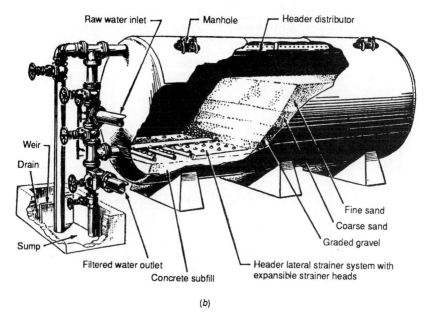

Raw water inlet   Manhole   Header distributor

Weir

Drain

Fine sand

Coarse sand

Sump   Graded gravel

Filtered water outlet   Header lateral strainer system with
Concrete subfill   expansible strainer heads

(*b*)

FIGURE 4.2   (*Continued*) Pressure filters may be of the vertical or horizontal tank type. (*b*) horizontal pressure filter. (*Courtesy, The Permutit Co.*)

The *hot-lime and soda softening process* operates at 212°F and over and uses steam as a heat source. The heat causes a faster chemical reaction to take place. Calcium hydroxide, lime and soda, or sodium carbonate are also used in the hot lime method of softening (see Fig. 4.3*b*).

The *zeolite softening process* uses a sandlike substance called zeolite, which can be of synthetic or natural derivation. This substance is also arranged inside a tank as a filter bed (see Fig. 4.3*c*). Zeolite has the remarkable property of base exchange. When the hard water passes through the bed of zeolite, the calcium and magnesium compounds pass into the zeolite and are replaced by sodium from the zeolite. The calcium bicarbonate becomes sodium bicarbonate, and magnesium sulfate becomes sodium sulfate. These sodium compounds do not form scale; thus, this ionic exchange makes the water soft by freeing it from the hardness compounds. Eventually, the zeolite loses its sodium concentration because the sodium is combining with the calcium and magnesium compounds, and this causes the zeolite to lose its exchange power.

Zeolite regeneration involves irrigating the zeolite with a strong brine solution (sodium chloride). A reverse reaction causes the sodium from the brine to replace the calcium and magnesium in the zeolite. Figure 4.3*c* shows a zeolite softener that uses automatic controls. In the softening cycle, water flows downward through the zeolite bed. As the water-softening capacity of the zeolite declines to a set point, automatic valves cut off the downward flow of water and *backwash* upward to loosen the material and also remove deposited dirt. In the third step, a measured quantity of common salt brine is admitted at the top of the bed. After a timed interval, a stream of rinsing water is introduced to remove excess salt and to clean the zeolite, after which the bed is ready for another softening cycle.

In a boiler water application, zeolite-treated water shows zero hardness by a soap test, but the water now has soluble sodium salts in solution. Blowdown is necessary to

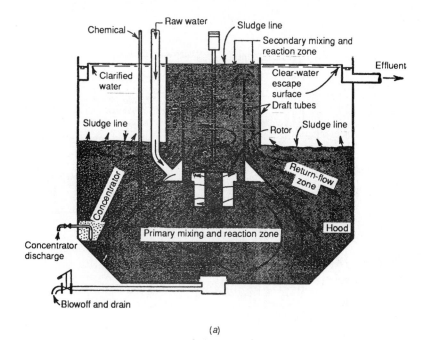

Chemical — Raw water — Sludge line
Secondary mixing and reaction zone
Effluent
Clarified water
Clear-water escape surface
Draft tubes
Sludge line
Sludge line
Rotor Sludge line
Concentrator
Return-flow zone
Concentrator discharge
Primary mixing and reaction zone Hood
Blowoff and drain

(a)

Steam inlet
Condensate inlet
Make-up deaerator
Steam inlet
Condensate heater and deaerator
Condensate outlet
Deaerated make-up outlet
Wash-water outlet
Softener
Wash-water return
Filter bypass
Raw-water supply
Chemical mixer
Agitating pump
Filter
To service
Backwash pump
Chemical-feed pump
Suction box

(b)

FIGURE 4.3 Three types of water softeners. (*a*) Continuous cold-lime softener; (*b*) hot-lime softener also has a deaerating section for return condensate. (*Courtesy, The Permutit Co.*)

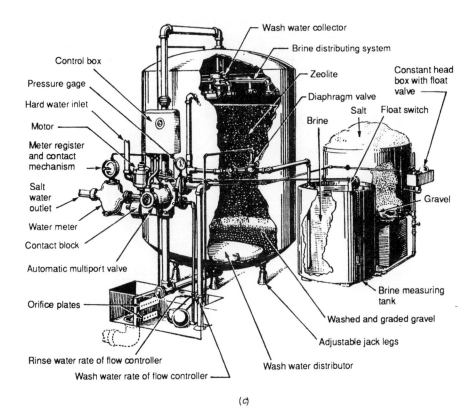

FIGURE 4.3   (*Continued*) Three types of water softeners. (*c*) automatic zeolite-type water softener. (*Courtesy, The Permutit Co.*)

limit the concentration of these salts so that priming and foaming do not occur in steam boilers. A high percentage of sodium carbonate may cause steel embrittlement under certain conditions; thus, zeolite softeners are better suited to handle magnesium sulfate hardness. To remove the carbonate hardness, the hot-process lime treatment is used ahead of the zeolite softener.

The amount of salt required for regeneration depends on the grain of hardness. The usual figure cited is $\frac{1}{4}$ to $\frac{1}{2}$ lb of salt per 1000 gal of water per *grain hardness*. For example, regeneration of 50 ft^3 of a high-capacity synthetic zeolite requires 300 lb of salt. If the water has a 10-grain hardness and $\frac{1}{4}$ lb of salt is required per 1000 gallons, how many gallons of water can now be softened?

$$\text{Gallons softened} = \frac{4 \times 300 \times 1000}{10} = 120,000 \text{ gal}$$

Hot-process chemical softening and zeolite softening can be combined into a *hot-process-hot-zeolite treatment* system that will deliver, especially to high-pressure boilers, a hot feedwater with zero hardness.

Synthetic *ion-exchange resins* are polymers or copolymers of various organic chemicals, with the resins being in the form of beads. They have replaced zeolites in the normal sodium cycle of softening water and can be regenerated with sodium chloride. If regenerated with solutions of sulfuric or hydrochloric acids, it is called the *hydrogen cycle,* in which case the sodium in the water, in addition to the calcium and magnesium, is taken up by the resin with all three being replaced by hydrogen in the regeneration cycle of water softening. In the hydrogen cycle, it is necessary to control the acidity by mixing the treated water with a suitable alkali, such as caustic soda, soda ash, or phosphate.

Demineralization

Demineralization of water in certain industries requires the water to be completely free from mineral salts. This also applies to central power station's boiler water. Distillation is one method, but it is costly. Ion exchange is a two-step method, the first being the hydrogen cycle, the *cation* exchange, followed by the second step, the *anion* exchange. The anion exchangers are further divided into weakly and strongly basic anion exchangers. The weakly basic unit will not remove weak acids, such as carbonic or silicic, and therefore the treated water may contain silica and carbon dioxide. Weak base resins are regenerated by alkali, such as ammonia, caustic, or soda ash.

The strongly basic anion exchanger can remove both strong and weak acids, producing water that is free of silica and carbon dioxide, but it is more expensive to operate. Regeneration is with caustic.

Aeration

Aeration of water is performed for the following reasons:

1. Removal of gases such as carbon dioxide and hydrogen sulfide.
2. Improvement in taste and removal of odors.
3. Oxidation of any iron present so that it can be filtered or removed chemically in the water treatment steps. The same is applicable to other minerals that combine with air to form oxides.

Spray aeration is the most common method used in municipal systems. The water is broken up into fine droplets as it is discharged from spray heads. The water and air mix as the water drops down to a collection basin. In industrial plants where space may be restricted, coke-tray aerators may be used, or a *mechanical draft* type as shown in Fig. 4.4. The water drops down over a series of slot trays as the air flows upward in a counterflow arrangement.

Deaerators are extensively used to remove oxygen and carbon dioxide from solution, where it is desirable to accomplish this without the use of chemicals, or internal boiler water treatment (see Chap. 9).

Wells

Well waters may also need treatment for hardness and in many localities have heavy concentrations of iron rust, which is noticeable when the water stands for a while. For industries where clear water is essential, the removal of iron from solution is accomplished by aeration and sedimentation. Coagulants are used to speed up the process of precipitation, followed by a final filtering. The iron content can be reduced to acceptable levels of concentration by aeration, sedimentation, and filtering.

There are two types of wells. *Gravity* wells start at the surface and descend into an aquifer, which is at atmospheric pressure. In contrast, the *artesian* well descends to an

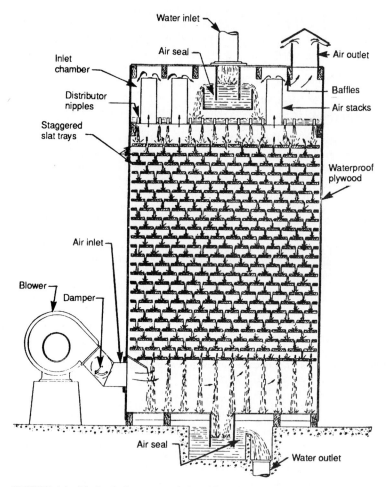

FIGURE 4.4 Mechanical aerator and degasifier of water uses blower to force air through water spray.

aquifer under pressure greater than atmospheric. The surface of the underground water exposed to atmospheric pressure is designated as the *groundwater table.*

WATER COOLING AND TREATMENT

The primary function of cooling water in refrigeration, air conditioning, and industrial plants is to remove heat; a secondary function is to conserve water by recirculating it after it has cooled and sending it back to the equipment requiring heat removal. Many problems can arise as the cooling water is recirculated, and it is necessary to test and treat cooling water to avoid equipment problems, such as corrosion, scale, and algae growth.

Some cooling water is not recirculated but is taken out of a river to be passed through a condenser, for example, in a power plant, and then returned to the river downstream of the plant. However, environmental regulations now require plants to meet certain temperature standards in discharging heated water back to a river, and the trend is to install water-saving devices, such as cooling towers.

Water can be cooled by *spray ponds,* where jets of water are sprayed into the air by connecting the nozzle heads to pumps. Spray ponds are used where a large water reservoir is available and the drift of water mist that is generated is not objectionable to the surroundings. Efficiency of cooling is low in comparison to cooling towers.

Water is cooled by air circulating through sprays by *sensible heat* removal where cooler air picks up some of the heat from water that is warmer than the air. The second and more prevailing cooling is by *evaporative cooling,* where moisture sprayed into the air evaporates, or becomes a vapor. In transferring from a liquid to vapor, it picks up about 1000 Btu/lb of heat, thus lowering the water temperature. This change of state from water to vapor by heat removal is referred to as the heat of vaporization. The rate of evaporative cooling depends on how moisture-laden the cooling air is and the relative humidity of the air. Both *dry-* and *wet-bulb* temperatures must be considered in evaluating the rate of evaporative cooling.

The dry-bulb temperature is that displayed on an ordinary thermometer, while the wet-bulb temperature is obtained by covering the bulb of an ordinary thermometer with *wetted* silk gauze. When the instrument is placed in a moving air stream, some water in the gauze evaporates, which in turn takes heat from the remaining water in the gauze, thus dropping its temperature. The amount the water temperature drops depends on what the initial temperature was, and air dryness, or humidity. Finally, an equilibrium or balance is reached, which is the wet-bulb temperature.

With both dry- and wet-bulb temperatures known, the relative humidity is determined from psychrometric charts. The latter is also important in air conditioning. For example, when the humidity is 100 percent, dry- and wet-bulb temperatures are equal. In evaporative cooling devices, heat will be absorbed from the water as long as the wet-bulb temperature is *lower* than the water temperature.

In addition to humidity, rate of cooling is also determined by (1) the amount of water surface in contact with the air, (2) the velocity of air and water during their contact, (3) the length of air to water contact, and (4) the difference between wet-bulb temperature of air and inlet temperature of water.

Mechanical-draft-type cooling towers use fans to promote air circulating through the water spray and are classified as (1) forced draft type, pushing air through the tower, (2) induced draft, pulling air from the tower, (3) counterflow to the water flow, and (4) crossflow, where the air flows from the sides, while the water flows downward (see Fig. 4.5). These types of towers remove about 25 percent of the heat by sensible heat and the rest being by evaporative cooling. Most towers are of the induced-draft type.

Cooling towers lose water by evaporation, by the drift of droplets with the wind around a tower, and by the blowdown that is required to control the quality of the water. This makeup water can average from 2 to 5 percent of total flow of water through the cooling tower, depending on water quality and local atmospheric air conditions. An increase in the humidity of the air passing through the tower may require a steady addition of makeup water, but this is small compared to the water that would be needed if a nonrecirculating system was used. Where more than one circulating fan is used for draft purposes, these can range to 18 ft in diameter, with each cooling or draft section being called a *cell.*

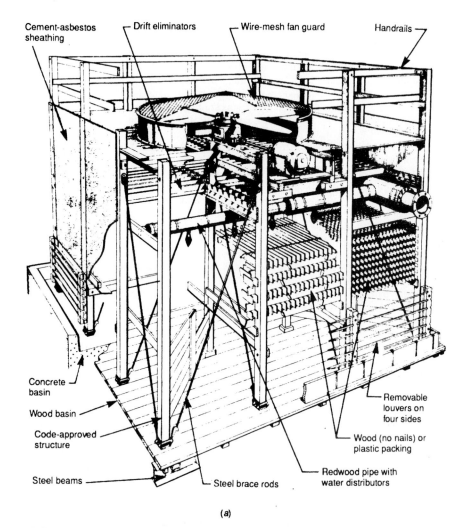

Cement-asbestos sheathing — Drift eliminators — Wire-mesh fan guard — Handrails

Concrete basin — Removable louvers on four sides

Wood basin —

Code-approved structure — Wood (no nails) or plastic packing

Redwood pipe with water distributors

Steel beams — Steel brace rods

(a)

FIGURE 4.5   Mechanical-draft cooling towers. (*a*) Induced draft pulls air upward while water spray drops downward. (*Courtesy, Power magazine.*)

Maintenance of Mechanical-Draft Towers

The following are guidelines for maintaining mechanical-draft towers:

1. Machinery on the tower should be inspected for proper mechanical and electrical operation each 8 h of operation.

2. The alignment of motors, drive shafts, gearboxes, and fans needs special attention because any problem here can cause excessive vibration, abnormal wear and tear, and some form of failure that can cause a shutdown of the system. This includes checking oil levels in the gearbox.

3. Drive motors should be of the totally enclosed type to prevent wind-driven rain from affecting the stator insulation. Otherwise, sheet metal covers are recom-

mended to protect the motor from water splashing off plates and entering the openings of the motor, which are usually located on the bottom. It is also essential to make insulation resistance tests annually or whenever it is suspected that water is entering the motor and affecting the winding insulation. Space heaters can be installed to go on automatically during periods when the motor is idle.

4. Gearboxes and fans should be checked annually for wear and tear and proper meshing and pitch of blades per manufacturer's instructions.

5. Winter operation, if required, creates icing problems on louvres, decks, gearboxes, and even fans primarily from water spray. A strategy for winter operation is usually provided by the tower manufacturer. For example, this may include reducing the velocity of air flow by the use of two-speed motors, throttling water flow, because cooler water in winter requires less cooling water through heat exchanger apparatus. The strategy, basically, is to use the heat picked up in cooling process water to prevent icing of the cooling tower equipment as much as possible.

Evaporative Condensers

Evaporative condensers are used to cool smaller flows than are possible with mechanical-draft cooling towers (see Fig. 4.6). The other distinguishable difference is that the gas or liquid being cooled is piped into the evaporative condenser, whereas with a cooling tower, the water from the tower is piped to the apparatus needing cooling water. In Fig. 4.6, note that condensing coils are within the unit and the evaporative cooled water is sprayed over the coils. Thus, an evaporative condenser combines evaporative water cooling with heat transfer to the gas or fluid that is to be cooled. Evaporation is aided by a fan, thus cooling the water. Air is pulled upward from an inlet and through the water spray, with the fan assisting in discharging the air back to the atmosphere. The cooled water droplets, or sprays, drop to a catch basin, and a

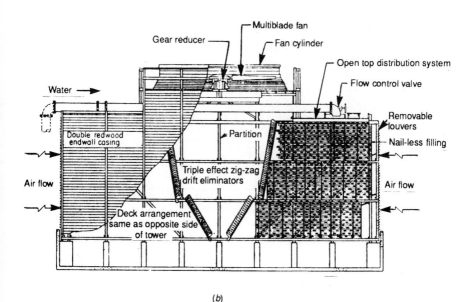

(b)

FIGURE 4.5   (*Continued*) Mechanical-draft cooling towers. (*b*) Cross-flow of air from both sides is classified as double-flow.

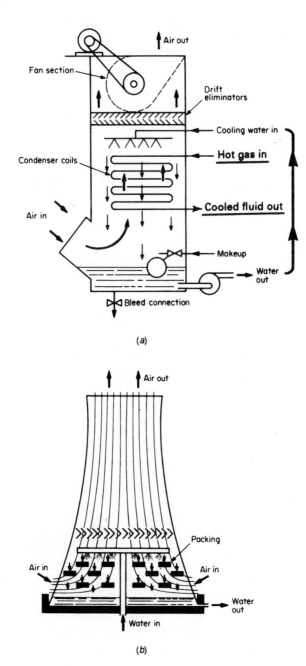

(a)

(b)

FIGURE 4.6 (*a*) Evaporative condenser combines cooling of water
and the subsequent cooling within the unit of the gas/liquid to be
cooled or condensed; (*b*) hyperbolic cooling tower has no fans and
depends on the chimney effect to obtain large air circulation from
bottom to the top of the tower.

pump recirculates the water up to a spray header. Automatic makeup of water lost by evaporation and drift is by a float-operated makeup valve that goes on and off with the rise and fall of the water reservoir on the bottom of the evaporative condenser. These units handle relatively small water flows and can be installed indoors; however, if this is done, a fresh air intake from the outside is required, as is a means to discharge the humid air to the outside.

Natural-Draft Towers

Natural-draft towers depend on hot air rising to produce a draft condition that can cool water. The most prominent is the *hyperbolic cooling tower* illustrated in Fig. 4.6. This tower has no fan to promote air circulation but depends on its hyperbolic shape to create a venturi nozzle-type effect, creating an increase in air velocity as the width of the tower narrows toward the top, commonly referred to as chimney effect. These hyperbolic towers can be designed to handle large water and air flows by increasing the diameter and height and are extensively used by utilities to cool steam in large condensers that are part of steam turbo-generators.

While the hyperbolic tower may be over 600 ft high, the water is introduced and cooled in the bottom 30 to 50 ft of the tower. Here wooden bars or waterproof synthetic bars are spaced and staggered in rows with the sprayed water splashing from row to row, exposing the droplets to the air flow and thus achieving evaporative cooling as in a mechanical-draft tower. Utilities rate these towers per the megawatt size of the turbo-generator for which the tower cools the condenser water.

Cooling Water Treatment

Cooling water treatment is especially necessary on recirculated water because evaporation and makeup water can increase the concentration of mineral solids. Water analysis and treatment depend on local conditions but must consider (1) corrosion, (2) deposits, (3) biological or algae growth, and (4) chemical attack of tower components, such as wood. *Chromates* are used to provide a tough film on the metal and thus restrict the corrosion process. Inhibitors such as polyphosphates, silicates, and alkalies have also been used for corrosion protection.

Deposits, or scale, are prevented by eliminating the scale-forming substances such as calcium, magnesium carbonates, and sulfates by softening the water with lime or soda ash, zeolite, or some of the phosphates supplied by water treatment companies. Testing the treated water is important because if it is overtreated and/or if blowdown procedures are not followed, concentrations may affect the equipment that the cooling water is pumped to, such as heat exchangers. Too high a concentration of soluble solids in the cooling tower water can affect wet-bulb temperature and thus cause warmer water to leave the tower.

Biological or algae growth may be controlled by chemical treatment with chlorine, copper sulfate, and potassium permanganate, but the dosage requires review by an experienced water treatment specialist. Chemical attack on wood usually causes *delignification.* Chemicals in the water, such as sodium carbonate, dissolve the lignin that binds the wood fibers together. Evidence of this is wood surfaces becoming white and fibrous, thus losing structural strength. High pH values are considered to contribute to delignification, and pH value for the tower water of 7 to 7.5 are usually recommended to avoid this wood deterioration. Neutralization of cooling water with an alkali or an acid is the usual method of adjusting pH.

Caution is required on the use of chromates for corrosion control. Care must be taken that spray from the cooling tower does not contact equipment or personnel because the chromates will stain paint and are irritating to skin, eyes, and nose. Local regulations may prevent their use because of the danger of discharging them to waste. They should never be used on lines or connections that may contaminate drinking water.

Water for Fire Control

Water for fire fighting is generally taken from the municipal supply of potable water, and the major concern of fire underwriters is the adequacy and availability of the water at all times. Where private water supply is available, such as in large industrial plants, the filtering of the water to remove suspended solids is of major concern, as is capacity. The standard grading schedule of the National Board of Fire Underwriters requires the following capacity for fire fighting purposes at all times, even during maximum periods of water use for domestic or industrial process requirements:

Population	Required maximum fire flow, gpm	Hours of duration of capability
1,000	1,000	4
2,000	1,500	6
3,000	1,750	7
4,000	2,000	8
5,000	2,250	9
6,000	2,500	10
10,000	3,000	10
13,000	3,500	10
17,000	4,000	10
22,000	4,500	10
27,000	5,000	10
33,000	5,500	10
40,000	6,000	10
55,000	7,000	10
75,000	8,000	10
95,000	9,000	10
120,000	10,000	10
150,000	11,000	10
200,000	12,000	10
Over 200,000	12,000 plus 2,000 to 8,000 gpm for a possible second fire	10

The values provided in above table are at 20 psi residual pressure. For fire protection, excess volume is desired over and above the maximum industrial and domestic demand. This is why fire underwriters request periodic flow tests from one or more fire hydrants at staggered intervals of time to coincide with maximum demand of domestic and industrial users, whose demands may vary through a 24-h period. This is really a test of the adequacy of water storage in reservoirs, storage tanks, and pumping facilities to make sure that the distribution storage system is able to maintain normal demands as well as the emergency demands of fighting a fire with water.

WASTEWATER AND TREATMENT

Federal legislation has been increasing to clean up water waste streams since the passage of the Clean Water Act in 1972. The aim is to clean water streams of pollutants and prevent underground water supplies from also becoming polluted. *Water pollution* occurs when waste matter is introduced in sufficient amounts to make the water quality unsuitable for domestic or industrial use *and* unsuitable for the proper support and growth of marine life.

Plant service operators will become more involved with waste water treatment methods as federal laws become more prominent. At present, most facility waste waters are discharged into publicly owned water treatment works (POTWs) or directly to surface and groundwater. Direct discharges are now regulated under the National Pollutant Discharge Elimination System. Federal and state regulations are encouraging POTWs to have industrial plants institute their own pretreatment programs before discharging into the public system or else pay a fee or surcharge to the public wastewater facility based on the amount of effluent and the concentration of the pollutants.

In addition to process wastewater, recent EPA regulations also include rain runoff that picks up plant contamination on the ground from spills, washing of equipment, and leaks, to name a few of the sources of plant site pollutants.

Industry is taking several approaches in reducing water pollution:

1. Reviewing process flows and aiming for zero discharge of polluted wastewater. This includes identifying the source of the contaminant, for example, identifying leakages or sloppy operational practices. A further aim in many industries is to recycle the polluted water and recover the pollutant. For example, acids in wastewater sometimes can be recycled by returning the wastewater to a recovery unit concentrator. With the acid reconcentrated to a usable value, it can be used again in the facility or sold to others.

2. Source reduction depends on reengineering process flows and may include process modifications to reduce the pollution at the source. In many cases, it is more economical to follow a source-reduction strategy than a so-called back-end strategy in which the total wastewater is treated before it is discharged.

3. Since 1990 EPA regulations have required POTWs to implement testing, reporting, and enforcement programs for any significant industrial discharges that they may handle in their systems. The EPA regulations also affect significant industrial users of the POTWs by requiring legally binding permits that govern pretreatment of wastewater and its discharge into POTW facilities. Included in federal regulations is sludge discharge; the regulations are intended to limit the potential release of toxins from the sludge into the ground. In addition, limits for metals, organics, and pathogens in water disposal have been established.

Facilities are required to establish a *wastewater management plan* in order to avoid federal fines or citations. Industrial process plants must determine the quantity of discharge to be treated, the temperatures involved, and, of course, its composition. Technical details on flow rates, pounds, gallons, and similar data will determine the size and composition of the wastewater pretreatment program before the waste is discharged. The objectives of treatment are:

1. Remove suspended solids
2. Remove oils and greases
3. Remove heavy metals
4. Eliminate or reduce volatile organics
5. Eliminate or reduce toxic compounds to acceptable levels
6. Neutralize the wastewater's acidity and alkalinity to an acceptable pH level

As with water treatment, wastewater treatment requires testing, identification of the pollutants, determining the method of pollutant removal, establishing controls and testing to make sure that the treatment is producing the desired results, and documenting the test results for regulatory review. Wastewater treatment, the equipment required to remove pollutant, and managing the system are rapidly expanding as government regulations and the public strive for clean water and its preservation as a national asset. A brief description of methods used in wastewater treatment follows.

Suspended Solids

Suspended solids removal depends on the material in the wastewater and its size. For large suspended matter, 25 mm in diameter or larger, methods used include:

Straining or screening

Gravity separation, including the use of coagulants

Filtration with granular media or precoat filtration

For suspended matter under 25 mm in diameter, methods used include:

Dewatering by pressure filtration

Use of centrifuges

Sand-bed drying

Vacuum filtration

Particles that will not settle unaided are called *colloids* and are of very small diameter. These are removed by chemical coagulants and filtering or by dissolved-air flotation and skimming.

Oils and Grease

Oils and grease, if free and not emulsified, float to the surface of the water and agglomerate. This permits the oil and grease to be mechanically skimmed off the surface.

Oils and greases of low concentration can be removed by dissolved-air flotation. In this process, the effluent is pressurized with excess air and then pumped into a flotation tank. The dissolved air comes out of solution with the oil and smaller particles clinging to air bubbles, which float to the surface, and again the oil and grease can be mechanically skimmed off.

With oil that is emulsified, coagulants must be employed in the wastewater so that the oil and grease coalesce into agglomerates large enough for removal.

Heavy Metals

Heavy metal removal is usually done by precipitating out the metal from the wastewater by making the metal form its hydroxide. Lime or a caustic is added to the wastewater to raise the pH value until it reaches the metal's minimum solubility. This helps the formation of the metal's hydroxide, which will precipitate out of the water. Coagulants help in flocculating the hydroxides into larger, heavier particles which can be filtered out. The resulting sludge can be dried out or vacuum filtered. The sludge can be used for landfill; however, it may be classified as hazardous waste under the Federal Resource Conservation and Recovery Act, depending on contaminant contact.

Water that contains several metals requires further analysis for treatment, with several other options available such as ion-exchange resins, activated carbon, and reverse osmosis.

Volatile Organic Compounds (VOCs)

Volatile organic removal is usually performed by stripping the vapors out of the wastewater by air or steam in packed distillation towers. Air or steam flow countercurrently to the wastewater, removing the VOCs in the vapor phase. VOCs in low concentrations may also be removed from the wastewater by activated carbon.

Toxic Compounds

Toxic compound reduction is generally by oxidation of the compounds by chemical agents. Because of the varieties of toxic agents, the chemical treatment of these toxics is beyond the scope of this book. As an example, oxidation of cyanide to cyanate is the method used to destroy cyanide's toxic properties, and this is performed by treatment with sodium or calcium hypochlorite.

Acids and Alkalies

Acid and alkali neutralization are required because pH values below 5.0 adversely affect the more delicate fish life, and below pH 4.0 nearly all type of fish will die. The food supply of fish is also affected by low pH values. High pH values above 9.0 also have a detrimental effect on fish life. Therefore, discharge of wastewater is required to have a pH value between 6.0 and 9.0. Acidic or low pH waters are neutralized by the addition of proper dosages of lime, limestone, or caustic solutions. Alkaline or high pH wastewater is neutralized by sulfuric, hydrochloride, or carbon dioxide gas.

Nonvolatile Organics

Nonvolatile organics must be treated if the loading of organics in the wastewater is sufficient to cause serious depletion of the oxygen in the water. At oxygen values below 5 ppm, the finer fish life is affected, and at 4 ppm nearly all fish life is adversely affected. If the oxygen content is low, aeration can be employed as described earlier in this chapter. The effect of organic matter in the depletion of dissolved oxygen in water is referred to as the "biochemical oxygen demand," or BOD. Many organic compounds can be oxidized catalytically with chemicals. A water treatment specialist should be consulted for more details.

QUESTIONS AND ANSWERS

1. How are grains per gallon and ppm related?

   Answer. One grain per U.S. gallon = 17.1 ppm. This makes 1 ppm = 0.058 gr/gal.

2. How is hardness expressed?

   Answer. In U.S. practice, it is expressed as parts of calcium carbonate per ppm of water or grains of calcium carbonate per gallon of water.

3. What is the chief chemical compound in temporary hard water?

Answer. Calcium bicarbonate, $Ca(HCO_3)$.

4. What two methods are used to soften water with temporary hardness?

Answer. Temporary hard water may be softened by *boiling* the water, with the calcium carbonate precipitating out as soft sludge and carbon dioxide being liberated. Temporary hardness can be eliminated by *adding lime* in the right quantity, with calcium carbonate being precipitated and water being a by-product of the chemical reaction.

5. What chemical compounds in water cause permanent hardness?

Answer. Calcium sulfate and magnesium sulfate.

6. How is hard water softened?

Answer. Permanent hard water can be softened by the correct addition of sodium carbonate or by passing the water through a zeolite softener.

7. What is the chemical constituent of zeolite?

Answer. Sodium aluminum silicate.

8. What impurities are removed from potable water?

Answer. (*a*) Suspended matter, (*b*) pathogenic bacteria, (*c*) microscopic plants or algae, (*d*) color, (*e*) odor, and (*f*) inorganic gases.

9. What is the soap test for hardness?

Answer. This test is commonly performed in boiler rooms. Fill a burette with a standard soap solution called APHA and add 50 mL of water. Mix the two in a bottle by shaking until a permanent lather forms and lasts about 5 min. Subtract 0.30 mL from the lather formed, and multiply the remainder by 20 to find the hardness of the sample, which represents the parts per million of calcium carbonate.

10. What is activated carbon and its use in water filtering?

Answer. Activated carbon is used to remove tastes and odors from water and also to improve its color. Charcoal is heated in a closed retort in an atmosphere of steam, air, carbon dioxide, or other gas to improve the ability of the carbon or charcoal to adsorb gases. One method is to pass water that has been filtered through a bed of activated carbon at a rate similar to a sand filter. Gases from the water are adsorbed by the carbon bed. The bed will require periodic cleaning depending on the amount of contamination in the water. It has a fairly long life before it has to be revivified.

11. What is a clarifier?

Answer. See Fig. 4.7. This is a mechanical type of sediment remover, used extensively in sewage treatment plants to remove suspended matter. The stream with suspended matter enters in the center of a small circular wall or baffle, which helps in distributing the laden stream both vertically and horizontally. The heavier solids deposit near the center of the clarifier, while finer deposits occur at the outer periphery. The weir shown carries the water out to be further filtered downstream. The settled solids are pushed slowly to a central sludge hopper for pumping out. In sewage treatment applications, the pumped-out sludge then goes to sludge-digestion tanks.

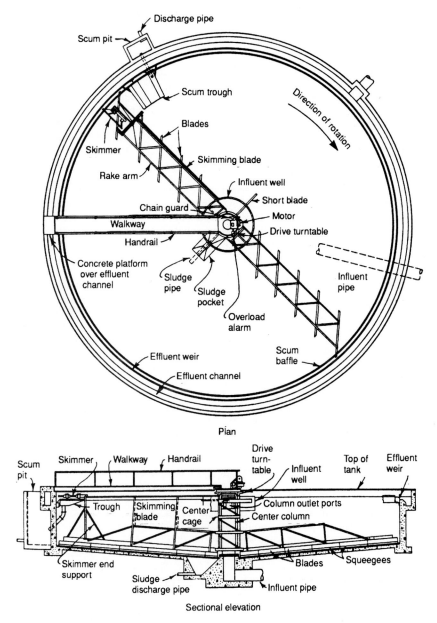

FIGURE 4.7 A clarifier is a mechanical-type sedimentation apparatus, extensively used to remove heavy sludge matter in a water stream. The unit shown is used in sewage treatment plants at the beginning of the sewage treatment cycle.

12. What is considered the required pressure at the base of a fire fighting nozzle?

    Answer.   A pressure of 50 to 75 psi is required for adequate fire fighting streams to be developed. Therefore, hydrants and fire pumpers must have 100 psi or more to deliver the pressure required by the nozzles at the end of hoses.

13. What does the term *coagulation* mean?

    Answer.   Coagulation assists in gathering fine suspended particles into larger "clumps" so that the suspended particles are easier to remove by sedimentation or filtering. Gelatinous or jellied substances are added to water containing fine particles, and this helps to coalesce the particles into larger ones for eventual removal by sedimentation or filtering.

14. How is acidity or alkalinity expressed?

    Answer.   This is expressed by the pH scale, which ranges from 1 to 14. A pH of 7 is considered neutral; anything below this value is acidic, and anything above this value is alkaline. The pH values are related to the laws of dissociation of water into OH and hydrogen ions. It is an exponential function with 10^{-7} being the neutral point. This means that a pH of 5 is 10 times more acidic than a pH of 6.

15. What is chemical precipitation?

    Answer.   When certain chemicals are added to a solution containing dissolved minerals, the chemicals react with the dissolved mineral and form solid particles that drop or precipitate out of the solution and can then be removed by filtering and blowdown in boilers, for example. Lime-soda softening of water, which causes the dissolved minerals to drop out of solution, is a good example of chemical precipitation.

16. What is ion exchange?

    Answer.   An ion is an electrically charged atom or radical from a molecule, and the electric charge can be positive or negative. In ion exchange, ions of different solids in solution are interchanged. For example, in a zeolite softener, a base exchange takes place with the calcium and magnesium compounds passing into the zeolite. They are replaced by sodium from the zeolite. The calcium bicarbonate that was in solution becomes sodium bicarbonate, and magnesium sulfate becomes sodium sulfate. Neither of these compounds are scale formers.

17. What is *anion* and *cation* exchange in demineralizers?

    Answer.   This is an ion exchange process which requires two steps of exchange. The first step is hydrogen zeolite, or *cation,* exchange which removes calcium, magnesium, and sodium from water but substitutes hydrogen in their place, making the water acidic. The second step is the use of weak or strong base resins; the acidity can be corrected in an *anion* exchanger. The strong base anion exchanger will remove both strong and weak acids in the water, but it is more costly to operate.

18. What is the advantage of hot-lime softening over cold-lime?

    Answer.   The reaction time is faster at higher temperature.

19. What impurities are removed from water by aeration?

    Answer.   (*a*) Carbon dioxide; the pH value (less carbonic acid) is increased, (*b*) hydrogen sulfide gas, (*c*) foul taste and odor, (*d*) ferrous iron, and (*e*) manganous manganese.

20. What precautions are required when using chromates for cooling water treatment against corrosion?

    Answer. This treatment may not be allowed by authorities because it may contaminate water runoff. Care must be taken that spray from cooling towers treated with chromate does not contact equipment or personnel. Chromates stain paint and irritate skin, eyes, and the nose. Chromate lines must be completely isolated from potable water for health reasons.

21. In neutralization, what chemicals are used to raise the pH?

    Answer. Low pH indicates an acidic condition, which can be corrected by the addition of base substances, such as lime, soda ash, or caustic.

22. How is a hyperbolic cooling tower classified?

    Answer. This is considered a natural-draft tower for cooling water.

23. At what humidity level do wet- and dry-bulb thermometers read the same?

    Answer. At 100 percent humidity.

24. What are two methods of heat transfer in a cooling tower?

    Answer. One is by the temperature difference between outside air entering the tower and the hot water to be cooled. This is called *sensible* heat transfer. The other is by *evaporative* cooling in which the fine water droplets in the tower may vaporize and, in doing so, cool the remaining water by absorbing out of the water the heat of vaporization, roughly equal to 1000 Btu/lb of water vaporized.

25. Why is makeup water needed in air-water cooling devices?

    Answer. (*a*) Water is lost by evaporation in cooling the water, (*b*) there is always some water drift out of the tower, and (*c*) the concentration of chemical treatment that is applied to the cooling water must be controlled. Blowdown may be needed, which must be replenished.

26. What encourages algae growth in cooling towers?

    Answer. Sunlight will cause more algae growth. If the algae on walls and reservoirs grows too thick, it can break off and clog pipes going to the heat-transfer equipment.

27. What are some impurities that can be present in wastewater?

    Answer. (*a*) Dissolved salts, (*b*) organic matter, (*c*) material in suspension, and (*d*) floating oils, acids, alkalies, and toxic substances.

28. What does the term *BOD* mean?

    Answer. This stands for "biochemical oxygen demand" and refers to organic matter in water using up the dissolved oxygen in it, which eventually can affect fish life. For example, at oxygen levels below 4 ppm, fish will die in the water so affected.

29. What pollutants must be removed from wastewater?

    Answer. Suspended solids, oil and grease, heavy metals, reduction of volatile organics, reduction of toxics to permissible threshold limits, correcting the acidity or alkalinity to acceptable levels, and nonvolatile organic treatment if oxygen level in the wastewater is low.

CHAPTER 5
FLUID MOVERS—PUMPS

All types of pumps are used in manufacturing process flows, power plants, and irrigation systems; however, this chapter will concentrate on service pumps that supply water to a plant for domestic, process, or fire fighting purposes, as well as pumps for chilled water flow, condensing water flow, condensate return, boiler feedwater, and drainage. Pumps cause flow by adding pressure to the fluid handled, and this in turn produces velocity to overcome friction or to lift the water to a certain head.

Pump types usually found in providing plant services are:

1. Reciprocating pumps
   a. Direct-acting steam type; simplex and duplex
   b. Power machine driven; single-acting simplex and triplex
2. Centrifugal pumps
   a. Single and multistage
   b. Volute and turbine type
3. Rotary pumps
   a. Gear and screw pumps
   b. Propeller pumps
4. Jet pumps
   a. Steam jet injectors and ejectors
   b. Water jet ejectors

SELECTION OF PUMPS

Pump selection is determined by the service if corrosive or abrasive fluids are to be handled; however, it usually starts with determining basic data such as head in feet, capacity in gallons per minute, horsepower required, and the speed and direction of rotation. The properties of the fluid to be handled require attention to viscosity, temperature, corrosiveness, and abrasiveness.

Pumping Terms

The terms discussed below can be useful when solving pump problems or specifying pumps.

Head. Liquid pressure in pump applications is considered equivalent to a column of liquid of a height sufficient because of the weight of the column to produce this pressure. For example, for water at atmospheric temperature, 2.31 ft of water = 1 lb/in^2 pressure. *Static head* is the height in feet of the fluid above a designated gage point. *Pressure head* is the static head *plus* the gage pressure expressed in feet plus the friction head (if the fluid is flowing). *Velocity head* is the vertical height or feet required to produce a certain speed of flow, expressed by the equation

$$\text{Velocity head, } h_v = \frac{v^2}{2g}$$

where g = 32.2 ft/s^2, or the acceleration of a free falling body at sea level. *Pump operating head* is the difference between the pressure and suction heads.

Except for water velocities well above average, or for large volumes at low heads, the velocity head is often left out in calculations.

Friction head is the feet of liquid required to overcome the resistance to fluid flow in pipes and fittings.

Velocity.   Velocity of flow per the velocity head equation is expressed in feet per second and is important in flow calculations. In order to determine the flow past a given point, use $Q = Av$, where A = cross-sectional area in square feet of the pipe or fluid conduit, v = velocity of flow in feet per second and Q = ft^3/s. Velocity heads are calculated from pressure heads on both sides of a venturi nozzle or orifice plate installed in the pipe per hydraulic standards. Some approximation is possible for a circular pipe with an inside diameter d, in:

$$v = \frac{0.4085\ (\text{gpm})}{d^2}$$

Example.   A test on a 3-in inside diameter pipe showed that the flow velocity was 68.08 ft/s as measured by an orifice plate. What is the gpm flow?
Substituting

$$68.08 = \frac{0.4085(\text{gpm})}{3^2}$$

and solving for gpm:

$$\text{gpm} = \frac{68.08(9)}{0.4085} = 1500\ \text{gpm}$$

Work in Pumping.   The work required of a pump is influenced by the amount of head the liquid will be raised, the force required to pump it into a higher-pressure system, and that required to overcome friction. This work is called the hydraulic horsepower, or theoretical pump horsepower, expressed as follows:

$$\text{Theoretical hp} = \frac{\text{gpm}(H)(s)(\text{lb/gal})}{33,000}$$

where gpm = flow rate
   H = total head of liquid, ft
   s = specific gravity of fluid   For water = 1

For water, which weighs 8.33 lb/gal at ordinary temperature, the equation becomes

$$\text{Theoretical hp} = \frac{\text{gpm}(8.33)H}{33,000} = \frac{\text{gpm}(H)}{3962}$$

The brake horsepower (bhp) is the above theoretical horsepower divided by the pump efficiency E, or

$$\text{bhp} = \frac{\text{gpm }(H)}{3962\ (E)} \qquad \text{Note that } E = \frac{\text{Theoretical hp}}{\text{bhp}}$$

Example. If a pump has an efficiency of 70 percent and is delivering 1500 gpm of water from ground level against a total head of 1000 ft, what size motor will be required? Substituting in the bhp equation,

$$\text{bhp} = \frac{1500(1000)}{3962(0.70)} = 541\ \text{hp}$$

Viscosity. Viscosity is a term used to indicate the internal friction of a fluid. In fluid mechanics, dynamic viscosity is expressed as

$$\frac{\text{Shearing stress of fluid}}{\text{Rate of shearing strain}}$$

Another term used in fluid mechanics is kinematic viscosity, which is the dynamic (also called absolute viscosity) divided by the density of the fluid. Kinematic viscosity can be expressed as ft^2/s or cm^2/s, which is called a *stoke*. As can be noted, viscosity is important in pump design and in some fluid flows, such as oils in pipes, in order to determine internal friction to flow. Viscosity varies considerably from one fluid to the other and decreases with rising temperatures. This is why viscous liquids are heated when they are pumped from one point to another. Viscous liquids require more horsepower to pump and reduce the pump efficiency and capacity because of this internal resistance of some fluids.

Suction. Total suction lift is the reading of the gage at suction flange of pump, which is converted to feet minus the velocity head in feet at that point. Total suction head is the same as lift except the velocity head is *added.*

From experience, pumps have suction limitations even though in theory they should lift a liquid to the feet height represented by atmospheric pressure, or 14.7 atmospheric pressure = 2.31 × 14.7 = 34 ft of lift possible. Factors which reduce the possible lift are internal friction to flow, vapor pressure of the fluid, pump speed, capacity, and internal pump design. Vibration and possibly cavitation occur when a pump is trying to operate with a suction lift that it cannot handle.

Cavitation. Cavitation can occur on pumps when the fluid pressure equals the vapor pressure at the existing temperature, and as a result vapor bubbles alternately form and collapse. The fast formation of bubbles causes the liquid at high velocity to fill the void with impact force on the internal parts of the pump. These surges into the voids are equivalent to explosions on small areas, but the forces produced on the internal parts can exceed the tensile strength in that part of the pump where they are occurring. This causes particles to be knocked off, and rapid pitting and erosion take place, even to the extent that pieces break off internally, producing serious damage to the pump.

To prevent cavitation, most pump manufacturers stamp their pumps with a *net positive suction head, or NSPH,* which should not be exceeded in order to avoid cavitation damage. The fluid being pumped and the corresponding temperature need to be known since vapor pressure has an effect on the NSPH. In power plant applications, feedwater pumps have been ruined due to cavitation damage because the feedwater temperature control went astray, and this resulted in the water "flashing," which affected the permissible NSPH. Many pumps, especially those designed to handle fluids above atmospheric temperatures, have a bypass valve from a stage of the pump back to the suction side. This bypass is activated if the NSPH allowed on the pump is approached in operation.

Useful Pump Data.   Other useful data (all will be for water) when calculating pump problems or performance are:

$$1 \text{ psi gage} = 2.31 \text{ ft of water}$$

$$1 \text{ ft of head} = 0.434 \text{ psi}$$

$$1 \text{ ft}^3 \text{ of water} = 62.4 \text{ lb in weight}$$

$$1 \text{ ft}^3 \text{ of water} = 7.48 \text{ U.S. gal} = 6.24 \text{ imperial gal}$$

$$1 \text{ U.S. gal} = 8.33 \text{ lb in weight}$$

$$1 \text{ imperial gal} = 10 \text{ lb in weight}$$

$$1 \text{ U.S. gal} = 231 \text{ in}^3$$

$$1 \text{ imperial gallon} = 277 \text{ in}^3$$

$$1 \text{ hp} = 33,000 \text{ ft} \cdot \text{lb/min of work}$$

RECIPROCATING PUMPS

Reciprocating pumps are positive displacement pumps and can be used to obtain very high pressures by staging the cylinders or by using more than one pump. In staging, the discharge pressure from one cylinder is the suction pressure for the next cylinder and this can be carried out with each cylinder boosting the pressure to the desired result.

Steam-Type Reciprocating Pumps

Steam-type reciprocating pumps are classified as *direct acting* if a steam cylinder is in line with the pumping cylinder or steam *power driven* if the steam engine has a crank, flywheel, and crosshead. The term *simplex* means it has one water cylinder. The term *double acting* means it pumps water from both ends of the piston or plunger of the pump. See Fig. 5.1 for pump arrangements.

A *duplex pump* of the steam direct-acting type has two water cylinders whose operation is coordinated to obtain the final desired pressure. A duplex pump can also be driven by a steam engine with a crank, flywheel, and crosshead arrangement. *Triplex pumps* have three water cylinders in parallel, can be driven by steam, and are either direct acting, steam power driven, or have a motor drive (see Fig. 5.1a).

Reciprocating pumps are also driven by electric motors, diesels, and gas, and steam turbines either directly through one shaft or coupling or through gearing. Figure 5.1d illustrates a triplex, single-acting reciprocating pump driven through gears by an electric motor.

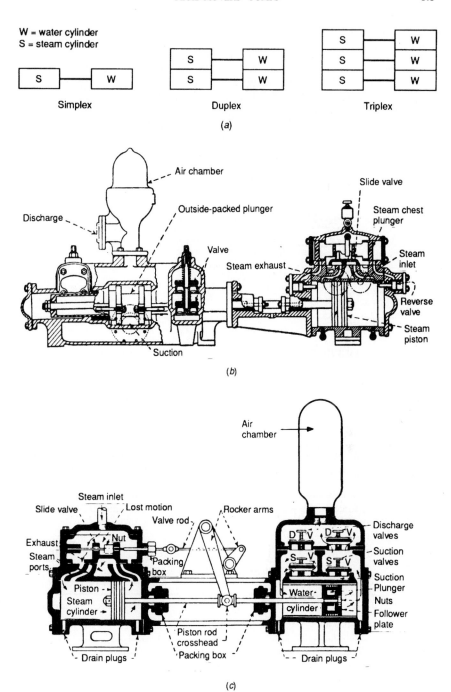

FIGURE 5.1 (*a*) Direct-acting steam-type reciprocating pump cylinder arrangements; (*b*) simplex outside-packed plunger pump; (*c*) duplex direct-acting steam pump.

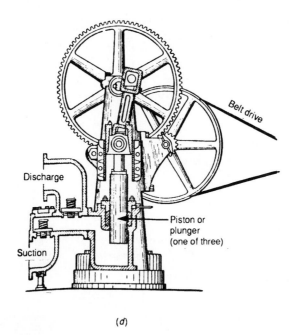

(*d*)

FIGURE 5.1   (*Continued*) (*d*) triplex single-acting power driven pump.

Water Pressure Calculation.   A common question about reciprocating pumps of the direct-acting steam type is how low-pressure steam can produce high discharge water pressure on the pump.

Example.   A simplex pump has a 10-in diameter steam cylinder operating at 100 psi, with a single-acting water pump with a diameter of 5 in. If efficiency is neglected, approximately what water pressure would be generated if a final 20 percent was added for pump slip?

The easiest way to solve this is to compare the force on the steam piston and equate it to the force on the water piston. Assume the cylinder diameter is the same as piston diameter. Then

$$\text{Steam piston area} \times \text{steam pressure} = \text{water piston area} \times \text{water pressure}$$

$$\text{Steam piston area} = \frac{\pi(\text{diameter})^2}{4} = \frac{\pi(10)^2}{4} = 78.54 \text{ in}^2$$

$$\text{Water piston area} = 0.7854(5)^2 = 19.64 \text{ in}^2 \qquad \left(\text{Note: } \frac{\pi}{4} = 0.7854\right)$$

Equating forces by multiplying each area by corresponding pressure,

$$100(78.54) = \text{water pressure}(19.64)$$

$$\text{Water pressure} = \frac{7854}{19.64} = 400 \text{ psi}$$

From this must be deducted 20 percent of 400 for pump slip, giving a final water pressure of 400 − 80 = 320-psi water pressure.

Horsepower Calculation. Another calculation that is quite often needed is the *indicated horsepower* (ihp) that is developed by the steam cylinder of steam-type reciprocating pumps. Reference is made to the foot · pounds of work per minute performed by the piston in terms of the following:

$$A = \text{piston area being acted on by the steam} = 0.7854(d)^2$$

where *d* is the steam piston diameter, inches.

$$P = \text{mean effective pressure of the steam}$$

This is usually determined by a steam engine indicator card in order to get the average pressure during a stroke of the engine, in psi.

$$L = \text{length of stroke, ft}$$

$$N = \text{number of strokes per minute}$$

$$\text{ihp} = \frac{PLAN}{33,000}$$

Example. What is the indicated horsepower for the steam cylinder in the previous example if the stroke is 20 in, the mean effective steam pressure is 90 psi, and the number of strokes per minute is 60? Using the above equation and substituting the known data,

$$\text{ihp} = \frac{90(20/12)(78.54)(60)}{33,000} = 21.42 \text{ hp}$$

Reciprocating steam pump sizes are always given with the steam cylinder diameter in inches listed first, followed by the water piston size and then the stroke in inches.

Use of Reciprocating Pumps

The reciprocating pump type to use is ordinarily dictated by service needs. The advantage of a reciprocating pump is its flexibility in achieving desired pressures, capacity, and speed. Reciprocating pumps are generally more expensive than centrifugals, require more floor space, can be noisy, and need more maintenance.

Simplex double-acting pumps are used for water service and moderate capacity boiler feed. Duplex double-acting pumps, when made of the correct material, are used extensively to pump viscous-type fluids such as oils and tars. Triplex pumps are used for constant-type loads, such as pumping water to a higher pressure.

The *air chamber* shown in Fig. 5.1*b* is used where water lines are long in order to avoid the shock of water hammer, and it reduces surges due to the reciprocating motion of the pump.

Reciprocating pumps must have *packing* to prevent water from leaking past the piston or plunger and also where any rod comes out of the cylinder, called gland packing. Material for packing depends on the fluid being handled, temperatures, pressures, and pump material. The manufacturer's maintenance and installation instructions should be followed when the packing requires replacement to avoid excessive leakage. Strainers should be installed on the suction side of reciprocating pumps in order to prevent foreign substances from damaging the valves or cylinders. *Foot valves* are used for pumps under suction pressure to prevent backflow on the suction line. *Aspirators* are used to drain air out of suction lines to prevent the pump from becoming air bound. It is a form of priming the pump by making sure only water flows to the pump from suction lines.

CENTRIFUGAL PUMPS

The most frequently used pumps are centrifugal pumps; they are used for all types of service, including industrial manufacturing applications, in addition to the traditional ones of pumping water, condensate, boiler feed, and chilled water and similar plant service applications. Sizes can go to 20,000 hp, as used for utility boiler feed. The growth of centrifugal pump application is due to lower cost, smaller floor space, non-pulsating flow, relatively quiet operation, and the adaptability to drive the pump by motors and gas and steam turbines at considerable variations from low to high speed. This type of pump has as its main component a casing in which an impeller rotates (see Fig. 5.2). The fluid to be pumped is directed through the inlet pipe to the center of the pump, called the eye. The impeller throws the liquid out radially through pump passageways, and this develops pressure by converting the kinetic energy. *Volutes* convert the velocity energy to pressure energy. In *diffusor-type* or turbine pumps, guide vanes are placed between the impeller and casing chamber, but the transformation of velocity energy to pressure follows the volute casing design (see Fig. 5.2).

Centrifugal pumps may be of the single- or double-suction type. In the double-suction type (Fig. 5.3) the fluid enters from both sides of the pump.

Multistage Pumps

Multistage centrifugal pumps are used for service pressures not attainable from single-stage pumps and are found in such service as water supply, fire, boiler-feed, and charge pumps in the refinery and petrochemical industry. Multistage pumps can be of the volute or diffusor type. Volute-type pumps usually have single-suction impellers with half the impeller inlets facing one direction and half in the opposite direction in order to balance the thrust forces. In the diffusor-type pump impeller inlets generally face one direction with the thrust force neutralized by a differential pressure arrangement, or *balancing piston* or drum. The unbalanced pressure differentials across each impeller create an axial thrust toward the suction end, which is counteracted by a balancing piston located near the end of the last impeller at the outboard end of the rotor. Kingsbury thrust bearings at the outboard bearing are also used to take any thrust fluctuation caused by abnormal operations.

The centrifugal pump is not considered a positive displacement pump, as is the reciprocating pump. For example, if the discharge valve on a centrifugal pump is closed completely, the pressure will only rise to a limited pump value with the rotating impeller churning the fluid and the work on the fluid being converted to heat. There will not be a rise in the head on the discharge end of the pump. In contrast, if the discharge valve on a reciprocating pump is closed, pressure continues to build up unless a pressure control stops the pump or a relief valve opens or the overload protection on the driver stops the pump. If none of the above work and there is unlimited horsepower drive available, something would have to burst from overpressure.

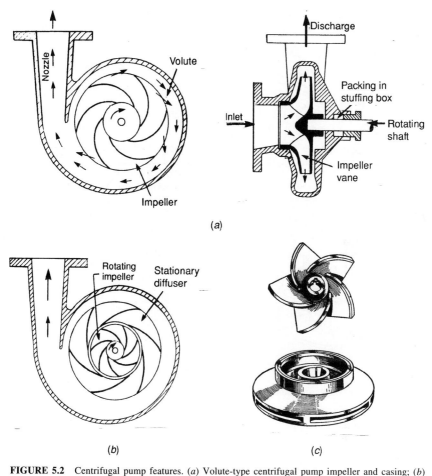

FIGURE 5.2 Centrifugal pump features. (*a*) Volute-type centrifugal pump impeller and casing; (*b*) diffuser-type centrifugal pump and casing; (*c*) impeller types, closed and open.

Performance Curves

Centrifugal pump performance curves are used to show the various relationships of head, power input, and efficiency at various speeds. Figure 5.4*a* shows the performance characteristics of a centrifugal pump with water as the fluid and operating at constant speed. Note how the head developed drops off after or near rated capacity. Efficiency rises to 91.7 percent and then drops. Figure 5.4*b* shows the pump performance curves at *different* constant speed tests. The analysis of centrifugal pump behavior can be quite complex when one considers the complicated, turbulent flow inside the pump.

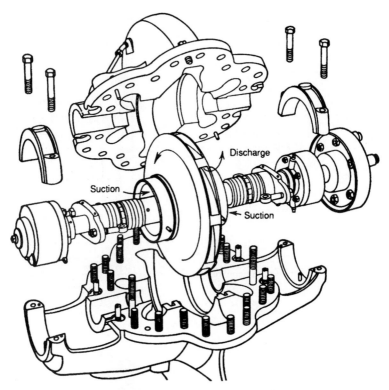

Discharge

Suction

Suction

FIGURE 5.3   Single-stage double-suction centrifugal pump details.

Pump Laws

Centrifugal pump laws have been established for approximate relations, which are helpful in indicating performance trends. The derivation of these centrifugal pump laws is treated in texts on fluid mechanics. For a pump operating at *constant speed,*

Total head varies directly as the pump impeller diameter *squared, D^2*

Capacity of the pump varies as the pump impeller *cubed, D^3*

Fluid power developed by the pump varies by D^5 with D = pump impeller diameter

For a centrifugal pump operating at *different speeds,*

Pump capacity varies directly as the speed, N

Total head varies directly as N^2

Fluid power developed by the pump varies directly as N^3

Example of Pump Laws.   A centrifugal pump operating at 1150 r/min produced a total head of 37.6 ft and a capacity of 800 gpm. Using pump laws, what would be the

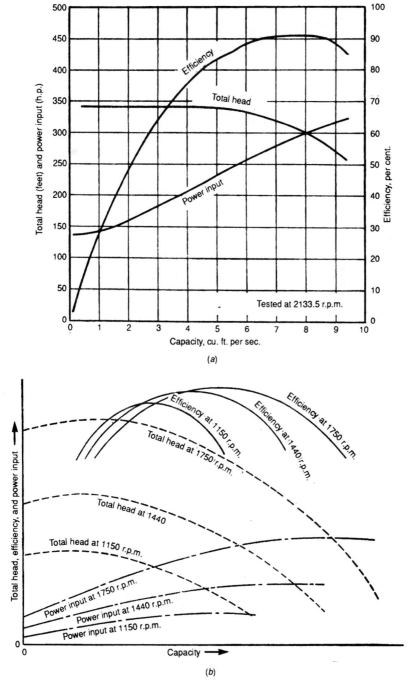

FIGURE 5.4 Centrifugal pump performance curves. (*a*) Performance characteristics of a single-stage, single-suction centrifugal pump operating at *constant* speed; (*b*) performance characteristics of a centrifugal pump at *different* speeds.

head and capacity at 1750 r/min? This involves the relationships of a pump at different speeds. By proportion, to find head,

$$\frac{37.6}{(1150)^2} = \frac{H}{(1750)^2}$$

$$H = \frac{37.6(1750)^2}{(1150)^2}$$

$$= 87.1 \text{ ft}$$

To find capacity,

$$\frac{800}{1150} = \frac{C}{1750}$$

$$C = \frac{800(1750)}{1150}$$

$$= 1217.4 \text{ gpm}$$

Some *operating problems* on centrifugal pumps are:

1. Low water flow can be caused by improper speed, plugged suction strainer, air-bound pump, open air vent valves used to prime the pump, worn wearing rings, damaged impeller.
2. Vibration can be caused by misalignment, bearing wear, impeller unbalance due to wear, and corrosion on pump parts.
3. Progressive shaft thinning and cracking can be caused by improperly installed shaft packing or chemical attack on the material from the fluid handled that may cause stress-corrosion cracking to occur.

Barrel-Type Pumps

Barrel-type centrifugal pumps do not have their casing split horizontally as shown in Fig. 5.3 but consist of a double-case cylinder with access to the pump internals being made through removable end heads. These barrel-type pumps are used for high-pressure boiler feed of up to 6000 psi and 600°F water temperature. For the common 2600-psi service, the pump runs at 3600 r/min and has 12 stages or more.

ROTARY PUMPS

Rotary pumps are considered positive displacement pumps because they can build up pressure by the action of two rotating meshing pump components (see Fig. 5.5). The flow from a rotary pump is not pulsating as it is for a reciprocating pump but is fairly steady. Rotary pumps are designed for certain services up to a medium head. Some are designed with stainless steel components and have large cavities for the fluid between

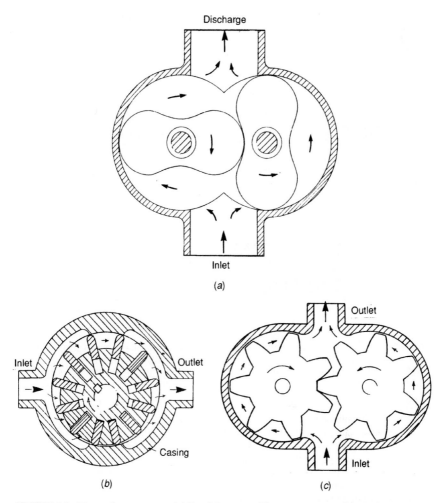

FIGURE 5.5   Types of rotary pumps. (*a*) Two-lobe rotary; (*b*) vane-type rotary; (*c*) gear-type rotary.

the rotating rotors which permit pumping of high viscous fluids (see Fig. 5.6). The absence of entry and discharge valves is another advantage of this type of pump. Generally, rotary pumps are not well suited for pumping fluids containing grit or abrasives because of the close clearance between the rotating elements and the pump casing.

The two-lobe type pump shown in Fig. 5.5*a* has two lobes mounted on parallel shafts that rotate in opposite directions. Timing gears that maintain the proper relation between the lobes throughout the rotation are used at one end of the shafts. Fluid is drawn into the space between the lobes and casing and is then pushed by rotation from inlet to outlet, while increasing the pressure. Figure 5.5*b* shows a vane-type pump, which is also a positive displacement pump. Note that the rotating member with its sliding vanes is set off center in the casing. The entering fluid on one side is trapped between the vanes, which ride on the inside of the casing, and is carried to the dis-

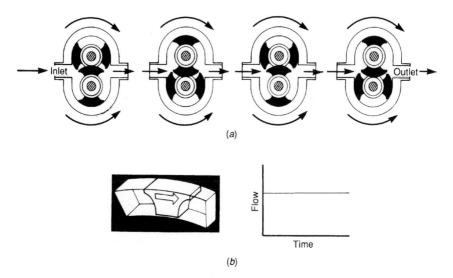

(a)

(b)

FIGURE 5.6 Large-cavity rotary pump principle of operation. (*a*) Fluid flows from inlet to outlet in one revolution of the two parallel shafts; (*b*) arc-shaped rotor wings or tips travel in annular-shaped "cylinders" that are machined into casing to produce less slip. (*Courtesy, Waukesha Pump Co.*)

charge opening as shown. A gear-type rotary pump is shown in Fig. 5.5*c*; its operation is similar to the lobe pump.

The Waukesha rotary pump shown in Fig. 5.6 has special shaped lobes that provide large fluid cavities. The rotor's tips travel in annular-shaped "cylinders" that are machined into the pump's casing as shown in Fig. 5.6. This produces low slip operation.

Some other rotary-type pumps are two- and three-lobe pumps, which are still mounted on two parallel shafts, and screw-type pumps with up to three screws intermeshed to provide positive displacement.

Air or gas entrainment in rotary pumps will cause reduced displacement and thus reduced capacity. Some liquids give off gas in the pumping cycle, so it is important to obtain from the manufacturer the permissible air or gas in the fluid to be handled so that the correct pump capacity will be obtained. Air or gas can also cause noisy and rough vibration operating conditions, with even cavitation damage possible.

Excess clearance between rotating parts and/or the casing reduces capacity. If the clearance becomes excessive, a point will be reached where the entire capacity of the pump will back up through the clearance from discharge to suction. Excessive clearance will show up in low discharge pressure. Another check is to close the suction valve and note suction pressure. A pump in good condition will produce 27 to 29.5 in of vacuum. Permissible clearances increase with increased viscous fluids. Like all pumps, they should not be operated dry because excessive wear will immediately occur on the designed clearances for the pump. When pumping highly volatile liquids such as butane, propane, and hot oils, it is necessary to have a suction static head, at the temperature of the fluid, in order to prevent vaporization of the fluid within the suction side of the pump.

On the discharge side of the pump, excessive pressure may result in cold weather due to pipe friction or by accidental closing of the discharge valve. It is necessary to have a relief valve that is piped to a safe discharge point, is set at or below the maximum allowable pressure of the pump, and has a capacity slightly greater than the pump capacity. Overpressure protection is required on all positive displacement pumps.

PROPELLER PUMPS

Propeller pumps are usually immersed in the liquid to be pumped and are used for drainage or for large-volume circulating service such as condensers in steam or refrigeration service (see Fig. 5.7). The propeller pump is considered an axial-flow pump. Note in Fig. 5.7a that fluid enters the pump at the bottom, and then the rotating

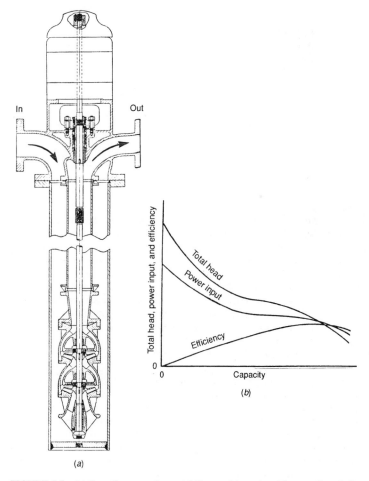

FIGURE 5.7 (a) Propeller pump has axial flow and is enclosed in a suction shell; (b) propeller pump characteristic curves at *constant* speed.

impellers on the shaft lift the fluid along the pump length to be discharged axially from the impellers. Note also that the total head that is developed comes from the dynamic lifting action between the vanes or impellers and the fluid. Figure 5.7*b* shows the characteristics of a propeller pump operating at constant speed. The efficiency rises, reaches a maximum, and then decreases. These pumps are generally used for low heads of less than 40 ft but with capacities greater than 300 gpm.

INJECTORS AS PUMPS

Pumping of water is possible by using the venturi nozzle effect (see Fig. 5.8*a*). Steam is introduced into a venturi nozzle. At the small diameter of the venturi, pressure is low, but velocity can be high. If a water suction line is introduced at the throat of the venturi, it will be pulled into the throat and carried by the high-velocity steam into the discharge line. When the device is used to lift water as shown in Fig. 5.8*a*, it is called an ejector. If properly designed, vacuum conditions exist at the throat so that water from the suction line flows into the venturi nozzle. Figure 5.8*b* shows a single-tube steam-water injector that is used to force water into a steam boiler. It is usually limited to standby service on small- and medium-size boilers, with pumps being used during normal operation. However, it can be very useful in supplying water to a boiler in emergency conditions, such as power failures on motor-driven boiler feed pumps. As shown in Fig. 5.8*b,* when the injector steam is admitted to the venturi nozzle and pressure energy is converted to velocity energy with low pressure, but high velocity, existing at the throat of the nozzle. If properly designed, a partial vacuum in the throat will cause water from an atmospheric tank to be pulled into the throat of the nozzle. The water will now flow with the steam into the divergent part of the nozzle, where velocity energy is converted back to pressure energy. This should be higher than the boiler pressure so that the condensed steam and water will be forced into the boiler water feed system. Note that the pressure generated by the injector must be high enough to lift the check valve on the feed system to the boiler. The overflow valve shown is to let steam and water escape through the small holes in the sides of the combining tube. As soon as the water pressure is sufficient to lift the boiler feed check valve, the water passes through the venturi in an unbroken stream, thus creating a partial vacuum in the overflow chamber, which in turn lifts the sliding washer up against its seat. This prevents any in-rush of air through the overflow opening that would break up the water jet flow.

Steam-type injectors with properly designed nozzles and tubes can develop *water pressure* up to 50 psi above the steam pressure used, but the incoming water temperature must be less than 150°F to prevent flashing in the throat of the nozzle. The added benefit of a steam injector is that it also heats the feedwater to the boiler; however, its pumping efficiency is in the 5 percent range because of the low expansion of the steam. The main advantages of the steam-water injector are simplicity, compact construction, and no moving parts.

Figure 5.8*c* shows a double-tube injector, which is designed to handle water hotter than the single-tube injector previously described. Two nozzles are used, one for lifting the water and the other to force the water into the boiler. A lever handle is used to operate the main steam valve and the so-called forcer valve. In operation, the lever handle is pulled just enough to open the main steam valve, which then begins the lifting part of the cycle. Once lift is established, the lever is pulled further to open the forcer steam valve and to close the overflow valve as in the single-tube design. Steam from the forcing nozzle then forces the water from the lifting tube into the forcing tube, where it is discharged at a pressure sufficient to lift the boiler check valve and enter the boiler water.

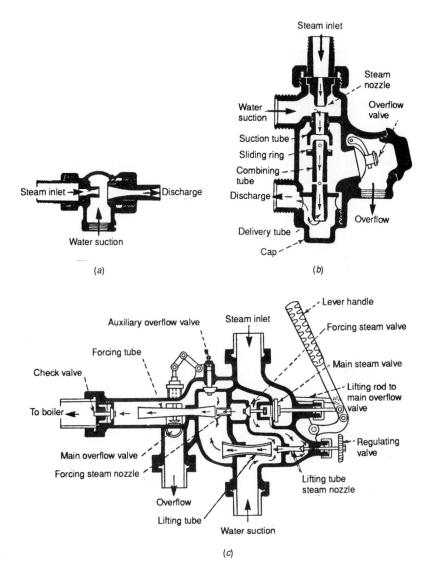

FIGURE 5.8 Steam-water injectors used as pumps. (*a*) Venturi-type nozzle using steam lifts water at throat of nozzle where partial vacuum exists due to high-velocity steam flow through the nozzle; (*b*) single-tube steam-water injector is used for moderate pressure boiler feed; (*c*) double-tube steam-water injector can handle warmer water than single tube in forcing water into a boiler.

Injector troubles include the following: feedwater too hot; suction lift too high; leaky boiler check valve, which would allow steam to blow back into the injector discharge line; leaks in suction pipe destroying vacuum so that water cannot be lifted; steam supply pressure too low for nozzle design, thus impairing lift; steam too saturated, thus not allowing steam to condense properly per design of the nozzle; and obstructions, scale, dirt, and rust plugging in suction lines, tubes, and nozzles.

The manufacturer's maintenance instructions should be followed so that this emergency-type pump is available for use when needed.

DIAPHRAGM PUMPS

Diaphragm pumps are primarily used for controlled volume pumping and for precise feeding of chemicals in manufacturing or as the equivalent of a meter. These pumps are considered to be the positive displacement type with low-volume flows and essentially consist of a flexible diaphragm enclosed in a pressure casing called a head with the pump being considered as single acting. The diaphragms can be pushed back and forth mechanically or by air or liquid on the other side of the fluid being pumped. The diaphragms can be flat discs, tubes, or cones. As long as the diaphragm is not cracked, these type of pumps are considered leakproof. Overpressure protection is usually provided by an internally installed relief valve. Another advantage of diaphragm pumps is that they require no packing. Many manufacturers use two diaphragms to prevent leakage to the outside. If the pressure rises in the space between the diaphragms, there is a defect on the pumping side of the pump.

Because there are no seals, these pumps are often used to pump hazardous liquids and, if the proper diaphragm material is used, high-temperature liquids and high-viscosity fluids also. Strokes are limited to 100 per minute to extend the life of the diaphragm. The manufacturer's NPSH guidelines should be adhered to in order to avoid severe pulsation and vibration. Pulsation dampers are quite often used in these pumps. The material for the diaphragms must be suitable for the fluids handled. API 675 lists the diaphragm materials that should be used with the different fluids with which this type pump may encounter in process applications.

GENERAL PUMP PROBLEMS

There are many common pump problems associated with centrifugal and rotary pumps. Some of the symptoms are: fails to deliver fluid, pump loses its prime after starting, fails to deliver rated capacity, discharge pressure too low, driver is overloaded, pump vibrates, stuffing boxes overheat, bearings overheat, excessive bearing wear, excessive pitting of shaft under packing or seals, and cavitation.

Problems may be combinations of the above, such as vibration and cavitation. The listing below can be used as a guide for troubleshooting and also can be used as a guide to perform maintenance and upkeep of the pump:

Fails to deliver fluid

1. Wrong direction of rotation
2. Pump not primed
3. Suction line not filled with liquid
4. Air or vapor pocket in suction line

5. Inlet to suction pipe not sufficiently submerged
6. Available NPSH not high enough
7. Height from liquid level to centerline of pump too great
8. Difference between suction pressure P and vapor pressure too small
9. Pump not up to rated speed
10. Total head greater than head for which pump is designed

Loses prime after starting

1. Suction line not filled with liquid
2. Air leaks in suction line or through stuffing boxes
3. Gas or vapor in liquid
4. Air or vapor pockets in suction line
5. Inlet to suction line not submerged far enough
6. Available NPSH not high enough
7. Height from liquid level to centerline of pump too great
8. Difference between suction and vapor pressure too small
9. Liquid-seal piping to lantern ring plugged
10. Lantern ring not properly placed in stuffing box

Fails to deliver capacity

1. Wrong direction of rotation
2. Suction line not filled with liquid
3. Air or vapor pocket in suction line
4. Air leaks in suction line or through stuffing boxes
5. Suction-pipe intake not submerged far enough
6. Available NPSH not sufficient
7. Height from liquid level to centerline of pump too great
8. Difference between suction pressure P and vapor pressure too small
9. Pump not up to rated speed
10. Total head greater than head for which pump was designed
11. Foot valve too small
12. Foot valve clogged with trash
13. Viscosity of liquid greater than that for which pump was designed
14. Mechanical defects, such as wearing rings worn, impeller damaged, or internal leaks caused by defective gaskets

Discharge pressure too low

1. Gas or vapor in liquid
2. Pump not up to rated speed
3. Greater discharge pressure needed than that for which pump was designed
4. Liquid thicker than that for which pump was designed
5. Wrong rotation
6. Mechanical defects such as wearing rings worn, impeller damaged, or internal leaks caused by defective gaskets

Driver overloaded

1. Speed too high
2. Total head higher than rated head of pump
3. Either the specific gravity or viscosity of liquid or both different from that for which pump is rated
4. Mechanical defects, such as misalignment, shaft bent, rotating element dragging, or packing too tight

Pump vibrates

1. Starved suction, because of gas or vapor in liquid, available NPSH not high enough, inlet to suction line not submerged far enough, or gas or vapor pockets in suction line
2. Misalignment
3. Worn or loose bearings
4. Rotor out of balance because of the impeller being plugged or damaged
5. Shaft bent
6. Control valve in discharge line improperly placed
7. Foundation not rigid

Stuffing box overheats

1. Packing too tight
2. Packing not lubricated
3. Wrong grade of packing
4. Not enough cooling water to jackets
5. Stuffing box improperly packed

Bearings overheat

1. Oil level too low
2. Improper or poor grade of oil
3. Dirt in bearings
4. Dirt in oil
5. Moisture in oil
6. Oil cooler clogged or scaled
7. Any failure of oiling system
8. Not enough cooling water
9. Bearings too tight
10. Oil seats fitted too closely on shaft
11. Misalignment

Excessive bearing wear

1. Misalignment
2. Shaft bent
3. Vibration
4. Excessive thrust resulting from mechanical failure inside the pump
5. Lack of lubrication
6. Bearings improperly installed
7. Dirt in bearings
8. Moisture in oil
9. Excessive cooling of bearings

Shaft pitting

1. Reduce packing or seal leaks
2. Renew packing or seals
3. Renew shaft pitted area by installing sleeve or metallizing pitted area

Cavitation

1. Check design temperature of fluid
2. Check vapor pressure of fluid at this temperature
3. Check upper temperature limit allowed on pump
4. Check minimum NPSH required on pump per manufacturer's instructions, and do not operate below this head

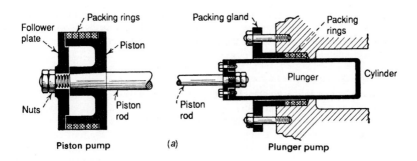

Piston pump   (*a*)   **Plunger pump**

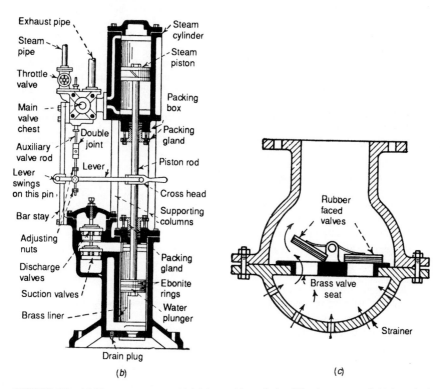

(*b*)   (*c*)

FIGURE 5.9 (*a*) Plunger-type pump (right) is outside packed, while piston pump (left) is packed inside the cylinder; (*b*) foot valve acts as check valve on suction side of pump and also features an incorporated strainer; (*c*) vertical simplex steam pump takes up less floor space.

QUESTIONS AND ANSWERS

1. What is the difference between a plunger and piston-type pump?

Answer. See Fig. 5.9*a*. In the plunger pump, the plunger does not have piston rings or packing inside the cylinder, but instead it is kept tight against water leakage by passing through outside the cylinder-packed packing box as shown in Fig. 5.9*a*. In contrast, the piston pump has the piston entirely within the cylinder and has rings or packing fitted into grooves of the piston (circumferentially), and these are held in place by a follower plate as shown in Fig. 5.9a. Note that one advantage of a plunger pump is that the packing can be replaced without having to open the cylinder as with a piston pump.

2. What is the function of a foot valve?

Answer. See Fig. 5.9b. The function of the foot valve is twofold: (1) To act as a check valve in two directions, when water is lifted in a suction pipe to a water pump and to prevent water from flowing back out of the suction pipe to the pump. (2) As Fig. 5.9b shows, a strainer is also incorporated into the foot valve construction to stop foreign objects from entering the pump suction. Another advantage of a foot valve is that it keeps the pump suction line above the valve full of water, making it easier for the pump to start up with water on the suction side.

3. What steam-type pump takes up the smallest floor space and is used for moderate pressure boiler-feed service?

Answer. See Fig. 5.9c. The vertical simplex steam-type pump is used extensively in smaller moderate pressure boiler systems, where the exhaust from the steam cylinder can be used for hot water heating or similar lower-pressure service. The pump essentially acts as a reducing valve while feeding water to a boiler. Another advantage is that there is less wear on the cylinders and pistons because the weight of the pistons does not rest on the cylinders. Note that the valves are in a separate chamber attached to the cylinders on both the steam and water ends. This makes them more readily available for inspection and repair.

4. What does the term *valve deck* mean?

Answer. A valve deck is a plate that contains either the suction or discharge valves of a direct-acting pump.

5. What do *air bound* and *steam bound* mean?

Answer. Air bound indicates air is leaking into the suction side of the pump, incapacitating the ability of the pump to draw water on the suction side of the pump. Steam bound indicates the water is too hot for the suction of the pump, and the water flashes into steam with reduction of pressure. Air binding usually requires venting the pump and then priming it with water to get the pumping action started. Steam flashing can cause pump seizure, vibration, and cavitation damage. Check the upper permissible temperature limits for the water as established by the pump manufacturer and do not exceed these limits.

6. What can cause knocking on a steam-type reciprocating pump?

Answer. There are several items that can cause these conditions. (1) Condensate in steam lines or steam cylinder. Drain both, especially when starting, to avoid a head being blown off or other internal damage to the steam side of the pump. (2) Make sure there is no air on the water side of the pump. Use vent valves and, if required, prime the pump. (3) Make sure no valves have been closed on the suction or discharge side of the pump and that water is flowing properly with no obstructions. (4) Check the operation of the valves to make sure no lost motion exists. (5) Check all linkages, piston rods, bearings, and similar mechanical parts for excessive wear.

7. What are static suction and dynamic suction lift?

Answer. Static suction lift is the vertical distance from the surface of a water supply below the pump to the center line of the pump suction entrance. Dynamic suction lift is the static suction lift *plus* friction and velocity head, or losses, in the suction line.

8. What is the purpose of a flinger ring in a centrifugal pump?

Answer.   A flinger ring is fitted to the pump shaft between the pump gland and the bearings. Any leakage past the packing or seals will reach the flinger ring, which throws the water by centrifugal force outward and away from the bearings, thus preventing water from entering the bearing oiling system.

9. What is the purpose of wearing rings on centrifugal pumps?

Answer.   Wearing rings are used to seal off the tip of impellers and the casing so that a pressure differential exists from the entrance to the impeller to the discharge side, thus assuring pumping action by the impeller. The rings provide a tight seal and are a wearable item, thus preventing casing wear and replacement. They should be checked during any dismantled inspections.

10. What is the danger of operating a reciprocating pump with the discharge valve closed? How does this compare to a centrifugal pump?

Answer.   Since a reciprocating pump is a positive displacement pump, excessive discharge pressure can result unless a relief valve is installed *ahead* of any discharge valve, set at a pressure not to exceed the maximum pressure allowed on the pump, and with sufficient capacity to at least equal the capacity of the pump. A centrifugal pump may not cause excessive pressure; however, the work done on the fluid will be converted to heat by the churning action of the pump. This will cause high vapor pressure, seizing, bearing damage from the thrust forces produced, and excessive internal wear, requiring the rebuilding or replacement of the pump. Centrifugal pumps with flat head-type casings have had this flat portion blown off from overheating damage.

11. What is the water cylinder diameter of a steam-type reciprocating pump that has the following name plate: 10" × 18" × 12"?

Answer.   The water cylinder is 18 in. The steam cylinder size is always listed first, being 10 in, while the stroke of 12 in is listed last.

12. What does priming a pump mean?

Answer.   Priming a pump means filling the suction side of the pump with water before starting the pump. This is especially applicable when the source of water is below the suction centerline of the pump. If a check valve has been placed in the suction line, the pump can be primed by filling the suction line and pump casing with water from another source, such as a city water line or even from the discharge side of the pump if this line has been standing full of water. It is necessary to first vent the air out of the pump by means of the air vent valve. Most pump manufacturers also provide priming instructions for their particular pump, and these instructions should be followed.

13. What is the difference between a volute and turbine-designed centrifugal pump?

Answer.   The casing shape of the volute pump is a gradually increasing cross-sectional area, which causes reduced velocity flow but increased pressure on the fluid. The casing sections are volutes shaped, hence the name. In a turbine centrifugal pump, the *impeller* is surrounded by a stationary diffusion ring containing the passages of increasing cross-sectional area. Water leaves the impeller rim at a high velocity, which is then converted to pressure as the water passes through the diffusion ring. The casing surrounding the diffusion ring is usually circular in shape and has a constant cross-sectional area.

14. What four features are common to rotary pumps?

Answer. (1) No suction or discharge valves. (2) They are positive displacement pumps, requiring relief valves on discharge side to avoid accidental overpressure. (3) Rotating elements have tight clearances with the casing. (4) Fluid is trapped on one side of the casing in voids between the rotating elements and the casing and is discharged on the other side with no valves used.

15. What is the danger of pitting on pump shafts under gland packing?

Answer. The pits act as stress concentration points and also as a place for stress corrosion to develop on the shaft. Combined they can produce cracks on the shaft and ultimate shaft failure as the cross-sectional area of the shaft that resists the external load is reduced in size.

16. Why would you find a propeller pump for the following service: delivery 2400 gpm, total head 20 ft, pump speed 2600 r/min?

Answer. Propeller pumps are best suited for large-volume, low-head flows.

17. What are pump slip and volumetric efficiency?

Answer. Pump slip is the *difference* between the displacement of volume of a reciprocating pump, which is calculated as the theoretical discharge volume, and the actual discharge of the pump. Slip is expressed as a percentage of the theoretical displacement, and these have values ranging from 3 to 15 percent. The volumetric efficiency is the ratio of the volume of water actually delivered to the theoretical displacement of the pump.

18. What is the function of shaft sleeves?

Answer. These are replaceable sleeves placed in stuffing or gland boxes and in interstage glands so that the shaft is protected from abrasive wear caused by too tight packing and or corrosion from the fluid handled. Since they are wearable items, most are replaced during major inspections, which is far less costly than replacing a pump shaft (see Fig. 5.10).

19. What is the function of a lantern ring?

Answer. See Fig. 5.10. This is a ring of cage construction and as shown in Fig. 5.10 is inserted between the turns of the packing. Sealing liquid from the pump or other source is conducted through an opening in the stuffing box and directed to the lantern ring. This ring distributes the sealant around the shaft and thus prevents air from leaking into the pump and lubricates the packing around the shaft. To prevent the shaft from being scored, lantern rings are made of soft materials, such as white metal, brass, and even plastic. Be sure not to crush this ring when tightening the gland.

20. What are the two main types of mechanical seals as used on pumps?

Answer. Mechanical seals have two flat sealing faces at right angles to the axis of rotation. One face can move axially in order to permit the sealing faces to remain in contact despite shaft end play, face wear, and face run-out. This one face is known as the seal ring. In a stationary seal arrangement, the sealing ring is in the housing and does not move. In the second, or rotating, type, the sealing ring turns with the shaft. Mechanical seals are replacing the so-called jam-packed

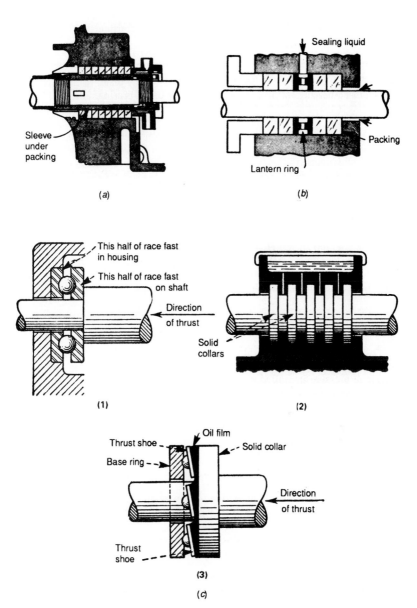

FIGURE 5.10 (*a*) Replaceable shaft sleeve under packing takes wear instead of shaft; (*b*) lantern ring adds lubricant to packing and prevents pump leakage; (*c*) thrust bearings used on pumps; top left, ball bearing type, top right, multicollar type, and bottom, Kingsbury thrust bearing features tapered bearing surfaces to establish an oil wedge.

seals. Mechanical seals require checking for wear and tear during any pump overhauls.

21. What are three methods used to balance end thrusts in centrifugal pumps?

Answer. See Fig. 5.10. (1) Ball bearings riding on end plate, (2) multicollar riding against a bearing surface, and (3) Kingsbury thrust bearing.

22. How is *specific speed* defined and used?

Answer. Pump manufacturers develop an index on a pump or pumps in order to determine what combinations of head, speed, and capacity are possible at the point of maximum efficiency for the pump to be employed; this is called specific speed. Mathematically, it is expressed as follows

$$N_s = N\frac{\sqrt{gpm}}{H^{3/4}}$$

where N = pump speed, H = total head per stage in feet, gpm = the capacity in gallons per minute, and N_s = specific speed. By using this relationship, it has been determined that impellers for high total heads have low specific speeds, and impellers for low total heads have high specific speeds.

Practical Pump Calculations

23. A pump originally manufactured for 3600 r/min has been operating with a 60-psi discharge pressure in chilled water service at 1800 r/min. Additions to the building requires more chilled water pressure. What would the theoretical discharge pressure be for this pump if it operated at the designed 3600 r/min?

Answer. Use pump laws that say pressure varies as the square of the speed, or

$$\frac{60}{P} = \frac{1800^2}{3600^2} \qquad \text{or} \qquad \text{new } P = 240 \text{ psi}$$

24. A centrifugal pump is operating with a 60-psi suction pressure and a 200-psi discharge pressure. What would the total head on this pump be if friction and velocity head totaled 25 ft?

Answer. Net increase in pressure performed by the pump is 200 − 60 = 140 psi. Converting to feet, 140/0.434 = 323 ft and total head = 323 plus 25 = 348 ft.

25. A steam condensate pump on a 50,000-kW turbo-generator has a friction and external head of 60 ft and suction of 29.5 in. The turbine has a steam rate of 12 lb/kW · h. The velocity head for the condenser pump is only ½ ft. What are the gallons of flow per minute through the pump and the theoretical horsepower required of the pump?

Answer. It is necessary to convert to workable units for use in pump equations.

$$\text{Converting steam rate to gpm} = \frac{50,000(12) \text{ lb/h}}{60 \text{ min/h} (8.33) \text{ no. of gal}} = 1200 \text{ gpm flow}$$

For the vacuum head, the vacuum on condensers is measured in inches of mercury, which weighs 0.49 lb/in^3. Therefore, 29.5 in $\times$ 0.49 = 14.455-psi negative pressure. Converting to feet of water = 14.455 $\times$ 2.31 = 33.4 ft of suction head.

$$\text{Total head} = 60 + 33.4 + 0.5 = 93.9 \text{ ft}$$

$$\text{ihp} = \frac{93.9(1200)(8.33)}{33,000} = 28.4 \text{ hp}$$

26. A reciprocating pump has the following nameplate information 12" $\times$ 6" $\times$ 14". What is the steam cylinder size, and if the pump is operating at 90 strokes per minute, what are the piston speed and total piston travel in ft/min?

Answer. The first number shown is the steam cylinder diameter: 12 in. The stroke is 14 in; therefore, the speed in feet per minute is 14/12 $\times$ 90/2 = 52.5 ft/min, and the distance traveled is 52.5 $\times$ 2 = 105 ft/min.

27. A boiler feed pump has a motor drive and delivers 1300 gal/min at 200-psi gage pressure on the discharge side with suction pressure at 0 psi. The feedwater weighs 62.3 lb/ft^3. The pump mechanical efficiency is 85 percent. What is the water horsepower, and what horsepower should the motor be.

Answer. The calculation is as follows:

$$\text{Head developed} = \frac{200 \times 144}{62.3} = 462.3 \text{ ft}$$

$$\text{Weight of gallon of water} = \frac{62.3 \times 231}{1728} = 8.33 \text{ lb/gal}$$

$$\text{Weight of water handled per minute} = 1300 \times 8.33 = 10,829 \text{ lb/min}$$

$$\text{Water hp} = \frac{10,829 \times 462.3}{33,000} = 151.7 \text{ hp}$$

$$\text{Motor hp} = \frac{151.7}{0.85} = 178.5 \text{ hp}$$

28. A simplex water pump of 8 $\times$ 4 $\times$ 10 in operates at 125-psi steam pressure. What theoretical water pressure can be developed?

Answer. Equate steam piston area times steam pressure to water piston area times water pressure and solve for water pressure as follows with $\pi/4 = 0.7854$:

$$0.7854(8)^2(125) = 0.7854(4)^2 P_w$$

$$P_w = \frac{64(125)}{16} = 500 \text{ psi}$$

29. If a reciprocating pump is 6 × 3 × 8 in, and the pump has a total head of 300 ft, what steam pressure would be required if the pump is of the simplex type?

Answer. Convert head to psi, or 300 × 0.434 = 130.2 psi. Using above example,

$$0.7854(3)^2(130.2) = 0.7854(6)^2 P_s$$

or

$$P_s = \frac{9(130.2)}{36} = 32.55 \text{ psi}$$

30. A flow test on a circular pipe showed 1500 gpm flowing past a 6-in diameter pipe point. What is the velocity of flow in feet per minute?

Answer. Use the equation $Q = AV$; however, watch the units since they must be matched. Convert gpm to ft³/s, or

$$Q = \frac{1500}{7.48(60)} = 3.34 \text{ ft}^3/\text{s}$$

Convert pipe internal cross-sectional area to square feet as follows:

$$\frac{0.7854(6)^2}{144} = 0.2 \text{ ft}^2 \quad \text{then} \quad v = \frac{3.34}{0.2} \times 60 = 1,002 \text{ ft/min}$$

31. The inside diameter of the discharge pipe of a condenser pump is 4 in, and the suction pipe is 6 in. Water delivered is 250 gpm at 60°F. The discharge pressure is 190 psi, and the suction is 12 in of mercury vacuum. The discharge gage is 3 ft above the pump centerline, while the suction gage is 2 ft below the centerline of the pump. Frictional resistance of the pump per the manufacturer is 2.2 ft. What is the total head that the pump must operate against?

Answer. The calculations are as follows:

$$\text{Water at } 60°F = 62.34 \text{ lb/ft}^3$$

$$12 \text{ in mercury height} = 12 \times 0.49 = 5.88 \text{ psi}$$

Then,

$$H_d = \frac{190 \times 144}{62.34} + 3 = 442 \text{ ft head}$$

$$H_s = \frac{5.88 \times 144}{62.34} + 2 = 15.6 \text{ negative head in ft}$$

where H_d is discharge head and H_s is suction head. To find velocity head, convert flow to ft³/s.

$$Q = \frac{250 \times 231}{1728 \times 60} = 0.557 \text{ ft}^3/\text{s}$$

The velocity in the discharge pipe is calculated with equation $Q = A \times V$:

$$V_d = \frac{0.557}{0.7854(4)^2/144} = \frac{0.557(144)}{12.56} = 6.38 \text{ ft/s}$$

$$V_s = \frac{0.557(144)}{0.7854(6)^2} = 2.83 \text{ ft/s} \qquad \text{velocity head} = \frac{V_d^2 - V_s^2}{2g}$$

where g is acceleration of gravity $= 32.2$ ft/s². Therefore, velocity head is

$$\frac{6.38^2 - 2.83^2}{2(32.2)} = \frac{32.69}{64.4} = 0.51 \text{ ft}$$

This is why velocity heads are usually ignored in ordinary flows. The *total head* is $442 + 15.6 + 2.2 + 0.51 = 460.31$ ft.

32. One pump can empty a tank in 8 h 15 min, while another takes 11 h. If both pumps are used together, how long will it take to empty the tank?

Answer. If $V =$ volume of tank, rate of pumping per pump is $V/8.25$ and $V/11$. Rate × time t for the two pumps should equal V pumped out:

$$\frac{Vt}{8.25} + \frac{Vt}{11} = V$$

then if we divide both sides by Vt, we obtain

$$\frac{1}{8.25} + \frac{1}{11} = \frac{1}{t}$$

By using the common denominator 33,

$$\frac{4 + 3}{33} = \frac{1}{t}$$

Therefore $t = 4.714$ h $= 4$ h, 42 min and 50 s for both pumps operating together to pump out the tank.

33. What is the relationship between head in feet and pounds pressure per square inch assuming the water is fluid?

Answer. The relationship is:

$$1\text{-lb pressure per in}^2 = \frac{1}{0.434} = 2.3 \text{ ft of head}$$

$$1 \text{ ft of head} = \frac{1}{2.3} = 0.434 \text{ psi}$$

CHAPTER 6
COMPRESSED AIR SYSTEMS

Plant services include air systems that are used extensively for the following applications:

1. To provide air for pneumatic controls for air conditioning, heating systems, and other such controls in process applications.
2. To operate punch presses; rotating hand tools such as grinders, wrenches, drills, screwdrivers, reamers; and similar tools powered by compressed air.
3. To provide air for cleaning machine tools and other work stations.
4. To aerate and agitate liquids, operate paint spray guns, and spray oils and insecticides.
5. To inflate tires, charge shock absorbers, and act as a cushioning device for many compressed fluids, such as expansion tanks on hot water heating systems.
6. To start diesel engines and for pumped air storage for operating combined-cycle electric power generating plants.

It is essential for plant service operators to be familiar with compressed air systems in order to inspect, service, and maintain them properly.

The chief advantage of using air is that it is available from the earth's atmosphere, which consists of a mechanical mixture of gases, mostly oxygen and nitrogen, as well as inert gases, such as argon, helium, neon, krypton, radon, and xenon as shown in the table below:

Composition of Dry Air

Substance	Volume, %	Weight, %
Nitrogen	78.0	75.32
Oxygen	21.0	23.0
Inert gases	0.94	1.42
Carbon dioxide*	0.03	0.03
Miscellaneous†	0.03	0.03

*Varies by area; however, winds keep the air in motion, so concentration is reasonably steady, except for locked-in basins such as Los Angeles.
†Miscellaneous substances include carbon dioxide and dust, which also vary by areas, as well as water vapor, which varies with temperature and humidity.

(a)

Gas	Mol. Weight	R	c_p	c_v	k
Air	28.967	53.34	0.240	0.1715	1.40
Carbon monoxide (CO)	28.000	55.18	0.249	0.1789	1.40
Carbon dioxide (CO_2)	44.000	35.10	0.199	0.153	1.30
Helium (He)	4.000	386.0	1.25	0.754	1.66
Hydrogen (H_2)	2.018	766.4	3.445	2.460	1.40
Methane (CH_4)	16.000	96.5	0.529	0.404	1.31
Nitrogen (N_2)	28.016	55.15	0.249	0.178	1.40
Oxygen (O_2)	32.000	48.28	0.218	0.516	1.40

(b)

FIGURE 6.1   (a) Air tank rupture caused by internal water corrosion thinning; (b) data on gases for use in perfect gas equations.

AIR FILTERING

In certain applications, dust and moisture must be at a minimum in compressed air systems. All air contains suspended solid particles, or dust, composed of various substances, such as clay, salts, soot, plant fibers, and other substances that may be released into the atmosphere at that particular locale. Dust can obstruct compressor passages, block heat transfer, and cause sludge to form in the compressor lubricating system; therefore, all air compression systems use filters to block out the dust from the intake of compressors. These filters must be periodically replaced or restored by cleaning in order to maintain their effectiveness.

Water vapor is compressed with the air as it goes through the compressor and then condenses in the piping or air receiver. Water in air systems can cause corrosion on metal parts if not removed and eventual thinning due to corrosion, which will weaken the pressure-containing part and in severe cases can cause explosions (see Fig. 6.1). Low points of piping should have moisture traps, and air tanks should be bled daily of the condensate that collects at the bottom of the tank in order to avoid deterioration due to water vapor in the air. *Air dryers* are used in certain compressed air applications in order to reduce the moisture content in the compressed air to a minimum. This is especially true on critical pneumatic control systems and where the air piping and controllers are outdoors and subject to freezing temperatures. Processes have stopped because the water in the air line froze, making the controls inoperable. This has occurred in the South where severe cold spells were not expected.

COMPRESSED AIR DYNAMICS

Compressed air is considered a working substance which carries energy to and from a machine. The working substance, or air, may absorb or reject energy, do work, or have work done on it. Because of energy transfers, the properties of the working substance changes from one point to the next, and these properties are pressure, volume, weight, temperature, internal energy in the substance called enthalpy, and entropy. Thus, air has a specific volume at some particular pressure and also temperature and internal energy. The reader should review a few of these basics or thermodynamics of compressed air systems since they help in analyzing facility problems or needs.

Atmospheric pressure is the weight of air above a certain point as measured by a barometer, and the standard is usually 14.7 psia at 68°F, which is equal to the weight of a mercury column 1 in^2 and 29.92 in high. Note that gage pressure would be zero at atmospheric pressure. *Perfect gas laws* are used to solve many compressed air problems, and the reader is referred to texts on thermodynamics for them. However, Boyle's and Charles' law need mentioning.

The following units are used in the equations below:

P = absolute pressure = 144 (P_g+14.7) lb/ft^2

v = specific volume of gas, ft^3/lb

V = volume of gas, ft^3 at pressure P and temperature in °F

Also $V = M/v$ cu. ft where M = weight of gas, lb. v = specific volume.

Boyle's law states that at *constant* temperature, the volume of a given weight of gas varies inversely as the absolute pressure, and in equation form, for two points of a gas flow,

$$\frac{P_1}{P_2} = \frac{V_2}{V_1}$$

Note that since T_1 equals T_2 and is a constant temperature process, it is not in the equation. This is called an *isothermal* relation.

Charles' law states that at *constant* volume, the *absolute* pressure and *absolute* temperature vary directly with the heat added or subtracted from a perfect gas, where absolute temperature $T = 459.7+°F$. In equation form,

$$\frac{P_1}{T_1} = \frac{P_2}{T_2} \quad or \quad \frac{P_1}{P_2} = \frac{T_1}{T_2}$$

The two laws are combined to form the following useful equation:

$$\frac{P_1 V_1}{T_1} = \frac{P_2 V_2}{T_2}$$

Also useful is the following:

$$PV = MRT$$

where R = constant, for a gas $R = Pv/T$, and R = ft · lb/lb/°F absolute. See Fig. 6.1 for some R values.

An example of gas law applications is that air has a specific volume of 12.4 at atmospheric pressure and 32°F.

AIR COMPRESSORS

The most prominent compressor used has been the *reciprocating piston* type, which produces air pressures of 100 psi and above. The *centrifugal* and *axial* types are used for large volumes of air flow, such as blast furnace applications in the steel industry. Figure 6.2 illustrates the many types of air compressors in use. Figure 6.2b illustrates a screw-type compressor, which is extensively used for low-volume flow and generally requires less maintenance than reciprocating units. A brief description of the types shown in Figs. 6.2 and 6.3 follows:

1. *Reciprocating.* (Figure 6.3) A piston moves back and forth in a cylinder with suction valves opening to admit atmospheric air, which is compressed by the piston to discharge pressure. These compressors may be double acting, which requires the cylinder to have suction and discharge valves at each end of the cylinder. They can also be single- or multistage units.

2. *Sliding-vane rotary.* Air is trapped between the vanes as it passes the suction openings and is compressed as the volume between the vanes decreases to the designed discharge pressure. The vanes slide outward due to centrifugal force in slots and ride against the inner casing.

3. The *two-lobe* type consists of two impellers mounted on parallel shafts that rotate in opposite directions, drawing air into the pockets between the impellers and the casing and then carrying this compressed air to the outlet at design pressure.

4. The *liquid-piston* rotary units have a multiblade rotor revolving in an elliptical casing that is partly filled with a liquid which acts as a piston. The blades form a series of buckets. The liquid follows the contour of the casing and alternatively

leaves from and returns to the space between the blades twice per revolution. When the liquid leaves the buckets, air is drawn in, and when the liquid returns to the bucket, it compresses the air to the designed discharge pressure.

5. The *centrifugal* compressor takes air at the eye of the impeller and then accelerates it outward radially. In the diffuser section of the casing, the velocity of the air is converted to static pressure. These units can be staged to obtain different discharge pressures.

6. The *axial* compressor takes air in at one end and discharges it at the other axial end after it passes through a series of rotary and stationary blades, similar to a steam turbine of the reactionary type. Each row on the rotor increases the pressure; thus, the number of stages of compression required will be influenced by the final pressure needed, as well as by the volume of flow.

7. The *screw* compressor illustrated in Fig. 6.2b is a twin-shaft positive displacement unit. The air is slowly compressed from the suction end to the discharge port by ever-diminishing space between the convolutions of the two helical rotors as the air is compressed to the final design pressure before leaving the discharge port. The space containing the trapped air is formed between the casing walls and the interlocking convolutions of the two helical rotors.

Selection of Compressor Type

Compressor selection is determined by the *pressure, capacity,* and *control* that may be needed for the facility operation. For example, if the air is to be used for pneumatic controls, 100 psi may be needed, plus an extra 25 psi if the lines are long. Where one or two operations require air at higher pressure, it is usually more economical and convenient to install a separate smaller high-pressure unit just for that service. Where small amounts of air are required at pressures lower than that carried by the main distribution line, a reducing valve from the main line might be appropriate. On the other hand, if a large volume of low-pressure air is needed, it is more economical to install a separate unit just for that purpose.

Capacity of the units depends on load and whether it is steady or intermittent. This requires careful review, and assistance can be obtained from compressor manufacturers or from plants with similar operations. Capacity is measured in cubic feet per minute (cfm) and the corresponding pressure that is needed. Approximations are possible as shown in the following table:

Approximate Driver Horsepower Needed to Compress 100 cfm of Air

| Altitude, ft | One-stage | | | Two-stage | | | |
| | Discharge pressure, psig | | | Discharge pressure, psig | | | |
	60	80	100	60	80	100	125
0	16.3	19.5	22.1	14.7	17.1	19.1	21.3
1,000	16.1	19.2	21.7	14.5	16.8	18.7	20.9
2,000	15.9	18.9	21.3	14.3	16.5	18.4	20.5
3,000	15.7	18.6	20.9	14.0	16.1	18.0	20.0
5,000	15.2	17.9	20.3	13.5	15.5	17.3	19.2
10,000	14.1	16.5	18.6	12.3	14.1	15.6	17.2

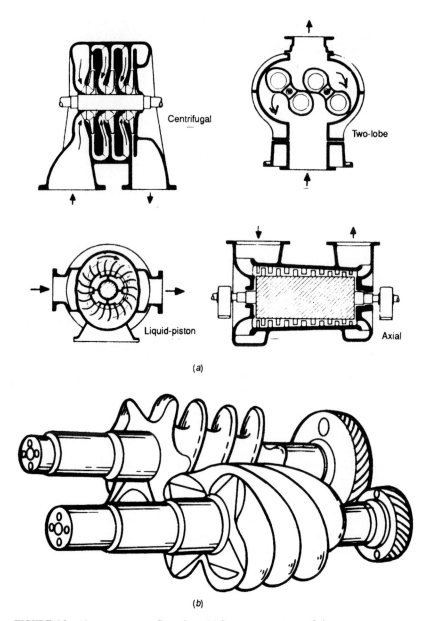

Centrifugal

Two-lobe

Liquid-piston

Axial

(a)

(b)

FIGURE 6.2 Air compressor configurations. (a) Some common types of air compressor arrangements (*Courtesy,* Power Magazine); (b) screw-type compressor features two helical screws rotating in opposite directions (*Courtesy, Aerzen U.S.A. Corp.*).

Example. What size motor would be needed if a two-stage compressor were to be installed with a total capacity of 2500 cfm at sea level and a discharge pressure 125 psig and if the electric motor had a 90 percent efficiency? For 2500 cfm, from above, the driver horsepower needed is 2500/100 × 21.2 = 530 hp. The electric motor hp = 530/0.9 = 589 hp, with nearest standard size being 600 hp. The compressor selected would probably be a water-cooled reciprocating type.

Capacity Regulation

The intermittent demand for air requires regulating the output of a compressor. For small machines with low demand, the easiest and simplest means is a pressure switch, which turns a compressor on at a certain lower pressure and then shuts it off at the selected higher pressure, with the usual spread being about 10 percent between on and off operation.

In constant speed control, which is used when air demand is fairly constant, the compressor runs continuously but compresses air only when the pressure drops to a set point. Two methods, called unloaders, are used: (1) closed-suction unloaders and (2) open-inlet valve unloaders. The closed-suction unloader has a pressure-actuated valve in the compressor intake, which prevents compression from taking place while the compressor is operating at full speed, by shutting off the air intake. In the open-inlet valve unloader method, the inlet valves are kept open, thus preventing compression.

Large motor-driven compressors, usually those over 100 hp, are usually equipped with capacity step control. It is a variation of constant-speed control, but the unloading is programmed in steps from full load to no load. For example, three-step control of full load, three-fourths load, and no load are accomplished with free air unloaders. Many compressors use clearance pockets as small air reservoirs that are opened when unloading occurs. The air is compressed into the clearance pockets on the compression stroke and then fills the cylinder on the return stroke, thus preventing the compression of additional air.

Compressor Efficiencies

It is perhaps appropriate to list and calculate the many efficiencies used before reviewing compressor types:

1. Volumetric efficiency = actual air delivered, ft^3/theoretical air delivery, or piston displacement
2. Isothermal compression efficiency = theoretical isothermal hp/indicated hp
3. Adiabatic compression efficiency = theoretical adiabatic hp/indicated hp
4. Mechanical efficiency = indicated hp/actual hp input (brake hp)
5. Overall efficiency = compression efficiency × mechanical efficiency (State if compression efficiency is isothermal or adiabatic)/actual hp input (brake hp)

RECIPROCATING COMPRESSORS

Most compressors are reciprocating (Fig. 6.3), ranging in horsepower from small garage units of 2 to 10 hp to several thousand horsepower in large industrial plants. The arrange-

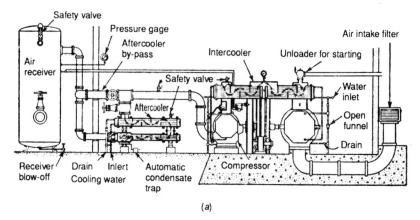

(a)

FIGURE 6.3   Two types of reciprocating compressors. (a) Two-stage motor-driven reciprocating compressor.

ments can range from single-acting, double-acting, single-stage compression to multistage, air-cooled, water-cooled, vertical, and horizontal cylinder arrangements; V or Y types with one crankshaft; and motor, steam turbine, diesel, and gasoline engine drive.

Figure 6.3a illustrates the layout for a two-stage reciprocating air compressor driven by a synchronous motor directly between the low- and high-pressure cylinder. The unit is equipped with safety valves on the intercooler, on the discharge side of the compressor *before* any shutoff valve and on the air receiver.

Figure 6.3b illustrates a two-stage *steam*-driven air compressor. This unit has one crankshaft with a flywheel mounted on the shaft between the steam engine and compressor. Usually, control is by speed regulation of the engine in step with the pressure on the discharge side or in the receiver, with the governor limiting the maximum speed. Unloaders are also used while the unit runs at constant speed. Both compressors shown in Fig. 6.3 are referred to as cross-compound machines because the cylinders are across from each other and not in tandem or in-line with each other.

As Figs. 6.3b and 6.4a show, the key elements of a reciprocating compressor are the cylinders, heads, pistons, inlet and discharge valves, crankshaft, connecting rods, crosshead, flywheel, cooling system, lubrication system, and controls for the compressor capacity and pressure and for the driver.

Whether a compressor is *air* or *water cooled* is determined by the amount of heat generated in compressing the air. Air cooling is accomplished by means of fins on the cylinders and may include a fan which blows cooling air over the fins and also the intercoolers if the compressor is of the multistage type. In general, air-cooled machines are used for pressures up to 125 psi and 100 hp size. Water-cooled units have water circulated in cylinder jackets, and if they are multistage units, the water is circulated in intercoolers and aftercoolers as shown in Figs. 6.3b and 6.4a. Water-cooled units are used for sizes over 100 hp because of the high heat of compression that must be removed.

Staging in a reciprocating compressor depends on the pressure that is needed. In single-stage units compression takes place from suction to discharge in one cylinder. Pressure is limited to 125 psi, and up to 3 hp and above this, multistaging is used. In multistaging, pressure is raised as the air travels from the initial inlet cylinder to the next. Each step is called a stage. Intercoolers are used to cool the air between stages as close to inlet temperature as may be possible. This reduction in temperature reduces the power required to compress the air to its final desired pressure.

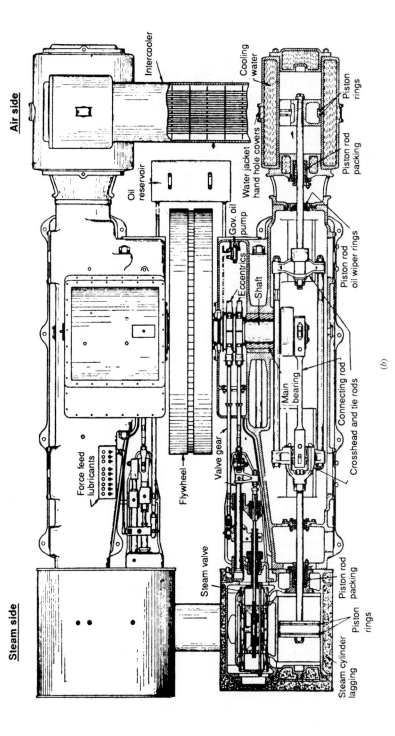

Steam side

Air side

Intercooler

Cooling water

Piston rings

Water jacket hand hole covers

Piston rod packing

Oil reservoir

Gov. oil pump

Eccentrics

Shaft

Piston rod oil wiper rings

Force feed lubricants

Flywheel

Valve gear

Main bearing

Connecting rod

Crosshead and tie rods

Steam valve

Piston rod packing

Piston rings

Steam cylinder lagging

(b)

FIGURE 6.3 (*Continued*) Two types of reciprocating compressors. (*b*) two-stage steam-driven reciprocating air compressor.

6.9

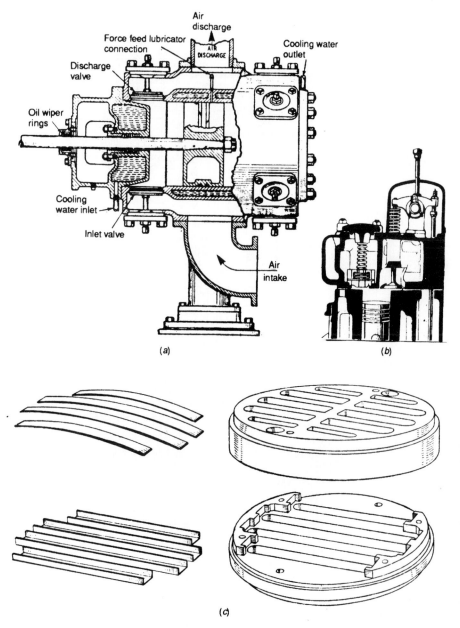

FIGURE 6.4 (*a*) Water-cooled air compressor cylinder; (*b*) poppet intake valve uses push rod rocker arm from cam shaft; (*c*) automatic channel valve parts.

Valves

Most valves for air compressors are of the automatic type that depend on pressure difference for operation instead of mechanical linkage. For small air compressors, cam-operated poppet-type inlet valves are used as shown in Fig. 6.4*b*. Larger units use lightweight plate or strip valves as shown in Fig. 6.4*c*. The *channel valve* consists of bowed stainless-steel strips A, valve channels B, stop plate C, and valve seat D. On the suction stroke, each channel lifts off its seat without flexing, thus opening the full length of each port. The bowed spring strip within each channel controls the vertical movement of the channel and helps to cushion the valve lift. The design of the controlled cushioning requires a good fit of the spring in the valve channel and with the height of the sides of the channel. In closing, the spring strips contract by reassuming the bowed condition, and the channels are forced to return to their seats. This downward movement occurs automatically with the slightest increase of cylinder pressure above the intake pressure. The same valve arrangement is used on the compressor discharge valves but is designed for discharge pressure operation.

CENTRIFUGAL COMPRESSORS

Centrifugal compressors are used for large-volume air flows at moderate pressure. The pressure obtained depends on the speed of operation and the number of stages or impellers. A four-stage unit is shown in Fig. 6.5*a*. The rotor consists of high-tensile-strength alloy steel impellers of the radial type. At high speed, usually over 3000 r/min, the impellers accelerate the air radially to speeds of 300 ft/s and over. The velocity of the air from each impeller is converted into pressure as the air passes into the larger passages leading to the next stage, or impeller. Each stage adds to the pressure of the air. This design must have curved guide vanes mounted between the impellers to direct the air discharged from one impeller to the next stage at the proper angle in the direction of rotation. Water cooling is used on multistage units to remove the heat of compression.

Surging must be avoided on this type of compressor because it can cause unstable operation, usually accompanied by excessive vibration. Surging is caused by the air pulsating between the impellers because of insufficient flow for the design of the machine. The air or gas cannot be accelerated to keep flowing downstream, and it pulsates in the compressor passages, similar to a churning-type action. Each compressor has a lower limit of speed and volume flow that is called the surge point. Surging usually occurs at low load, and the manufacturer's instruction should be followed to avoid operating a centrifugal compressor below this surge point.

Figure 6.5*b* shows a centrifugal compressor arrangement in which the staging or impellers are not on one shaft but on separate shafts with a gear drive interconnecting the stages through pinions connected to a bull gear. The advantage of this arrangement is that different speeds per pinion are possible, thus making it possible to drive each impeller near its optimum specific speed. Speeds may reach 70,000 r/min. The air coolers are outside the compressor casing, usually in the machine's base.

Control of centrifugal air flow is by (1) varying the speed, (2) controlling the flow by movement of internal vanes, and (3) throttling the suction or discharge. Speed control is obtained directly by using a steam turbine, wound rotor motor, or dc motor. If a constant-speed driver is used, variable speed can also be obtained by using hydraulic or magnetic couplings.

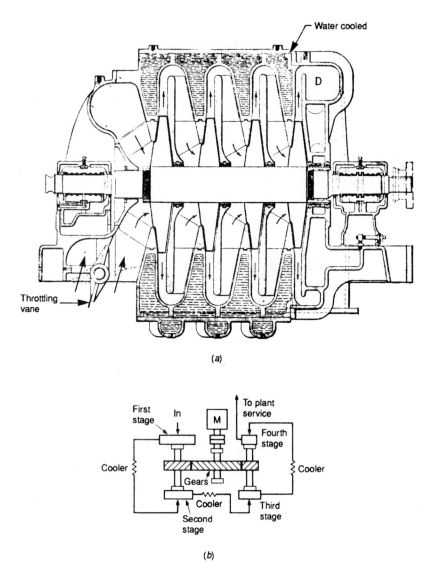

FIGURE 6.5 (*a*) Four-stage centrifugal air compressor with inlet throttling; (*b*) four-stage centrifugal air compressor arrangement features separate shaft impellers with integral gear drive, allowing different impeller speeds.

Compressed air auxiliaries are air filters, inter and after air coolers, moisture separators or traps, and air receivers. Some of these auxiliaries are illustrated in Fig. 6.6.

Air filters are used to remove solid particles in the incoming air and may be of the dry type or of the viscous-impingement type, where the air passing through impinges on fine wire coated with oil, and this traps the dust. In traveling screen filters, the dust

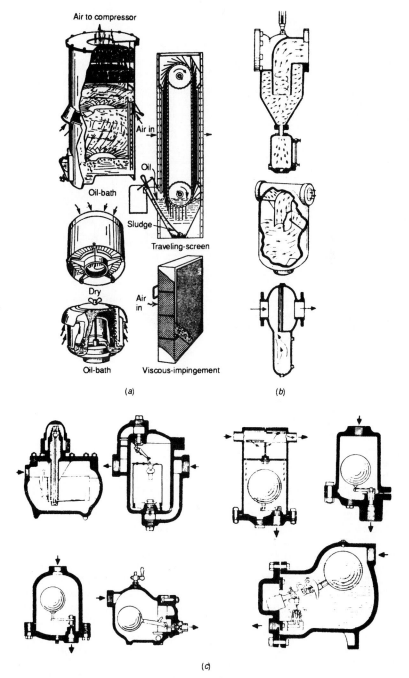

FIGURE 6.6 Compressed air auxiliaries. (*a*) Air filters; (*b*) moisture separators; (*c*) traps. (*Courtesy,* Power Magazine.)

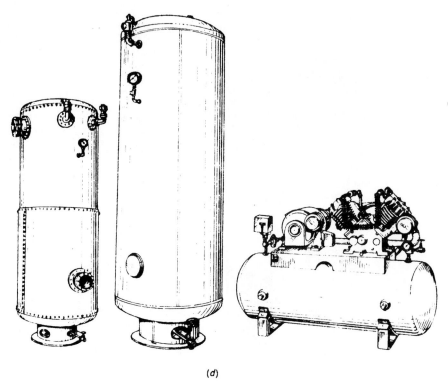

(*d*)

FIGURE 6.6 (*Continued*) Compressed air auxiliaries. (*d*) air receivers. (*Courtesy,* Power Magazine.)

is trapped on a traveling oil-coated screen and then washed off in an oil bath that must be replenished or filtered periodically.

Inter- and aftercoolers not only cool the air for better efficiency but also remove oil and water vapor by condensing them out of the air. These coolers are essentially air-to-air or air-to-water heat exchangers. The water-cooled units have water in the tubes with the air to be cooled flowing over the tubes. *Separators* are used to remove oil and moisture in the air flowing to the receiver after being discharged from the aftercooler. *Traps* are used to remove moisture from air lines but need special attention if they are in an unheated area to avoid frozen condensate, which would make the trap inoperable. Receivers are tanks to store the compressed air and also dampen the pulsations in compressor discharge flows due to pressure waves. Receivers also help to cool the air and remove some of the moisture. They should be constructed according to the ASME code and stamped with National Board number to show that code construction has been met. The allowable pressure stamped on the tank should not be exceeded; therefore, install a safety valve set for the allowable pressure with a volume equal to the compressor's volume of delivery to the tank. Tanks should be drained periodically, preferably daily, in order to remove the oil and moisture from the bottom.

Diesel action can occur if oil is not drained from the intercooler, aftercooler, and air receiver. Explosions have occurred when the hot air under pressure combined with the oil in the system and ignited, sending pressure waves into the piping and tanks. If severe enough, rupture of the weakest element in the system can occur.

TURBO-BLOWERS

Turbo-blowers are similar to centrifugal compressors but are used for low-pressure service (see Fig. 6.7). The turbo-blower is a constant-pressure machine in which the power consumed is almost directly proportional to the volume delivered. Controls can

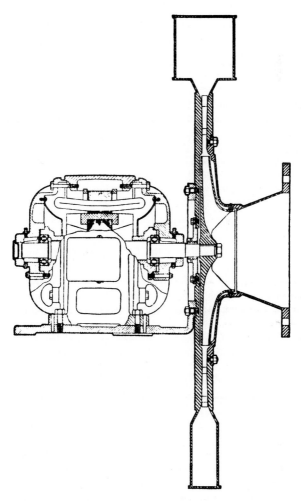

FIGURE 6.7   Direct-drive single-stage turbo-blower.

be provided for constant discharge pressure, constant suction pressure required in some services, or constant volume. Control can be by vanes installed on the suction or discharge side that are actuated by sensors measuring pressure, volume, or other process needs. Steam turbine drives or variable-speed motor drives may use speed control.

Surging is a problem at about one-third to one-half of rated capacity and is accompanied by whining noise and unstable operation, including vibration. Blowers should be seized and operated above the surge point to avoid these problems. The mechanical maintenance required on these machines is similar to centrifugal compressors.

SAFETY PRECAUTIONS

Compressed air has some inherent hazards that facility service operators should recognize. Among these are:

1. Compressed air released through a nozzle travels at high speed and should not be directed at the human body or used to play pranks on others. External and internal injury and even death can result.

2. Air piping and tanks, intercoolers, and aftercoolers should be of the right material, installed with adequate support, built to ASME code or ANSI standards, and stamped with the allowable pressure.

3. Intercoolers, aftercoolers, and air receivers should have safety relief valves with an NB or ASME nameplate showing set pressure and capacity and should be of material suitable for the service. The set pressure should not exceed the allowable pressure for the equipment the safety valve is protecting. The capacity should prevent the pressure from rising above that allowable.

4. Safety valves should be tested periodically to make sure they are not frozen in a closed position.

5. Make sure there is a safety valve on the discharge from the compressor if there is a closing valve between the compressor and the receiver.

6. Oil and moisture should be drained manually from air line traps, separators, inter- and aftercoolers, and receivers to avoid "diesel" actions and to limit internal corrosion and deterioration. Automatic drains also require inspection and testing at regular intervals.

7. Valves or other parts of a compressor should *never* be cleaned with gasoline or other highly volatile and inflammable liquids. People have been severely burnt when these highly flammable substances combined with the air in the compressed air system and caused fires and explosions.

8. Investigate high air temperatures immediately. This can indicate valve problems or blocked cooling water flow, to name a few. High intercooler temperatures point to problems in the high-pressure cylinder, while low intercooler pressure indicates problems with the low-pressure cylinder.

9. Avoid excessive carbon buildup on valves and piston rings by periodic overhauls and cleaning. Avoid excessive oiling of the air cylinders.

10. Periodically check the water flow on water-cooled machines. Open funnels are usually provided to make this water flow check. Accidental shutoff of water is always possible, and alarms should be installed for this possibility.

MAINTENANCE

Most manufacturers provide good maintenance and operation instructions for their machines, and these should be followed. With time, these instructions may be filed in a forgotten corner of the plant. It is, therefore, appropriate to list the following compressor problem areas and the possible causes so that corrective actions can be taken by plant service operators:

Malfunction symptom	Possible causes
Noise or knocking	Loose or burned-out bearings, loose valve or unloader, loose flywheel, motor rotor shunting back and forth from unlevel mounting or belt misalignment
Squeal	Motor or compressor bearings tight, belts slipping, lack of oil, leaking gasket or joint
Intercooler safety valve blows while running unloaded	Broken or leaking horsepower discharge valve, or suction unloader, defective or stuck low pressure unloader, blown gasket
Intercooler valve blows while running loaded	Broken or leaking high-pressure discharge or suction valve, defective high-pressure unloader held in unloaded position, blown gasket
Sudden capacity drop	Bad leak in air-operated equipment or air lines, discharge piping clogged, suction filters blocked, broken, or badly leaking valves, blown gaskets, leak in intercooler
Gradual capacity drop	Accumulation of small leaks in air lines, poorly seating valves, restricted suction filters, worn rings or cylinders
Receiver safety valve blows	Defective pop valve, defective pressure switch or pilot valve, leak in control line, inoperative suction unloaders
Unit blows fuses	Fuses too small, low voltage, pressure-switch differential setting too close, unit starting against full load, electrical trouble, motor or compressor tight on moving parts
Unit will not start	Blown fuse or tripped overload relay, motor or electrical trouble, defective pressure switch, motor or compressor binding
Roughness and vibration	Base too light, improper shimming under unit, foundation bolts loose, unbalance from one cylinder not pumping
Excessive oil consumption	Oil level too high, oil viscosity too light, too high oil pressure (if forced-feed lubricated), worn rings and cylinders

QUESTIONS AND ANSWERS

1. A 14- by 12-in double-acting single cylinder compressor has a 3-in diameter piston rod and operates at 250 r/min. What would be the piston displacement?

Answer. Since it is double-acting, both sides of the piston area are compressing and both travel the length of the stroke. Therefore,

Piston area without rod $= 0.7854 \times 14^2 = 153.94$ in^2

Rod area $= 0.7854 \times 3^2 = 7.07$ in^2

Piston area with rod $= 153.94 - 7.07 = 146.87$ in^2

Total piston area $= 153.94 + 146.87 = 300.81$ in^2

Displacement is then $\dfrac{300.81}{144} \times \dfrac{12}{12} \times 250 = 522.2$ ft^3/min

2. What are three functions of an air receiver?

Answer. An air receiver (1) dampens air pulsation, especially on reciprocating machines, (2) acts as a reservoir of reserved stored air, and (3) eliminates some moisture and oil from the compressed air, which must be drained periodically.

3. Why are check valves necessary on centrifugal or rotary machines that operate in parallel and discharge into a common header?

Answer. Check valves should be installed on the discharge line of each compressor in order to avoid backflow of air into an idle machine that has not been properly isolated. If the machine is equipped with a separate motor-driven oil pump, damage will occur to the compressor so driven by backflow because the bearings will have no lubrication with the oil pump shut down.

4. What have been the problems with small compressors mounted on top of air tanks?

Answer. Because the units were not balanced properly, and not mounted on a firm enough foundation, severe vibration of the assembly caused cracking to occur on the air tank, especially on saddle welds to the air tank. Several explosions of air tanks in California caused that state to require compressors to be mounted *off* the air tank unless the design included an analysis of the vibration forces imposed on the assembly to show that the assembly was adequate to resist the vibratory forces imposed on the assembly.

5. What is the difference between isothermal and adiabatic compression?

Answer. In isothermal compression, the temperature is constant throughout the cycle by removing the heat of compression. In adiabatic compression, no heat is added or subtracted from the compression cycle; thus, the discharge pressure is higher, and more work is needed to compress the air than with isothermal compression.

6. What does the term *compression ratio* mean?

Answer. This is the ratio of absolute discharge pressure of a compressor to *absolute* intake or suction pressure. If the discharge pressure is 200 psig and intake pressure is 0 psi at sea level, the compression ratio would be 214.7/14.7 = 14.6.

7. How does intercooling affect compression?

Answer. By dividing the compression of the air into two or more steps, air can be cooled between the steps, and this reduces the work that the compressor has to perform, and it makes it possible to compress more weight of air. Another benefit is that some of the water vapor in the air is condensed in the intercoolers.

8. How is mechanical efficiency defined for an air compressor?

Answer. This is a ratio of the power necessary to compress the air in the cylinders to the brake horsepower (bhp) required to drive the compressor. The power necessary to compress the air in the cylinder can be obtained by using indicator cards to obtain the mean effective pressure (mep). For reciprocating compressors, this mep is used in the equation

$$\text{ihp} = \frac{PLAN}{33,000}$$

Chapters 5 and 10 discuss how this equation is used for steam cylinders; the methods and principles are the same. Then mechanical efficiency E_m = ihp/bhp. It is also necessary to state whether the ihp is for isothermal or adiabatic compression.

9. What is compression efficiency?

Answer. This is the ratio of the *isothermal* horsepower required to compress the amount of air required for the ihp as explained above.

10. An instruction book for a compressor shows 170 hp as the isothermal horsepower, a brake horsepower of 200 hp, and an indicated horsepower (ihp) of 185 hp. What is the mechanical efficiency and compression efficiency?

Answer. They are:

$$E_m = \frac{\text{iph}}{\text{bhp}} = \frac{185}{200} = 92.5\%$$

Compression efficiency for isothermal compression = 170/185 = 91.9%

11. Why is the indicated horsepower not the same as brake horsepower?

Answer. The ideal or indicated horsepower departs considerably from the actual work required by the compressor due to the additional work required to compensate for leakages, friction of moving parts, imperfect valve action, inefficient cooling, obstructed intake, and similar losses.

12. What is the displacement of a compressor?

Answer. The *displacement* of a compressor is the volume displaced or swept through by the net area of a piston for a reciprocating machine during the compression stroke and is expressed in ft³/min. Remember, for multistage compressors, *only* the low-pressure cylinder is considered.

13. An 8- by 6- by 3-in compressor has a single-acting compression cycle with an 8-in stroke and operates at 450 r/min. What is the compressor displacement?

Answer. Only the low-pressure cylinder is used, or

$$\text{Displacement} = \frac{0.7854(8)^2}{144} \times \frac{8}{12} \times 450 = 104.8 \text{ ft}^3/\text{min}$$

14. What do capacity and volumetric efficiency mean?

Answer. Capacity is the actual volume of air compressed and delivered and is usually expressed in ft³/min. Capacity will always be less than displacement due to the clearances needed to prevent mechanical interference, friction, compression heat, and even altitude of operation. The volumetric efficiency then is

$$\frac{\text{Actual displacement or volume}}{\text{Theoretical displacement or volume}}$$

15. What kind of marking is required on a safety valve?

Answer. It should have the ASME or National Board stamping to indicate it was built and tested to nationally recognized standards. It must be of the direct, spring-loaded type and be of material suitable for the service. This includes non-corrosive material construction and the seat being fastened to the body so that it cannot lift with the valve disk. No failure of any part should interfere with limiting the full discharge capacity of the valve. The following markings are required in such a manner that they cannot be obliterated in service: (1) Identifying name or trademark of the manufacturer of the valve, including manufacturer's design or type number, (2) inlet size of the valve in inches, (3) set pressure, psig, (4) capacity in cubic feet per minute of air at 60°F and 14.7 psi absolute, and (5) year the valve was built or coding to identify the year.

16. What is the set pressure tolerance for code safety valves?

Answer. The set pressure tolerance is 2 psi for pressures up to and including 70 psi and 3 percent for pressures above 70 psi.

17. Why is it desirable to have cool intake air in a compressed air system?

Answer. Cooler air in the compressor takes up less volume, and the unit compresses a greater weight of air per stroke. For every decrease of 5°F in the incoming air, there is a 1 percent increase in the mass of air compressed, or in the volume compressed. The intake air should be clean, cool, and dry as much as possible. The best air usually comes from the outside of the building, preferably on the shady side. Dust can be blocked by installing and maintaining the filters recommended by the manufacturer.

18. What precautions are needed in starting a steam-driven air compressor?

Answer. Always follow the manufacturer's instructions, but pay particular attention to opening the drain cocks on the steam cylinder to blow out condensed steam. Also, open the drain lines in the steam line above the throttle valve until the condensate is blown out and the line warms up. This will avoid water hammer. Open the throttle valve a little and slowly, and let the steam cylinder warm up. This will avoid too-rapid expansion stresses. Turn the engine over so that steam blows from both of the cocks. Slowly warm up the engine and slowly bring it up to speed with the throttle valve until the governor takes over. Make sure this happens. Test the overspeed device, if there is one, manually. Some old units have a belt-driven governor with an idler pulley arrangement. Make sure there is a trip mechanism on the throttle valve to shut off the steam if the governor belt breaks. Do not forget to check the oil level in the bearings, and make other mechanical checks for knocks, leaks, loose linkages, and unusual noises, as you would any mechanical piece of machinery.

19. What causes moisture in air lines?

Answer. Water vapor in the air pulled in by the compressor depends on the humidity, and this in turn affects the water vapor pressure. Initially as the air is compressed and picks up heat from the work done on it by compressing the air, the water vapor remains vaporized but starts to condense out of the air in the intercooler and aftercooler, if so equipped. The remaining water vapor may condense in the air receiver and even in the air distribution lines, especially if the condensed water is not properly removed along the air travel path. Air dryers are now extensively used to give a "final" polish in moisture removal before the air goes to controls and process use. Moisture in the air may cause rusting, wash away lubricants from compressed-air-using tools, make controls operate erratically, cause trapped moisture in unheated air lines to freeze, or contaminate some process applications.

20. What shell equation should be used for calculating allowable pressure on an air tank that has been reduced in thickness?

Answer. Use ASME Code, Section VIII as follows:

$$P = \frac{SEt}{R + 0.6t}$$

where S = allowable code stress, for ordinary steel, use 13,750 lb/in^2
E = longitudinal joint efficiency. This factor depends on the type of welded joint and extent of radiographic examination of the welded joints. Use 65 percent to be conservative.
R = inside radius of shell, inches
P = gage pressure
t = shell thickness, inches

Example. An air tank was found corroded due to outside installation and the fact the outside of it never had been painted. A thickness test with an ultrasonic instrument showed an average thickness of 0.70 in. The outside diameter was 48 in. The tank had a stamping showing 300 psig as allowable pressure. What is the present AWP?

$$R = \frac{48 - 2(.7)}{2} = 23.3 \text{ in}$$

$$P = \frac{13,750(0.65)(0.70)}{23.3 + 0.6(0.7)} = 263.8 \text{ psi allowable}$$

Pressure must be reduced, and new safety valve installed, or else the tank should be replaced with a new one. Another possibility is to check the manufacturer's data sheet to obtain true design stress and to see if full radiography was used on the welded joint. This would increase the E factor in the equation.

CHAPTER 7
AIR MOVERS: FANS AND BLOWERS

Fans and blowers are extensively used in all types of occupancies and service requirements, such as draft for boilers, ventilation, supplying conditioned air to offices, and supplying combustion air for ovens and similar industrial equipment. Plant service operators can be expected to operate and maintain some type of fan service as part of their duties. Fans and blowers move air at desired volumes and pressure; however, the pressure is not measured in psi but rather in inches of water. Blowers usually deliver large volumes of air at higher pressure than fans, which are considered low-pressure air deliverers.

DRAFT

Draft can be natural, such as a chimney in a furnace, or mechanical, where a fan is used to move the air or gas. The movement is caused by a pressure difference between inlet and outlet and is measured in terms of the height in inches of water of a water column that is equivalent to the air pressure acting on the water column. Figure 7.1 shows two types of draft gages. In U-tube manometers, one side is connected to the duct, chimney, or furnace where the draft is to be measured, while the other end is open to the atmosphere. The difference in the water level is a measure of the pressure difference. For example, to convert 3 in of draft to pressure, remember that a cubic foot of water weighs 62.4 lb and by dividing this by the cubic inches in a cubic foot, we obtain $62.4/1728 = 0.036$ psi/in^3 of water. Thus for a 3-in draft, the pressure would be $3 \times 0.036 = 0.108$ psi.

PRESSURE

Pitot tubes are used to measure *static* pressure in a duct, *velocity* pressure, and *total* pressure in order to determine the velocity and volume of flow in the duct. Pitot tubes have one leg shorter than the other, which is at a right angle (see Fig. 7.2). The short leg is inserted into the duct parallel to the longitudinal axis of the duct, and the open end faces the air or gas flow. Since the velocity of flow in a duct varies from the walls to the center of the duct, standards have been developed to take a profile of readings

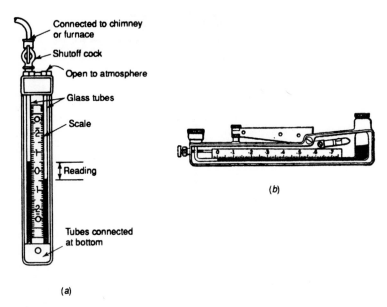

(b)

(a)

FIGURE 7.1 Gages used for draft measurement. (a) U-tube measures draft by indicating pressure differential in inches of water; (b) an inclined-draft gage uses a light oil as the gage fluid; however, the scale divisions are so positioned that the readings show the equivalent inches of water.

so that an average can be obtained for the particular duct position that is being measured. To convert velocity pressure h_v from inches of water to feet per second, use

$$V = \sqrt{\frac{2g\,Dh_v}{12d}}$$

where V = velocity of flow in duct, ft/s
  g = acceleration of gravity, 32.17 ft/s^2
  D = weight of 1 ft^3 of water at the gage temperature; in most cases, use 62.4
  h_v = velocity pressure measured, in of water
  d = density of air or gas flowing, lb/ft^3

Example.   Air flowing in a duct has a velocity pressure of 0.13 in of water and an air temperature of 170°F, with the water temperature in the gage being 80°F. Tables show the air density at 170°F is 0.06304 lb/ft^3 and that of water is 62.19 at 80°F. What is the velocity of flow in feet per minute? Substituting known quantities in above equation,

$$V = \sqrt{\frac{2(32.17)\,62.19 \times 0.13}{12 \times 0.06304}} = 26.22 \text{ ft/s} = 1{,}573.2 \text{ ft/min}$$

The *volume* of flow is calculated by using

$$Q = AV_a$$

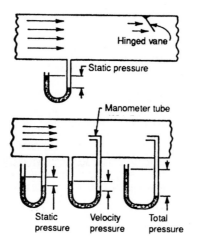

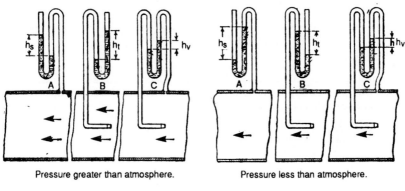

FIGURE 7.2 (a) Three air pressures as measured by pitot tubes and manometers; (b) position of fluid in manometer for pressures greater than and below atmospheric pressure.

where Q = ft³/s flow
V_a = average velocity of flow, ft/s
A = duct cross-section area, ft²

Example. If the average flow is as determined in the previous problem, and the duct is 16 in in diameter, what would the volume of flow be in cubic feet per minute? The cross-sectional area of duct is $0.7854 \times (16/12)^2 = 1.396$ ft²:

$$Q = 1.396 \times 26.22 = 36.6 \text{ ft}^3/\text{s} = 2196 \text{ ft}^3/\text{min}$$

The *total pressure* as shown in Fig. 7.2 is used to calculate the fan horsepower required to move the air. Use

$$\text{ahp} = \frac{Q \times 60 \times h_t \times D}{12 \times 33000}$$

where Q = volume of air handled, ft^3/s
  h_t = total pressure difference created by the fan, in of water
  D = density of water (ft^3) at the gage temperature, lb

Example. Assume total pressure difference in the previous flow example was 9.226 in of water, the density of water was 62.2 lb/ft^3, and the flow was 36.6 ft^3/s. What is the air horsepower required? Substituting known quantities in above equation

$$\text{ahp} = \frac{36.6 \times 60 \times 9.226 \times 62.2}{12 \times 33000} = 3.18 \text{ hp}$$

$$\text{Fan mechanical efficiency} = \frac{\text{air horsepower}}{\text{horsepower input}}$$

It should be remembered that the work done by a fan or blower is the total average pressure difference between inlet and outlet for the fan or blower. If the inlet has no inlet duct, this normally would be zero inlet pressure, while if an inlet duct is present, the inlet pressure must be considered.

Example. A blower has an inlet duct for a gas from a process that shows 0.25 in of water static pressure and a velocity pressure of 0.4 in of water. The discharge static pressure measured 18 in of water with a velocity pressure of 1.0 in of water. What is the total pressure difference produced by the blower?

$$\text{Pressure difference} = \text{outlet total pressure} - \text{inlet total pressure}$$

$$= (18.0 + 1.0) - (-0.25 + 0.4)$$

$$= 19 - 0.15 = 18.85 \text{ in of water}$$

Another term quite often noted in fan literature is manometric efficiency, designated as

$$\text{Manometric efficiency} = \frac{\text{draft actually produced}}{\text{theoretical draft}}$$

FAN TYPES AND CHARACTERISTICS

A basic type of classification for fans is by flow, *radial or centrifugal* and *axial.* Figure 7.3 shows the different types. The propeller fan is used mostly for ventilation and exhausting service. The vane axial fan is a propeller fan set inside a cylinder.

The centrifugal fan is the most common type of fan, and this is divided into two classes: (1) *steel-plate* as shown in Fig. 7.4a and (2) *multiblade* as shown in Fig. 7.4b. In the steel-plate fan, the wheels consist of one or two spiders, each having 6 to 12

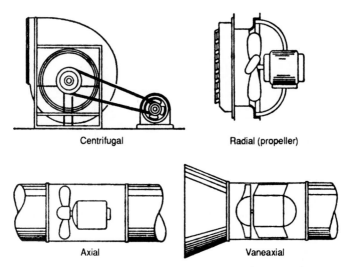

Centrifugal   Radial (propeller)

Axial   Vaneaxial

FIGURE 7.3   Basic types of fans can be classified by gas flow into radial or centrifugal and axial.

arms. Each pair of arms has a flat blade extending radially as shown in Fig. 7.4*a*. The blades may be straight or curved either forward or backward. These types of fans are usually used with engine drives.

The *multiblade* fan is built up of two or more annular rings and has many narrow curved blades riveted or welded in place between the annular rings as illustrated in Fig. 7.4*b*. A double inlet conoidal fan is shown in Fig. 7.5*a*.

Designation of centrifugal fans includes the following: (1) number of inlets—single or double, (2) width of wheel, (3) diameter of wheel, (4) discharge—horizontal, vertical, top, bottom, or angular, (5) housing—full in which the fan scroll is completely above the base as shown in Fig. 7.5*c*, seven-eighths, and three-quarters, where the scroll is located below the top of the supporting base, and (6) rotation of clockwise or counterclockwise as noted when looking at the fan from the drive side.

Steel-plate and forward-curved blade fans are generally used for slower rotation speeds, while conoidal and backward-curved multiblade fans are used for high speeds, which makes them suitable for direct connection to mechanical-drive steam turbines and high-speed motors.

Fans that are in high-temperature and abrasive service need special attention. For example, induced draft fans on coal-burning boilers should have water-cooled bearings because of the high temperature of the exhaust gases that the fan is exposed to. The abrasive action of fly ash requires more rugged construction and material such as stellite tips on blades in order to prevent rapid wear of the blades and perhaps a sudden rupture of a part of the fan.

Volume and pressure control on the gas flow can be manual or automatic. There are several methods of controlling the volume and pressure: (1) varying the speed of the driver, (2) louvre dampers placed at the outlet of the fan with constant speed drive, (3) adjustable vanes placed at the inlet of the fan as shown in Fig. 7.6*a*, (4) variable-pitch blades on vane axial fans, and (5) bypass or spillover where some of the gas is directed back to the suction side or discharged to the atmosphere.

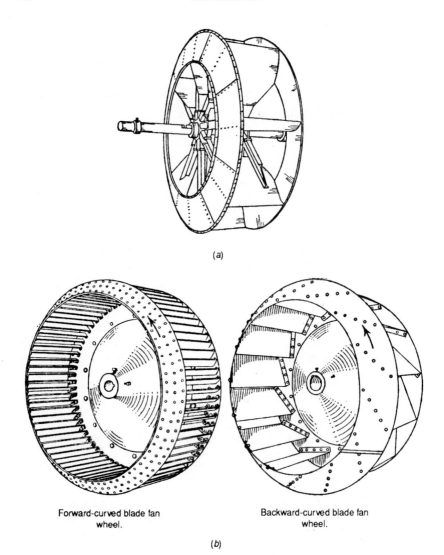

(a)

Forward-curved blade fan wheel.

Backward-curved blade fan wheel.

(b)

FIGURE 7.4 Steel-plate and multiblade fan wheels. (*a*) Steel-plate fan wheel is used for lower speeds; (*b*) multiblade fans may have forward- or backward-curved blades.

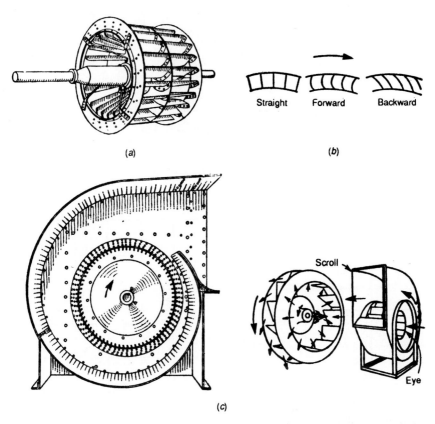

FIGURE 7.5   (*a*) Double-inlet, double-width conoidal fan; (*b*) multiblade fans may have straight, for-ward-, or backward-slanted blades; (*c*) fan housing designation depends on scroll position in relation to base.

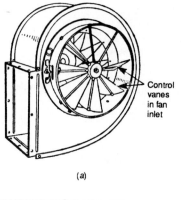

(a)

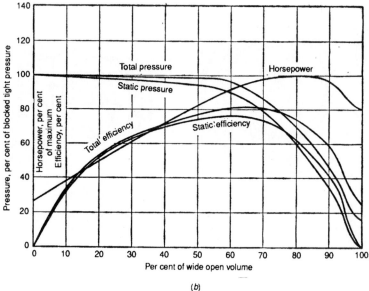

(b)

FIGURE 7.6 (*a*) Control is by vanes in fan inlet for this design; (*b*) typical performance curves for a backward-curved multiblade fan.

Testing of fans for performance is based on standards as established by the National Association of Fan Manufacturers and the American Society for Heating and Ventilation. Figure 7.6*b* shows characteristic performance curves for a backward-curved blade fan.

CENTRIFUGAL FAN PERFORMANCE LAWS

Similar to centrifugal pumps, for a given fan size, duct system, and air density, the following relationships exist on performance:

1. Capacity varies directly as the speed ratio.
2. The static pressure varies as the speed ratio.
3. The speed and capacity vary as the square root of the static pressure.
4. The horsepower varies as the cube of either the speed or capacity ratio.
5. The horsepower varies as the 3/2 power of the static pressure ratio.
6. The air velocity varies directly as either the speed or the capacity ratio.

Where the static pressure is *constant* at the fan outlet, the following rules apply:

1. The capacity and the horsepower vary as the square of the wheel diameter.
2. The speed varies inversely as the wheel diameter ratio.
3. At constant static pressure, the speed, capacity, and power vary inversely as the square root of the ratio of the air densities.
4. At constant capacity and speed, the horsepower and static pressure vary directly as the ratio of the densities of the air.

Example. For conditions 1 and 2 just above, one fan has a wheel diameter of 22.75 in with a capacity of 7890 cfm at 6 ½-in static pressure and a speed of 2064 r/min. Assume each fan operates at the same static pressure of 6 ½ in of water. What would be the capacity and speed of a 45.5-in diameter fan? Per rule 1,

$$\text{Capacity} = 7890 \times \left(\frac{45.5}{22.75}\right)^2 = 31{,}560 \text{ cfm}$$

Per rule 2, the speed is:

$$2064 \times \frac{22.75}{45.5} = 1032 \text{ r/min}$$

Fan Drives

Most fans are driven by squirrel-cage induction motors. Wound-rotor motors give some speed control as do dc motors, but the trend is to use single- or two-speed induction motors with inlet vane control. Vane control provides an initial spin to the air, thereby reducing inflow to the blades, which is more efficient than damper control on the fan outlet. Mechanical-drive turbines are extensively used where the exhaust steam from the turbine can be used in the process. Variable speed can be obtained on induction motors by solid-state frequency changers, and this also permits electronic control of pressure and volume flow in any integrated control such as for boiler combustion controls.

Induced-draft fans pull the products of combustion in a boiler and direct them to the chimney for atmospheric discharge. The *forced-draft fan* takes air from the atmosphere and delivers it through air ducts to air preheaters and burners and even directly into the furnace, depending on the size of the boiler and its arrangement. Forced-draft fans produce some air pressure, and if the boiler's casing is not tight, furnace gases may escape into the boiler room through leaking joints or cracks. For coal-fired plants, the boiler room may start to be covered by unburnt coal and ash. This may be the penalty for operating with a *positive-pressure* furnace. In a *negative*-pressure furnace, the induced-draft fan creates a partial vacuum condition in the furnace, and atmospheric air rushes in through leaking joints or cracks of the furnace casing. This reduces the efficiency of burning by introducing excess air into the furnace. To make the effects of leaks minor, *balanced draft* is used in most large boiler systems; this keeps the furnace slightly vacuum at about 0.1 in of water vacuum.

OPERATION AND MAINTENANCE

Fans and blowers should be started with minimum flows and gradually brought up to full load to prevent shocking the machine's components and to also avoid seriously overloading the driver. As with all machinery, it should be maintained clean and free of deposits; otherwise possible corrosion attack and imbalance with resultant vibration may occur. Fans and blowers have been torn apart by the destructiveness of prolonged operation with abnormal vibration. For this reason, a daily operational check should include checking the vibration.

Critical fans and blowers, and those located away from operator attention, should have vibration pickups that display this vibration in control rooms. Vibration causes can include misalignment, poor or loose foundation attachments, imbalance due to erosive wear, such as fly ash on induced fans, cracks from sharp corner designs, and stress-corrosion cracking, to name a few. Therefore, fans should be internally inspected in order to detect wear and tear on blades and their attachments. This is especially applicable to critically assigned operational fans and those with high monetary repair or replacement costs.

Proper lubrication and the maintenance of oil quality are other important functions of service operators who have fans and blowers under their care and control.

QUESTIONS AND ANSWERS

1. What are some causes of fan or blower vibrations?

Answer. Vibration may be caused by: (1) bearings that are worn or poorly lubricated, (2) a defective thrust bearing, (3) excessive deposits on blades, (4) unevenly eroded blades, (5) rotating parts rubbing on or binding with stationary parts, (6) unit misaligned with gear set and driver, (7) bent shaft or some other bent component, (8) unit that is loose from foundation, (9) one-side running hotter than other side, causing misalignment, and (10) cracks in rotating element.

2. How is draft measured?

Answer. Air and gas under flow conditions are measured in inches of water by instruments such as manometers, where one side of a tube is connected to the duct or furnace in which the draft is to be measured and the other is connected to the atmosphere or inlet duct. The difference in the water level in the two columns indicates the inches of water of draft, which is a measure of the difference in pressure (see Fig. 7.2).

3. Air flows in a duct 24 × 30 in with an average velocity of 20 ft/s when its static pressure is 0.75 in of water. The air is at 50°F and weighs 0.07788 lb/ft^3, while the water temperature in the gage is at 60°F and weighs 62.34 lb/ft^3. What is the *total pressure* of the air in the duct and the volume of flow in cubic feet per minute?

Answer. Use the equation

$$V = \sqrt{\frac{2gDh_v}{12d}}$$

where $V = 20$, $g = 32.17$, $D = 62.34$, and $d = 0.07788$ and solve for h_v. Substitute

$$20 = \sqrt{\frac{2(32.17)\,62.34h_v}{12 \times 0.07788}}$$

and by solving, we obtain

$$20 = 65.51\,h_v$$

thus, $h_v = 0.305$ in of water. The *total pressure* is the sum of the static and velocity, or

$$0.75 + 0.305 = 1.055 \text{ in of water.}$$

And the flow Q = duct cross-section area × average velocity of flow, or

$$Q = \frac{24 \times 30}{144} \times 20 = 100 \text{ ft}^3/\text{s} \qquad \text{or 6000 ft}^3/\text{min.}$$

4. What is the relationship of horsepower to speed?

Answer. Horsepower changes directly as the cube of the speed. To double the speed, you must have a motor 8 times as big.

5. What is the relationship of pressure to speed?

Answer. Pressure changes directly as the square of the speed of the fan. By doubling the speed, you get 4 times more pressure.

6. What is the relationship of volume to speed?

Answer. Volume changes directly in proportion to the speed of the fan. To get more flow, speed up the fan.

7. Which type of blading (straight, front- or back-tipped) is best for a centrifugal fan used in an air conditioner?

Answer. Back-tipped blading is your best bet because it is the least noisy at its maximum efficiency.

8. What is a lobe blower or Root's blower?

Answer. A lobe blower works on the same principle as a rotary lobe pump. It may have two or more shafts cut with two or more lobes on each rotor. External gears time the lobes for positive rotation. Lobe blowers are used on air conveyors and gas pumps or as scavengers or superchargers on internal-combustion engines.

9. How are fans controlled?

Answer. Fans are controlled by (1) varying the speed of the drive unit (turbine, motor fluid drive, variable-pitch sheaves), (2) dampers, to throttle the discharge, (3) inlet vanes, which can be adjusted to spin the air entering the fan, (4) variable-pitch blades on the vane axial fans, and (5) spillover, which diverts part of the discharge to the atmosphere or back to the suction of the fan.

10. What are two mechanical draft systems?

Answer.   Two systems are *forced* draft and *induced* draft.

11. What are some attractive features of backward-curved blades?

Answer.   Backward-curved blades in fans permit high operating speeds and produce high volumetric efficiencies and a wide range of capacities at constant speed with small changes in power requirements. They also operate at lower noise levels than forward-curved blades.

12. Why is static pressure in a fan important?

Answer.   Any fan which develops a large static pressure in comparison with its total pressure has a greater capacity to overcome the resistance offered by air or gas flow due to duct friction, baffles, and any other obstructions in the air or gas flow.

13. What measurement precautions are required in using pitot tubes?

Answer.   Measurements are not adaptable to velocity pressures corresponding to flows less than 6 ft/s unless used with delicate pressure-measuring gages. Pressure measurements should be made in a straight section of duct at least 20 times the duct diameter in length with at least a 10-duct diameter space on either side of the pitot tube location. Unstable data will be obtained if readings are taken in bends or near a fan outlet.

14. Why is it necessary to have high stacks or chimneys in power plants equipped with mechanical draft fans?

Answer.   To disperse and dilute the discharge from the chimneys, such as smoke and flue gas as well as cinders and fine ash, in order to comply with air pollution requirements and to avoid offending adjacent property owners.

15. A fan discharges air into a duct 36 × 48 in at a velocity of 3000 fpm against a total pressure of 1.5 in of water. The static pressure at the fan inlet is 1.30 in of water, and the corresponding velocity pressure is 0.65 in of water. The gage fluid has a density of 62.3 lb/ft³. What is the power required to drive the fan if the mechanical efficiency is 75 percent?

Answer.   In order to use the air horsepower equation, it is first necessary to calculate Q, or the volume of air handled:

$$Q = AV$$

where A = (36 × 48)/144 = 12 ft²
  V = 3000/60 = 50 ft/s
  Q = 12 × 50 − 600 ft³/s

$$\text{ahp} = \frac{Q \times 60 \times h_t \times D}{12 \times 33000}$$

where $D = 62.3$ and $Q = 600$; therefore, it is necessary to determine $h_t = total$ pressure difference created by the fan:

$$h_t = \text{total outlet pressure} - \text{total inlet pressure}$$

$$= 1.5 - (-1.30 + 0.65) = 2.15 \text{ in of water}$$

$$\text{aph} = \frac{600 \times 60 \times 2.15 \times 62.3}{12 \times 33000} = 12.18 \text{ hp}$$

$$\text{Power required} = \frac{12.18}{0.75} = 16.24 \text{ hp}$$

16. What fan types are suitable for low- and high-speed applications?

Answer. Steel-plate and forward-curved blade fans are best suited for slow speeds, while fans of the conoidal, radial-tipped, and backward-curved multiblade types are best suited for high-speed applications.

17. What is the difference between a fan, a blower, and a compressor?

Answer. There is no definite difference. However, a very-low-pressure machine is often called a *fan.* A medium-pressure machine is usually called a *blower.* When a machine delivers in *pounds per square inch,* it is generally regarded as a *compressor.*

CHAPTER 8
HEATING, PROCESS, AND POWER BOILER SYSTEMS

Plant service operators are generally assigned the responsibility of operating heating, process, or power boilers in a safe, efficient, and continuous manner, even though many boilers are operated automatically. These devices must be periodically tested, and the controls require attention if set limits are no longer maintained. The boiler system and automatic controls must be analyzed and understood so that corrective actions can be taken by the operator. At times, manual operation is required until the boiler can be shut down to make the necessary repairs. Skilled facility operators are also required to inspect and maintain the boiler system. This requires understanding the wear and tear that takes place in service, such as cracking, thinning, overheating, and scale coatings on heat-transfer surfaces, to name a few. The wear and tear can cause unexpected failures, even explosions, and/or bulging of pressure-containing parts. Most jurisdictional examinations stress the importance of recognizing dangerous conditions that may be developing in a stored energy system, which a boiler is, and thus require operators to be familiar with the forces acting on boiler components, the stresses that are produced by pressure, and the weakening effect of overheating metal parts.

PRESSURE, FORCE, AND STRESS

Pressure

Gage pressure is the pressure in pounds per square inch that exists inside of a pressure-containing part; it is 14.7 psi less than *absolute pressure* or atmospheric pressure at sea level. Most code calculations are based on gage pressure, except where thermodynamic principles are involved. The standard atmospheric pressure of 14.7 psi is also defined as the weight of a mercury column of 760 mm (29.92 in) in height at 32°F, where mercury weighs 0.491 lb/in^3. As can be noted, each inch of mercury height exerts 0.491 psi. Pressures below atmospheric are defined as vacuum conditions and are measured in inches of mercury. Atmospheric pressure that is expressed in inches of mercury is called barometric pressure.

Example. A condenser on a steam turbine reads 27.5 in of mercury, while the barometric reading is 30 in of mercury. What is the absolute pressure in the condenser? Absolute pressure in inches of mercury = 30 − 27.5 = 2.5 in of mercury. Pressure = 2.5 × 0.491 = 1.23 psia.

Force

The force acting on a flat surface of a boiler is equal to pressure times exposed area and is expressed in pounds. For example, if a flat tube sheet in a fire-tube boiler has 361 in^2 of unstayed surfaces and the boiler is operating at 125 psi, the force on the tube sheet will be 361 × 125 = 45,125 lb, or over 22 tons. The tube sheet material must resist this force and thus becomes internally stressed.

Tensile stress is defined as

$$S_T = \frac{F}{a}$$

where F = external force, lb
  a = cross-sectional area of the *material* resisting F, in^2
  S_T = internal tensile stress of the material, lb/in^2

Example.   A 2-in diameter steel bar is tested in a tension machine and starts to tear apart at 92,500 lb force. What is the tensile stress? The area resisting the force is

$$\pi \times \frac{2^2}{4} = 3.14 \text{ in}^2$$

$$S_T = \frac{92,500}{3.14} = 29,458.6 \text{ lb/in}^2$$

Strain

Strain is defined as the stretch per unit length as a material is stressed under load and is expressed as inches per inch. For example, if the bar in the above example was 15 in long and it stretched to 15.020 in total length, the strain would be 0.020/15 = 0.00133 in/in. A useful ratio is the modulus of elasticity E = stress/strain; it is used to calculate stress by measuring strain. For steel, $E = 30 \times 10^6$ lb/in^2.

The reader is referred to strength of materials texts for further study on how materials resist the loads imposed on them. (See also McGraw-Hill's *Boiler Operator's Guide* by Spring and Kohan.)

Stress

Stress concentration increases the normal calculated stress due to abrupt changes in the component's shape, such as corners, holes, or discontinuity in the metal's structure. Normal calculated stresses may be increased by 2 to 3 times due to stress concentration factors, and this is quite often manifested by fatigue cracks developing in sharp corners or other abrupt material configurations.

Crack growth due to fatigue has been quantified by new knowledge of material behavior at the tip of a crack, and this new stress analysis technique is called *fracture mechanics*. It has been established that plastic deformation takes place about the crack tip with each cycle of applied stress and that the distance the crack advances with each cycle depends on the stress intensity factor applied to the tip of the crack. Equations are available in the literature to determine crack growth and cycles required for it to reach critical size.

Stress-corrosion cracking occurs when a repetitive stress is being experienced in an environment that is attacking the metal's grain structure. It is most pronounced in high-stressed areas. The failures may be intergranular or transgranular. Intergranular cracking occurs when the cohesion between the crystallographic grains of a metal is reduced by environmental chemical attack, with the cracks following the grain boundaries. Transgranular failure splits the metal's grain structure. Most stress-corrosion cracking is intergranular. Some corrosives affecting the following metals are:

Carbon and low-alloy steels. Hydrogen sulfide, sodium hydroxide, nitrate solutions, hydrogen cyanide solutions, and mixed acids such as sulphuric and nitric

Stainless steels. Acid chloride solution and hydrogen sulfide

Copper alloy. Ammonia vapors and solutions, mercury compounds, and amines

Aluminum. Sodium chloride solutions

Nickel alloys. Caustic soda (sodium hydroxide), hydrofluoric acid, and fluosilic acid

Caustic embrittlement of ferrous metals was one of the first well-known examples of stress-corrosion in which alkaline solutions or concentrations attacked the carbon steel in highly stressed areas, such as riveted joints.

High temperature is a common cause of boiler part failure from overheating. The ASME stress tables for boiler material amply demonstrate that above permissible temperatures the allowable stress declines very rapidly. For example, SA 178A is a common carbon steel tube material, which has an allowable stress from -20 to $650°F$ of $11,800$ lb/in^2. At $800°F$, the allowable stress drops to 7700 lb/in^2. At the same pressure, it is obvious that the tube would fail from *overstress* as the strength of the metal declines with temperature over a specified range as shown by the code. Causes of overheating include low water in a boiler, excessive scale on heat-transfer surfaces, poor water circulation in tubes, flame impingement on pressure parts, and loss of power, causing overheating from inherent heat on grate fired boilers; these are only a few of the reasons for overheating damage on boiler pressure containing parts.

Creep can occur in steel above about $800°F$. Permanent deformation starts to take place even at constant load due to the plastic nature of the material. The slow stretching with time at high temperature is called creep. Long exposure to creep will cause unacceptable deformation and a sudden rupture of the part so affected. The boiler code has established criteria for stress, temperature, and time for creep to occur. High stress and temperature expedite the plastic deformation. The material may act in a brittle manner and fail suddenly due to loss of ductility at elevated temperature.

Charpy V-notch tests are used to measure the resistance of a material to impact or brittleness. *Steel has transition temperatures* where it changes from ductile behavior to brittle behavior. The drop-weight test is used to determine the fracture resistance against brittle failure of steel that is 3/4 in thick and over, with the specimen having a notch that can start a fracture. The tests are run at different temperatures to get a profile on the transition zone with temperature changes. Impact testing details are available from ASTM E-23 standard on the subject.

NONDESTRUCTIVE TESTING AND EXAMINATION (NDT)

NDT is used in boiler construction to assure good-quality welding and also in repairs. The purpose of NDT is to detect flaws such as voids, cracks, inclusions, porosity, lack of fusion, laminations, undercut, lack of weld penetration, and similar defects. NDT methods used include visual examination, hydrostatic or water leak detection under pressure, radiography, magnetic particle, dye penetrant, ultrasonic, and eddy current testing. Also being developed for code inspection application are acoustic emission and holographic testing. NDT is a method of testing or examining material for defects, or lack of same, without affecting the material physically or chemically. Note that tensile tests to obtain a material's property are destructive to the material. The same is true of welder bend test of prepared welded specimens.

BOILER DEFINITION AND TYPES

Most jurisdictions *define a boiler* as a closed pressure vessel in which a fluid is heated for use external to itself by the direct application of heat resulting from the combustion of fuel (solid, liquid, or gaseous) or by the use of electricity or nuclear energy. A *water-tube* boiler has the water in the tubes and the heat source outside or surrounding the tubes. *Fire-tube* boilers have the heat source inside the tubes and the water in the shell surrounding the tubes. The third classification is *cast-iron* boilers; here the boiler components are made of cast iron, and the boiler has a different configuration from steel boilers, which the other two types are.

Further code classification is based on pressure and use. A *power* boiler is a steam or vapor boiler operating above 15 psig. The same definition applies to a *high-pressure* steam boiler. A *low-pressure* boiler has two pressure limits:

1. A *steam boiler* operating up to 15 psig. Above this, it is classified as high pressure.
2. A *hot-water boiler* operating up to 160 psig or 250°F. Hot-water boilers operating above this pressure and temperature level are classified as high pressure and must be designed to the Power Boiler Code, Section I of the ASME Boiler Code. Low-pressure boilers are built to Section IV requirements of the ASME Code.

Fire- and water-tube steel boilers can be constructed for high and low pressure. Cast-iron boilers are considered low-pressure boilers and are usually used for heating system service. Most large-capacity boilers over 50,000 lb/h are of the water-tube type. This is also true for pressures over 300 psi. Thus, most boilers for large power generation are of the water-tube type because the fire-tube type is restricted by shell size and thickness requirements of the shell for large capacities and pressure.

Boiler capacity can be expressed as pounds per hour, horsepower, Btu per hour, and in power generation, megawatts of electricity produced from the boiler.

A *boiler horsepower* is defined as the energy required to convert 34.5 lb/h of water to steam at a temperature of 212°F. This rating dates to the initial introduction of the steam engine as a power source, and the boiler was rated by the steam engine it was driving. Ten square feet of *heating surface* was considered the equivalent of 1 hp. The heating surface of a boiler is that area that is exposed to the products of combustion.

Example. What is the pounds per hour rating of a boiler rated at 750 hp, and what heating surface rating would the boiler have?

1. Output $= 750 \times 34.5 = 25{,}875$ lb/h
2. Estimated heating surface $= 10 \times 750 = 7500$ ft^2

Heat transfer in a boiler is by *conduction, convection,* or *radiation,* such as exists in the furnace of a boiler. Water flow in natural circulation boilers is induced by the

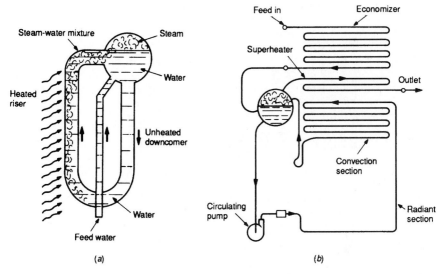

FIGURE 8.1 Two types of water circulation in tubes. (*a*) Natural circulation has cold, more dense water, while lighter hot water and steam rises; (*b*) forced circulation uses a pump to circulate water through tubes.

difference in density of cold water and heated water-steam mixtures (see Fig. 8.1*a*). The heavier cold water will flow to the bottom of the heat-transfer component such as a tube, while the lighter heated water-steam mixture rises in the boiler water or hydraulic circuit. The higher the steam pressure, the more dense the steam becomes, and a point is reached where natural circulation must be supplemented by pump circulation for circulation between cold and hot fluid to continue in the boiler (Fig. 8.1*b*). Insufficient flow or flow that cannot match heat input can cause tube overheating in the hydraulic circuit of the boiler.

Fire-Tube Boilers

The early boilers built during the industrial revolution were fire-tube boilers, also called shell boilers, because the tubes are within the shell, as shown in Fig. 8.2*a,* which shows a horizontal return tubular boiler (*HRT*). The fire-tube boiler has its tube ends exposed to the products of combustion, so these tubes are rolled and beaded over or rolled and welded to prevent the tube ends from being burnt off by the hot gases going into the tubes. Other fire-tube boiler types are the *economic* or firebox type (Fig. 8.2*b*), *locomotive* firebox type (Fig. 8.3), *scotch marine* (SM) type (Fig. 8.3*b* and *c*), the *vertical tubular* (VT) (Fig. 8.4), and not shown, the *vertical tubeless,* a small boiler used in dry cleaning plants for pressing clothes.

The *SM* boiler is the most used fire-tube boiler today for industrial and commercial occupancies. It gained its popularity because it is completely factory assembled and packaged with the burner, fans, and controls factory fire-tested, thus requiring very little on-site work except for hooking up steam and water piping and electrical power. It has been adapted from marine service, where its compactness required small ship space for the capacity produced. SM boilers are built with up to four passes of flue gas traveling through the tubes in order to gain more heat transfer (see Fig. 8.5).

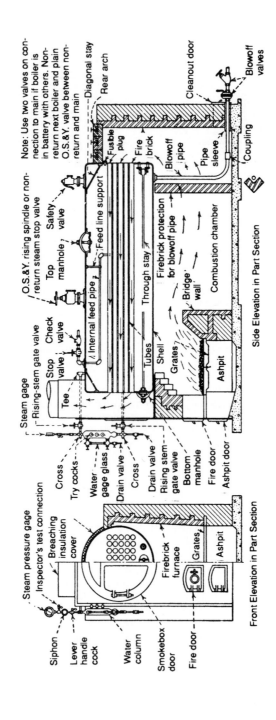

FIGURE 8.2 (a) HRT-type fire-tube boiler details.

8.6

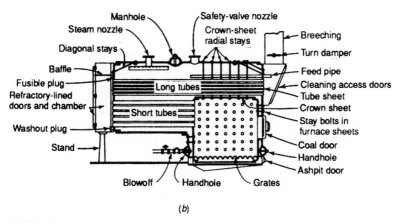

FIGURE 8.2 (*Continued*) (*b*) economic-type fire-tube boiler is externally fired.

Heat Release

Overfiring has caused some problems in the past with tube leakage in the tube sheet on SM boilers as manufacturers attempted to boost output per unit size. Some jurisdictions have imposed limits as a result of consumer complaints. Recommended guidelines are the following:

1. *Volumetric* heat release rate:

$$\frac{\text{Maximum fuel input at boiler rating} \times \text{fuel's heat value}}{\text{Cubic feet of furnace volume}}$$

Recommended volumetric ratios are:

For oil fired. 35 MBtuh/ft^3, with a maximum of 45 MBtuh/ft^3
For gas-fired. 40 MBtuh/ft^3, with maximum of 50 MBtuh/ft^3

2. *Effective projected radiant surface* heat rate:

$$\frac{\text{Fuel release rate at boiler rating}}{\text{Square feet of furnace radiant surface}}$$

Recommended (some jurisdictions have adopted this ratio) radiant surface ratios are:

For oil fired. 180,000 Btuh/ft^2
For gas fired. 200,000 Btuh/ft^2

Most designers try to limit the heat release rate to below 150,000 Btuh/ft^3 of furnace volume; otherwise difficulties in maintaining proper air-fuel ratios start to occur. With higher firing rates, the fuel is still burning at entry to the first tube pass, overheating this area of the boiler and causing cracking of tube ends and also in the weld connecting the tube sheet to the furnace. Any scaling or sediments in the boiler can aggravate the overheating damage with overfiring.

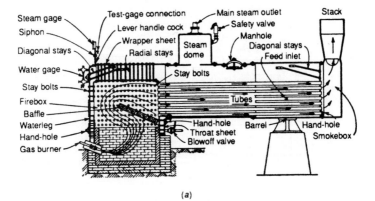

(a)

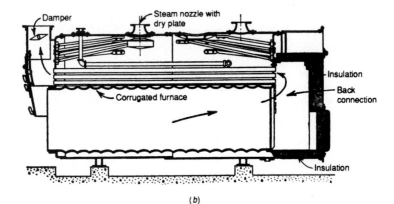

(b)

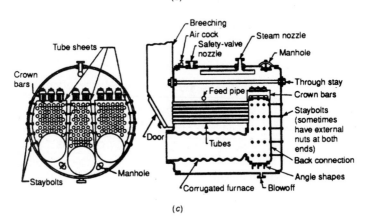

(c)

FIGURE 8.3 Locomotive and SM fire-tube boilers. (a) Locomotive firebox fire-tube boiler; (b) SM fire-tube boiler has dry-back insulation at the rear of the furnace; (c) SM boiler has three corrugated furnaces and a wet-back stay-bolted design at the rear of the furnace for additional heating surface.

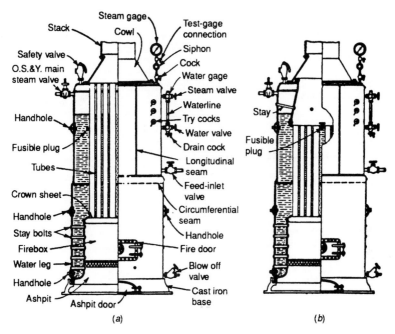

FIGURE 8.4   Two types of VT boilers. (*a*) Dry-top; (*b*) wet-top.

The VT boiler is used for moderate pressure and capacities, such as in laundries and dry cleaning plants, and also for portable service, such as cranes and pile driving. Capacities seldom exceed 10,000 lb/h and 300 psi pressure. Figure 8.4 shows a dry-top and wet-top unit. Other VT boiler names or types are: *Manning,* with enlarged firebox; *tapered-course* bottom, with enlarged firebox; and a tubeless unit for small capacity service. This essentially consists of a cylindrical vertical furnace with a larger cylinder surrounding the furnace with the space in between containing the water and steam. The ends are closed with welded-in flat plates to confine the pressure.

Water-Tube Boilers

The water-tube boiler has water in the tubes with the products of combustion sweeping the outside of the tube for heat transfer. However, the tubes are not located within a shell as is the case with the fire-tube type but are connected to drums or shells or headers. This makes it possible to design many parallel tube circuits to increase the capacity and also to raise the pressure, which is not possible with tubes confined within a shell.

Water-tube boilers can be classified by tube arrangement as *straight, bent,* and *coil tube.* The early water-tube boilers were inclined straight tubes rolled into *sinuous* or *box headers.* Drums above the header contained the water and steam. Tubes were usually 4 in in diameter, providing good natural circulation. A *vertical straight-tube* water-tube boiler is shown in Fig. 8.6*a.* This boiler is still in use and consists of an upper and lower cylindrical drum with bumped heads, with each side containing tubesheets into which the tubes are rolled. The furnace is located in a separate "Dutch

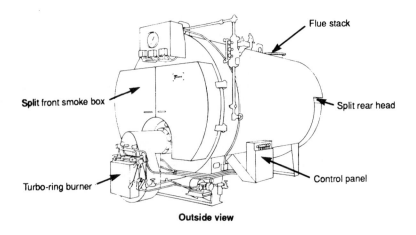

Flue stack

Split front smoke box

Split rear head

Turbo-ring burner

Control panel

Outside view

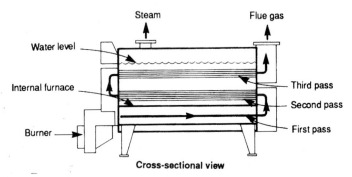

Steam   Flue gas

Water level

Internal furnace

Third pass

Second pass

Burner

First pass

Cross-sectional view

FIGURE 8.5   SM boiler has three passes for the products of combustion in order to gain more heating surface and more output in a compact, factory-assembled, fire-tested package that is quickly placed into service at a location.

oven" with the hot gases sweeping the tubes for heat transfer. The boiler is supported by four brackets welded to the mud drum and resting upon plates fastened in the walls of the brickwork.

Figure 8.6*b* shows an early *bent-tube boiler* of the *Sterling* type. It consists of three transverse steam and water drums above that are set parallel and connected to one mud drum by tubes so curved that their ends enter the drum at right angles to the surface. The curvature of the tubes permits expansion and contraction with temperature changes.

The advantages of bent-tube boilers are the large heating surface that can be developed and, thus, large capacities. They can also respond to sudden demands for high-quality steam, and the Sterling-type boiler had a large reserve water capacity in the three upper drums.

Waterwalls, Preheaters, Economizers, and Superheater

With increased demand for larger-capacity boilers, *furnace waterwalls* were introduced to extract radiant heat from the furnace. This was followed by preheating boiler water

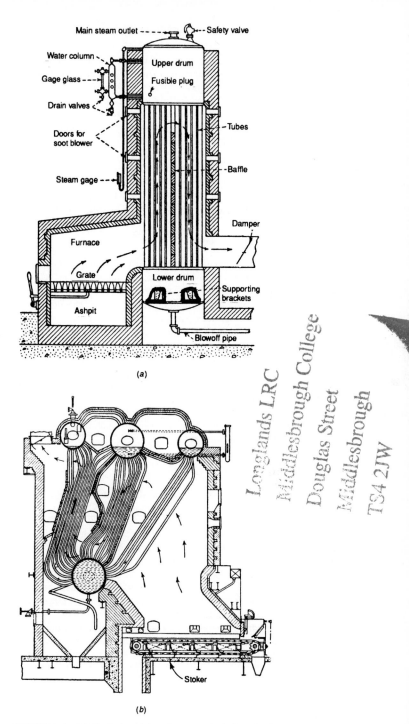

FIGURE 8.6 (*a*) Vertical straight-tube watertube boiler; (*b*) the Stirling boiler was an early bent-tube design.

in *economizers,* superheating steam in *superheaters,* and preheating combustion air in *air heaters,* all of which have improved efficiency and raised pressures and capacities to where one boiler can now supply one 1,300,000-kW turbo-generator of steam energy. Such large utility boilers have ratings of 9,775,000 lb/h at 3845 psi and 1000°F steam temperature. This is a *supercritical* pressure boiler, or one that operates above the critical pressure of 3206.2 psi and an operating temperature above the critical temperature of 705.6°F for water. The *critical pressure* is the pressure at which water and steam have the same density, while the *critical temperature* is the temperature above which water cannot exist as a liquid, no matter how high the pressure. There is no latent heat of vaporization above the critical pressure because water and steam are at the same density. Supercritical units are used to generate electricity in utility plants and are a development from the simple water-tube boilers of the last century.

Industrial Water-Tube Boilers

Most modern industrial water-tube boilers operate at moderate pressure and have the *A-, D-,* or *O*-type configuration of bent tubes as shown in Fig. 8.7. The furnace walls are water cooled as are the floor and roof of the furnace, usually with tangent tubes to give a tight furnace. These boilers come packaged from the factory with ignition and combustion flame safety devices, safety interlocked starting with limit switches for steam pressure, low water, fan failure, low fuel pressure (oil and gas) and for oil-fired units, low oil temperature. The D boiler shown in Fig. 8.7*b* is equipped with a radiant superheater.

The *coil-type* water-tube boiler shown in Fig. 8.8*a* is a controlled circulation unit with a separate pump that forces the water from a drum through the three coil sections shown in the illustrations. At the outlet from the coils, steam flows back to the drum to be used in process applications. This type of boiler is a quick steamer and may reach capacities up to 15,000 lb/h and up to 900 psi. The disadvantage is the small reserve capacity since the drums are small.

Cast-Iron Boilers

Cast-iron boilers are extensively used for steam and hot-water heating service. The maximum pressure allowed is 15 psi for steam and 30 psi for hot water heating per ASME Section IV Construction Code, except for special cast-iron boilers used for hot-water supply service and not for heating. Sections for the heating boilers may be stacked front to back vertically, or horizontally like pancakes. The sections are held together in two ways. One is with internal tapered push nipples that are inserted into holes of the vertical section, and by means of tie rods, the sections are pulled together by tightening nuts against washers (see Fig. 8.9). In the other way the external header type has the individual sections assembled into supply and return headers by means of threaded nipples, locknuts, and gaskets

Cast-iron boilers are prone to cracking failures from thermal shock, such as rapid introduction of cold water into a hot boiler. Internal surfaces are difficult to clean without dismantling the boiler. Frequent blowdown of the boiler will assist in minimizing sludge buildup.

Electric Boilers

Electric boilers are simple cylinders with electric heating elements within the shell (see Fig. 8.12*b*). There are no tubes in this type of boiler. In the *resistance* type, heat

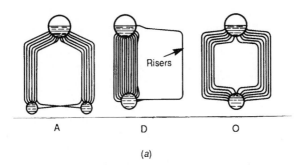

(a)

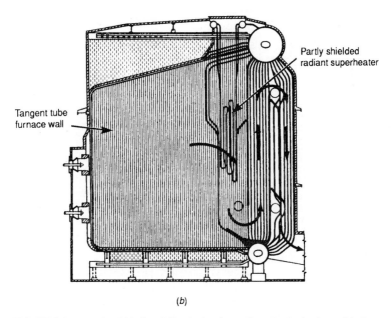

(b)

FIGURE 8.7   (a) Industrial boilers follow basic tube configuration in the shape of A, D, or O and are factory packaged for fast field erection; (b) D-type water-tube boiler equipped with superheater.

is generated by current flowing through resistance elements, which are wire encased in an insulated metal sheath and submerged in water. The electric heat generates moderate steam pressures at low capacities. The second type, called *electrode boilers,* has the current flowing through the water and not through wires to generate heat and steam. High-voltage units with 2300 to 15,000 V have been developed for capacities above 10,000 lb/h for use in low-cost electricity areas.

Capacity of the safety valves for these type boilers is based on kilowatt input, whether high or low pressure. The minimum capacity must equal 3.5 lb/h/kW input.

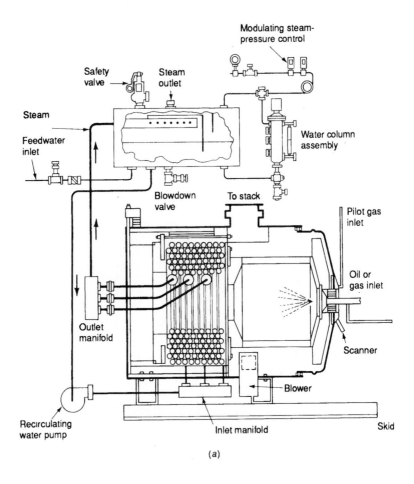

Modulating steam-
pressure control

Safety
valve

Steam
outlet

Steam

Feedwater
inlet

Water column
assembly

Blowdown
valve

To stack

Pilot gas
inlet

Oil or
gas inlet

Outlet
manifold

Scanner

Blower

Recirculating
water pump

Inlet manifold

Skid

(a)

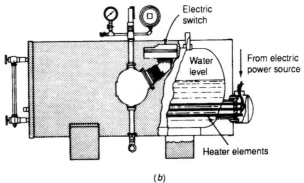

Electric
switch

Water
level

From electric
power source

Heater elements

(b)

FIGURE 8.8 (a) Coil-type water-tube boiler has three interconnected coils and forced circulation; (b) resistance-type electric boiler has no tubes and is a simple shell-type pressure vessel.

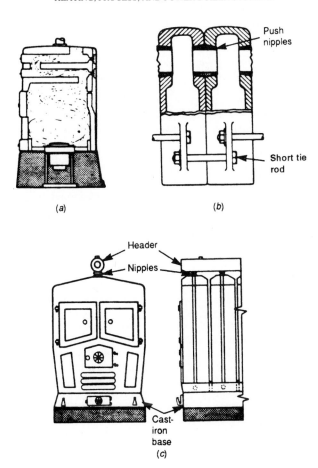

FIGURE 8.9 Three types of cast-iron boilers. (*a*) Round (Hy-test) for hot water supply; (*b*) vertical sectional, tie-rod assembled for heating service; (*c*) external header type for heating service.

HEATING SYSTEMS

Boilers used for heating systems are usually the low-pressure type, except where a central heating plant has high-pressure boilers, which distribute the steam or hot water to various buildings, where the pressure is reduced for the heat-transfer apparatus that is in the building. This could consist of radiators; unit heaters with circulation fans, commonly called hot-blast heating coils; finned-pipe radiation, usually located along baseboards; coils in air conditioning ductwork; and hot water circulation to baseboard heaters.

Types of heating systems are *steam* and *hot-water heating.* In *steam heating* the condensate is returned to the boiler. A pound of steam will produce about 950 Btu of heat as the steam gives off its heat of vaporization. A *gravity return system* has all the heating equipment above the boiler, and no pumps are used to return the condensate, while the *mechanical return system* uses traps and/or pumps to aid in the return of condensate. If a vacuum is pulled on the return lines, it is called a vacuum return system. Steel and cast-iron boilers are used for steam heating.

Hot-Water Heating

Hot-water heating systems pump hot water to the spaces to be heated, where radiators or finned pipes are used to heat the air in the space by convection. Hot-water systems require expansion tanks to permit the water to expand as it is heated; otherwise rapid pressure rise would occur due to the almost incompressibility of water. The relief valve would release the pressure. Steel and cast-iron boilers are used for this service.

Hot-water supply boilers are used to furnish hot water for washing and cleaning; therefore, there is no returning water or pumps. Generally, jurisdictions classify domestic hot-water units as boilers if input exceeds 200,000 Btu/h, a water temperature over 200°F, and nominal water containing capacity over 120 gal. Hy-test cast-iron boilers are used for hot-water supply, as are steel boilers.

Expansion Tanks

Figure 8.10 shows a hot water boiler with expansion tanks, sometimes called compression tanks. As water is heated in a tightly closed container, its pressure rises very rapidly. The expansion tank has an air cushion to permit the water to expand without raising the pressure too high. A common problem with hot-water heating systems is that eventually the air is absorbed into the water, and the system becomes water-logged. Usually this can be noted by a sudden pressure rise in the system. In extreme cases, the relief valve opens as the pressure rises to its setting. It is necessary to drain the system and reestablish the air cushion. Some hot-water systems use expansion tanks that are equipped with a "balloon" of air inside the tank. This prevents the air from being absorbed by the water, unless the balloon cracks inside the tank.

Because heating boilers are considered low pressure, automatic operation was an early goal of manufacturers; however, after several unforeseen accidents, the ASME has required pressure and temperature controls to have redundant features. For *steam heating boilers:*

1. Two pressure controls are required, one an operating control that cycles the boiler on and off and the other an upper limit control set no higher than 15 psi as backup for the operating control.

2. Low water fuel cutoffs must be installed on the boiler.

3. Each boiler must have a spring-loaded safety valve set no greater than the allow-able working pressure for the boiler, sized to boiler output.

4. Automatically fired boilers must have flame safeguard controls to shut off the fuel if improper combustion occurs in order to prevent fireside fuel explosions.

For *hot-water heating boilers:*

1. Two temperature controls are required, one an operating control that cycles the boiler on and off and the other an upper limit cutout set no higher than 250°F.

2. For hot-water heating boilers with heat inputs over 400,000 Btu/h, low water fuel cutoffs are required.

3. Each boiler must have a spring-loaded ASME-approved relief valve set at or below the maximum stamped allowable pressure of the boiler and with a capacity in Btu per hour greater than the stamped Btu output of the boiler.

4. Automatically fired boilers must have flame-safeguard controls to shut off the fuel if improper combustion occurs to prevent furnace explosions.

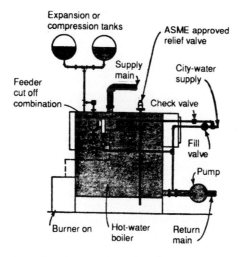

FIGURE 8.10 Expansion tanks on hot-water systems permit water to expand when it is heated without caus-ing excessive pressure rise.

PROCESS BOILERS

Process boilers are high-pressure boilers used in various process applications ranging from steam being used as a sterilizer in dairies to large refinery boilers in crude oil cracking operations. The smaller boilers are generally the fire-tube type, while high-pressure and large-capacity units are of the water-tube type. In some applications, the boiler is an integral part of the process, burning waste streams, such as bagasse in sugar mills and black liquor recovery boilers in the paper mills. In the latter application, the boiler recovers some of the liquid used in cooking wood chips to make pulp.

UTILITY BOILERS

Utility boilers are of the water-tube and reheaters type and feature high-pressure and capacities that include economizers, superheaters to raise the steam temperature so that it is dry before being admitted to a steam turbo-generator, and air preheaters. The reheaters make the steam dry again after it passes through several rows of turbine blades. Utility boilers operate in a closed loop of water to steam, with very little makeup required. In process work, on the other hand, makeup water can be as high as 100 percent.

STEAM CHEMISTRY

Steam chemistry control is more critical on utility boilers because of the high pressures and temperatures and because poor-quality steam may have an adverse effect on the connected steam turbine blades, such as loss of capacity from deposited scale on the blades, blade erosion, corrosion of internal turbine parts, and stress-corrosion cracking.

COMBINED-CYCLE GENERATION

Figure 8.11 shows a flow diagram for a *combined-cycle* generating plant that uses a gas turbine to drive a generator, with the exhaust gases from the gas turbine going to a waste heat boiler. The steam from the waste heat boiler is used to drive a steam turbo-generator. Supplemental firing is used in the waste heat boiler to start steam flow or to boost boiler output. These waste heat boilers may have three pressure ratings, with the high-pressure steam used for driving the steam turbo-generator and the lower pressure used for feedwater heating or for process use. *Independent power producers* (IPP) have developed the combined cycle as a supplementary way to add to the nation's generating capacity. The overall thermal efficiency is greater than with the traditional boiler-steam turbo-generator method of producing electricity. Because of the need to integrate the steps in the operation of a combined-cycle plant, computers are used to guide the system as it is brought on-line and when it is secured during planned or emergency shutdowns.

FUEL COMBUSTION

Combustion of fuel in a boiler involves the rapid chemical combination of oxygen with the combustible elements in fuel, carbon, hydrogen, and sulfur. The three elements needed for combustion to proceed are (1) mixing of oxygen with the fuel to obtain a combustible mixture, (2) ignition temperature to start the combustion process so it is self-sustaining, and (3) sufficient time for the chemical combination of oxygen and fuel elements to proceed to completion; otherwise unburned fuel particles will leave the furnace.

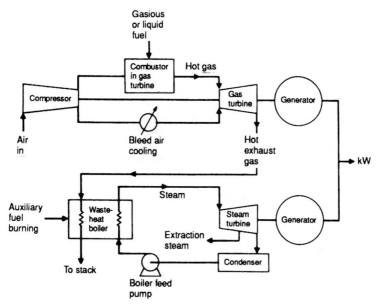

FIGURE 8.11   Combined-cycle plant uses gas turbine to drive a generator, and the waste heat from the exhaust of the turbine is directed to a waste-heat boiler for steam generation. The steam is used to drive a steam turbo-generator and for process use.

The following steps are followed in burning fuel in a boiler:

1. Preparing the fuel for burning, such as grinding solid fuel to small particle size for better mixing with oxygen, preheating the fuel and air, and atomizing or suspending these fuels as they are forced into the furnace

2. By heat, converting the fuels into a gas by thermal decomposition or by hydroxylation (Oxygen penetrates the hydrocarbon compound to form oxygenated compounds)

3. Producing the right air-fuel ratio in the correct proportions at the proper temperature for ignition to occur and combustion to be self-sustaining

4. Transferring the heat from combustion to the heat-absorbing surfaces of the boiler while maintaining ignition and volatilization temperatures in the furnace

Burning fuels in a boiler now requires the fuel gas to be as free as possible of toxic gases or products that can affect the environment, such as SO_2 and NO_x, and in some areas, even the temperature of the exiting gas is limited. Solid wastes generated in burning must be as inert as possible before being disposed of. Environmental factors have complicated burning fuels in a boiler and generally require additional treatment and equipment to do so, such as fluidized bed burning to reduce SO_2 emissions, flue gas scrubbers, and bag house dust collectors, to name a few. Technological changes have resulted in new controls and methods in burning fuels to limit environmental effects.

Solid-fuel burning methods are manual, stoker-firing, suspension or pulverized fuel burning, and atmospheric-combustion fluidized-bed systems. Coal is the largest solid fuel used in industrial and utility boilers. Coal classification is into *anthracite,* a hard, noncoking coal, and *bituminous,* which is a soft coal with a high percentage of volatile matter. Semibituminous coal is softer than anthracite, burns with little smoke, and can be easily broken into small pieces. Subbituminous, or *black lignite,* is a low grade of bituminous coal with heating value between 9000 and 11,000 Btu/lb. Lignite is between peat and subbituminous coals; it has a wood structure, a claylike appearance, and a heating value from 7000 to 11,000 Btu/lb.

Coal analysis by *ultimate and proximate* methods is one of the functions in operating coal-fired plants because of the variance in coal properties. Proximate analysis determines the percentage of moisture, volatile matter, fixed carbon, and ash in a sample of coal. The ultimate analysis is more extensive and is conducted in a laboratory to determine the percentages of carbon, oxygen, hydrogen, sulfur, nitrogen, and inert matter in the coal. The heating value of the coal can then be calculated under the following criteria; as received, air-dried, moisture-free, moisture and ash-free, and moisture and mineral-free. The heating value is determined by burning a coal sample in a calorimeter filled with oxygen under pressure.

Stokers are classified into (1) overfeed, in which the fuel is carried in the furnace above the stoker, with the spreader and chain-grate stoker being of this classification, and (2) underfeed, where the fuel is carried by the stoker underneath. Further classification is by the mechanism used to move the coal, such as single retort, multiple retort, screw feed, or ram feed. Figure 8.12 shows a sprinkler and chain-grate stoker.

Pulverizers (Fig. 8.13a and b) are used to grind the coal into a fine dust for better mixing of the fuel with oxygen in the furnace so that burning is equivalent to gas burning (i.e., faster and more complete than with grate burning). Preparing fuel properly and then feeding it into the furnace with the proper amount of combustion air (Fig. 8.14) assures more complete combustion, thus preventing excess unburned carbon and carbon monoxide from escaping into the stack.

Waste fuels include municipal waste, wood chips, sawdust, bagasse, all types of hulls such as from shells of nuts, corn cob, and so on. Liquid wastes include tar and pitch and the black liquor from the cooking of wood chips in the paper industry. All require special consideration in fuel preparation, fuel burning, boiler arrangement, and waste disposal, whether gaseous or ash, to comply with clean air and water legislation.

Air for combustion is divided into three parts: *Primary air* is used to transport the pulverized coal to the burner nozzle and then blow it into the furnace. *Secondary air,* the major portion of the total air, is admitted into the furnace through a large number of air ports to supply the air needed to ensure complete and uniform combustion of the fuel in the furnace. The *tertiary air,* in about the same percentage of the total air as the primary, is admitted under damper control through the burner housing around the fuel nozzles to the furnace.

Manufacturers of burner systems are concentrating on developing *low-excess* air burners in order to reduce NO_x emissions to the atmosphere.

Fire-protection for coal-fired plants must consider the coal-handling area prior to coal burning in a boiler's furnace. This includes the fire hazard of *spontaneous combustion* in coal storage areas. The strategy is to compact the stored coal in 1-ft layers. This keeps the oxygen out and minimizes the chance that exposed areas will spontaneously combust. Enclosed-area coal storage requires a program of inspection to detect developing hot spots, usually by checking for the odor of burning coal. Emergency capability to transport the coal out of enclosed storage areas is needed so that the fire in a pile cannot be sustained.

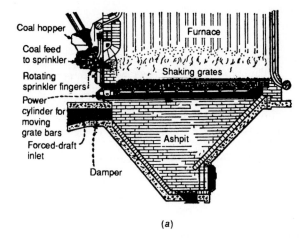

(a)

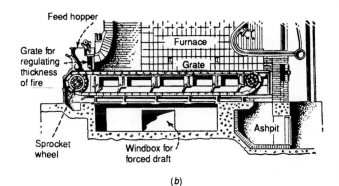

(b)

FIGURE 8.12   Sprinkler and chain-grate stoker for burning coal. (a) sprinkler-stoker uses shaking grates to mix fuel and air; (b) in the chain-grate stoker air comes in from below the grate.

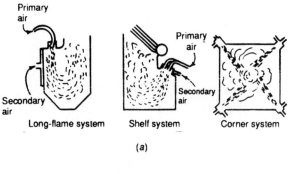

Primary air

Primary air

Secondary air

Secondary air

Long-flame system   Shelf system   Corner system

(a)

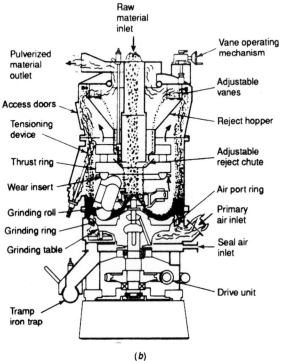

Raw material inlet

Vane operating mechanism

Pulverized material outlet

Adjustable vanes

Access doors

Reject hopper

Tensioning device

Thrust ring

Adjustable reject chute

Wear insert

Air port ring

Grinding roll

Primary air inlet

Grinding ring

Grinding table

Seal air inlet

Drive unit

Tramp iron trap

(b)

FIGURE 8.13 (a) Three methods of suspension firing of pulverized coal: long-flame system, shelf system, and corner system; (b) bowl-mill pulverizer has grinding stationary rolls that crush the coal against a rotating grinding ring.

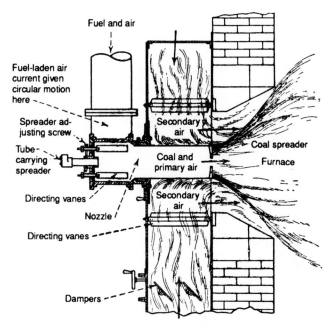

Fuel and air

Fuel-laden air current given circular motion here

Spreader adjusting screw

Tube-carrying spreader

Directing vanes

Nozzle

Directing vanes

Dampers

Secondary air

Coal and primary air

Secondary air

Coal spreader

Furnace

FIGURE 8.14 In pulverized-coal burning, primary, secondary, and tertiary (not shown) air is used to obtain complete combustion of the pulverized coal.

Enclosed-type conveyors are now required to limit the spread of dust into the environment. Conveyors should have sprinklers and fire stops to prevent the spread of fire.

Enclosures with large coal storage require some form of instant *pressure relief opening* to minimize structural damage from a coal dust explosion.

Magnetic separators are required at dump and reclaim hoppers and before any coal crushers to remove any scrap metal in the coal being processed that might be a source for spark ignition of the coal or its dust.

Some other fire insurance company recommendations and NFPA standards, which facility service operators in coal burning plants should be familiar with, are *ventilation systems* in indoor coal storage areas to draw off methane gas and *smoke and heat vents* at the roof of enclosed conveyor galleries and associated structures.

Fluidized-bed coal burning has received attention as a means of emission control. Particles of coal are burned in a bed of limestone through which air is passed (see Fig. 8.15). The velocity of air through nozzles is maintained so that the limestone becomes suspended over the burning fuel, with the bed resembling a boiling fluid. The temperature of the fluidized bed is kept low enough to minimize the formation of nitrogen oxides. The limestone combines with the sulfur dioxide, thus eliminating the need for a flue-gas desulphurization system.

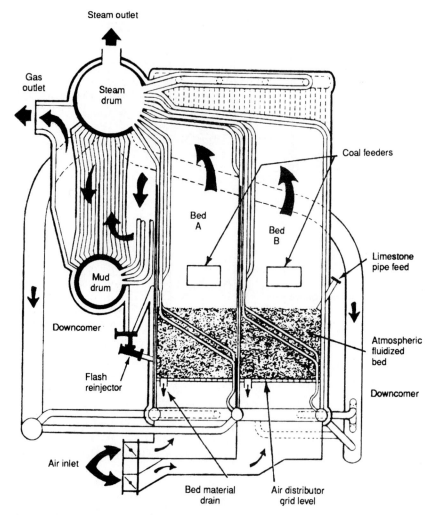

FIGURE 8.15 Atmospheric fluidized-bed burning, coal-fired boiler is used to control SO_2 and NO_x stack emissions.

BURNING FUEL OILS

Fuel Oil Properties

Fuel oils are sold in six standardized grades, and the properties of specific gravity, heating value, viscosity, flash and fire point, sulfur content, ash content, and congealing point are all important in determining the preheating that may be needed in pumping the oil from storage to burner, the type of burner, and pressure of the oil that may be needed for good atomization and burning. API's standard formula for degrees API is based on the density of fuel oil at 60°F and refers to water at 60°F:

$$\text{Degrees API} = \frac{141.5}{\text{oil specific gravity at 60°F}} - 131.5$$

Table 8.1 shows some properties of fuel oil per degree API.

$$\text{Actual specific gravity} = \frac{141.5}{131.5 + \text{API deg}}$$

Fuel oils are sold in six standardized grades, under the numbers or grades of 1, 2, 3, 4, 5, and 6. Grades 1, 2, and 3 are light, medium, and heavy domestic fuel oils. These usually do not require heating prior to burning in a furnace. Grades 4, 5, and 6 correspond to federal specifications for Bunkers A, B, and C, respectively. These oils are heavy and viscous and thus require heating before being sprayed into a furnace.

TABLE 8.1   Some Properties of Fuel Oil per API Degrees

Degrees API	SP gr	Lb/gal	Btu/lb	Btu/gal	Lb/42-gal bbl	Lb/ft³
3	1.0520	8.76	18,190	159,340	368.00	65.54
5	1.0366	8.63	18,290	157,840	362.62	64.59
7	1.0217	8.50	18,390	156,320	357.37	63.65
9	1.0071	8.39	18,490	155,130	352.46	62.78
11	0.9930	8.27	18,590	153,740	347.71	61.93
13	0.9792	8.16	18,690	152,510	342.88	61.07
15	0.9659	8.05	18,790	151,260	338.22	60.24
17	0.9529	7.94	18,890	149,980	333.64	59.42
19	0.9402	7.83	18,980	148,610	329.23	58.64
21	0.9279	7.73	19,060	147,330	324.91	57.87
23	0.9159	7.63	19,150	146,110	320.71	57.12
25	0.9042	7.53	19,230	144,800	316.59	56.39
27	0.8927	7.44	19,310	143,670	312.60	55.68
29	0.8816	7.35	19,380	142,440	308.70	54.98
31	0.8708	7.26	19,450	141,210	304.92	54.31
33	0.8602	7.17	19,520	139,960	301.18	53.64
35	0.8498	7.08	19,590	138,690	297.57	53.00
37	0.8398	7.00	19,650	137.550	294.04	52.37
39	0.8299	6.92	19,720	136,400	290.64	51.76
41	0.8203	6.83	19,780	135,090	287.23	51.16

Viscosity is the relative ease or difficulty with which an oil flows. It is measured by the time in seconds a standard amount of oil takes to flow through a standard orifice in a device called a *viscosimeter.* The usual standard in this country is the Saybolt Universal, or the Saybolt Furol, for oils of high viscosity. Since viscosity changes with temperature, tests must be made at a standard temperature, usually 100°F for Saybolt Universal and 122°F for Saybolt Furol. Viscosity indicates how oil behaves when pumped and, more particularly, shows when preheating is required and what temperature must be held.

Flash point represents the temperature at which an oil gives off enough vapor to make an inflammable mixture with air. The results of a flash-point test depend on the apparatus; so this is specified as well as temperature. Flash point measures an oil's volatility and indicates the maximum temperature for safe handling.

Pour point represents the lowest temperature at which an oil flows, under standard conditions. Including pour point as a specification ensures that an oil will not be difficult to handle at expected low temperatures.

All test properties above are covered by ASTM standards, which should be consulted for details of apparatus and methods. A copy of *ASTM Standards of Petroleum Products and Lubricants* can be obtained from the American Society for Testing and Materials at 1916 Race St., Philadelphia, PA 19103.

The *sulfur content* of fuel oil is determined in percent by weight. The allowable limit is 4 percent. If you buy oil with excessive amounts of sulfur, sulfurous or sulfuric acids are apt to form in the boiler breaching and uptakes that corrodes the economizer, air heater, and stack. The EPA is strict about the amount of sulfur released to atmosphere from a stack.

Moisture or water in fuel oil, if allowed to enter the atomizer (burner), will cause sputtering. This may extinguish the flame, reduce flame temperature, or lengthen the flame. As for sediment, it is troublesome because it clogs strainers and burner sprayer plates. If there's a lot of it, frequent strainer and atomizer clearing will keep you busy, or you will lose steam pressure.

If water or sediment or both are present to any great extent, they can be separated from oil by heating. This is done while oil is in settling tanks or by passing it through a centrifuge.

OIL BURNERS

In addition to proportioning fuel and air and mixing them, oil burners must prepare the fuel for combustion. Two ways (with many variations) are (1) oil may be vaporized or gasified by heating within the burner or (2) oil may be atomized by the burner so vaporization can occur in the combustion space.

First step in atomizing oil is heating it enough to get the desired viscosity. This temperature varies slightly for each grade of oil. Lighter oil, numbered 1, 2, and 3, usually needs no preheating. The commercial grades 5 and 6 give best combustion results when heated enough to introduce the oil to the atomizer tip between 150 and 200 Saybolt Seconds Universal (SSU) viscosity. Then, after oil is heated to the right temperature, it must be pumped into the burner at the right pressure.

Steam-atomizing burners possess the ability to burn almost any fuel oil of any viscosity at almost any temperature. Air is less extensively used as an atomizing medium because its operating cost is apt to be high. These burners can be divided into two types:

1. Internal-mixing or premixing oil and steam (see Fig. 8.16*a* and *b*) or air mix inside the body or tip of the burner (8.16*c* and *d*) before being sprayed into the furnace.

2. External mixing, where oil emerging from the burner is caught by a jet of steam or air (see Fig. 8.16*d*).

Steam consumption for atomizing runs from 1 to 5 percent of the steam produced, with the average around 2 percent. The pressure required varies from about 75 to 150 psi.

In the burner in Fig. 8.16*d*, oil reaches the tip through a central passage, flow being regulated by the screw spindle. Oil whirls out against a sprayer plate to break up at right angles to the stream of steam, or air, coming out behind it. The atomizing stream surrounds the oil chamber and receives a whirling motion from vanes in its path. When air is used for atomizing, it should be at 10 psi for lighter oils and 20 psi for heavier. Combustion air enters through a register, shown in Fig. 8.16*e*. Vanes or shutters are adjustable to give control of excess air.

A burner in which the oil is atomized by pressure alone is usually called a *mechanical* burner to distinguish it from the steam-atomizing burner. The mechanical burner breaks up the oil into very fine particles by forcing it under high pressure through a number of very small passages in the tip. Oil pressures run from 30 to 300 psi, depending on the burner, and the oil must be heated to temperatures ranging from 100 to 300°F so that it will flow freely. Air for combustion is usually admitted through the burner casing, and the burners are blown out with compressed air when not in use to prevent the oil from hardening and clogging the passages.

A mechanical oil-burner tip is shown in section in Fig. 8.17*a*. In this burner, oil is forced through tangential slots in the nozzle against a sprayer plate; this causes the oil spray to issue from the burner in the form of a hollow cone.

Figure 8.17*b* shows the complete burner installation. Air for combustion enters through adjustable louvers, which give it a rotary motion. This motion is further accentuated by the diffuser or impeller plate at the tip of the burner. The diffuser plate also splits up the airstream so that air and oil spray mix thoroughly before burning. Burners of the type shown in Fig. 8.17*b* are also made to burn either oil or gas. There are even triple-service burners that burn oil, gas, or pulverized fuel.

A combination oil and gas burner is shown in Fig. 8.17*c*. The oil-burning part is of the mechanical type, somewhat like that of Fig. 8.17*b*. The gas burner is a ring with many small jets on the inside. The air supply is controlled by opening or closing the movable blade registers.

The horizontal *rotary cup* atomizes fuel oil by literally tearing it into tiny droplets. A conical or cylindrical cup rotates at high speed (usually about 3500 r/min) if motor-driven. Oil moving along this cup reaches the rim, where centrifugal force flings it into an air stream (see Fig. 8.18). This system of atomization requires no oil pressure beyond that needed to bring oil to the cup. But high oil preheat temperatures must be avoided since gasification may develop. The rotary cup can satisfactorily atomize oils of high viscosity (300 s SSU) and has a wide range of about 16:1 turndown ratio.

Oil Burner Maintenance

Some aspects of oil-burner maintenance for operator's attention are: Make sure that the burner gets uniformly free-flowing oil, clear of sediment that clogs burner nozzles. This means avoiding sludge buildup in storage tanks and keeping strainers in good condition. The preheat temperature must be right for the fuel and burner type and must be uniform. Watch for wear caused by ash abrasion in the fuel and for carbon buildup. In rotary-cup burners, worn rims cause poor atomization. If cups are not properly pro-

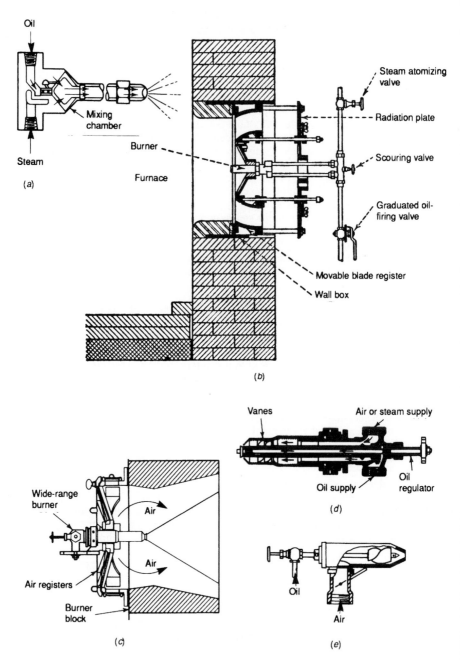

FIGURE 8.16 *Steam* and *air-type atomizing* oil burners. (*a*) and (*b*) Steam-atomizing burner with oil and steam being mixed inside the burner; (*c*) air-atomizing burner employs low pressure air; (*d*) steam or air-atomizing burner with mixing done externally; (*e*) air registers control excess air.

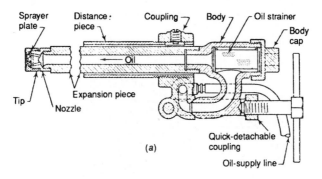

(a)

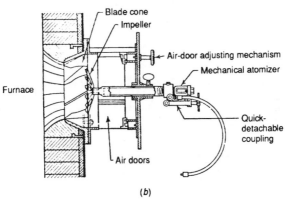

(b)

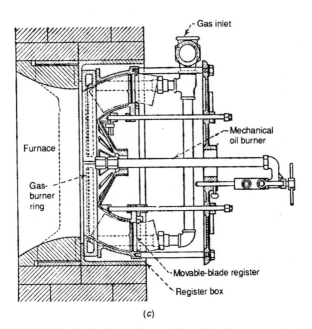

(c)

FIGURE 8.17 *Mechanical oil burners* force oil under pressure through tiny nozzles to obtain a fine oil spray into the furnace. (*a*) Burner details; (*b*) sketch of assembly; (*c*) combination oil and gas burner of the mechanical-atomization type.

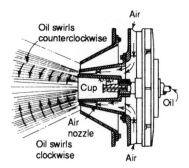

FIGURE 8.18 Rotary-cup burner atomizes oil into fine spray by centrifugal force.

tected after being turned off, carbon forms on the rim. When the burner is shut down, always take out the cup and insert a flame shield. *Remember:* Worn or carbonized mechanical-atomizing nozzles give trouble. Always replace worn nozzles and keep them clean.

GASEOUS FUELS AND BURNING GAS

Natural Gas and Coal Gas

Of many gaseous fuels, only natural gas has any commercial importance in steam generation because manufactured gases cost too much. By-product gases usually have low heating values and are produced in relatively minor quantities; so they are ordinarily used at the production point and not distributed. Natural gas is colorless and odorless. The heating value averages about 1000 Btu/ft^3 (20,000 Btu/lb) but may run much higher. Natural gas is usually sold by the cubic foot but may be sold by the therm (100,000 Btu).

Coal gas and coke-oven gas (manufactured gases) are produced by carbonizing high-volatile bituminous coal in retorts that exclude air and are heated externally by producer gas. Usually a number of by-products result. Cleaned of impurities, these gases are roughly one-half hydrogen and one-third methane, plus small amounts of carbon monoxide, carbon dioxide, nitrogen, oxygen, and illuminants (C_2H_4 and C_6H_6). The heating value runs around 550 Btu/ft^3.

GAS BURNERS

Figure 8.19*a* shows a single-port *atmospheric burner.* A needle valve controls the gas flow through a spud; air is drawn in around the shutter at the end. This gas-air mixture passes through the tube and burns at the end. Single-port burners are often grouped, several banks high and wide, to serve larger furnaces. Each burner pulls in primary air for combustion by the action of a stream of low-pressure gas expanding through an orifice.

With a fixed burner-port size and shape, the nature of burning depends a lot on the amount of primary air or premix. With the premix low, the flame is long and pale blue. It may have a yellow tip, which indicates some cracking and the presence of free carbon. When the primary air is increased, the burning becomes more rapid; the flame shortens and a greenish inner cone appears. When the speed of burning, or flame propagation, exceeds that of the gas issuing from the port, the flame flashes back into the mixing tube. About 30 to 70 percent premix is good; some burners use 100 percent primary air. Secondary air, which is often drawn in around the burner, can be controlled by varying the draft or by adjusting the opening area by shutters.

Figure 8.19*b* shows a *high-pressure burner* that mixes air and gas in a short distance. Most high-pressure burners use gas at 20 to 30 psig and air at atmospheric pressure, but some use compressed air.

Refractory Burner

Figure 8.19*c* shows a refractory burner that premixes the gas and air needed for combustion in a chamber outside the furnace. Multiple gas jets discharge into the airstream, causing a violent agitation in the short mixing tube, or tunnel, of the refractory.

In figure 8.19*d,* turbulence vanes impart a swirling motion to the air entering the *tunnel burner.* Each small jet of gas issuing from the multiple-jet orifice entrains with the air and impinges it outward against the tunnel walls; the result is turbulent, thorough mixing.

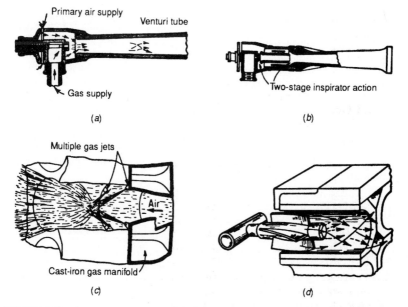

FIGURE 8.19   Gas burner types. (*a*) Atmospheric gas burner pulls in primary air; (*b*) two-stage burner is for high-pressure air; (*c*) refractory burner premixes gas and air in a mixing chamber; (*d*) turbulent vanes impart a swirling motion.

Low Excess-Air and Low NO_x Burners

There is a decided trend to trim air-fuel ratios on burners as a means of saving fuel and, even more important due to government regulations, as a means to control air pollution. Boilers produce NO_x from burning or combining nitrogen in the fuel with oxygen, called *fuel-bound NO_x*, and by the reaction of nitrogen and oxygen in the air at high temperature to form thermal NO_x. This has resulted in several burner strategies to trim excess air and to lower the combustion temperature so as to limit thermal NO_x formation. Trimming excess air as low as 3 percent of the theoretical air required for combustion reduces NO_x formation by limiting the amount of air going into the boiler. There is an element of risk involved, and this low excess-air burning requires good stack gas monitoring for CO, CO_2, and oxygen as a redundant check that combustion with low excess-air is proceeding safely. This requires a well-engineered installation, installed by organizations specializing in this type of work. If O_2 is trimmed to too low a level, NO_x will be reduced, but very high levels of CO and unburned hydrocarbons can result, which will produce unstable and resonating boiler flames. The biggest risk, however, is unburnt fuels in the furnace suddenly igniting and causing a furnace explosion. Therefore, operators should seek technical assistance from burner manufacturers or combustion control firms and not try low excess-air manipulation without having redundant features to warn of too rich a mixture existing that can endanger boiler operation.

FIRESIDE BOILER CONSIDERATIONS

There are many conditions on the firing side of boilers that affect the operation of the boilers by facility operators. *Cold-end corrosion* is the corrosion that may occur on the tail end of the boiler system, namely, economizer, air heaters, and induced-draft fans. The primary cause for this fireside or flue-gas-side corrosion is the sulfuric acid vapors that may be in the flue gas when sulfur dioxide combines with water. This acid condenses in the cooler part of the boiler passages. The acid dew point is the temperature at which sulfuric acid vapor is in equilibrium with liquid sulfuric acid so that the vapor condenses at that temperature.

Prevention of cold-end corrosion can start with design, such as the use of corrosion-resistant boiler passage material and metal-spraying parts with similar material. The usual method is to operate the boiler in a manner so that the back-end temperatures are maintained above the acid dew point in order to prevent acid vapors from condensing in the flue passages. Fuels free of sulfur will, of course, also prevent this type of corrosion.

Particulate Emission Control

Methods used to control particulate emission may consist of *mechanical separators,* which depend on centrifugal force to separate out the particles from the gas stream. *Baghouses* use filters or glass-fabric material to block the particulate matter but not block the flow of gases through the fabric. *Wet scrubbers* are used to wash out the particles from a gas stream. *Electrostatic precipitators* use the principle that an electric charge between two plates will attract particles to the plate out of a passing gas stream. When the plates are rapped, the deposited particles are dropped into a collection bin for disposal.

The term *flue-gas desulfurization* (FGD) is applied to the various methods used in removing sulfur from flue gas. The removal of SO_2 from flue gas is receiving great attention because of the so-called acid rain threat to the environment, which is being blamed on coal-firing plants. Among the FGD systems under investigation are:

1. Scrubbing the flue gas through limestone beds
2. Fluidized-bed burning by limestone injection
3. Scrubbing the flue gas through magnesia beds
4. Scrubbing with sodium and regenerating the sodium
5. Scrubbing with sodium and throwing the resultant products away in approved landfill areas
6. Catalytic oxidation of the SO_2
7. Carbon absorption
8. Manganese oxide absorption

Operators can control emission by following the maintenance, inspection, and operating instructions of the emission-control manufacturers on their installed systems so that jurisdictional requirements are complied with on such items as permissible threshold limits on SO_2, NO_x, noise levels, solid disposal requirements, and similar regulations that involve the environment. This requires paying attention to on-line monitoring systems as well as extracting periodic samples for manual testing on parts per million or similar threshold readings.

COMBUSTION AND BURNER CONTROLS

Combustion controls in modern boiler systems incorporate operating and safety controls as follows:

1. Maintain proper fuel and air feed for combustion as load on the boiler system varies.
2. Incorporate the necessary interlock system on pressure, water feed, draft, and similar associated boiler flows in order to maintain load and safe operating conditions.
3. Provide alarm systems to notify the operator when unsafe operating conditions are being approached or have been reached.
4. Provide safety trips so that if established safe operating conditions are exceeded, the trips will secure the boiler system or safely shut it down.

Manual and Automatic Operation

Some large-city ordinances define manual and automatic boiler controls. They relate to the degree of automation and may involve operator skill requirements specified by the jurisdiction. The ordinances usually refer to the terms *manual* and *automatic*. There are four classifications, two manual and two automatic:

1. *Manual.* A boiler which is purged, started, modulated, and stopped manually.

2. *Supervised manual.* A boiler which is purged and started manually, modulated automatically, and stopped manually.

3. *Automatic nonrecycling.* A boiler which, when actuated manually by a push button, is purged, started, modulated, and stopped automatically but does not recycle automatically.

4. *Automatic recycling.* A boiler which is purged, started, modulated, and stopped automatically and which recycles on a preset pressure or temperature range automatically.

FURNACE EXPLOSIONS

A furnace explosion is the ignition and almost instantaneous combustion of explosive or highly inflammable gas, vapor, or dust accumulated in a boiler setting. Often it is of greater expansive force than the boiler setting can withstand. In minor explosions, called puffs, flarebacks, or blowbacks, flames may blow suddenly for a distance of many feet from all firing and observation doors. Thus anyone in the flame path may be seriously or fatally burned. Such minor explosions indicate dangerous conditions, even if no real damage is done. Heavier explosions may shatter gas baffles, bulge setting walls, loosen the refractory, blow brick tops off boiler settings through roofs, blow the sidewalls out from under the boiler, break connecting piping, and even demolish boiler housings.

Three conditions are usually necessary to cause a furnace explosion: (1) accumulation of unburned fuel, (2) air and fuel in an explosive mixture, and (3) a source of ignition, such as hot furnace walls, improper ignition timing, faulty torch, and dangerous light-off procedures on manually started boiler combustion systems.

The common types of furnace explosions are gas and coal-dust explosions; each results from the presence of unburned fuel and its delayed ignition. Furnace explosions are of the primary and secondary types. For example, a quantity of unburned fuel in the primary combustion chamber, or furnace, may distill off a large volume of gas during a period of interrupted ignition. If this gas does not ignite promptly, it may fill the furnace and circulate back to a secondary pass.

Continued ignition delay may cause the volatile content of the unburned fuel to be exhausted and the major part of the gas to pocket in the second or third pass of the setting. When the diluted gas mixture remaining in the furnace ignites (because the burner is relighted or for some other reason), a furnace explosion of minor intensity occurs in the primary chamber. The blast of flame may then follow through to ignite the gas in the succeeding passes, resulting in a secondary, and more violent, explosion.

Causes of Explosions

Some of the contributing factors that cause furnace explosions are:

1. Flame failure due to liquids or inert gases entering the boiler fuel system
2. Insufficient purge before the first burner is lighted
3. Human error
4. Faulty automatic fuel regulating controls
5. Fuel shutoff valve leakage
6. Unbalanced fuel/air ratio
7. Faulty fuel supply systems

8. Loss of furnace draft

9. Faulty pilot igniters

It is important to *purge a furnace* before light-off of fuel.

For safe operation, the furnace should be purged before light-off in order to reduce the flammability limits well below those at ambient temperature of the fuel being burned. The purging of a furnace will make the air/fuel ratio lean so that ignition may not occur.

Pilot Proving

Burners using an auxiliary pilot to ignite a main burner always present the hazard of the pilot going out accidentally before the main burner goes on. By then the entire setting may be filled with fuel. Fire departments and insurance company requirements usually stipulate that pilots be equipped with automatic shutoffs to stop the fuel flow if the flame on the pilot goes out. Pilot-proving flame detectors are now becoming mandatory on larger boilers and also on smaller gas-fired units.

Three types of gas pilot ignition for industrial boilers are:

1. *Interrupted gas-electric ignition.* This type uses a pilot for seconds only. The burner fires after ignition without the pilot. It is used mostly for firing residual fuel oil but also for natural gas, depending upon gas-line valving and vent line to prevent leakage of gas into the furnace during off-firing cycle.

2. *Intermittent pilot.* This type uses a pilot to ignite fuel and continues to burn during the firing cycle. The burner and pilot go off simultaneously.

3. *Continuous pilot.* Once ignited, this type of pilot burns continuously whether the burner is firing or is off. Thus protection is provided against unburned gases entering the furnace. This pilot will ignite and burn off any leakage of gas that may enter the furnace.

Mechanical and electrical interlocks play a vital role in operating equipment in proper sequence. They immediately shut down a unit if any of the components in a combustion system is not operating within set limits.

Operator Procedures to Prevent Explosions

The operating precautions needed to prevent furnace explosions include the following:

1. Check the operation of the boiler periodically.

2. If a burner goes out accidentally, shut off the igniter and fuel supply and thoroughly scavenge the furnaces and gas passes before again attempting ignition. Always, always determine and remedy the cause of the stoppage.

3. Keep burners and all allied equipment clean.

4. On boilers using both forced- and induced-draft fans, test the interlock periodically.

5. Do not attempt to secure excessively high CO_2 by using too rich a fuel/air ratio or by an inadequate secondary air supply.

6. Keep the temperatures and pressures for preheated air, drying air, fuel oil, etc., at the right levels.

7. Never allow an unstable flame condition to continue uncorrected.

BURNER FLAME-SAFEGUARD SYSTEMS

Flame safeguard systems were developed to detect unsafe flame conditions in a furnace. It is an arrangement of flame-detection systems, interlocks, and relays which will sense the presence of a proper flame in a furnace and cause fuel to be shut off to the furnace if a hazardous (improper flame or combustion) condition develops. Modern combustion controls are closely interlocked with flame-safeguard systems and also pressure-limit switches, low-water fuel cutoffs, and other safety controls that will stop the energy input to a boiler when a dangerous condition develops.

Flame Characteristics

Some of the characteristics of a fire that can be used to monitor a flame are based on the physical characteristics of a flame:

1. A flame produces an *ionized zone,* meaning that it can conduct a current through it. The amperage is low, as the resistance through the flame is high, being on the order of 250,000 to 150,000 Ω. The currents are in microamperes when a voltage is impressed across a flame, but electronic devices can amplify this current and make it a control signal for proving a flame. Conductivity flame rod detectors use the principle of a conducting flame for flame-detection monitoring.
2. A flame can *rectify an alternating current.* This is done by making one electrode across a flame larger than the other, thus making electrons flow through a flame much more readily in one direction than in the opposite direction. When an alternating voltage is applied to these electrodes, the resulting current is, in effect, an intermittent direct current.
3. *Radiation of light* is a known phenomenon of any fire. A flame radiates energy in the form of waves which produce heat and light. Three types of radiation from a flame are:
   a. *Visible light* that can be seen by the human eye. The wavelengths of visible radiation extend only from 0.4 to 0.8 μm. Visible radiation for flame detection has the limitation of low intensity. This intensity varies with different fuels, burner types, and methods of mixing. Refractory-material radiation approaches visible radiation, making flame detection difficult. Visible-radiation flame detection is found to be more suitable for oil flames than for gas flames. *Visible-radiation flame detection* is by means of the oxide of the metallic element cadmium. When this metallic element is exposed to visible light, it emits electrons with the strength of the visible light. Thus, if a cadmium phototube is designed in an appropriate electronic circuit, electricity will flow through the circuit when the cadmium is exposed to sufficient light. This electricity can be used to trigger relay circuits for flame detection.
   b. *Infrared radiation* covers most of the useful band of wavelengths and also covers most of the radiation strength. Infrared detectors are suitable for gas and oil flames. Because hot refractories also radiate infrared, scanners must avoid hot refractories. Lead sulfide cells are used in photocells to sense infrared radiation. Unlike cadmium phototubes, they do not emit electrons but have the property of having their electrical resistance reduced while exposed to infrared radiation. The greater the strength of the radiation, the lower the resistance of the lead sulfide. This principle is used for flame-detection purposes when such cells are connected to a designed electronic circuit.

c. *Ultraviolet radiation* is the most widely used flame detector based on the phenomenon of sensing the strength of ultraviolet radiation in a flame. It is insensitive to visual and infrared radiation and is not affected by hot refractories since these usually do not give off appreciable ultraviolet radiation. When radiation from a flame passes through the typical quartz viewing window of one of these detectors into the flame-sensing tube, the tube becomes electrically conductive. The strength of the detector signal, or current passed through the sensing tube, depends on the kind of fuel, size and temperature of the flame, and distance between the flame detector and the flame. Figure 8.20*a* shows some typical wavelengths in a flame and the response percentage of total wavelengths of typical flame pickup devices.

Types of Detectors

The types of flame detectors used on boilers depends on the fuel used and size and method of firing, the type of burner, and the size and arrangement of the boiler. Flame detectors vary from those used on small domestic boilers to those on large boilers. Types of detectors and the task each does are:

1. Stack switch, heat-sensing (small boilers)
2. Rectifying flame rod, heat-sensing (see Fig. 8.20*b*)
3. Rectifying phototube, visible-light-sensing
4. Lead sulfide photocell, infrared-light-sensing
5. Cadmium sulfide photocell, visible-light-sensing
6. Ultraviolet flame-detector tube, ultraviolet-light-sensing
7. Visible-light detectors using fiber-optic light guides

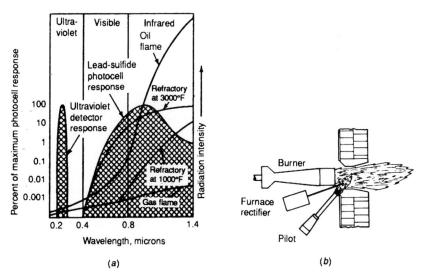

FIGURE 8.20 (*a*) Various flame detection systems showing percentage wavelength light response versus wavelength; (*b*) flame-rod-type detector is used to prove pilot and main flame exist.

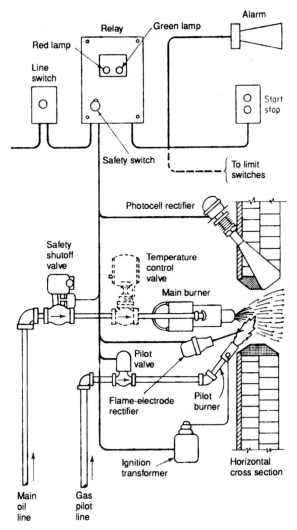

FIGURE 8.21 Oil-fired boilers generally use a combustion safe-guard system that is based on the rectified impedance principle.

For example, the rectified impedance system (see Fig. 8.20b) operates on the principle that either a flame or a photocell sighted at a flame is capable of conducting, as well as rectifying, an alternating current. This alternating current is applied to either a flame electrode inserted in the flame or a photocell sighted at the flame. The resultant rectified current, which can be produced only when a flame is present, is in turn detected by the relay.

The actual flame-detecting units consist of flame-electrode and photocell rectifier assemblies. The flame-electrode type is generally used for nonluminous flames, such as gas flames, whereas the photocell type is usually used for luminous flames, such as oil flames. The system shown in Fig. 8.21 provides gas-pilot flame supervision on start, and oil main-flame supervision on run, for an oil-fired burner. The flame-electrode rectifier is used to supervise the gas-pilot flame (constant pilot) during start-up only, while the photocell rectifier is used to supervise the oil main flame. If the oil-fired burner is ignited by an intermittent pilot, the flame-electrode rectifier is not needed.

ASME, NFPA, AND JURISDICTIONAL REQUIREMENTS

The ASME has adopted requirements for combustion safeguards for automatically fired boilers per their CSD-1 standard first issued in 1988. NFPA Standard 85C titled, "Prevention of Furnace Explosions in Multiple Boiler Burner Boiler-Furnaces," is an excellent reference for plants with multiple burner furnaces firing oil, gas, or pulverized coal. It details requirements for design, installation, operation, and maintenance of multiple burner firing systems to help in preventing furnace explosions or implosions.

Many jurisdictions now include flame safeguard systems on automatically fired boilers as an installation requirement. For example, one jurisdiction has the following requirements:

1. *Gas-fired boilers.*   (a) Pilot has to be proved, whether manual or automatic, before the main gas valve is permitted to open, either manually or automatically, by completing an electric circuit. (b) A timed trial for the ignition period is established based on the input rating of the burner. For instance, for input rating of 400,000 to 5,000,000 Btu/h per combustion chamber, the trial for the ignition period for the pilot of automatically fired boilers cannot exceed 15 s. And the main burner trial for ignition also cannot exceed 15 s. (c) The burner flame-failure controls must shut off the fuel within a stipulated time, again depending on the fuel input of the burner. For a burner rated with an input of 400,000 Btu/h or more, the electric circuit to the main fuel valve must be automatically deenergized within 4 s after flame failure. And the deenergized valve must automatically close within the next 5 s.

2. *Oil-fired boilers.*   Similar provisions have been adopted, with requirements on response time for controls to shut off the burner based on fuel input in gallons per hour, instead of Btu per hour. The flame must be continuously supervised by the controls.

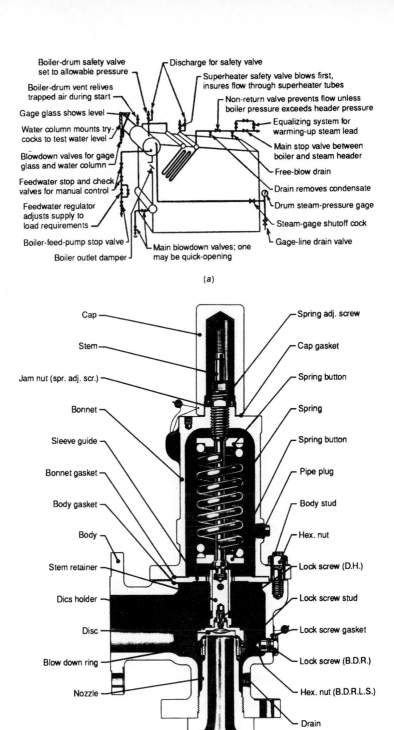

FIGURE 8.22 (*a*) Fittings and appurtenances for high-pressure boilers; (*b*) safety valve nomenclature. (*Courtesy, Teledyne Faris Co.*).

8.40

CODE APPURTENANCE REQUIREMENTS

All boilers require certain fittings and devices for safe operation, and the broad term *appurtenances* is used to denote such devices. Figure 8.22*a* shows some of the fittings that are installed and in most cases required by ASME code and jurisdictions. Not shown in the illustration is a low-water fuel cutoff, which is mandatory on automatically fired boilers.

The code requires the following as a minimum on high-pressure boilers: (1) Pressure gage and test connection, (2) safety or relief valve, (3) blowdown valve, (4) gage glass and gage cocks, (5) stop valve in steam line, (6) stop and check valve in the feedwater line, (7) low-water fuel cutoff on automatically fired boilers as defined in CSD-1, (8) combustion safeguards on boilers firing fuel in suspension, (9) pressure and/or temperature controls and cutouts, including upper-limit settings as detailed in CSD-1 of the ASME for smaller automatically fired boilers.

The function of these appurtenances can be briefly summarized. *Safety valves* open up at set pressures, per code at the maximum allowable pressure for the boiler, and thus relieve the pressure on the boiler and prevent overstressing and possible rupture from taking place on pressure-containing parts. Figure 8.22*b* shows a typical safety valve. The code provides details on popping tolerances and reseating of safety valves after they open at the set pressure. The questions and answers at the end of the chapter will stimulate further reading about code requirements for many code tolerances or size requirements for fittings and appurtenances.

Water gages are required on steam boilers in order to show if the boiler has water in it per design requirements. *Pressure gages* are a must so that the pressure on the boiler can be checked at all times in order to make sure it is within the maximum allowable range; they are also important for maintenance purposes. *Stop valves* control the steam flow from the boiler, assist in isolating a boiler for maintenance purposes, and are also required when cutting a boiler into an active steam line supplied by other boilers. *Feedwater* connections include stop valves on feed lines, check valves to prevent back flow, and control connections for regulating water input per load. *Blowoff* valves and connections are required to control concentrations of solids in the boiler water, to blow out mud and sediments, and to drain the boiler for maintenance purposes. On boilers operating over 100 psi, two blowoff valves are required in series to prevent accidental draining of a boiler in operation and for personnel safety.

WATER GAGES AND WATER COLUMNS

A water column is a hollow casting, or forging, connected by pipes at top and bottom to the boiler's steam and water spaces (see Fig. 8.23). The steam-pipe connection to the top of the water column must not be lower than the top of the glass, and the water-pipe connection to the column must not be higher than the bottom of the glass. The minimum size of these connecting pipes must not be less than 1 in. Use plugged tees or crosses at right-angle turns so that all piping may be easily examined and cleaned by removing the plugs. Valves, if used on steam and water connections to the water column, must be outside screw-and-yoke, lever-lifting gate valves or stop cocks with a level handle. Or they must be other valve types that offer a straightway passage and show by position of the operating mechanism whether they are open or closed. Always lock these valves or cocks open, or seal them open. If this is not done, the whole purpose of the water column and gage-glass connection will be destroyed, and the true level of water in the boiler cannot be determined.

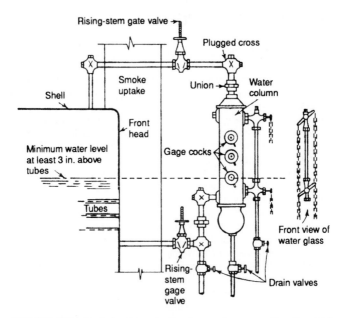

FIGURE 8.23 Typical water-column and gage-glass connection for a horizontal fire-tube boiler.

The water-gage glass with its steam, water, and drain valves is placed on the water column as shown and also on the required number of gage cocks. Damper regulators, feedwater regulators, steam gages, and other pieces of apparatus that do not require or permit escape of an appreciable amount of steam or water may be connected to the pipes leading from the water column to the boiler. Cast-iron water columns may be used for pressures not exceeding 250 psi, and malleable-iron columns for pressures not exceeding 350 psi. Above that pressure, steel columns are used.

BLOWDOWN VALVES

Blowdown is necessary on boilers because blowing down does three jobs:

1. Rapidly lowers the boiler-water level in case it accidentally rises too high. This action reduces the possibility of slugs of water passing on with the steam and wrecking machinery.
2. Permits removal of precipitated sediment or sludge while the boiler is in service. Otherwise it might be necessary to take the boilers off the line frequently to wash out sludge accumulations.
3. Controls the concentration of suspended solids in the boiler. The solids would settle on metal parts, reducing heat transfer and causing metal overheating where the scale is located. Rupturing of tubes, shells, and tube sheets may then occur.

On all boilers, except those used for traction or portable purposes or both, when the allowable working pressure exceeds 100 psi, the bottom blowoff pipe must have two

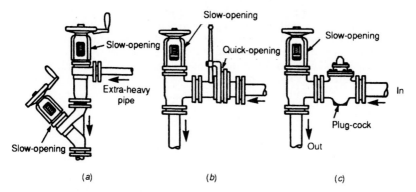

(a)   (b)   (c)

FIGURE 8.24   For boilers operating over 100 psi, the code requires two blowdown valves in series. (a) Two slow-opening valves; (b) one slow-opening and one quick opening valve; (c) one slow-opening and one code-approved plug cock.

slow-opening valves or one slow-opening valve and one quick-opening valve or a cock complying with the code. The blowdown connection to the boiler must be at the lowest point of the boiler so as to drain it properly (see Fig. 8.24).

Avoid accidental opening of a blowoff valve, especially of the quick-opening type. Remove the handle or lock it in a closed position when the valve is not in use. Open and close the valve slowly to prevent water hammer and possible rupture of pipes, valves, or fittings. If a blowoff valve appears to leak when closed, open it again so that boiler pressure will remove whatever is holding it open. Forcing only damages the valve. When boilers are in battery and one boiler is open for cleaning, always "break" the connection between the idle boiler and the blowdown line (if installed). Otherwise blowing down the operating boiler will blow back into the idle boiler, scalding personnel inside.

A *blowoff tank* is necessary when there is no open space available into which blowoff from the boilers can discharge without danger of accident or damage to property. For example, discharging to a sewer would probably damage the sewer by blowing hot water under high pressure directly into it. A good blowoff-tank installation is always nearly full of water (Fig. 8.25).

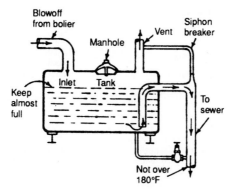

FIGURE 8.25   Blowoff tank details. Use of blowoff tanks will prevent possible scalding accidents.

Low-water fuel cutoffs (LWCO) on automatically fired boilers are second only to the safety valve for protecting boilers against serious accidents. All plant operators should be familiar with their method of operation and the testing that should be performed periodically to make sure they are functional at all times. Figure 8.26a shows what can happen to low-water fuel cutoffs if they are not tested periodically and maintained per manufacturer's instructions.

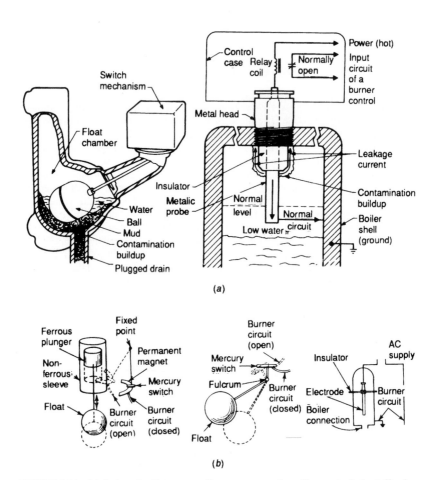

FIGURE 8.26 (*a*) Scale and sediment may affect proper operation of low-water fuel cutoffs when needed. Left, float-type LWCO stuck in sediment, and right, probe-type LWCO has leakage current through scale to complete circuit, thus providing no low-water protection; (*b*) three common types of low-water fuel cutoffs. Left, float-magnet type, center, flow-linkage type, and right, electrode probe type.

Common Types of LWCO

Figure 8.26*b* shows three common types of low-water fuel cutoffs; they are:

1. The float-magnet type (see Fig. 8.26*b,* left) has a ferrous plunger on one end of a float rod. The plunger slides within a nonferrous sleeve. A permanent magnet, with a mercury switch affixed, is supported by a pivot adjacent to the nonferrous sleeve. Under normal water conditions, the ferrous plunger is above and out of reach of the magnetic field. In this position the mercury switch is in a horizontal plane, keeping the burner circuit closed. But if the boiler-water level drops, the float also drops, bringing the ferrous plunger within the magnetic field. Then the magnet swings through a small arc toward the plunger; the mercury switch tilts, opening the burner circuit.

2. The float-linkage type (Fig. 8.26*b,* center), has a float connected through linkage to a plate supporting a mercury switch. Because the plate is horizontal in the normal water-level position, the switch holds the burner circuit closed. If the water level drops, the float drops, tilting the plate so the switch opens the circuit.

3. The submerged-electrode type (Fig. 8.26*b,* right) uses boiler water to complete the burner circuit. If the water level drops below the electrode tip, current flow is interrupted, shutting down the burner. On FT boilers, the low-water cutoff generally includes an intermediate switch that controls the feed pump.

AUXILIARIES

Boiler auxiliaries are external to the boiler proper, but include equipment that supplies (1) water to the boiler, such as feed pumps, (2) fans that supply combustion air and forced-draft fans, (3) those that help exhaust flue gas to the atmosphere, such as induced-draft fans. Other auxiliaries could be (1) coal unloaders and conveyors, (2) feedwater heaters, (3) fuel oil heaters, (4) condensate tanks, (5) evaporators, and (6) other such equipment that is necessary to operate a boiler system. This is influenced by the size and pressure of the boiler system, the use of the system, the fuel used, the water supply, and the condensate returns, to name a few. Operators should carefully check this equipment for proper operation since a malfunction of any auxiliary equipment will, or could, immediately affect boiler operation. Most manufacturers supply good information on maintaining the auxiliary equipment in a boiler plant, which if followed, will help in maintaining the reliability of the boiler system. Some auxiliaries are described in other chapters under pumps, fans, turbines, etc.

QUESTIONS AND ANSWERS

1. What are four types of fire-tube boilers?

   Answer.   Horizontal return tubular (hrt), locomotive firebox (lfb), scotch marine (sm), and vertical tubeless and tubular (vt).

2. What does *creep* mean?

   Answer.   It is the tendency of metals to stretch under load of long duration in high temperatures until the metal starts to develop crack failures.

3. What is the blowdown of a safety valve?

Answer. The blowdown is the pressure differential between the popping pressure and the pressure at which the valve reseats.

4. What are two general classifications of superheaters?

Answer. The general classifications are radiant and convection types; they depend on whether heat is absorbed in the furnace area (radiant) or in the convection sections of the boiler (convection type).

5. What is the purpose of an air heater?

Answer. Air heaters recover heat from boiler flue gas, and this preheats the air to be used in burning a fuel in the boiler furnace. The preheater thus saves the fuel that would otherwise be used to heat the incoming air in the furnace.

6. What is a downcomer?

Answer. This is a large vertical pipe outside the boiler casing of water-tube boilers that takes water from the water space of the steam drum and brings it to water-wall headers to be heated in the furnace or circulation plan for the boiler. The downcomer is placed outside the casing so that the downward natural gravity circulation of the cooler water is not disturbed, which it would be if the downcomer was placed in a heat-absorbing zone.

7. Steam boilers over 15 psi are classified as high pressure per the ASME code. What are the limitations for hot water boilers?

Answer. If the pressure exceeds 160 psi and 250°F, the boilers are classified as high pressure.

8. What is the purpose of telltale holes in solid staybolts?

Answer. Three-sixteenth-in holes are drilled and extend $\frac{1}{2}$ in beyond the inside surface of the plate supported by the staybolt. If a staybolt cracks from overheating, expansion, or contraction, steam or water will start to leak out of the telltale hole, thus warning an operator of a weakened stayed area in the boiler. If the warning is ignored, the usual flat surface requiring staying can bulge and rupture.

9. How many pounds of steam will a 1250-hp boiler deliver?

Answer. Use $34.5 \times 1250 = 43{,}125$ lb/h.

10. What is the purpose of water screen tubes in water-tube boilers?

Answer. Water screen tubes are placed above the furnace *before* the convection section to lower the temperature of any particulate in the flue gas stream below the fusion point. This prevents slagging from occurring in the heat-transfer areas of the boiler *beyond* the furnace area. These areas are more difficult to deslag.

11. What are dry- and wet-bottom furnaces?

Answer. On pulverized coal burning units, the furnace can be designed to burn coal of any fusion range. If the resulting ash in the furnace (bottom ash) that is removed is in a dry state, this furnace is called a dry-bottom type. For low-fusion coals, the bottom ash may be in a liquid form, and the furnace is called a wet-bottom furnace.

12. On what principle does a low excess-air oil burner operate?

Answer.   By refinement, or better atomizing of the oil, it is possible to obtain complete combustion of the oil by using only the theoretical amount of air as developed by combustion equations, thus improving combustion efficiency by not heating excess air.

13. What do the terms *primary* and *secondary* air mean?

Answer.   These terms refer to the combustion process on large boilers, where the primary air is the air mixed with the fuel at or in the burner, while the secondary air is introduced around the burner or through openings in a furnace wall or floor. From a combustion perspective, primary air ensures instant combustion as the fuel enters the furnace, while the secondary air provides additional air for combustion to be completed inside the furnace.

14. What is the flash point of oil?

Answer.   The flash point is the lowest temperature at which the oil can be heated so that the vapor given off flashes momentarily in the presence of an open flame. Temperature of any oil going to a burner should be 10°F below the oil's flash point to prevent preignition. Lighter oils numbered 1, 2, and 3 require no preheating. Burner temperatures recommended for no. 4 is 150°F, for no. 5 is 175°F, and for no. 6 is 275°F.

15. What does a Charpy V-notch test measure?

Answer.   This test and the drop weight test are used to measure the resistance of a material to impact forces or brittleness. Since certain steels have transition temperatures from ductile to brittle behavior, tests are conducted at different temperatures as detailed in ASTM Standard E-23.

16. What are some NDT methods?

Answer.   Visual examination, hydrostatic testing or leak detection, radiography, magnetic particle, dye penetrant, ultrasonic, eddy current, acoustic emission, and holographic tests.

17. What is the critical pressure, and what is this quantity for water/steam?

Answer.   The critical pressure is the pressure at which water and steam have the same density, while the critical temperature is the temperature above which water cannot exist as a liquid, no matter how high the pressure. The critical pressure for water/steam is 3206.2 psi, and the critical temperature is 705.6°F.

18. What is the term used for a boiler that burns solid fuel in a bed of limestone through which air is passed?

Answer.   This is a fluidized-bed burning boiler.

19. What is the difference between "fuel-bound NO_x" and "thermal NO_x"?

Answer. NO_x formed from burning nitrogen in the fuel with oxygen is fuel-bound NO_x, while the reaction of nitrogen and oxygen from the air introduced into a high-temperature furnace is called thermal NO_x.

20. On what sections of a boiler does cold-end corrosion occur?

Answer.   The primary cause of this type corrosion is sulfur dioxide that may be in the flue gas, and as it enters the cooler sections of the boiler, it may combine with dew point water in the passages to produce sulfuric acid. This acid attacks the metals in the cooler section of large boilers such as economizers, air pre-heaters, and induced draft fans, hence the term *cold-end corrosion.*

21. What are five methods of flue-gas desulfurization?

Answer. They are scrubbing the flue gas through limestone beds, fluidized-bed burning by limestone injection, scrubbing the flue gas through magnesia beds, catalytic oxidation of the sulfur dioxide, and using carbon absorption or manganese oxide absorption.

22. What is a flame-safeguard system and what is its purpose?

Answer. A flame-safeguard system is an arrangement of detectors, interlocks, and relays for sensing the presence of a proper flame at a burner and automatically shutting off the fuel if an improper flame is detected.

23. What is measured in a flue-gas analyzer?

Answer. It measures the percentages of volume of carbon dioxide, carbon monoxide, and oxygen. It is based on the principle that air contains 21 percent oxygen and 79 percent nitrogen and that the maximum percentage of CO_2 in flue gas that is possible is the same as the percentage of oxygen in the air, namely, 21 percent. Thus, if there is no excess air, the flue gas analysis will be: oxygen, 0 percent; nitrogen, 79 percent; and carbon dioxide, 21 percent.

24. What would excess air of 50 percent show?

Answer. It would show oxygen, 7 percent; nitrogen, 79 percent; and carbon dioxide, 14 percent. Note that the total percentage of oxygen and carbon dioxide is 21 percent, but the oxygen is 50% of the carbon dioxide.

25. What is the range of pressure setting permitted by the code rules for two or more safety valves on a boiler?

Answer. Three conditions must be met: (1) One or more safety valves must be set at or below the maximum allowable pressure, (2) the highest pressure setting of any valve cannot exceed the maximum allowable pressure by 3 percent, and (3) the range of pressure settings of all the saturated steam safety valves cannot exceed 10 percent of the highest pressure setting to which any valve is set. For example, a boiler has three safety valves and an allowable pressure of 200 psi. What are the permissible settings of the safety valves? One at 200 psi (maximum allowable pressure), the second at $200 + 0.03 \times 200 = 206$ psi, and the third at $206 - 0.10 \times 206 = 185.4$ psi.

26. What is the proper sequence in opening and closing blowoff valves when blowing down a boiler?

Answer. When a boiler is equipped with both a blowoff valve and cock or quick-opening valve in the same blowoff connection, always open the cock or quick-opening valve first and the blowoff valve second. To close, always close the blowoff valve first and the cock or quick-opening valve second. Caution the boiler operator to open and close blowoff valves and cocks slowly to reduce shock as much as possible. And *never* take your hands off the blowoff valve while it is open or your eyes off the gage glass.

27. Why is proper venting important in feedwater heaters?

Answer. As steam condenses in the feedwater heaters, condensate falls to the bottom of the shell and forms a seal against the steam blowing through the shell.

This prevents noncondensable gases from leaving the heater. Thus, an air blanket forms and puts the heater out of action. This condition is most pronounced in low-pressure heaters operating below atmospheric pressure. The only protection against the forming of an air blanket is to vent the heaters properly. Make sure a vent is placed in each end of the steam space. Besides removing gases, the vents help to distribute steam throughout the shell, thus making better use of the heat-transfer surface.

28. Why are evaporators used in power plants?

Answer. Evaporators produce makeup water for boilers from raw water that has a high solids content. Modern boilers need evaporated water to prevent scaling. Aboard ships and in areas where fresh water is scarce, evaporators are used to distill potable water (drinking water) from sea water.

29. Give five primary reasons for deaerating.

Answer. Deaerators remove oxygen, remove carbon dioxide, raise feedwater temperature, improve heat-transfer efficiency, and improve overall boiler operation.

30. Why must oxygen be removed from feedwater?

Answer. Oxygen is extremely corrosive to boilers. The maximum safe oxygen content for most boilers is about 0.005 cc/L (7 ppb). As steam condenses, any oxygen carried over with the steam will be highly corrosive to the condensate lines and the process equipment. Carbon dioxide may also be present in the steam and condensate; it is extremely corrosive when it condenses and combines with water.

31. Why must carbon dioxide be removed from raw water?

Answer. Most raw water contains free carbon dioxide, which is corrosive when combined with water to form carbonic acid. In steam, carbon dioxide is not corrosive, but when the steam condenses, the carbon dioxide combines with water and forms very corrosive carbonic acid, which will dissolve return lines and process equipment rapidly.

CHAPTER 9
BOILER OPERATION AND MAINTENANCE

Boiler operation is being automated with the advancement of semiconductor technology, commonly called the computer on a chip, or microprocessor. This technology is permitting advanced or retrofitted boiler plants to boost plant performance with more responsive control technology by using more sophisticated measuring instruments or pickups on pressure, flow, draft, temperatures, flue gas content, air-fuel ratios, and similar operating variables. This information is then used to calculate control settings needed to reach predetermined set points. In larger plants, microprocessors are installed at key process flow locations, such as feedwater flow control, but the information on proper or improper control at this process location is also supplied to a centralized boiler control room for overall management control. CRT displays show the operator at workstations the present level or quantity of unit performance. Printouts of data are also available from control room printers. Data are also stored for retrieval as needed. On-line diagnostics have been developed to assist operators in large-boiler plants to avoid boiler damage during startups, shutdown, steady-state operation, and cyclic conditions. The skill level needed when something goes wrong, however, is much greater and includes instrumentation and computer technicians in addition to the traditional boiler operator.

BASICS

Plant operators should be familiar with certain boiler operating and maintenance procedures that are still applicable, especially for commercial and industrial plant locations. These can be summarized as follows by considering a high-pressure steam boiler plant:

1. Use gage cocks to check the *water level* at least once per shift or during any water-level disturbance, such as a pump tripping. Use test gage cocks to assure that connections to the water gage and water column from the boiler are not obstructed and that the water-level reading reflects true boiler level.

2. Know the actions to take if *low-water alarm sounds* or low-water fuel cutoff trips the boiler out of service.

3. Maintain the gage glass free of leaks. Keep gage cocks clean and dry.

4. Maintain firing equipment clean and free of leaks. *Check the flame* to make sure it is not striking the boiler side walls, tubes, or shell in a blowtorch effect since this can cause local overheating.

5. Keep the *outside of the boiler* clean and dry to avoid soot and oil damage to controls, electrical wiring, and actuators and to eliminate binding, which would make them inoperative.

6. Check *flame safeguards* on each shift in order to make sure they function properly in order to avoid furnace explosions.

7. Test *safety valves,* usually once per month, by lifting the test lever. If the valve does not lift, rust or scale may be binding the mechanism. Follow the safety valve manufacturer's instructions to free the valve. Secure the boiler and have the valve repaired by a qualified repair concern if it cannot be freed.

8. *Steam and water leaks* cause corrosion, endanger personnel, and can add to fuel costs. They are a sign that something on the boiler system has been weakened, and possible unexpected failures may occur as a consequence. They should be repaired as soon as possible since they are dangerous to operation.

9. When starting a boiler, make sure it is purged properly so that accumulated fuel from leaks cannot cause a furnace explosion. Also make sure the boiler has water at the proper level before firing the unit.

10. Check all *appurtenances and auxiliaries* for cleanliness and tightness before firing and, thereafter, for leaks, vibrations, and similar abnormal operating conditions. Problems found should be corrected as soon as possible since any malfunction could affect boiler operation.

11. When taking a boiler out of service, *avoid thermal stresses* by cooling the boiler slowly, and avoid blowing off the boiler under pressure to accelerate cooling. Slow cooling will prevent any mud or scale from baking onto internal surfaces.

12. Boilers that are to be *internally inspected* should have all internal surfaces accessible for inspection and cleaning; this means removing manholes and handholes as well as drum internals that may interfere with checking drum surfaces, tube holes, and ligaments for erosion, wear and tear, and cracking.

13. Check the low-water fuel cutoff on each shift. Serious damage can result from low water. Follow the manufacturer's instructions on the testing of the low-water fuel cutoff for the boiler under your control.

14. It is important to study the boiler manufacturer's instructions that normally are provided with the boiler with respect to operation, testing of controls and safety devices, maintenance to be applied, and inspections that are required on a periodic basis, including jurisdictionals.

15. Remember to maintain a log book and enter any unusual operating problem as well as the date of testing and the results. This includes conditions found during inspections of all appurtenances and boiler auxiliary equipment under your care and control.

16. Perform necessary blowdowns per shift, or more often, on water columns, low-water fuel cutoff drains, gage-glass drains, and the boiler proper to control sludge accumulation and to maintain boiler water quality per water treatment guidelines.

Proper operation will maintain design efficiency for the boiler with a minimum number of forced outages, which management of enterprises expect.

PERSONNEL SAFETY

Some personnel safety precautions to follow when working in a high-pressure boiler plant include the following:

1. Follow OSHA rules for entering a confined space by making sure the space has been cooled and all water and steam valves, including drain and blowdown valves on drums to be entered, have been closed and locked or tagged. This is especially important where more than one boiler is connected to the system. By following common precautions, injuries from backflows will be avoided. Make sure all gases have been purged from the confined space by actually testing the inside with a meter per OSHA requirements to make sure the vessel has the required oxygen level inside. A person should be stationed by any entry point so that emergency help can be provided if so needed.

2. Use low-voltage lamps and cords (12 V or less) when inspecting the inside of a drum or shell. Make sure the equipment is grounded. The electrical equipment should be explosionproof.

3. Make sure there is no pressure on the boiler system when opening any manhole, handhole, or similar opening on a boiler that has been in service to prevent possible scalding accidents.

4. Wear tinted glasses or a tinted shield when looking at furnace flames in order to protect your eyes from harmful light rays or flying ash and slag particles.

5. Do not stand in front of furnace openings when the boiler is operating unless proper heat-resistant clothing is worn and your face, eyes, and hands are protected from possible furnace pulsations, tube failures, soot-blower operation, and back drafts that can blow hot gases from the furnace through the furnace openings.

6. Never tighten connections while the boiler is at full pressure and temperature. A sudden rupture of the connection can cause serious injury. Make connections tight with the boiler idle and with no pressure on the system.

7. Entering idle furnaces that burn solid fuel requires suitable clothing, including hard hats, in order to avoid being struck by any falling slag or dust that may come loose from overhead tubes.

8. Be extra careful when opening or entering any boiler auxiliary machinery. Make sure circuit breakers are locked open and tagged so that equipment cannot be started. Fans and similar equipment may start rotating from backflow or differential air pressure. The rotor should be locked with a suitable device to prevent rotation.

Practice common safety practices that are applicable for all types of locations.

BOILER START-UP

As design pressures rise, so does the amount of care required in starting boilers properly to prevent abnormal stresses. In general, a heating rate of a 100°F change in saturation temperature per hour is a good rate to follow in start-ups.

In filling a boiler to be operated, use treated water as specified for the boiler pressure and temperature. Make sure all internals of the drum, headers, and tubes have been properly cleaned and are free of tools, rags, and similar foreign objects before closing a boiler for operation. During the filling operation, open all vents on top of the boiler to displace all air with water. This will prevent oxygen attack on heat-transfer surfaces and pockets that may interfere with water circulation.

When filling the boiler, and before firing, it is essential to establish the correct water level in the gage glass. Since water expands with heat, it is generally accepted that the level in the gage glass should be 1 in above the minimum in the glass, unless a special design has different instructions.

Boilers with economizers, superheaters, and reheaters require special attention during start-up because on initial firing, there is no water and steam flow in them as pressure and temperature is raised.

Superheaters must have steam flow to avoid overheating damage. Since there is no steam flow on start-up, several methods are used to prevent overheating damage to the superheaters during start-up cycles; they include:

1. Controlling the flue-gas temperature across the superheater tubes to below 900°F metal temperature by controlling start-up firing rate and flue-gas recirculation and by using excess air. Thermocouples are strategically placed to measure superheater metal temperatures.

2. For drainable superheaters, the header drains are left open to establish flow through the superheater.

3. By the use of a start-up boiler, steam is supplied to the superheater on start-up from this boiler until the superheater flow can be switched to the regular boiler at about 10 percent of normal superheater flow. Steam from the start-up boiler is fed to auxiliary low-pressure steam lines or a condenser.

Reheaters also must have steam flow on start-up, and the same strategy as for superheaters is followed to avoid overheating damage.

Purging of furnaces is an important procedure when starting a boiler that burns fuel in suspension. Make sure any automatic controls go through a furnace purging cycle. For small, single-burner units, purging of the furnace is normally programmed with 70 percent of full-load air flow until eight furnace volume changes are made. Some burners specify a time frame to achieve this purging volume. For large boilers with multiple-burner units, purging generally recommended by boiler manufacturers is 25 percent of full-load air flow until five volume changes are achieved.

Hydrostatic testing is often applied after repairs or on new boilers in order to make sure all parts are tight. The ASME hydrostatic test is $1\frac{1}{2}$ times the allowable working pressure, and this pressure is maintained long enough for the inspection to be made for leaks. It is necessary to apply gags to safety valves during the test unless the valve is flanged; then blanks are installed in the flange. Use treated water for the hydrostatic test. The water temperature must be above 70°F per code rules. In general, the water and metal temperature must be above the dew-point temperature of the surrounding air in order to prevent condensation on metal parts, which would prevent the detection of small leaks during the hydrostatic test. The water cannot be too hot either (about 120°F) because this would prevent touching metal parts for close inspection. Too high a water temperature may also cause the hot water to evaporate at leak points.

It is also important to vent the unit of any air when filling the boiler for the hydrostatic test. Open all vents until water appears.

Chemical cleaning is often applied to remove the accumulation of oil, grease, paint, and scale or corrosion products. The chemicals to be applied may depend on the deposits found in the boiler. Waterside cleanliness is important since it improves boiler efficiency. Waterside impurities can coat tubes and cause tube failures. On power machinery using steam, carryover of solids can affect turbine blades. Oil deposits can cause *foaming,* which can cause carryover of wet steam and a false indication of water level. Usually the boiler is filled with a solution of the chemicals in the water. Follow the filling procedure that was previously described. A soaking period follows, and the chemical solution is drained and the boiler flushed. In some chemical cleaning, caustic and phosphate solution boil-out is used, and then the boiler is drained and flushed. Bent superheater and reheater tubes need special attention by blowing out all water used for flushing. In general, operators must work with the firm performing the chemical cleaning. However, it is important to check by inspection whether the chemical cleaning was effective and that the boiler was properly flushed so that no chemical residue remains in it.

SHUTTING DOWN

Procedures for shutting down are influenced by the size of the boiler and the fuel that is burned. Follow the instructions for the type of firing system installed for the boilers under your control to avoid overheating damage, excessive thermal stresses, and a furnace explosion from an accumulation of unburned fuel.

Slow cooling of the boiler is required to prevent thermal stresses from developing. Manufacturers provide recommended cooling rates, sometimes called ramp rates, in bringing a boiler up to pressure and in cooling a boiler on shutdown. Superheater drains must be opened during the shutdown procedure to make sure flow is maintained while the boiler is cooling. When water feed to the boiler is stopped per instructions, the economizer and reheater may require similar attention in order to maintain circulation while the boiler is cooled.

For units that are to be out of service after shutting down, it is necessary to prevent a vacuum from developing inside the steam/water space from the condensation of steam. This applies even to steam-heating boilers. Vacuum conditions can cause leaking oil heaters that use steam to fill a boiler with oil since the leak will follow the path with the least pressure. An alert operator may detect this condition by noting the discoloration in the gage glass or when performing drain tests. To avoid a vacuum condition on shutdown, open the steam drum vent valves as the boiler pressure drops to near atmospheric pressure. Vacuum conditions may also cause leaks to appear at gasketed joints.

INSTRUMENT AND CONTROL CHECKING

Modern boilers have various degrees of automation; however, periodic checking of instruments and controls for proper operation must be performed by operators.

Some basic instruments are needed for large power boilers. As a minimum, the ASME boiler code recommends the following: (1) steam pressure gage, (2) feedwater pressure gage, (3) furnace draft gage, (4) an outlet pressure gage on the forced-draft fan and an inlet pressure gage on the induced-draft fan, (5) steam flow recorder for checking boiler output, (6) CO_2 recorder to check on combustion, (7) superheater inlet and outlet temperature recorder, (8) inlet and outlet temperature recorders for air heaters, (9) thermometers indicating inlet and outlet steam temperatures for boiler reheaters, (10) feedwater temperature recorders for checking degree of deaeration and economizer operation, (11) pressure gages on pulverizers to check differential pressure for fuel-air mixtures to burners, (12) pressure gages for oil-fired boilers on oil lines to burners and temperature gages before and after any oil preheaters, and (13) pressure gages for gas-fired boilers on the main gas line to burners and on individual burners.

The two main types of pressure gages are the Bourdon tube and the diaphragm type. Figure 9.1a shows the interior mechanism of the single-tube Bourdon gage with the dial removed. The bent tube has an oval cross section and is closed at one end and connected at the other to boiler pressure. The closed end is attached by links and pins to a toothed quadrant, which in turn meshes with a small pinion on the central spindle.

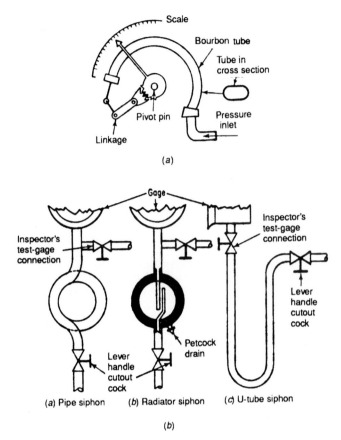

FIGURE 9.1 (*a*) Bourdon-type pressure gage principle is based on tube expanding with pressure; (*b*) types of siphons used on steam-pressure gages.

As pressure builds up inside the oval tube, it attempts to assume a circular cross section, thus tending to straighten out lengthwise. This action turns the spindle by the links and gearing, causing the needle to move and register the pressure on a graduated dial.

The siphon is simply a pigtail or drop leg in the tubing to the gage for condensing steam, thus protecting the spring and other delicate parts from high temperatures. Three forms are shown in Fig. 9.1*b*. If there is danger of freezing during long periods of shutdown, the siphon should be removed or drained.

Safety controls generally are those that limit energy input and thus shut down the equipment when unsafe conditions develop. They are (1) pressure-limit or temperature-limit switches, (2) low-water fuel cutoffs, (3) flame failure safeguard systems, (4) automatic ignition controls, (5) oil and gas fuel shutoff valve controls, (6) air and fuel pressure interlock controls, and (7) feedwater regulating controls.

The safety valve (or relief valve) is the most important safety device. While not considered a control in the usual sense, it is the last measure against a serious explosion.

Critical interlock-control trips are needed where there is a danger of serious damage to the boiler and where personnel safety may be jeopardized. Generally, trips would be initiated for the loss of power, water, air for combustion, and fuel.

Loss of combustion air can result if the forced-draft fan fails; and if the fuel supply is not tripped, the furnace may fill with fuel, which could result in a serious furnace explosion from delayed ignition.

Lack of feedwater should also trip the fuel supply to avoid dry firing the boiler.

On balanced-draft boilers, an induced-draft fan failure should also trip the fuel supply to avoid a possible furnace or implosion incident, and the forced-draft fan should also be tripped to prevent the products of combustion from being forced out of the boiler settings owing to pressurized furnace operation.

ABNORMAL OPERATION

Abnormal operation requires emergency actions to be taken by the operator, and these operations can include tube failures, furnace puff-backs or explosions, fire around fuel-conveying or fuel-handling equipment, low-water alarms, and trips, to name a few.

Tube failures are the most common causes of boiler operational problems. Many of these are caused by the way the boiler is operated. For example:

1. *Neglected tube leaks* can cause erosion or steam-cutting of adjacent tubes, leading to a low-water failure. Leaks add to feedwater treatment loads. Leaks also disturb circulation. It is best to secure the boiler and repair the leaks.

2. *Improper water treatment* can cause corrosion or scale buildup inside tubes of the water-tube type. With time, tube failures can result from tube thinning due to corrosion or from overheating if scale develops. Water treatment should be checked per prescribed schedules, and this should be supplemented by internal inspection or turbining tubes and then analyzing the deposits for feedwater adjustments if found necessary. Selective tube cleaning can be used to obtain samples of deposit.

3. On larger boilers, it is important to follow *permissible temperature rise and cooling rates* to avoid placing additional stresses on boiler components from too rapid expansion and contraction. If the recommended ramp rates are exceeded due to wide-load swings, thermal fatigue can result on rolled joints or on U-bends of suspended tubes.

4. *Cyclic service* can also lead to thermal fatigue failures. This is a common problem with boilers that were operated with a base load and are now operated only on a demand basis.

5. Sudden *loss of boiler feed pumps* may develop into a low-water condition on the boiler with resultant tube failures. Periodic checking of low-water alarms and low-water fuel cutoffs will help avoid this type of abnormal operational problem.

6. Coal-burning boilers may have *passage plugging* from fly ash deposits that may concentrate heat on tubes producing short-term overheating. Frequent soot-blower operation to eliminate fireside plugging presents a tube erosion problem that shortens tube life and can lead to unexpected failures.

7. *Low-temperature corrosion* at the cold end of the boiler can lead to economizer tube failures due to temperatures falling below the acid dew point (primarily sulfuric acid). Operators must watch exhaust flue gas and economizer inlet temperatures to avoid this type of attack and possible economizer tube failure. It is usually recommended that exhaust temperatures to the stack be kept above 300°F to avoid low-temperature corrosion attacks unless it is known that special alloyed tubes have been used that can resist the acid attack at the cold end of the boiler.

8. *Exfoliation* of magnetite layers inside water-tube boiler tubes spall off during start-up or shutdown and is prevalent in utility boilers that are being cycled either ON-OFF or

due to wide load swings, such as day to night loads. It is believed the principal cause of exfoliation of tube metal is the difference in the coefficient of expansion between the oxide coating on the tubes and the solid tube metal. Exfoliated material is carried by the steam to connected steam turbines at very high velocity, and this causes turbine blade erosion wear. It also causes long-term tube thinning on superheaters and reheaters, which have the most tube exfoliation. New tube materials are being used to prevent exfoliation and are one of the options available during any boiler retrofitting.

EMERGENCY ACTIONS

The actions to be taken during emergencies such as tube failures or low water should be clearly understood and even rehearsed by reviewing written instructions and "dry runs" so that prompt steps can be taken during the emergency.

Investigation as to cause of tube failure is very important since it may require adjustment in operations, starting procedures, water treatment review, frequency of inspections, and type of inspections. Visual inspections may provide evidence of the reason for the failure, as may a review of past inspections, including NDT or thickness checks, scale and corrosion history, method of operation, and similar variables. Large-boiler plants generally also subject a failed tube to metallurgical failure analysis because outages can be so expensive in terms of lost income.

Low water also requires a well-planned procedure for safely shutting down the boiler to avoid more overheating damage. The procedures follow almost the same steps as those required for tube failures, namely, preventing thermal shocking the boiler by an orderly shutdown.

Inspections after low-water incidents must be more thorough than with tube failures and include drums or shells, supports, baffles, and any other location that may have suffered damage from excessive temperature excursion. Since incipient cracking could have occurred during low water, all repairs will have to be approved by a commissioned jurisdictional boiler inspector.

WATER TREATMENT

Many boiler problems arise as a result of scale, corrosion, and even cracking that have their origin in poor water or steam conditions. The term *water treatment* is applied to the prevention of these problems by the external and internal treatment of water that is used in boilers. As pressures and temperatures rise in a boiler system, so does the care required in assuring that only properly treated water is used in them.

Boiler operations can be affected by impurities in the boiler water. Most problems can be grouped into one or more of the following as a result of water problems that were not properly controlled:

1. *Deposits,* or scale. Metal parts that transfer heat to the water become coated, and this interferes with the heat transmission through the metal to the boiler water or through steam for superheaters and reheaters. The loss of heat transfer adds to the amount of fuel that must be additionally burned to obtain the same output. There is a loss of efficiency when converting the fuel energy into heat. An additional result is the danger of overheating metal parts as the scale becomes thicker. Bulged or bagged furnaces in internal-fired boilers may result due to the weakening of the metal from overheating damage. Another effect of scale is the blocking of water circulation, which causes overheating damage and loss of efficiency.

2. *Corrosion* has many forms and may be caused by impurities, such as oxygen, which causes pitting attack on metals and caustics, which cause grooving or gouging of metal and may also cause caustic cracking in stressed areas.

3. *Carryover* is caused by impurities that cause the steam to be "wet" as it leaves the boiler. This can affect steam-using machinery by heavier wet steam having an impact on machinery components and possibly by overstressing the components from the shock loading. Wet steam can also affect processes. The impurities carried over with the steam can deposit out as scale and can interfere with controls and proper heat transfer in the process and can affect the quality of the product in the process flow.

Water may have the following type of impurities (see also Chap. 4):

1. Suspended matter, such as sand, mud, and organic wastes.

2. Dissolved or in-solution solids that precipitate out of the water at higher temperatures, such as limestone (calcium carbonate), dolomite (magnesium carbonate), gypsum (calcium sulfate), epsom salt (magnesium sulfate), sand (silicate), common salt (sodium chloride), Glauber salt (hydrated sodium sulfate), and traces of iron, manganese, fluorides, and aluminum, depending on the source of water.

3. Dissolved gases, such as oxygen, carbon dioxide, hydrogen sulfide, and even methane.

WATER TREATMENT TERMS

Terms

Some common terms used in water treatment are:

1. *Acid.* A compound that yields hydrogen ions, such as sulfuric acid.

2. *Alkali.* A substance or salt that will neutralize an acid.

3. *Base.* A compound which can react with acids to form a salt and in water solutions yields hydroxyl ions.

4. *Colloidal.* A gelatinlike substance which appears to be in a dissolved state but is actually in suspension. It is made of very tiny particles.

5. *Concentration.* This is applicable to solutions and expresses the ratio of the dissolved substance to the amount of water, usually expressed in a weight ratio.

6. *Corrosion.* The chemical action of a metal being combined to form an oxide of the metal by the action of oxygen, an acid, or alkali on the metal.

7. *Grain.* A unit of weight as used in water treatment, with 7000 gr equaling 1 lb.

8. *Hardness.* A measurement of the amounts of calcium and magnesium compounds in solution in water because the chemical action of these compounds forms an insoluble product, or scale, in the water.

9. *Hydrogen ion.* An ion formed from hydrogen with the symbol H^-, which forms an acid when combining with certain impurities in water.

10. *Hydroxyl ion.* This has the symbol OH^- and forms a base compound when combined with some impurities in water.

11. *Ionization of water.* The breaking up of a molecule of water into hydrogen ions and hydroxyl ions with the reaction proceeding at a higher rate at higher water temperature.

12. *Oxide.*   The chemical combination of oxygen with a metal to form the metal oxide, such as iron oxide, commonly called rust.

13. *Parts per million, or ppm.*   The ratio of impurities on a *volume basis* to a million units of pure water.

14. *pH.*   A measurement of the hydrogen-ion concentration in order to measure the relative acidity or alkalinity of a solution. It is the logarithm to the base 10 of the reciprocal of the hydrogen-ion concentration, with 7 being neutral. Below 7, the solution is acidic; above 7, it is base.

15. *Precipitate.*   The solid substance that is separated from a solution by chemical reaction in the solution.

16. *Reagent.*   A substance of a known strength which is used for the detection and measurement of another unknown substance.

17. *Salts.*   Those substances which in solution yield ions other than hydrogen or hydroxyl. They are also the product of an acid and base combining or an acid and certain metals chemically reacting.

18. *Scale.*   An adherent deposit on metal surfaces in a boiler, which is caused primarily by impurities precipitating out of the water and cementing on the metal as temperatures rise in the boiler.

19. *Soluble.*   The ability of a substance to go into solution by dissolving.

20. *Condensate.*   Steam that is condensed in the steam loop and is returned to the boiler system.

21. *Makeup water.*   Water that must be replenished in the boiler system as a result of leakage, blowdown, and steam process use.

22. *Feedwater.*   The combination of condensate and makeup water that is supplied to the boiler for evaporation.

23. *Blowdown.*   The bleeding of a portion of the water in the boiler in order to remove suspended solids.

24. *Condensate polishing.*   The purification of returned condensate by passing it through demineralizers.

External Treatment

The source of makeup water must be analyzed by water treatment specialists to determine the pretreatment that will be effective in maintaining good water quality. See Chap. 4 for some water treatment methods that are applied to raw water and can also be applied to boiler water.

Evaporators can be used to distill makeup water required in boilers as a result of leakage, process, or other unavoidable losses. They are classified by the method of vaporization used as:

1. *Flash type.*   Hot water is pumped or injected into a chamber under vacuum, where the water flashes into steam.

2. *Film type.*   Water in a thin film is passed over steam-filled tubes.

3. *Submerged type.*   Steam-filled tubes are submerged in the water to be evaporated.

Distillation of the water in this way removes dissolved gas, but all solid impurities are left in the evaporator, which must be cleaned regularly to remove scale from the outside surface of the tubes.

Deaerators are used to preheat boiler feedwater, but their main function is to remove oxygen and carbon dioxide. Steam deaerators break up water into a spray or film and then sweep the steam across and through it to force out dissolved gases such as oxygen and carbon dioxide. The removal of carbon dioxide increases the pH value of the water. Oxygen becomes less soluble as the water temperature is increased. Therefore, it is removed by bringing the water near its boiling point for the operating pressure involved with the boiler system. Vacuum units are used where the boiler water is to be further heated after the deaerator, and low-pressure steam is used.

Since deaerators liberate oxygen from the water, internal inspections should be made on them at the same time the boiler receives its internal inspection in order to make sure no serious corrosion is taking place from the release of oxygen. Highly stressed areas such as welded seams and nozzle connections require inspections for stress-corrosion cracking. Wet, fluorescent, magnetic-particle testing is recommended for welds that were not stress relieved during tank fabrication. As a result of several severe cracking incidents on deaerators, the Heat Exchange Institute (HEI) has recommended new guidelines for fabricators; they are:

1. Corrosion allowance to be increased from $\frac{1}{16}$ to $\frac{1}{8}$ in
2. Full x-ray of shell and head welded seams
3. Stress relief of welds in storage tanks
4. Smooth grinding of the internal weld seams to prevent rust pockets from forming in the weld ripples
5. Wet, fluorescent, magnetic-particle testing of nozzle-to-shell welds

The National Board has also issued bulletins on deaerator inspections as a result of in-service failures. The new standard is available from HEI, 1300 Summer Ave., Cleveland, Ohio 44115. It can be used as a guide for checking existing deaerators.

Demineralizers are used in boiler plants that operate over 1000 psi. Demineralizers resemble the zeolite process because it also involves ion exchange. The cation exchange is operated on the hydrogen cycle where the metal ions are replaced with hydrogen ions by using the appropriate resin in the bed. The anion exchanger operates on the hydroxide cycle using specially prepared resins saturated with hydroxide ions. The salt anions, such as bicarbonate, carbonate, sulfate, and chloride, are replaced by hydroxide ions. The final effluent consists essentially of hydrogen ions and hydroxide ions or pure water.

In mixed-bed demineralizers, the two types of resins are mixed together in a single tank. Regeneration in a mixed bed can be performed because the two resins can be hydraulically separated into different beds.

Condensate polishing is used to purify returned condensate, and demineralizers are used in power plants to remove corrosion products and ionized solids that come from connected piping, turbines, heaters, or condensors. This improves turbine-generator efficiency, protects the steam loop from the effects of condenser leakage, and avoids having harmful deposit or corrosion products in the boilers.

Internal Treatment

Internal boiler water treatment may be all that is required for boilers operating at low pressure and which have a large condensate return with very little makeup or where the makeup water is of good quality. In high-pressure boiler plants, internal treatment complements external treatment by reducing impurities further and controlling the pH of the boiler water. The purpose of internal treatment is:

1. For the chemicals to react with any feedwater hardness introduced into the boiler and prevent it from precipitating on boiler metal parts as scale.
2. To condition suspended matter in the boiler water such as sludge or iron oxide with chemicals so that it does not adhere to metal.
3. To prevent foam carryover by providing antifoam protection.
4. To eliminate oxygen from the water and also to provide enough alkalinity to prevent corrosion.

Chemicals that are used to prevent some of the above conditions include:

1. To remove hardness and soften the water: soda ash, caustic, and various types of sodium phosphates.
2. To prevent foam carryover, certain organic materials are used as antifoam agents.
3. To condition sludge, various organic materials are also used, such as tannin, lignin, or alginates.
4. To eliminate or scavenge oxygen, sodium sulfite and hydrazine are used. To prevent corrosion, various combinations of polyphosphates and organics are used. These also prevent scale in feedwater systems. For preventing condensate source corrosion, volatile neutralizing amines and filming inhibitors are employed.

The pH control in high-pressure boilers is also important. To reduce the corrosion of preboiler equipment such as feedwater heaters, feed pumps, and feed lines, feedwater is treated to control the pH level and the amount of free carbon dioxide and oxygen gases.

Maintaining the condensate pH between 8.0 and 9.5 is generally accomplished by one of two methods. The first is the addition of neutralizing amines (ammonia, morpholine, cyclohexylamine, and hydrazine) which neutralize the acids present in the condensate. The alternative method, the injection of filming amines, is used in situations where high carbon dioxide content in the steam is causing increased corrosion rates. The filming amines are a waxlike substance which coats the tube walls, protecting them from the corrosive condensate. Their major drawback is that control of the amine feed rate is critical to ensure coating without causing flow restrictions.

MAINTENANCE

Boiler maintenance can be assigned three objectives:

1. *Maintaining a safe and reliable plant.* This includes periodic testing to make sure the low-water alarms and cutoffs function properly; proper water levels are maintained to avoid overheating damage; the water treatment is preventing scale and rust from forming dangerous deposits on heat-transfer surfaces; the flame safeguard system is functional so that no accumulated fuel in the furnace can cause a furnace explosion; all auxiliaries are operating properly so that boiler operation is not endangered from faulty draft, faulty feedwater flow, or faulty fuel flow; pressures are within the AWP; and the safety valve is functional.

2. *Maintaining an efficient plant.* Heat loss up the stack is a significant percentage of loss in efficiency. Some causes of this are under the control of the operator, such as maintaining proper draft and air-fuel ratios as well as stack temperatures. Above-normal stack temperatures may be due to scale buildup on heat-absorbing surfaces in the boiler. Too much excess air wastes fuel by heating the air that is uncombined with the fuel. Poor draft may prevent all fuel from being burned in the furnace.

3. *Inspection and logging of items that need to be corrected during the next mainte-nance period.* This includes auxiliaries. During outages, internal inspections will indicate if corrosion or harmful deposits are occurring, and this may require adjustments in operation or feedwater treatment.

INTERNAL INSPECTIONS

Internal inspections include inspections for maintenance purposes and also to satisfy jurisdictional requirements, especially on high-pressure boilers, and to check on the structural soundness of the pressure-containing parts, so as to note any conditions that can affect its strength to confine the pressure. Wear, deterioration, corrosion, scale, oil, cracks, grooving, thinning, and other such weakening conditions require attention. Most boilers develop their own areas of trouble spots, depending on design, operating conditions, and maintenance practices. Check all exposed metal surfaces inside the boiler for effectiveness of water treatment and scale solvents and also for oil or other substances that enter with feedwater.

NONDESTRUCTIVE TESTING

NDT equipment can be used in boiler inspection to locate potential areas of failure. There are five major *nondestructive tests* being used: *ultrasonics, radiography, mag-netic particle, dye penetrant,* and *eddy current.* Ultrasonic equipment is now portable for field use and is extensively used for plate and tube thickness checks. These instru-ments become useful as tracing instruments for determining causes of failure of a repetitive nature. For example, tube failures in waterwalls are a common problem. After one tube failure, adjacent tubes can be checked ultrasonically and thinned tubes replaced prior to failures.

A similar practice is followed on tubes subject to fly-ash erosion. The thickness of the tubes in a suspected area is checked by ultrasonic equipment, and those found thinned are replaced during normal outages. Plate thickness around access hole and handhole openings, water legs, shells, and heads is checked ultrasonically for thick-ness in order to determine allowable pressure.

Flaw detection, such as checking for laminations, cracks, porosity underneath plate surfaces, or welds on inaccessible visual parts, is playing a more important role. Pulse-echo instruments are now available for field testing to do flaw detection.

Radiography, so important in new construction, is extensively used in field testing. Welded repairs on high-pressure boilers are tested by x-ray or other radiographic equipment.

Magnetic-particle inspection finds its chief use in surface crack detection. Its main use is on piping and joints of boilers.

Eddy-current testing is finding its chief use searching for defects in tubes.

FIRESIDE INSPECTIONS

Inspection of the fireside of boilers is important too. Carefully inspect the plate and tube surfaces that are exposed to the fire. Look for places that might become deformed by bulging or blistering during operation. Solids in the waterside of lower generating tubes cause blisters when sludge settles in tubes and water cannot carry away heat.

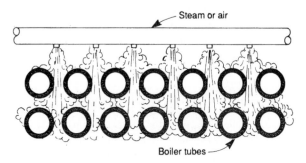

FIGURE 9.2   Stationary soot blower. Soot blowers improperly aligned or being used excessively can erode tubes and cause unexpected tube failures.

The boiler must be taken out of service until the defective part or parts have been properly repaired. Blistered tubes usually must be cut out and replaced with new ones.

Test stay bolts by tapping one end of each bolt with a hammer. For best results, hold a hammer or heavy tool at the opposite end while tapping. A broken bolt is indicated by a hollow sound.

Tubes in horizontal return tubular boilers deteriorate faster at the ends toward the fire. Tapping the outer surface with a light hammer shows if there is serious thinness. Tubes of vertical tubular boilers usually thin at the upper ends when exposed to the products of combustion. Lack of water cooling is the cause. Tubes subject to strong draft often thin from erosion caused by impingement of fuel and ash particles. Soot blowers, improperly used, will also thin the tubes (see Fig. 9.2). A leaky tube spraying hot water on nearby sooty tubes will corrode them seriously from an acid condition. Short tubes or nipples joining drums or headers lodge fuel and ash, then cause corrosion if moisture is present. First clean, then thoroughly examine all such plates.

Baffles in water tube boilers often move out of place. Then combustion gas, short-circuiting through baffles, raises the temperature on portions of the boiler, causing trouble. Heat localization from improper or defective burners, or operation causing a blowpipe effect, must be corrected to prevent overheating.

Waterwall tubes at air ports should be examined closely for loss of metal. The wall thickness of the tubes in this area should be determined periodically by using a thickness-measuring instrument or, if considered necessary, by removing tube samples.

Finned waterwall tubes should be checked for evidence of cracking at or adjacent to the fins. It may be necessary to sand or grit-blast questionable areas and use liquid dye penetrant to find the extent of the cracks if any are observed. All waterwall tubes should be carefully examined for any evidence of corrosion, erosion, or other defects. Any defects found should be repaired prior to returning the boiler to service.

The inspections generally made on *air-pollution-control* equipment should include the following: Look for plugging of bags, filters, and precipitator passages where the particulate matter is removed from the flue gas. Evidence of the corrosion and its effect on the components is essential to prevent an unexpected operating problem later. Damper freedom to move as intended is also an important check. Electrostatic precipitators need special attention on the following:

1. Are electrodes in place and not broken?
2. Are the hoppers plugged?

3. Are the insulators clean and not cracked or broken?

4. Are the rapping mechanisms free and operable?

BOILER REPAIRS

There are many repairs made on boiler auxiliaries, controls, and wiring that are not governed by jurisdictional code rules; however, all operators eventually become involved with some form of repair on the pressure parts of the boiler, such as tubes, thinned areas, leaking handhole, and manhole gaskets (see Fig. 9.3) and perhaps the appearance of cracks. It is good practice to always consult with the jurisdictional inspector who issued the operating certificate or the organization the inspector represents, which could be the jurisdiction or a licensed insurance company that insures the boiler before any pressure part repairs are made.

Repairs permitted are based on restoring the affected part or parts to as near the original strength as possible. They are governed by code requirements for new construction or by National Board rules on permissible repairs, where the state has adopted National Board rules for repairs. Follow the jurisdictional inspector's advice.

Gasket leakage is a common problem on some boilers, and their repair can usually be performed by an operator, unless the leakage has thinned the material around the opening; then the approval for metal repair is needed from the jurisdictional inspector.

For those operators who want to study the code on boilers, it is suggested the following be obtained from the American Society of Mechanical Engineers in New York City:

Section I—Power Boilers

Section IV—Heating Boilers

Section VI—Recommended Rules for Care and Operation of Heating Boilers

Section VII—Recommended Rules for Care of Power Boilers

Section II—Material Specifications—Ferrous Materials

Section V—Nondestructive Testing

Section IX—Welding and Brazing Qualifications

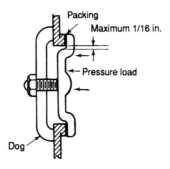

FIGURE 9.3 Gasket leakage around manhole and handhole cover plates should be corrected; otherwise the leaks may cause localized metal thinning.

QUESTIONS AND ANSWERS

1. What is an operator's first duty when taking over a watch?

Answer. First observe the water level in the gage glasses of all boilers. Blow down the water column and gage glass on each boiler and observe the return of water back into the glass. Then check the water level with gage cocks, if installed. Check the operating pressure and note if it is within the rated pressure of the boiler.

2. What test on a high-pressure boiler will indicate that the lower gage glass is obstructed even though the gage glass is half full of water?

Answer. Open the try cocks on the water column. If all show steam, it means the bottom connection is obstructed, permitting steam from the top connection to condense in the gage glass. The boiler should be shut down immediately and inspected for possible dry-firing damage. Naturally, the bottom connection of the water column and gage glass should be cleaned of all obstructions before the boiler is returned to service.

3. What should be done if water is not visible in the gage glass because of failure of the feedwater supply?

Answer. Immediately: (1) Shut off the fuel to the burners and secure the burners. (2) Check the water level by trying the try cocks and water-column drain. If definite low water is indicated below the gage-glass level, close the main steam valve and feedwater valve. (3) If the boiler is equipped with one, open the superheater drain. (4) Continue operating forced-draft and induced-draft fans until the boiler cools gradually. (5) Let the pressure reduce gradually, and when the furnace area is sufficiently cooled, check for leaking tubes and other signs of overheating damage. On fire-tube boilers, look for cracked or warped tube sheets, broken and leaking stay bolts in the water legs. On scotch marine boilers, check for cracked or leaking furnace-to-tube sheet welds. On cast-iron boilers, look for cracked sections. On steel boilers, check for leaking joints on longitudinal or circumferential welds or riveted joints. (6) If no leakage is evident, give the boiler a hydrostatic test of $1\frac{1}{2}$ times the allowable working pressure. Then again check for leakage at all critical parts of the boiler. If leakage is observed during the initial check or during the hydrostatic test, notify the authorized boiler inspector immediately for inspection of the boiler and advice on permissible repairs.

4. When you notice an unusual burning paint odor, especially from piping, or unusually higher steam temperatures, what may be wrong?

Answer. Low water is occurring in a boiler.

5. What should you look for on the first cold day?

Answer. Vapor revealing steam leaks.

6. What is used to remove fly ash and other particulate matter from flue-gas exhausts?

Answer. Coal-burning and other solid-fuel-burning boilers require auxiliary equipment to remove fly ash and other particulates so they are not emitted to the surrounding atmosphere. The equipment commonly used includes the following: (1) baghouse employing fabric filters now usually made of fiberglass that can withstand flue-gas temperatures of 275 to 550°F, (2) scrubbers that wash particulate emissions out of the flue gas and form a sludge that is disposed of in landfills, and (3) electrostatic precipitators.

7. How are superheat and reheat steam temperatures controlled?

Answer. Today steam temperature of 1000°F is common and units are being installed for 1050 and 1100°F. Because these high temperatures are limited only by metallurgy, steam temperatures must be held to close limits for safety as well as for economy. Six basic methods are used for controlling the temperatures of steam leaving the boiler: (1) Bypass damper control with a single bypass damper or series-and-shunt damper arrangement for bypassing flue gas around the super-heater as required. (2) Spray-type desuperheater control where water is sprayed directly into the steam with a spray-water control valve for temperature regulation. (3) Attemperator control where a controlled portion of the steam passes through a submerged tubular desuperheater and a control valve in the steam line to the desu-perheater or attemperator is used. (4) Condenser control with desuperheating con-denser-tube bundles located in the superheater inlet header and water-control valve or valves to regulate a portion of the feedwater flow through the condenser as required. (5) Tilting-burner control where the tilt angle of the burners is adjusted to change the furnace heat absorption and resultant steam temperature. (6) Flue-gas-recirculation control where a portion of the flue gas is recirculated into the furnace by means of an auxiliary fan with a damper control to change the mass flow through the superheater and the heat absorption in the furnace, as required to main-tain steam temperature.

8. What size draft-air opening is needed for a packaged boiler installed in a closed machine room?

Answer. There should be a fixed opening for fresh air, an average area of 2 ft^2 for each 100 boiler horsepower. Opening windows is not the answer because they are often closed in cold weather. Then the boiler starves for air. Remember that for each boiler horsepower, about 10 ft^3/min of air is needed.

9. Why must superheater safety valves be set to blow at a lower pressure than boiler safety valves?

Answer. The superheater safety valves should blow first in order to assure a flow of cooling steam through the superheater elements. If the boiler safety valves blow first, the steam flow to the superheater is seriously reduced, thus causing the superheater to be "starved" of cooling steam, and damage to the elements could result in the form of overheating.

10. What is the acid method of cleaning boilers?

Answer. Acid cleaning is often done to remove metallic oxide film or scale after the oil and grease boil-out operation. But this should be done only by experi-enced personnel using recommended dosages for the unit, followed by neutraliz-ing agents. Leaky tubes, especially at rolled joints in older boilers, often develop after improper acid cleaning.

The solvents used for chemical cleaning are varied. Some use hydrochloric acid, and others use phosphoric acid. The usual procedure is to refill the boiler until the solution overflows at the air vent (acid is added outside the boiler). The solution is *allowed* to soak the boiler from 4 to 6 h, followed by refilling with a neutralizing agent. If hydrochloric acid is used for soaking, a weak solution of phosphoric acid is used. After draining, fresh warm water is used for flushing; then the boiler is immediately filled with an alkaline solution and boiled again for several hours. This solution is drained, the boiler flushed again, and then refilled with normal service water, with proper feedwater treatment started immediately.

11. What is the wet-storage method?

Answer. This method is best when freezing is not a problem and if the unit will not be needed for at least a month. After it is prepared for storage, fill the boiler to the water level with deaerated water. If no deaerated water is used, open a top vent. Then build a light fire to boil the water for 8 h so dissolved gases are driven to the atmosphere. Use 1½ lb of sodium sulfite for each 1000 gal of water stored in the boiler to protect against oxygen. The concentration should be about 75 ppm. Use caustic soda to obtain alkalinity of 375 ppm. Keep the water temperature as low as possible and test the water weekly.

12. What is the dry-storage method?

Answer. The big problem with the dry method is keeping the insides dry. Airblast with independent outside hot air after draining. When dry, place shallow pans of quicklime inside, and then close all openings tight. Place trays between tubes and one in each steam drum (for the water-tube type) and bottom of shell (for the fire-tube type). Open the boiler every 30 days, and if the quicklime is saturated with water, replace it (or whatever material is used).

13. What are the four major boiler-water control tests used by stationary engineers? What water condition does each test determine?

Answer. See Fig. 9.4. The four major tests are for hardness, alkalinity, chlorides, and pH. (1) Hardness affects scale. (2) Alkalinity indicates required amounts of treatment chemicals (caustic soda or soda ash). (3) Chlorides control the concentration of solids and check on the surface condenser for leaks (especially where sea water is used, as in marine or tidal power plants). (4) pH (hydrogen ion) is a type of alkalinity or acidity to check to note if proper alkalinity control is maintained.

14. What are the two most common systems of reporting feedwater tests? How are they converted?

Answer. Feedwater tests are usually reported in parts per million (ppm) and in grains per gallon (gpg). To convert ppm to gpg, use the formula 17.1 ppm = 1 gpg. *Note:* Use 20 for very quick estimates, as an error of 3 ppm is a trifle. *Note*

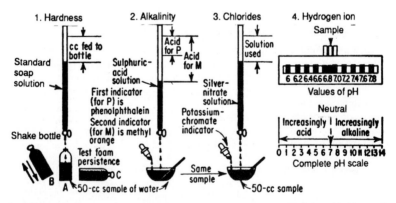

FIGURE 9.4 Four common boiler room water tests are hardness, alkalinity, chlorides, and acidity, or pH.

also: The SI metric system is also used, and tests may be reported in mL/L, meaning milliliters per liter. A milliliter is 1/1000 liter (L). If you are not sure which system is being used, watch for gr, which could be either grains or grams abbreviated. The SI metric system is easier to use and more scientific than grains and gallons; ppm is the same in any system.

15. What are acids and bases and the pH relation of each?

Answer. Acids are those ions in water that form hydrogen ions, H^+. For example, hydrochloric acid ionizes in water to form H^+ and Cl^-. An alkali forms hydroxide ions in water, OH^-. Also, sodium hydroxide (NaOH) ionizes in water to NA^+ and OH^-. When a base and an acid are put in a solution, a salt is formed. For example, $NaOH + HCl = NaCl + H_2O$, with NaCl being the salt formed by combining caustic soda with hydrochloric acid. The base OH^- neutralizes the acid H^+.

The pH value was derived from the hydrogen-ion concentration in *pure water,* which is

$$0.0000001 = \frac{1}{10,000,000} = \frac{1}{10^7}$$

To avoid the use of large decimal fractions, the exponential 7 in the above equation is used to denote neutral or pure water, without acid or alkaline properties. In this way it is easier to remember that 7 is neutral; values below 7 are acidic solutions, and above 7 they are alkaline solutions. The exponential relationship is brought out this way: If the hydrogen ion in pure water is increased 10 times, we have

$$10 \times 0.0000001 = 0.000001 = \frac{1}{10^6}, \text{ and the pH is 6.}$$

16. What compounds cannot be removed by boiling, making the water have *permanent hardness?*

Answer. Water containing calcium and magnesium sulfates cannot be softened by boiling; thus they have permanent hardness, which is solved by chemical treatment with the addition of sodium carbonate as noted by the following equations:

$$CaSO_4 + Na_2CO_3 = CaCO_3 + Na_2SO_4$$
$$MgSO_4 + Na_2CO_3 = MgCO_3 + Na_2SO_4$$

The carbonates precipitate out of solution.

17. What is zeolite softening?

Answer. Hardness in water is caused by calcium and magnesium compounds. A group of minerals called zeolites have the property of exchanging their sodium for the calcium and magnesium in the water, thus softening the water of the hardness compounds. The zeolites can be regenerated, usually by passing a strong solution of common salt through the beds of the zeolite.

18. What are some boiler deposits?

Answer. All of the following deposits can cause boiler parts to overheat if the scaling is too thick: Some features of deposits are (1) *calcium carbonates* are granular and porous. If dropped in an acid, bubbles of carbon dioxide are released.

(2) *Sulfate deposits* are hard, dense, and brittle but do not pulverize easily. (3) *Iron deposits* are dark colored and magnetic and, if placed in a hot acid, become soluble, forming a dark-brown solution. (4) *Silica deposits* are very hard, resembling porcelain, and are brittle but difficult to pulverize. They are not soluble in hydrochloric acid. (5) *Phosphate deposits* arise from internal boiler treatment and are soft brown or gray. They can be removed by normal cleaning methods (washed out).

19. How is the oxygen level controlled?

Answer. By external treatment in a deaerator and by chemical scavenging of oxygen by sodium sulfite and by hydrazine.

20. How is pH controlled?

Answer. When the makeup water is treated by the lime-soda or lime-soda-zeolite method, the effluent may be within the range of 8.0 to 9.5 pH, which is recommended by most boiler manufacturers. However, if excessive sodium bicarbonate is formed, this decomposes to form caustic soda, or excess alkalinity. The ABMA recommends that the alkalinity content should not exceed 20 percent of the total solids of the boiler water. *Split-stream* softening is used to reduce the alkalinity content.

21. What causes foaming?

Answer. This is when bubbles or froth form on the surface of the boiler water and are carried outside the boiler with the steam. It is caused by impurities such as alkalies, oils, fats, greases, and some organic matter being present as impurities in the water. Oil contamination can come from the oil used on steam-using equipment, such as pumps, compressors, and turbines, getting into the condensate returns. It is believed that soaps are formed by the oil reacting with water alkalies.

22. What is the purpose of blowdown?

Answer. Blowdown reduces any suspended solids or solids in solution. When the water blown out of the boiler is replaced by lower-solids feedwater, the boiler water is diluted to acceptable limits provided the amount of blowdown and its frequency is properly controlled.

23. How is the amount of blowdown determined?

Answer. It is based on the concentrations of the various impurities that can be tolerated. The term *feedwater concentration* is then related to feedwater flow. Assume a boiler has a set limit of 1500 ppm of dissolved solids that it must maintain, and the feedwater by tests shows 100 ppm of dissolved solids. The feedwater can be concentrated 1500/100 = 15 times. This means that for every 1000 lb of water fed into the boiler, 1000/15 = 66.7 lb of boiler water must be blown down to keep the dissolved solids from exceeding 1500 ppm. In high-pressure boilers, silica, alkalinity, and iron tests are also used in controlling blowdown.

24. What are some common water-treatment tests?

Answer. Titration, colorimetric, hardness, chloride, alkalinity titration, etc. Complete instructions on water testing are contained in (1) the manual on industrial water published by the American Society for Testing and Materials (ASTM) and (2) *Standard Methods for the Examination of Water, Sewerage, and Industrial Wastes,* published jointly by the American Public Health Association and the American Water Works Association.

25. What is the purpose of an electrical conductivity test of boiler water?

Answer. To measure the extent to which dissolved substances are concentrated in the boiler water. This test then helps in controlling carryover of dissolved solids, which condense in lines or equipment such as turbine blades, into the steam system.

26. What are some ways to reduce the oxygen content of boiler water?

Answer. Mechanical deaerator removal and the use of sodium, sulfite, or hydrazine.

27. Why is hydrazine used for oxygen control on high-pressure boilers?

Answer. Hydrazine (N_2H_4) is the oxygen scavenger most often used for high-pressure boilers. Its advantages over sulfite, which is used in most low-pressure units, are that it adds no dissolved solids to the boiler. Hydrazine scavenges oxygen according to the following reaction:

$$N_2H_4 + O_2 \rightarrow N_2 + 2H_2O$$

A major disadvantage to the use of sulfite in high-pressure boilers is that it may break down to corrosive H_2S and SO_2. The following reactions are involved:

$$Na_2SO_3 + H_2O + heat \rightarrow 2NaOH + SO_2$$
$$4Na_2SO_3 + 2H_2O + heat \rightarrow 3Na_2SO_4 + 2NaOH + H_2S$$

Another method used is *volatile treatment.* This system is based on the use of hydrazine and neutralizing amines or ammonia. Major advantages claimed of volatile treatment are that it adds no solids to the boiler and affords good preboiler protection. Boiler pH is controlled between 8.5 and 9.0, and the hydrazine residual in the feedwater is controlled between 0.01 and 0.10 ppm.

28. Why are chelants used in water treatment?

Answer. For boilers operating under 1000 psi, chelants have been used to control ion oxide deposits in boiler systems. The chelants react with metal ions to form soluble compounds which can then be flushed from the boiler water. The degree of combining is dependent on the system environment and the reactivity of the chelant to the metal ion that may be encountered. The chelant agents' reaction is considered to be applicable for soluble metal ions and not for insoluble metals. For the latter, dispersant polymer agents have been developed. The polymer is adsorbed onto the iron or insoluble metal oxide, thus controlling insoluble oxide deposits by altering their charge characteristics and preventing agglomeration. When used this way, the polymers are called *dispersants.*

29. What are major treatment problems with heating boilers?

Answer. Corrosion and pitting. Scale is not a problem because the same feedwater is used continuously, and the initial treatment usually lasts throughout the heating season.

30. What are the main elements of boiler-room management that require engineering attention and review?

Answer.   Boiler-room operation and related costs usually must consider the following: (1) Purchase of equipment and supplies needed to keep the equipment operating efficiently and reliably, (2) training and supervision of the staff required to operate the plant, (3) managing and controlling the chemical process of combustion in its varied forms from fuel purchase to burner control, (4) controlling and supplying load demands as reliably as possible, (5) controlling the impurities in the boiler water and supplying as much contaminant-free steam for process or power as possible, (6) managing routine cleaning, inspection, and periodic repairs of heat-transfer equipment and the auxiliaries involved, (7) conducting routine performance tests to check on efficiency of operation, and also safety tests such as low water, loss of flame, loss of power, and similar tests for emergency training of staff and to note if safety devices perform as intended, and (8) record keeping on output, fuel consumption, and supplies, and calculating efficiency of performance from recorded data.

Practical Calculations

31. What is boiler horsepower? How does it differ from engine horsepower?

Answer.   Boiler horsepower is an arbitrary unit involving 10 ft^2 of heating surface or 33,479 Btu (33,480 used in calculations) or the evaporation of 34.5 lb of water at 212°F. This is the potential of energy developed. Engine horsepower is 33,000 ft · lb of work per minute.

32. Give three formulas for calculating the horsepower rating of boilers.

Answer.   They are:

$$1. \quad \text{hp} = \frac{H}{33,480} \qquad \text{by Btu output method}$$

$$2. \quad \text{hp} = \frac{W}{34.5} \qquad \text{by steam flow output}$$

$$3. \quad \text{hp} = \frac{S}{10} \qquad \text{by heating surface}$$

where H = total Btu used
$\quad\;\; W$ = equivalent evaporation of water in pounds
$\quad\;\;\, S$ = ft^2 of heating surface of boiler

33. What is the horsepower output rating of a boiler containing 850 ft^2 of heating surface, and generating 3200 lb of steam per h at 95 psig, when the feedwater inlet temperature is 190°F and the steam quality is 97 percent? Use the Btu output method in Q. 32.

Answer.   Ten square feet of heating surface = 1 hp. Convert gage pressure to approximate absolute pressure by adding 14.696 or 14.7.

Enthalpy of evaporation (From steam tables in Chap. 10) at 109.7 psia = 883.2 Btu/lb

Enthalpy of liquid at 109.7 psi = 305.66 Btu/lb

883 × 0.97 steam quality = 856.5 Btu latent heat/lb

856.5 + 305.6 = 1162.1 Btu/lb at 97 percent dry

190°F − 32°F = 158 Btu/lb of feedwater

1162.1 − 158 = 1004.1 Btu/lb heat needed

3200 lb/h × 1004.1 = 3,213,120 Btu/h total output

$$\text{(Formula 1) hp} = \frac{H}{33,480}$$

$$H = 3,213,120 \text{ Btu}$$

$$\frac{3,213,120 \text{ Btu}}{33,480 \text{ Btu/hp}} = 96 \text{ hp output}$$

34. What is the meaning of the term *factor of evaporation?* What is the formula?

Answer. The factor of evaporation is the ratio of the actual amount of water evaporated at the boiler conditions to the equivalent evaporation of water from and at 212°F. The formula is:

$$F = \frac{H - q}{970.4}$$

when F = factor of evaporation
$\quad H$ = Btu required for steam under actual boiler operation
$\quad q$ = Btu heat of liquid of feedwater
$\quad 970.4$ = latent heat of steam at 212°F

Note: Always subtract 32°F from the temperature of the feedwater.

35. What is the equivalent evaporation of 4500 lb of water? Steam pressure is 100 psia and feedwater is 180°F.

Answer. The formula for calculating the factor of evaporation is

$$F = \frac{H - q}{970.4}$$

where F = factor of evaporation
$\quad H$ = total enthalpy in 1 lb of steam
$\quad q$ = enthalpy of liquid of feedwater
$\quad 970.4$ = latent heat of 1 lb of steam, Btu at 212°F

Consult the steam tables in Chap. 10 and note that 1 lb of steam at 100 psia has an enthalpy of 1187.2 Btu. This is H.

$180 - 32 = 148$ Btu/lb heat of liquid water (note that 32 is always subtracted).

$$F = \frac{1187.2 - 148}{970.4}$$

$$= 1.071$$

The formula for calculating the equivalent evaporation in pounds of water is

$$W' = WF$$

where $W' =$ equivalent evaporation, lb of water
$W =$ actual evaporation, lb of water
$F =$ factor of evaporation
$W' = 4500 \times 1.071$
$\quad = 4819.5$ lb of water

36. A boiler evaporates 5000 lb of water per hour under actual load. The factor of evaporation is 1.06. How many pounds of fuel are consumed to evaporate this water from and at 212°F if each pound of fuel has 18,000 Btu and the overall boiler efficiency is 80 percent?

Answer. The calculation is

$$W' = W \times F$$

$$W' = 5000 \times 1.06$$

Equivalent weight of water $= 5300$ lb

Btu available per lb of fuel at 80 percent efficiency $= 18,000 \times 0.80 = 14,400$

Latent heat of steam at 212°F $= 970.4$ Btu

$5300 \times 970.4 = 5,143,120$ total Btu's needed per hour

$$\frac{5,143,120}{14,400} = 357.16 \text{ lb of fuel/h needed}$$

37. How many pounds of steam at 15 psig are needed to heat 100 lb of feedwater from 60 to 180°F at atmospheric conditions?

Answer. The calculation is

$$180 - 60 = 120 \text{ Btu/lb needed}$$

$$100 \times 120 = 12,000 \text{ Btu needed}$$

$$15 \text{ psig} + 14.7 = 29.7 \text{ psia}$$

Total enthalpy of steam at 29.7 $= 1163.8$ Btu   (from steam table)

The steam temperature is lowered to that of a water-steam mixture as follows:

$$180 - 32 = 148 \text{ Btu/lb heat of liquid}$$

$$1163.8 - 148 = 1015.8 \text{ Btu available for heating water for each lb of 15 psig steam}$$

Now the Btu need is 12,000, therefore,

$$\frac{12,000}{1015.8} = 11.8 \text{ lb of 15 psig steam needed}$$

38. In a properly operated boiler the amount of blowdown is adjusted to some percentage of the steam flow. This is approximated by tests to keep the boiler-water solids concentration at or below some set limit. Suppose that 850-ppm solids is the limit and there is 30 percent makeup with 110-ppm solids. How many pounds of blowdown are needed for each 1000 lb of steam generated?

Answer. The calculation is

$$110 \times 0.30 = 33 \text{ ppm solids in feedwater}$$

$$\frac{850 \text{ ppm solids}}{33} = 25.75 \text{ concentrations}$$

$$\frac{1}{25.75} \times 100 = \text{percent blowdown} = 3.8 \text{ percent}$$

$$1000 \text{ lb} \times 3.8 \text{ percent} = 38 \text{ lb blowdown}$$

Note: Total dissolved solids (TDS) is more accurate than a chlorides test. TDS requires more skill and more equipment than the simple chlorides test.

39. What is the combustion safeguard system in Fig. 9.5?

Answer. This system is based on the rectified impedance principle for oil-fired burners, which states that either a flame or a photocell sited at a flame is capable not only of conducting an electric current but also of rectifying an alternating current. The system utilizes this principle by applying alternating current to either a flame electrode inserted in the flame or a photocell sited at the flame. And so the resultant rectified current, which can be produced only when flame is present, is in turn detected by the relay.

The actual flame-detecting units consist of flame-electrode and photocell rectifier assemblies and the protecting relay. The flame-electrode type is commonly used for nonluminous flames such as oil flames. A flame can conduct an electric current because when fuels are burned, a dissociation of the components of the fuel takes place and leaves free, within the flame boundary, a concentration of ions, which carry electric charges. These ions are produced by splitting the neutral particles which make up the gas into positive and negative parts. The splitting-up process takes place constantly during the burning of the fuel.

And since an electric current is a flow of charged particles through a substance, a flame which is made up of a concentration of charged particles may conduct an electric current.

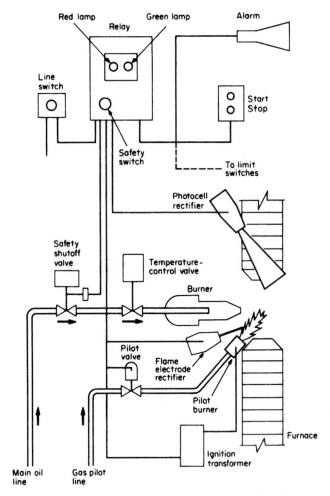

FIGURE 9.5 To prevent furnace explosions, combustion safeguard systems are used. The flame-electrode-type illustrated is based on the rectified imped-ance principle.

CHAPTER 10

PRIME MOVERS: STEAM ENGINES AND STEAM TURBINES

Prime movers are machines that convert some form of energy such as heat, wind, or hydraulic energy into useful mechanical energy. Steam plants use heat energy derived from burning fuel and convert it to mechanical energy in steam engines or steam turbines. The term *vapor cycle* is also used because heat is carried by a working medium, water and steam, which converts the fuel energy and then carries it to the prime movers to be converted to mechanical energy; it is then returned to a boiler to start the cycle again. In other prime movers, such as internal combustion engines, heat is obtained directly from the products of combustion; a portion is converted to mechanical energy, and the rest is exhausted to the atmosphere.

Plant service operators may come in contact with steam engines and steam turbines in their jobs and may also be required to have some knowledge of prime movers if they expect to secure a licensed operator status, which is required in many jurisdictions (see App. 4). Prime movers as energy converters can pose a threat to surroundings and life if the energy conversion is not controlled, and this is why many jurisdictions have licensing requirements to show that an operator has some basic knowledge about the energy conversion equipment in order to control it safely.

BASIC STEAM PROPERTIES REVIEW

Since steam is the working medium in steam engines and turbines, some basics require review. Boiling and vaporizing are defined in steam tables by temperature and absolute pressure, where *absolute pressure* is equal to gage pressure plus about 14.7 psi. For example, vaporization temperature for water is 212°F at 14.7 psi absolute pressure (atmospheric pressure).

Saturated steam is steam which exists at the temperature corresponding to its absolute pressure. *Superheated* steam is steam that has a temperature above the saturation temperature at that absolute pressure. The *quality of steam* is the amount of dry steam per pound of wet steam and is expressed as a percentage. Steam quality is determined by *steam calorimeters* as illustrated in Fig. 10.1. By measuring the weight of dry steam passing through the calorimeter and the weight of moisture that is separated from the incoming steam, the quality X is determined by the following equation:

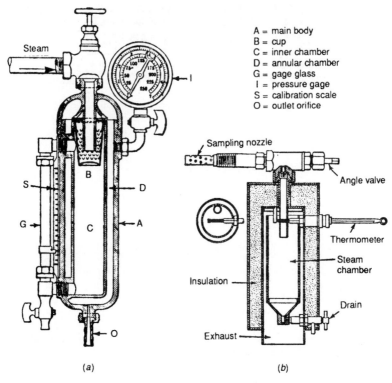

A = main body
B = cup
C = inner chamber
D = annular chamber
G = gage glass
I = pressure gage
S = calibration scale
O = outlet orifice

Steam

Sampling nozzle

Angle valve

Thermometer

Steam chamber

Insulation

Drain

Exhaust

(*a*)   (*b*)

FIGURE 10.1   Calorimeters are used to determine steam quality. (*a*) Separating calorimeter; (*b*) throttling calorimeter.

$$X = \frac{\text{weight of dry stream, lb}}{\text{weight of dry steam + weight of wet steam, lb}}$$

Enthalpy of steam indicates the amount of Btu per pound that may exist in steam at a defined absolute pressure and temperature and consists of two enthalpies that are combined to give the total heat content. (1) *Enthalpy of liquid* is the amount of energy in Btu that is required to raise 1 lb of water from a temperature of 32°F to a temperature of boiling at a given absolute pressure (see Table 10.1). The enthalpy of liquid is shown under the Sat. Liquid column and has the symbol h_f. (2) *Enthalpy of evaporation* is the heat energy, in Btu, necessary to convert 1 lb of liquid water into dry steam at a particular absolute pressure and temperature. It is shown in Table 10.1 in the Evap. column and has the symbol h_{fg}. The *total enthalpy* at a particular pressure and temperature is the sum of the enthalpy of liquid and the enthalpy of evaporation and is shown in Table 10.1 under the Sat. vapor column and has the symbol h_g. The temperature column in Table 10.1 provides the boiling temperature of water for that absolute pressure.

Example.   Steam at 600-psi gage pressure at saturated condition supplies a mechanical steam turbine. What is the corresponding saturated steam temperature and total enthalpy?

$$600 + 14.7 = 614.7 \text{ absolute pressure}$$

Go to Table 10.1 and interpolate:

$$\text{Temperature} = \left(16.89 \times \frac{14.7}{100}\right) + 486.21 = 488.69°F$$

$$\text{Enthalpy} = 1203.2 - \left(2 \times \frac{14.7}{100}\right) = 1202.9 \text{ Btu per lb}$$

Steam turbines expand steam through rows of blades as the size increases, and this requires "dry," or superheated, steam in order to avoid very moist steam in the lower rows of blading. Otherwise there could be blade material erosion from high-velocity flowing wet steam. Table 10.2 provides some of the properties of superheated steam, based on *degrees* of superheat. Note that the first column provides the saturation temperature for the corresponding absolute pressure.

Entropy is defined as the amount of heat in 1 lb of steam that is added or subtracted per *degree absolute* (459.6+°F) at which the addition or subtraction is made. The change in entropy is defined as heat content per degree of absolute temperature, and it is this ratio that is called entropy. The change of entropy is the *total* of the sum of the entropy of liquid s_f and the entropy of evaporation s_g, and these are shown in Tables 10.1 and 10.2. Entropy is used by engineers to predict the quality of the steam as it passes through boilers and heat-using engines. For example, if no heat is added to or subtracted from a machine (adiabatic conditions), there is no change in the total entropy. However, if work is done by the machine to lower the steam temperature, the enthalpy is lowered, and if the expansion is carried far enough, the quality of the steam will change, while the entropy will remain the same.

Example.   A steam engine receives steam at 100 psia at dry saturated conditions and discharges it at 30 psi absolute pressure. Measurements by calorimeter show that inlet steam had a quality of 0.98. Assuming adiabatic expansion through the engine, what will be the quality of the exhaust steam, and its enthalpy? The initial entropy equals the final entropy under adiabatic conditions. Using Table 10.1, the total entropy is modified by the steam quality factor affecting the entropy for evaporation, or s_{fg}. In equation form

$$\text{Entropy inlet} = \text{entropy outlet}$$

$$S_{fi} + X_i(S_{gi}) = S_{fo} + X_o(S_{go})$$

and from Table 10.1,

$$0.470 + (0.98 \times 1.1286) = 0.3680 + (X_o \times 1.3313)$$

and solving for X_o,

$$X_o = 0.91 \text{ quality at exhaust}$$

The enthalpy on the exhaust is affected by the quality factor of the exhaust steam as follows and using enthalpy designations from Table 10.1:

TABLE 10.1 Properties of Dry Saturated Steam Based on Absolute Pressure

Abs Press., Lb Sq In. p	Temp., F t	Specific Volume		Enthalpy			Entropy			Internal Energy		Abs Press., Lb Sq In. p
		Sat. Liquid v_f	Sat. Vapor v_g	Sat. Liquid h_f	Evap h_{fg}	Sat. Vapor h_g	Sat. Liquid s_f	Evap s_{fg}	Sat. Vapor s_g	Sat. Liquid u_f	Sat. Vapor u_g	
0.491	79.03	0.01608	652.3	47.05	1049.2	1096.3	0.0914	1.9473	2.0387	47.05	1037.0	0.491
0.736	91.72	0.01611	444.9	59.71	1042.0	1101.7	0.1147	1.8894	2.0041	59.71	1041.1	0.736
0.982	101.14	0.01614	339.2	69.10	1036.6	1105.7	0.1316	1.8481	1.9797	69.10	1044.0	0.982
1.227	108.71	0.01616	274.9	76.65	1032.3	1108.9	0.1449	1.8160	1.9609	76.65	1046.4	1.227
1.473	115.06	0.01618	231.6	82.99	1028.6	1111.6	0.1560	1.7896	1.9456	82.99	1048.5	1.473
1.964	125.43	0.01622	176.7	93.34	1022.7	1116.0	0.1738	1.7476	1.9214	93.33	1051.8	1.964
2.455	133.76	0.01626	143.25	101.66	1017.0	1119.4	0.1879	1.7150	1.9028	101.65	1054.3	2.455
5	162.24	0.01640	73.52	130.13	1001.0	1131.1	0.2347	1.6094	1.8441	130.12	1063.1	5
10	193.21	0.01659	38.42	161.17	982.1	1143.3	0.2835	1.5041	1.7876	161.14	1072.2	10
14.696	212.0	0.01672	26.80	180.07	970.3	1150.4	0.3120	1.4446	1.7566	180.02	1077.5	14.696
15	213.03	0.01672	26.29	181.11	969.7	1150.8	0.3135	1.4415	1.7549	181.06	1077.8	15
16	216.32	0.01674	24.75	184.42	967.6	1152.0	0.3184	1.4313	1.7497	184.37	1078.7	16
18	222.41	0.01679	22.17	190.56	963.6	1154.2	0.3275	1.4128	1.7403	190.50	1080.4	18
20	227.96	0.01683	20.089	196.16	960.1	1156.5	0.3356	1.3962	1.7319	196.10	1081.9	20
25	240.07	0.01692	16.303	208.42	952.1	1160.6	0.3533	1.3606	1.7139	208.34	1085.1	25
30	250.33	0.01701	13.746	218.82	945.3	1164.1	0.3680	1.3313	1.6993	218.73	1087.8	30
35	259.28	0.01708	11.898	227.91	939.2	1167.1	0.3807	1.3063	1.6870	227.80	1090.1	35
40	267.25	0.01715	10.498	236.03	933.7	1169.7	0.3919	1.2844	1.6763	235.90	1092.0	40
45	274.44	0.01721	9.401	243.36	928.6	1172.0	0.4019	1.2650	1.6609	243.22	1093.7	45
50	281.01	0.01727	8.515	250.09	924.0	1174.1	0.4110	1.2474	1.6585	249.93	1095.3	50
55	287.07	0.01732	7.787	256.30	919.66	1175.9	0.4193	1.2316	1.6509	256.12	1096.7	55
60	292.71	0.01738	7.175	262.09	915.5	1177.6	0.4270	1.2168	1.6438	261.90	1097.9	60
65	297.97	0.01743	6.655	267.50	911.6	1179.1	0.4342	1.2032	1.6374	267.29	1099.1	65
70	302.92	0.01748	6.206	272.61	907.9	1180.6	0.4409	1.1906	1.6315	272.38	1100.2	70
75	307.60	0.01753	5.816	277.43	904.5	1181.0	0.4472	1.1787	1.6259	277.19	1101.2	75
80	312.03	0.01757	5.472	282.02	901.1	1183.1	0.4531	1.1676	1.6207	281.76	1102.1	80
85	316.25	0.01761	5.168	286.39	897.8	1184.2	0.4587	1.1571	1.6158	286.11	1102.9	85
90	320.27	0.01766	4.896	290.56	894.7	1185.3	0.4641	1.1471	1.6112	290.27	1103.7	90
100	327.81	0.01774	4.432	298.40	888.8	1187.2	0.4740	1.1286	1.6026	298.08	1105.2	100
110	334.77	0.01782	4.049	305.66	883.2	1188.9	0.4832	1.1117	1.5948	305.30	1106.5	110

10.4

Abs. Press. (psi)	Temp (°F)	v_f	v_g	h_f	h_{fg}	h_g	s_f	s_{fg}	s_g	u_f	u_g
120	341.25	0.01789	3.728	312.44	877.9	1190.4	0.4916	1.0962	1.5878	312.05	1107.6
130	347.32	0.01796	3.455	318.81	872.9	1191.7	0.4995	1.0817	1.5812	318.38	1108.6
140	353.02	0.01802	3.220	324.82	868.2	1193.0	0.5069	1.0682	1.5751	324.35	1109.6
150	358.42	0.01809	3.015	330.5	863.6	1194.1	0.5138	1.0556	1.5694	330.01	1110.5
160	363.53	0.01815	2.834	335.93	859.2	1195.1	0.5204	1.0436	1.5640	335.39	1111.2
170	368.41	0.01822	2.675	341.09	854.9	1196.0	0.5266	1.0324	1.5590	340.52	1111.9
180	373.06	0.01827	2.532	346.03	850.8	1196.9	0.5325	1.0217	1.5542	345.42	1112.5
190	377.51	0.01833	2.404	350.79	846.8	1197.6	0.5381	1.0116	1.5497	350.15	1113.1
200	381.79	0.01839	2.288	355.36	843.0	1198.4	0.5435	1.0018	1.5453	354.68	1113.7
250	400.95	0.01865	1.8438	376.00	825.1	1201.1	0.5675	0.9588	1.5263	375.14	1115.8
300	417.33	0.01890	1.5433	393.84	809.0	1202.8	0.5879	0.9225	1.5104	392.79	1117.1
350	431.72	0.01913	1.3260	409.69	794.2	1203.9	0.6056	0.8910	1.4966	408.45	1118.0
400	444.59	0.0193	1.1613	424.0	780.5	1204.5	0.6214	0.8630	1.4844	422.6	1118.5
450	456.28	0.0195	1.0320	437.2	767.4	1204.6	0.6356	0.8378	1.4734	435.5	1118.7
500	467.01	0.0197	0.9278	449.4	755.0	1204.4	0.6487	0.8147	1.4634	447.6	1118.6
550	476.94	0.0199	0.8424	460.8	743.1	1203.9	0.6608	0.7934	1.4542	458.8	1118.2
600	486.21	0.0201	0.7698	471.6	731.6	1203.2	0.6720	0.7734	1.4454	469.4	1117.7
650	494.90	0.0203	0.7083	481.8	720.5	1202.3	0.6826	0.7548	1.4374	479.4	1117.1
700	503.10	0.0205	0.6554	491.5	709.7	1201.2	0.6925	0.7371	1.4296	488.8	1116.3
750	510.86	0.0207	0.6092	500.8	699.2	1200.0	0.7019	0.7204	1.4223	498.0	1115.4
800	518.23	0.0209	0.5687	509.7	688.9	1198.6	0.7108	0.7045	1.4153	506.6	1114.4
850	525.26	0.0210	0.5327	518.3	678.8	1197.1	0.7194	0.6891	1.4085	515.0	1113.3
900	531.98	0.0212	0.5006	526.6	668.8	1195.4	0.7275	0.6744	1.4020	523.1	1112.1
950	538.43	0.0214	0.4717	534.6	659.1	1193.7	0.7355	0.6602	1.3957	530.9	1110.8
1000	544.61	0.0216	0.4456	542.4	649.4	1191.8	0.7430	0.6467	1.3897	538.4	1109.4
1100	556.31	0.0220	0.4001	557.4	630.4	1187.8	0.7575	0.6205	1.3780	552.9	1106.4
1200	567.22	0.0223	0.3619	571.7	611.7	1183.4	0.7711	0.5956	1.3667	566.7	1103.0
1300	577.46	0.0227	0.3293	585.4	593.2	1178.6	0.7840	0.5719	1.3559	580.0	1099.4
1400	587.10	0.0231	0.3012	598.7	574.7	1173.4	0.7963	0.5491	1.3454	592.7	1095.4
1500	596.23	0.0235	0.2765	611.6	556.3	1167.9	0.8082	0.5269	1.3351	605.1	1091.2
2000	635.82	0.0257	0.1878	671.7	463.4	1135.1	0.8619	0.4230	1.2849	662.2	1065.6
2500	668.13	0.0287	0.1307	730.6	360.5	1091.1	0.9126	0.3197	1.2322	717.3	1030.6
3000	695.36	0.0346	0.0858	802.5	217.8	1020.3	0.9731	0.1885	1.1615	783.4	972.7
3206.2	705.40	0.0503	0.0503	902.7	0	902.7	1.0580	0	1.0580	872.9	872.9

SOURCE: Condensed from *Thermodynamic Properties of Steam* by Joseph H. Keenan and Frederick G. Keyes. John Wiley, N.Y.

TABLE 10.2　Properties of Dry Saturated Steam Based on Temperature

Tempera-ture F t	Absolute Pressure Lb per sq in. p	Specific Volume		Enthalpy			Entropy	
		Sat. Liquid v_f	Sat. Vapor v_g	Sat. Liquid h_f	Evap. h_{fg}	Sat. Vapor h_g	Sat. Liquid s_f	Sat. Vapor s_g
32°	0.08854	0.01602	3306	0.00	1075.8	1075.8	0.0000	2.1877
35	0.09995	0.01602	2947	3.02	1074.1	1077.1	0.0061	2.1770
40	0.12170	0.01602	2444	8.05	1071.3	1079.3	0.0162	2.1597
45	0.14752	0.01602	2036.4	13.06	1068.4	1081.5	0.0262	2.1429
50	0.17811	0.01603	1703.2	18.07	1065.6	1083.7	0.0361	2.1264
60°	0.2563	0.01604	1206.7	28.06	1059.9	1088.0	0.0555	2.0948
70	•0.3631	0.01606	867.9	38.04	1054.3	1092.3	0.0745	2.0647
80	0.5069	0.01608	633.1	48.02	1048.6	1096.6	0.0932	2.0360
90	0.6982	0.01610	468.0	57.99	1042.9	1100.9	0.1115	2.0087
100	0.9492	0.01613	350.4	67.97	1037.2	1105.2	0.1295	1.9826
110°	1.2748	0.01617	265.4	77.94	1031.6	1109.5	0.1471	1.9577
120	1.6924	0.01620	203.27	87.92	1025.8	1113.7	0.1645	1.9339
130	2.2225	0.01625	157.34	97.90	1020.0	1117.9	0.1816	1.9112
140	2.8886	0.01629	123.01	107.89	1014.1	1122.0	0.1984	1.8894
150	3.718	0.01634	97.07	117.89	1008.2	1126.1	0.2149	1.8685
160°	4.741	0.01639	77.29	127.89	1002.3	1130.2	0.2311	1.8485
170	5.992	0.01645	62.06	137.90	996.3	1134.2	0.2472	1.8293
180	7.510	0.01651	50.23	147.92	990.2	1138.1	0.2630	1.8109
190	9.339	0.01657	40.96	157.95	984.1	1142.0	0.2785	1.7932
200	11.526	0.01663	33.64	167.99	977.9	1145.9	0°2938	1.7762
210°	14.123	0.01670	27.82	178.05	971.6	1149.7	0.3090	1.7598
212	14.696	0.01672	26.80	180.07	970.3	1150.4	0.3120	1.7566
220	17.186	0.01677	23.15	188.13	965.2	1153.4	0.3239	1.7440
230	20.780	0.01684	19.382	198.23	958.8	1157.0	0.3387	1.7288
240	24.969	0.01692	16.323	208.34	952.2	1160.5	0.3531	1.7140
250°	29.825	0.01700	13.821	218.48	945.5	1164.0	0.3675	1.6998
260	35.429	0.01709	11.763	228.64	938.7	1167.3	0.3817	1.6860
270	41.858	0.01717	10.061	238.84	931.8	1170.6	0.3958	1.6727
280	49.203	0.01726	8.645	249.06	924.7	1173.8	0.4096	1.6597
290	57.556	0.01735	7.461	259.31	917.5	1176.8	0.4234	1.6472
300°	67.013	0.01745	6.466	269.59	910.1	1179.7	0.4369	1.6350
310	77.68	0.01755	5.626	279.92	902.6	1182.5	0.4504	1.6231
320	89.66	0.01765	4.914	290.28	894.9	1185.2	0.4637	1.6115
330	103.06	0.01776	4.307	300.68	887.0	1187.7	0.4769	1.6002
340	118.01	0.01787	3.788	311.13	879.0	1190.1	0.4900	1.5891
350°	134.63	0.01799	3.342	321.63	870.7	1192.3	0.5029	1.5783
360	153.04	0.01811	2.957	332.18	862.2	1194.4	0.5158	1.5677
370	173.37	0.01823	2.625	342.79	853.5	1196.3	0.5286	1.5573
380	195.77	0.01836	2.335	353.45	844.6	1198.1	0.5413	1.5471
390	220.37	0.01850	2.0836	364.17	835.4	1199.6	0.5539	1.5371
400°	247.31	0.01864	1.8633	374.97	826.0	1201.0	0.5664	1.5272
410	276.75	0.01878	1.6700	385.83	816.3	1202.1	0.5788	1.5174
420	308.83	0.01894	1.5000	396.77	806.3	1203.1	0.5912	1.5078
430	343.72	0.01910	1.3499	407.79	796.0	1203.8	0.6035	1.4982
440	381.59	0.01926	1.2171	418.90	785.4	1204.3	0.6158	1.4887
450°	422.6	0.0194	1.0993	430.1	774.5	1204.6	0.6280	1.4793
460	466.9	0.0196	0.9944	441.4	763.2	1204.6	0.6402	1.4700
470	514.7	0.0198	0.9009	452.8	751.5	1204.3	0.6523	1.4606
480	566.1	0.0200	0.8172	464.4	739.4	1203.7	0.6645	1.4513
490	621.4	0.0202	0.7423	476.0	726.8	1202.8	0.6766	1.4419
500°	680.8	0.0204	0.6749	487.8	713.9	1201.7	0.6887	1.4325
520	812.4	0.0209	0.5594	511.9	686.4	1198.2	0.7130	1.4136
540	962.5	0.0215	0.4649	536.6	656.6	1193.2	0.7374	1.3942
560	1133.1	0.0221	0.3868	562.2	624.2	1186.4	0.7621	1.3742
580	1325.8	0.0228	0.3217	588.9	588.4	1177.3	0.7872	1.3532
600°	1542.9	0.0236	0.2668	617.0	548.5	1165.5	0.8131	1.3307
620	1786.6	0.0247	0.2201	646.7	503.6	1150.3	0.8398	1.3062
640	2059.7	0.0260	0.1798	678.6	452.0	1130.5	0.8679	1.2789
660	2365.4	0.0278	0.1442	714.2	390.2	1104.4	0.8987	1.2472
680	2708.1	0.0305	0.1115	757.3	309.9	1067.2	0.9351	1.2071
700°	3093.7	0.0369	0.0761	823.3	172.1	995.4	0.9905	1.1389
705.4	3206.2	0.0503	0.0503	902.7	0	902.7	1.0580	1.0580

SOURCE:　Reprinted from *Thermodynamic Properties of Steam* by Joseph H. Keenan and Frederick G. Keyes. John Wiley, N.Y.

$$\text{Total enthalpy} = \text{Sat. liquid} + (X_o \times \text{Evap})$$

or at 30 psia,

$$\text{Total enthalpy} = 218.82 + (0.91 \times 945.3) = 1079 \text{ Btu per lb}$$

Superheated steam is produced by heating saturated steam above the saturation temperature, corresponding to the steam pressure. This raises the total enthalpy of the steam. Advantages of superheated steam include: (1) The additional heat added to the steam vapor causes it to act more like a gas, (2) steam condensation losses are reduced in pipelines and the steam-using equipment, (3) the temperature differential in steam-using equipment may be increased, and (4) erosive wear and tear from wet steam passing through steam-using machinery is reduced.

Example.   How many British thermal units per pound of steam are required to produce 600 psia at 900°F from a feedwater temperature of 200°F? From Table 10.2, h_f at 200°F is 168 Btu per lb. Using Table 10.3, (pp. 10.8 to 10.10) at 600 psia and 900°F superheat, $h = 1462.5$. The Btu's required are $1462.5 - 168 = 1294.5$ Btu per lb.

Heat Conversion Factors

Per the first law of thermodynamics, energy cannot be created or destroyed, but all forms are mutually convertible as follows:

$$\text{Heat and work} - 778 \text{ ft} \cdot \text{lb} = 1 \text{ Btu}$$

$$1 \text{ hp} \cdot \text{h} = 2545 \text{ Btu/h}$$

For heat and electricity,

$$1 \text{ kW} \cdot \text{h} = 3413 \text{ Btu's per h}$$

Vapor Cycles

The *Carnot cycle* expresses vapor cycles as an efficiency involving a heat source at temperature T_1 and heat rejection at temperature T_2. The Carnot ideal efficiency is $E_c = T_1 - T_2/T_1$. The Carnot cycle is theoretically the most efficient for an engine working between given temperature limits.

The *Rankine cycle* considers the expansion of steam and the amount of Btu's exchanged as can be noted by the following definition of Rankine efficiency:

$$E_R = \frac{h_1 - h_2}{h_1 - h_{f2}}$$

where h_1 = enthalpy of admitting steam at the throttle pressure, Btu/lb
$\quad\ h_2$ = enthalpy of steam after isentropic (no heat energy enters or leaves the working substance, steam) expansion from an initial pressure p_1 to exhaust pressure p_2, Btu/lb
$\quad\ h_{f2}$ = enthalpy of the liquid at the exhaust pressure p_2, Btu/lb

TABLE 10.3 Properties of Superheated Steam Based on Absolute Pressure

Abs. press, psi (sat. temp.)*		Sat. liquid	Sat. vapor	Temperature, °F							
				300	400	500	600	700	800	900	1000
15 (213.03)	v	0.02	26.29	29.91	33.97	37.99	41.99	45.98	49.97	53.95	57.93
	h	181.1	1150.8	1192.8	1239.9	1287.1	1334.8	1383.1	1432.3	1482.3	1533.1
	s	0.3135	1.7549	1.8136	1.8719	1.9238	1.9711	2.0147	2.0554	2.0936	2.1296
20 (227.96)	v	0.02	20.09	22.36	25.43	28.46	31.47	34.47	37.46	40.45	43.44
	h	196.2	1156.3	1191.6	1239.2	1286.6	1334.4	1382.9	1432.1	1482.1	1533.0
	s	0.3356	1.7319	1.7808	1.8396	1.8918	1.9392	1.9829	2.0235	2.0618	2.0978
40 (267.25)	v	0.017	10.498	11.040	12.628	14.168	15.688	17.198	18.702	20.20	21.70
	h	236.0	1169.7	1186.8	1236.5	1284.8	1333.1	1381.9	1431.3	1481.4	1532.4
	s	0.3919	1.6763	1.6994	1.7608	1.8140	1.8619	1.9058	1.9467	1.9850	2.0212
60 (292.71)	v	0.017	7.175	7.259	8.357	9.403	10.427	11.441	12.449	13.452	14.454
	h	262.1	1177.6	1181.6	1233.6	1283.0	1331.8	1380.9	1430.5	1480.8	1531.9
	s	0.4270	1.6438	1.6492	1.7135	1.7678	1.8162	1.8605	1.9015	1.9400	1.9762
80 (312.03)	v	0.018	5.472		6.220	7.020	7.797	8.562	9.322	10.077	10.830
	h	282.0	1183.1		1230.7	1281.1	1330.5	1379.9	1429.7	1480.1	1531.3
	s	0.4531	1.6207		1.6791	1.7346	1.7836	1.8281	1.8694	1.9079	1.9442
100 (327.81)	v	0.018	4.432		4.937	5.589	6.218	6.835	7.446	8.052	8.656
	h	298.4	1187.2		1227.6	1279.1	1329.1	1378.9	1428.9	1479.5	1530.8
	s	0.4740	1.6026		1.6518	1.7085	1.7581	1.8029	1.8443	1.8829	1.9193
120 (341.25)	v	0.018	3.728		4.081	4.636	5.165	5.683	6.195	6.702	7.207
	h	312.4	1190.4		1224.4	1277.2	1327.7	1377.8	1428.1	1478.8	1530.2
	s	0.4916	1.5878		1.6287	1.6869	1.7370	1.7822	1.8237	1.8625	1.8990
140 (353.02)	v	0.018	3.220		3.468	3.954	4.413	4.861	5.301	5.738	6.172
	h	324.8	1193.0		1221.1	1275.2	1326.4	1376.8	1427.3	1478.2	1529.7
	s	0.5069	1.5751		1.6087	1.6683	1.7190	1.7645	1.8063	1.8451	1.8817

Abs. Press. psia (Sat. Temp)										
160 (363.53)	v	0.018	2.834	3.008	3.443	3.849	4.244	4.631	5.015	5.396
	h	335.9	1195.1	1217.6	1273.1	1325.0	1375.7	1426.4	1477.5	1529.1
	s	0.5204	1.5650	1.5908	1.6519	1.7033	1.7491	1.7911	1.8301	1.8667
180 (373.06)	v	0.018	2.532	2.649	3.044	3.411	3.764	4.110	4.452	4.792
	h	346.0	1196.9	1214.0	1271.0	1323.5	1374.7	1425.6	1476.8	1528.6
	s	0.5325	1.5542	1.5745	1.6373	1.6894	1.7355	1.7776	1.8167	1.8534
200 (381.79)	v	0.018	2.288	2.361	2.726	3.060	3.380	3.693	4.002	4.309
	h	355.4	1198.4	1210.3	1268.9	1322.1	1373.6	1424.8	1476.2	1528.0
	s	0.5435	1.5453	1.5594	1.6240	1.6767	1.7232	1.7655	1.8048	1.8415
250 (400.95)	v	0.0187	1.8438		2.151	2.427	2.688	2.942	3.192	3.439
	h	376.0	1201.1		1263.4	1318.5	1371.0	1422.7	1474.5	1526.6
	s	0.5675	1.5263		1.5949	1.6495	1.6969	1.7379	1.7793	1.8162
300 (417.33)	v	0.0189	1.5433		1.7675	2.005	2.227	2.442	2.652	2.859
	h	393.8	1202.8		1257.6	1314.7	1368.3	1420.6	1472.8	1525.2
	s	0.5879	1.5104		1.5701	1.6268	1.6751	1.7184	1.7582	1.7954
400 (444.59)	v	0.0193	1.1613		1.2851	1.4770	1.6508	1.8161	1.9767	2.134
	h	424.0	1204.5		1245.1	1306.9	1362.7	1416.4	1469.4	1522.4
	s	0.6214	1.4844		1.5281	1.5894	1.6398	1.6842	1.7247	1.7623
500 (467.01)	v	0.0197	0.9278		0.9927	1.1591	1.3044	1.4405	1.5715	1.6996
	h	449.4	1204.4		1231.3	1298.6	1357.0	1412.1	1466.0	1519.6
	s	0.6487	1.4634		1.4919	1.5588	1.6115	1.6571	1.6982	1.7363
600 (486.21)	v	0.0201	0.7698		0.7947	0.9463	1.0732	1.1899	1.3013	1.4096
	h	471.6	1203.2		1215.7	1289.9	1351.1	1407.7	1462.5	1516.7
	s	0.6720	1.4454		1.4586	1.5323	1.5875	1.6343	1.6762	1.7147
800 (518.23)	v	0.0209	0.5687			0.6779	0.7833	0.8763	0.9633	1.0470
	h	509.7	1198.6			1270.7	1338.6	1398.6	1455.4	1511.0
	s	0.7108	1.4153			1.4863	1.5476	1.5972	1.6407	1.6801

*V = Specific volume, h = enthalpy, s = entropy.

SOURCE: Condensed from *Thermodynamic Properties of Steam* by Joseph H. Keenan and Frederick G. Keyes. John Wiley. N.Y.

10.9

TABLE 10.3 Properties of Superheated Steam Based on Absolute Pressure *(Continued)*

Abs. press, psi (sat. temp.)*		Sat. liquid	Sat. vapor	300	400	500	600	700	800	900	1000
1,000 (544.61)	v	0.0216	0.4456				0.5140	0.6084	0.6878	0.7604	0.8294
	h	542.4	1191.8				1248.8	1325.3	1389.2	1448.2	1505.1
	s	0.7430	1.3897				1.4450	1.5141	1.5670	1.6121	1.6525
1,200 (567.22)	v	0.0223	0.3619				0.4016	0.4909	0.5617	0.6250	0.6843
	h	571.7	1183.4				1223.5	1311.0	1379.3	1440.7	1499.2
	s	0.7711	1.3667				1.4052	1.4843	1.5409	1.5879	1.6293
1,400 (587.10)	v	0.0231	0.3012				0.3174	0.4062	0.4714	0.5281	0.5805
	h	598.7	1173.4				1193.0	1295.5	1369.1	1433.1	1493.2
	s	0.7963	1.3454				1.3639	1.4567	1.5177	1.5666	1.6093
1,500 (596.23)	v	0.0235	0.2765				0.2815	0.3719	0.4352	0.4893	0.5390
	h	611.6	1167.9				1174.5	1287.2	1363.8	1429.3	1490.1
	s	0.8082	1.3351				1.3412	1.4434	1.5068	1.5569	1.6001
2,000 (635.82)	v	0.0257	0.1878					0.2489	0.3074	0.3532	0.3935
	h	671.7	1135.1					1240.0	1335.5	1409.2	1474.5
	s	0.8619	1.2849					1.3783	1.4576	1.5139	1.5603
2,500 (668.13)	v	0.0287	0.1307					0.1686	0.2294	0.2710	0.3061
	h	730.6	1091.1					1176.8	1303.6	1387.8	1458.4
	s	0.9126	1.2322					1.3073	1.4127	1.4772	1.5273
3,000 (695.36)	v	0.0346	0.0858					0.0984	0.1760	0.2159	0.2476
	h	802.5	1020.3					1060.7	1267.2	1365.0	1441.8
	s	0.9731	1.1615					1.1966	1.3690	1.4439	1.4984
3,206.2 (705.40)	v	0.0503	0.0503						0.1583	0.1981	0.2288
	h	902.7	902.7						1250.5	1355.2	1434.7
	s	1.0580	1.0580						1.3508	1.4309	1.4874

*V = Specific volume, h = enthalpy, s = entropy.

SOURCE: Condensed from *Thermodynamic Properties of Steam* by Joseph H. Keenan and Frederick G. Keyes. John Wiley, N.Y.

Example.   What is the Rankine cycle efficiency for a steam engine operating with dry admission steam at 160 psia and exhausting at 2 psia? From Table 10.1, $h_1 = 1195.1$ Btu/lb. Since the steam is expanded adiabatically, it is necessary to determine the quality of the steam and how this affected h_2:

Total entropy at 160 psia = 1.5640   (Table 10.1)

Entropy of liquid at 2 psia = 0.1749

Entropy of evaporation at 2 psia = 1.7451

$$\text{Exhaust quality of steam} = \frac{1.564 - 0.1749}{1.7451} = 0.796$$

Total enthalpy of *exhaust* steam = enthalpy of liquid + enthalpy of evaporation, considering the quality of the steam

or

Total enthalpy of *exhaust* = 93.99 + (0.796 × 1022.2) = 907.7

$$\text{Then } E_R = \frac{1195.1 - 907.7}{1195.1 - 93.99} = 0.261 \quad \text{or 26.1 percent}$$

Regenerative cycles modify the Rankine cycle by extracting steam from steam turbines to preheat or regenerate feedwater condensate to a temperature near the water temperature inside the boiler. This saves heat and improves the efficiency, and it has proven economical even though feedwater heaters have to be added to the vapor cycle. Steam is extracted from steam turbines at designated points of the steam flow through the turbine to obtain the temperatures required to heat the feedwater in external heat exchangers.

STEAM ENGINES

The steam engine was a major prime mover well into the twentieth century when the advent of the steam turbine proved to be more advantageous in generating large amounts of power. Steam engines can be used for smaller power needs and, where the exhaust steam from the engine can be employed for heating water, for supplying heating loads and similar low-pressure applications. The theoretical steam consumption per horsepower · hour is

$$\frac{2545}{h_1 - h_2}$$

where h_1 is the enthalpy of the inlet steam, and h_2 is the enthalpy of the exhaust as described in the previous problem.

Example.   What is the theoretical steam consumption for the steam engine operating with the pressure depicted in the previous problem?

$$\text{Theoretical steam consumption} = \frac{2545}{1195.1 - 907.7} = 8.8 \text{ lb hp} \cdot \text{h}$$

The steam engine is a reciprocating machine with pistons, cylinder(s) and crankshafts that convert a portion of the heat energy to mechanical power. Different engines were perfected as the industrial revolution developed and grew in size, and as a result, the demand for power to drive machinery expanded.

Engine classification depends on many factors as can be noted by the following:

Item to be considered	Classification
Frame and cylinder	Horizontal, vertical
Valve gear	Slide-valve, Corliss, poppet
Steam flow in cylinder	Unidirectional, counterflow
Number of cylinders for expansion	Single, two (compound), three (triple expansion), and four (quadruple expansion)
Speed of rotation	Low, medium, high
Ratio of stroke to diameter	Long, short
Type of exhaust	Noncondensing, condensing

Some other steam engine terms that operators should be familiar with are: The *head end* of the cylinder is farthest away from the crankshaft, while *crank end* is the nearest to the crankshaft. *Dead center* is the position at which reversal of piston motion occurs at either end of the cylinder. This indicates there are two dead ends—head end and crank end. *Stroke* is the linear distance traveled by the piston from one dead-center position to the other. The *forward stroke* is between the head- and crank-end dead centers, with the *return stroke* being the motion in the opposite direction. *Long-stroke* engines have a stroke or piston travel greater than the cylinder diameter, while *short-stroke* engines have a piston travel equal to or less than the cylinder diameter. *Piston displacement* is the volume swept through by the piston, or the product of piston area times stroke length. *Piston speed* is equal to twice the length of the engine stroke times crankshaft revolutions per minute. Thus, if an engine has a stroke of 12 in and turns at 125 r/min, the piston speed is $[(2 \times 12)/12] \times 125 = 250$ ft/min.

Engines are also identified by the *valve mechanism* employed to admit steam into the cylinder and then exhaust it at the end of the stroke. Valves control the events of admission, cutoff, expansion release, exhaust, and compression of the steam during each working cycle. Figure 10.2 shows the ideal events as described and which normally would be obtained from a real engine by an indicator diagram as shown in the figure. An actual pressure-volume diagram is also shown in Fig. 10.2.

The *D-slide valve engine* is an early engine and is illustrated in Fig. 10.3. Steam is admitted to the cylinder by the D-slide valve, and the expansion of steam after valve closure moves the piston, piston rod, and crosshead, which produces a rotary motion of the main shaft through the turning effort exerted on the crankpin. Figure 10.4 shows a simple D-slide engine of the vertical type. This type of engine is simple and costs less than other types of valved engines, but it is less efficient because the valves do not permit steam cutoff within the engine cylinder at less than about five-eighths of the stroke without causing problems with other events in the expansion process.

Lead is the term applied to the amount in inches that a valve is open to admit steam to the cylinder prior to the dead-center position of the piston, and this helps to avoid sudden jolts as the piston comes to rest at the end of the stroke (cushions it).

The term *lap* is used to show the amount by which the admission (steam lap) or exhaust lap overlaps the ports when the valve is in midposition. Steam laps are provided to give cutoffs earlier than the end of the stroke.

The *Corliss* engine is an improvement over the D-slide valve engine and features semirotary motion of the valve gear that admits steam or exhaust through narrow cut grooves in the valves as shown in Fig. 10.5. The cylinder has four valve chambers—two

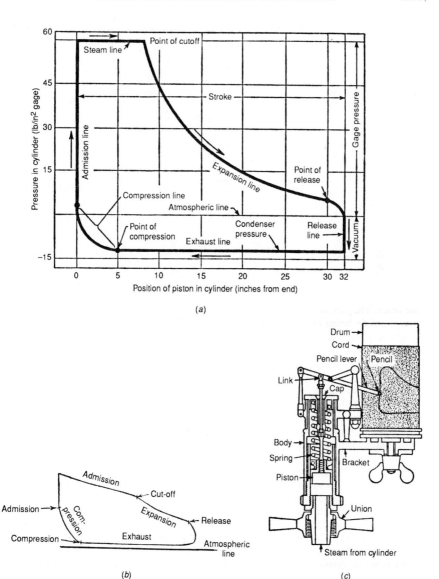

FIGURE 10.2 (*a*) Ideal indicator diagram shows admission, cutoff, expansion release, and exhaust in terms of steam pressure and piston stroke position; (*b*) actual pressure-volume diagram of an engine; (*c*) engine indicator apparatus.

steam valves and two exhaust valves. Very short ports exist between the cylinder and valve chambers. The steam and exhaust valves are illustrated in Fig. 10.5*b*. The valves are actuated by a single eccentric located on the engine shaft (see Fig. 10.5*c*). The eccentric motion is transmitted by eccentric straps and the eccentric rod to a vertical rocker arm pivoted at its bottom end. Motion from the upper end of the rocker is carried

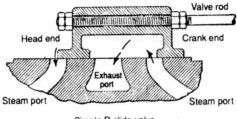

Simple D slide-valve

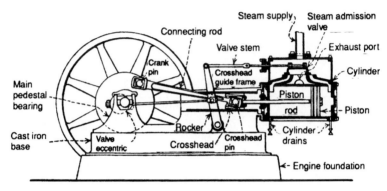

FIGURE 10.3   Horizontal D-slide valve engine.

by the reach rod to the wrist plate mounted on the side of the cylinder. It is the wrist plate which gives an oscillating motion that moves the valves. Fig. 10.5*d* shows a Murray steam engine valve gear.

Very rapid and sharp cutoff of valve admission and exhaust is obtained by the use of dash-pot plungers on Corliss engines. These plungers are lifted against the pressure of the atmosphere as a partial vacuum is produced under the plunger when it is lifted. By the intricate valve mechanism, the steam hook is disengaged from the catch block shown in Fig. 10.5*d,* and the valves are quickly closed by the air pressure exerted on top of the dash-pot plungers.

The governor for a Corliss engine is usually a loaded-pendulum type, belt-driven by a pulley. Speed changes may be made by increasing or decreasing the weights of the pendulum arms. If the governor belt breaks, the pendulum arms are not rotated, and the valve mechanism closes the engine valves.

The Corliss engine is more efficient than the D-slide valve engine because of reduced clearance volumes, less condensation losses, and good cutoff governing.

Poppet-valve Engines

Poppet-valve engines were developed for steam pressures up to 400 psi and 200°F superheat. The poppet valves are actuated by a cam shaft (see Fig. 10.6), which is driven from the main engine shaft. The poppet valves are mechanically opened by the cam shaft, and a spring is used to assist the valve in seating quickly at the steam cutoff point.

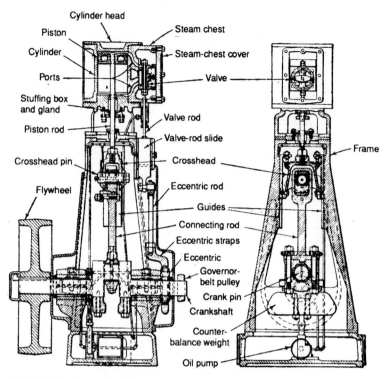

FIGURE 10.4   Vertical D-slide valve steam engine and parts.

All valves are of the double-seat type. The earlier engines had two admission and two exhaust valves. However, this produced counterflow, or all the expanded steam is returned to the inlet end of the cylinder during exhaust, thus cooling the cylinder walls. This produced condensation losses. In the *uniflow engine,* Fig. 10.6, admission is by poppet valves; however, the piston itself opens the exhaust port near the end of the stroke by uncovering the exhaust port's cut in the cylinder. This design reduces the condensation losses that occur in counterflow engines.

Compound engines (see Fig. 10.7) were built to expand the steam through larger pressure and temperature ranges. Two or more cylinders are used for the successive expansion of steam through high- and low-pressure cylinders. If the cylinders are in line, the arrangement is *tandem,* and if the cylinders are parallel to each other, it is a *cross-compound* engine.

Engine Power

Steam engine operators use an engine indicator, an instrument for drawing actual pressure-volume diagrams as is shown in Fig. 10.2b. Once this diagram is obtained, the areas of the indicator diagrams are obtained by a mechanical integrating device called a *polar planimeter.* From the area of the indicator diagram, the mean ordinate may be

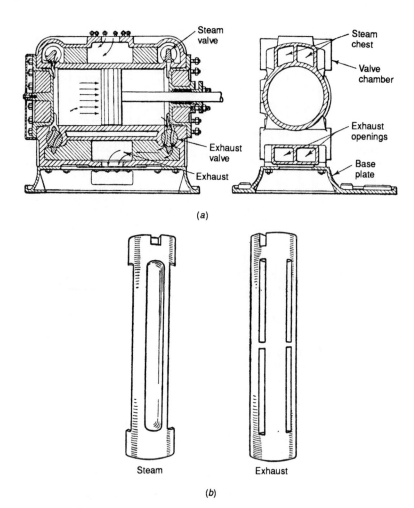

FIGURE 10.5   Corliss steam engine. (*a*) Cylinder has two admission and two exhaust valves;
(*b*) ported valves partially rotate to admit steam and exhaust it from the cylinder.

determined as provided in the instructions for the instrument. The mean ordinate in
inches times the indicator spring constant provides the *mean effective pressure,* mep, in
the cylinder. The indicated horsepower is calculated from

$$\text{ihp} = \frac{PLAN}{33,000}$$

where ihp = indicated horsepower
       P = mean effective pressure, psi
       L = stroke, ft
       A = effective area of piston, in^2
       N = engine r/min

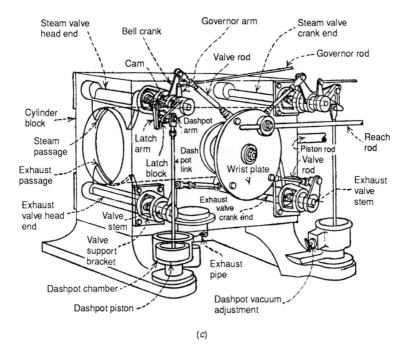

(c)

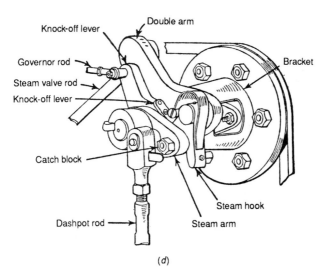

(d)

FIGURE 10.5 (*Continued*) Corliss steam engine. (*c*) intricate releasing-valve gear mechanisms are a feature of the Corliss engine; (*d*) steam valve-gear nomenclature.

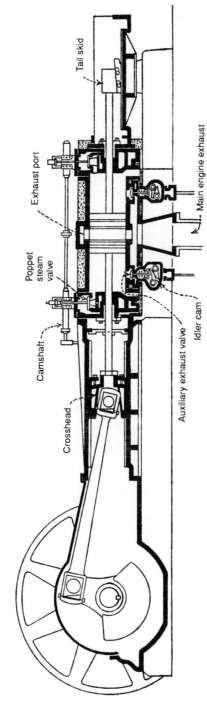

FIGURE 10.6 Poppet-valve uniflow steam engine has poppet valves for admission, but piston travel acts as exhaust valve in the middle of engine stroke, thus reducing condensation loss.

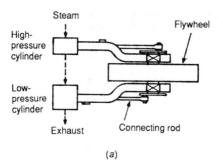

(a)

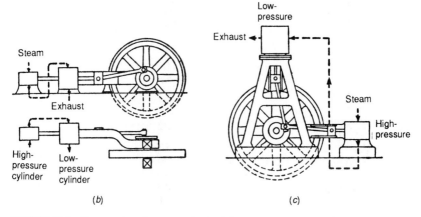

FIGURE 10.7   Compound engines come in the following arrangements. (a) Cross-compound engine; (b) tandem-compound engine; (c) angle-compound engine.

For double-acting pistons, it is usual practice to multiply the above by 2, or

$$\text{ihp} = 2 \times \frac{PLAN}{33,000}$$

Example.   A double-acting engine has a piston diameter of 14 in, a stroke of 30 in, a speed of 125 r/min, and an mep of 75 psi. What is the indicated horsepower? Substituting in above formula

$$\text{ihp} = \frac{2 \times 75 \times 30/12 \times 0.7854(14)^2 \times 125}{33,000}$$

$$\text{ihp} = 218.66$$

The actual power delivered by the crankshaft is the *brake horsepower,* bhp, the term being derived from the method of calculating the true output by Prony brakes. The *mechanical efficiency* of an engine is bhp/ihp.

Another term used in steam engine work is *thermal efficiency*. This is the ratio of the heat equivalent output (2545 Btu/hp · h) to the quantity of heat supplied by the steam to the engine to produce the power.

When superheated steam is supplied to the engine, the thermal efficiency E_t is

$$E_t = \frac{2545}{w_s(h - h_{f2})}$$

where w_s = actual pounds of steam supplied per horsepower · hour
h = total enthalpy of superheated steam at initial absolute pressure, Btu/lb
h_{f2} = enthalpy of liquid at absolute exhaust pressure, Btu/lb

Example. A steam engine uses 40 lb of steam per bhp · h with an initial pressure of 150 psia and exhaust pressure of 15 psia. Find the thermal efficiency. Using Table 10.1, for 150 psia, $h = 1194.1$, while for 15 psia, $h_{f2} = 181.11$. Then

$$E_t = \frac{2545}{40(1194.1 - 181.11)} = 6.28 \text{ percent}$$

Engine Governors

The main function of a governor is to control the speed with load changes, and this is done either to limit or automatically increase the steam input as the load increases or decreases. Governors must be stable within desired ranges of speed control. The term *coefficient of regulation* is the percentage of variation in speed that a governor is built for in controlling the speed from no load to normal full load, or

$$R = \frac{(N' - N)100}{N}$$

where N' = no-load speed, r/min, and N = normal full-load speed r/min. For electric generator drive, R is usually less than $1\frac{1}{2}$ percent. Two methods of mechanical governing have been used: (1) *throttling* to vary the steam pressure entering the cylinder as is shown in Fig. 10.8a and (2) the *cutoff* governor of the shaft or flywheel type which varies the volume of steam going into the cylinder. Figure 10.8b and c shows the cutoff governor types.

The throttling governor shown in Fig. 10.8a depends on its function by the weights G being thrown out by centrifugal forces as speed increases and at the desired speed setting. This outward movement is translated by linkages to start closing the valve A against the action of a coil spring J.

The flywheel or shaft governors control the volume of steam by controlling the duration of the steam admission period with cutoff for full or normal loads at one-quarter to one-third the stroke. A *centrifugal cutoff* governor of the flywheel type is shown in Fig. 10.8b. This mechanism has a leaf spring attached to one end of the spokes of the flywheel, while the free end of the spring has a weight connected to a level that is also weighted at one end. The lever is pivoted near the flywheel hub with an eccentric rod attached between the lever weight and the lever pivot point. It is the eccentric rod that transmits motion to a valve rocker to change valve positions with speed changes. The operating speed of the engine can be changed by moving the point of attachment of the link at the weighted lever.

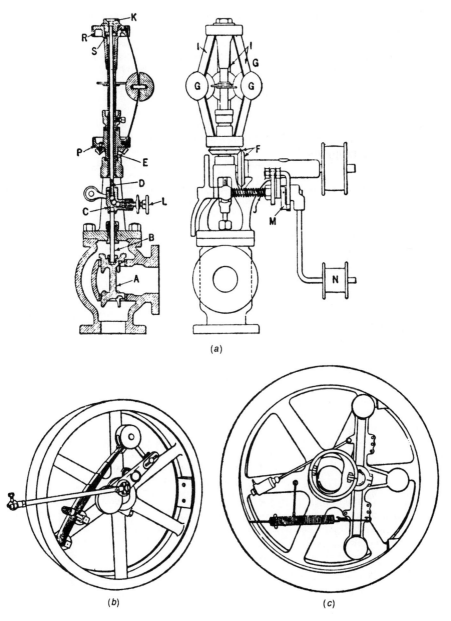

FIGURE 10.8   Steam engine governors. (*a*) Belt-driven throttling governor; (*b*) and (*c*) cutoff governors with (*b*) a flywheel centrifugal type and (*c*) a flywheel inertia type.

Another type of cutoff governor, an *inertia* governor, is shown in Fig. 10.8*c*. A weighted lever is pivoted at its center. A swinging eccentric is rigidly attached to the weight lever and rotates with the lever about the pivot point. When the engine is at rest, the eccentric is at a position to give the maximum cutoff. As the speed is increased, the rim of the flywheel tends to run ahead of the governor weights, which lag behind because of inertia forces, hence the term *inertia governor.* This causes the eccentric position to shift with respect to the engine crank, and the point of steam cutoff is made earlier. When the flywheel runs more slowly, the governor weights crowd ahead due to inertia, and the eccentric shifts to give later cutoff of steam to the cylinder.

The *efficiency* of *engines* is mostly affected by the heat losses in the exhaust steam, which are the largest amount of heat rejected by the engine and are carried away by the exhaust steam. The heat lost can vary from 70 to 90 percent of the total heat supplied. This is why engines were replaced by steam turbines, except where the exhaust heat can be used in a "down the line" process or heating equipment.

Operating and Maintenance

Operating a steam engine requires following certain basic principles and the instructions that the manufacturer provides. Some rules to follow are:

1. Make sure all drains before and after the throttle valve have been opened before attempting to start an engine in order to avoid slugging the cylinder head with condensate. This will also prevent water hammer on the steam line.

2. Check the oil wells, lubricators, and reservoirs for proper oil level and check the oil rings to make sure they turn freely.

3. Check the exhaust valves on condensing and noncondensing machines to make sure they are open to the condenser or the atmosphere. Start admitting steam slowly and bring engine up to speed while closing the previously opened drains.

4. Check the governor for proper speed control and, if so equipped, check the emergency overspeed trip at this time.

5. Check the lubrication system for proper operation while the engine is idling. Bearings that are loose or worn may start to knock. If bearings are running hot, it is an indication of insufficient lubrication, dirty oil or oil contaminated with moisture, bent shaft or misalignment, and/or loose parts.

6. When shutting down an engine after the load has been reduced, close the throttle valve slowly to reduce the speed at an equally slow rate, and try to prevent the engine from stopping at dead center to make starting easier. Make sure to open the cylinder, steam chest, and throttle valve drains after the engine stops to prevent buildup of condensation. Secure the steam and exhaust lines by shutting all valves so that the engine is isolated from any flows.

7. Piston ring wear requires periodic checking and replacement of rings. Solid pistons have to be pulled out of the cylinder to do this, while the rings on a built-up piston can be replaced without removing them from the cylinder by removing the cylinder head and progressively removing the parts with the rings, such as follower plates and bullring. While the cylinder is open, the walls should be checked for uneven wear and grooving. It is good practice to mike the cylinder bore to determine if the bore is still round and within tolerance in order to prevent blowby and uneven piston ring wear.

8. Engines with center cranks on the shaft should be periodically checked with a strain gage for crank alignment with the main shaft and also for cracks developing in the corners of the crank.

9. Valves may leak with time and may require resurfacing or replacement. Pressure-volume indicator cards in the cylinder can be used to detect faulty admission and cutoffs; valve leakage will show rippled lines on expansion and/or discharge. Another method to detect leaky valves is to maneuver the piston so that the suspected valve is directly over the ports and then opening the indicator card cocks. Slowly admit steam to the steam chest; if the valve is faulty, steam will flow out of the indicator card cocks.

10. Flywheels should be checked periodically to make sure the hub bolts are tight. Listen carefully when the engine is operating for any clicking noise coming from the flywheel since this is usually an indication that something on the flywheel is loose.

In general, as with all mechanical machinery, be alert to changes in the operation of the engine, which could include sounds, rhythm, smell, heat, odors, and increased steam consumption. All of these are indicators that something needs to be checked on the engine for correction, repair, adjustment, or replacement.

STEAM TURBINES

The steam turbine has certain advantages when compared with a reciprocating steam engine of the same size and operating conditions. The advantages include: turbines occupy less floor space, require less expensive foundations, provide oil-free exhaust steam, can be operated with superheated steam without any lubrication problems, have more uniform velocity of rotation, can be operated at higher speeds, and generally cost less than an equivalent steam engine. The turbine is primarily used to drive high-speed machinery, such as electric generators, and mechanical drives, such as pumps, compressors, fans and blowers, but it is not very applicable to direct drives where low speeds or large starting torques are needed. The high speed of the turbine can be reduced with intermediate gear set drive, but there is an increased cost, size, and weight involved and also a decrease in the mechanical efficiency.

Basics

The steam turbine converts potential energy of steam under pressure and temperature into kinetic energy that is directed onto blades to rotate a rotor of a turbine. Nozzles are used to expand steam from one pressure to a lower pressure, and the nozzles are usually of the convergent-divergent form shown in Fig. 10.9a. The steam pressure p_o at the throat of a nozzle drops to about 0.58 of the initial absolute pressure for saturated steam and to 0.55 for superheated steam. In most steam turbines, the blades on the rotor are moved by steam impulsing on the blades or by the reaction of impinging steam jets issuing from nozzles or guide passages as shown in Fig. 10.9b. In the impulse turbine, the steam jet issuing from the nozzle at high velocity is then directed against the blades of the turbine wheel, which pushes the rotor in the direction of the jet stream coming from the nozzle.

In the reaction turbine, steam pressure is further reduced in the moving blades with the resulting creation of kinetic energy that provides a reaction force against the blades,

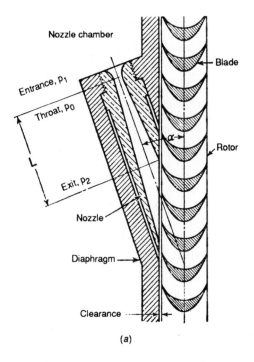

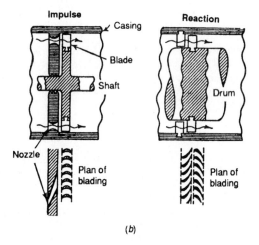

FIGURE 10.9 Steam turbines depend on converting the energy in steam to kinetic energy through nozzle-jet principles. (*a*) Convergent-divergent nozzle directs steam against moving blades at high velocity; (*b*) single-stage impulse and reaction turbine blade arrangement.

similar to a garden hose "kicking" back when high-pressure water flows out of a nozzle attached to the hose.

Many larger steam turbines combine the impulse and reaction blade design by staging, or having many rows of blades. Two terms used in staging are *velocity staging* and *pressure staging.* Figure 10.10 illustrates both staging methods. The velocity developed in the first-stage nozzles C is directed to two rows of moving blades D and F. Blades D absorb about one-half the velocity, with the exhaust from this row being reversed by the stationary blades E and then directed to the second row of moving blades F. This multivelocity absorption of energy from a nozzle discharge is *velocity staging,* or the *Curtis* principle, named after its developer.

In Fig. 10.10, after the steam goes through row F, it is expanded in the entrance nozzles of five successive *pressure stages,* each of which has one row of moving blades, or one velocity stage per pressure stage. Steam partly expanded in a set of nozzles, with the resulting steam velocity used in one row of moving blades, may be expanded further in a second set of nozzles, with another increase of velocity. Moving blades adjacent to the nozzles absorb the velocity. This expansion and increase in velocity may be repeated. This is called *pressure staging,* or the *Rateau* principle.

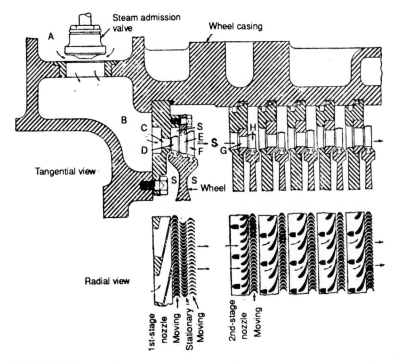

FIGURE 10.10   A multistage impulse steam turbine has six pressure stages with the first pressure stage at the beginning of flow having two velocity stages (Curtis), followed by five stationary pressure stages with moving velocity stages per pressure stage (Rateau).

Classification

Some turbine classifications are:

1. Mechanical or electric drive
2. Condensing or noncondensing
3. Extraction or nonextraction (see Fig. 10.11*a*)
4. Tandem and cross-compound (see Fig. 10.11*b*)
5. Single or multiple casing
6. Multiple casing such as high pressure, reheat, intermediate pressure, and low pressure on large utility units (see Fig. 10.12)
7. Single or double flow
8. Axial or radial flow (Sta-Laval)
9. Impulse, reaction, or mixed staging
10. Single and multiple staging
11. Straight connected, coupled, or gear drive
12. Topping and superposed

Turbine Performance

For the simpler turbine applications, such as mechanical drive, steam turbine performance is expressed as the *water rate,* or steam consumption in pounds per brake horsepower · hour. The theoretical water rate in pounds per horsepower · hour can be calculated as follows:

$$\text{Theoretical water rate per horsepower · hour} = \frac{2545}{H_2 - H_1}$$

where H_1 = total heat of steam in Btu/lb at turbine inlet conditions
 H_2 = total heat of steam in Btu/lb at turbine-exhaust pressure after an adiabatic expansion, or constant entropy, the same as for the steam engine

Rated water rates are what the turbine manufacturer specifies, so rated load and steam conditions must be used to compare a machine's water rate with the machine specifications.

Utilities use *heat rate* as a measure of boiler-turbine efficiency. This is the amount of Btu's required to make a kW of electricity, for example 11,000 Btu's/kW generated.

The *selection* of a steam turbine versus a motor for mechanical drive units depends on performing a heat balance in order to see if the exhaust steam can be efficiently used for process heating, boiler feedwater heating, and similar "downstream" applications. The turbine then must be selected so that excess exhaust steam is not vented to the atmosphere, thereby losing at least 1000 Btu's/lb steam flow to the turbine.

In steam turbine test work on performance, the *Willans* line is quite often referred to. For a machine with a constant speed governor, this line represents the throttle flow versus the load, usually expressed in kilowatts. The throttle flow is proportional to the load and is generally a straight line as shown in Fig. 10.13, except for low-load and overload conditions, where the proportionality differs. If the steam flow in pounds per hour is divided by the corresponding kilowatt · hour output, the *steam rate* in pounds per kilo-

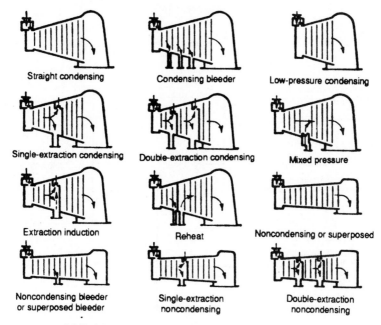

Straight condensing

Condensing bleeder

Low-pressure condensing

Single-extraction condensing

Double-extraction condensing

Mixed pressure

Extraction induction

Reheat

Noncondensing or superposed

Noncondensing bleeder or superposed bleeder

Single-extraction noncondensing

Double-extraction noncondensing

(a) Turbine classification by steam flow through unit

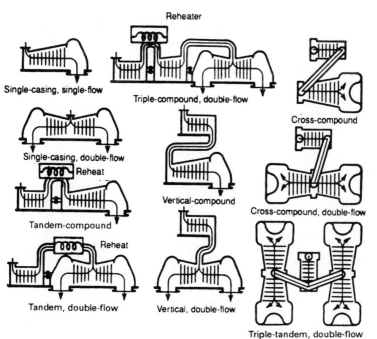

Reheater

Single-casing, single-flow

Triple-compound, double-flow

Cross-compound

Single-casing, double-flow

Reheat

Vertical-compound

Cross-compound, double-flow

Tandem-compound

Reheat

Tandem, double-flow

Vertical, double-flow

Triple-tandem, double-flow

(b) Turbine classification by single or double flow and casing arrangement

FIGURE 10.11 (*a*) Turbine classification by steam flow through unit; (*b*) turbine classification by single or double flow and casing arrangement.

10.27

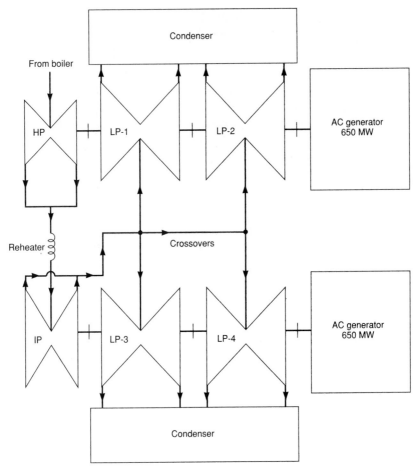

FIGURE 10.12   One of the largest utility units has six steam rotors arranged in cross-compound to drive two 650-mW ac generators. All turbines are double-flow.

watt · hour can be determined as shown in Fig. 10.13. The steam rate is usually high at low loads and starts to climb again at loads above the rated load. Note in Fig. 10.13 that steam and water rates for turbines connected to generators are the same thing. Large turbines over 100,000 kW use several valves to open in succession as the load increases, which assists in maintaining a flat water-rate curve.

Heat-rate curves are obtained by converting pounds per hour flow to the equivalent Btu content of the steam flow. If this is divided by the kilowatt · hour output, the Btu per kilowatt · hour is obtained.

Example.   What is the steam rate for a turbine operating at 600 psia and producing 20,000 kW with a steam flow of 190,000 lb/h at steam temperature of 700°F (superheated)? Also, what is the heat rate under these conditions?

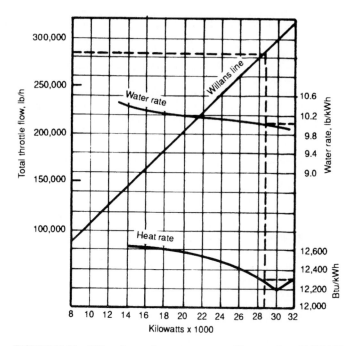

FIGURE 10.13 Willans line, water or steam rate, and heat rate for a 30,000-kW condensing turbo-generator. Dotted lines show steam flow, water rate, and heat rate when turbine is loaded to 29,000 kW.

$$\text{Steam or water rate} = \frac{190,000}{20,000} = 9.5 \text{ lb/kW} \cdot \text{h}$$

$$\text{Heat rate} = \frac{190,000 \times 1351.1}{20,000} = 12,835 \text{ Btu/kW} \cdot \text{h}$$

where 1351.1 is the enthalpy of steam at stated inlet steam pressure and superheat temperature.

Manufacturers usually test machines at the factory for power output, valve settings, speed regulation, overspeed trip setting, and running balance or vibration.

Acceptance tests are usually performed by large purchasers to verify guarantees and also to establish benchmarks of performance for future references, such as output, steam rate, speed regulation, vibration, valve operation, and overspeed trip function, to name a few.

Mechanical-drive turbines are used extensively in industry, power plants, and large office buildings to drive pumps, fans, mechanical machinery, and centrifugal compressors. Most units are of the back-pressure type with the exhaust steam used for heating, making hot water, or process applications. Many facility service operators come in contact with mechanical-drive turbines, and if the exhaust steam can be used in the facility, the turbine acts as a reducing valve with the power generated being actually a by-product. About 10 percent of the energy in the steam is used for power on back-pressure machines. The smaller turbines have been standardized with respect to horsepower, steam admission, exhaust pressure, and speed, with governors available for constant- and variable-speed operation. Most units are single stage in the smaller sizes, as is shown in Fig. 10.14a. This unit is equipped with an oil-relay governor and forced-feed lubrication, and it also has hand-valve-controlled nozzles for adding or reducing steam flow as the load changes. Hand valves are used to provide good starting power when beginning a plant operation at low pressure, low-speed operation control where the governor may not be suitable, and extra load capability when conditions of an operation require it.

Many mechanical-drive steam turbines are used for boiler-feed-pump drive, induced and forced fan drive, paper machine drives, and various types of high-speed centrifugal compressors. They are also used for emergency generator operation. As the power need becomes greater, *multistage* units are used, some even with bleeder or extraction steam provisions.

Dual drives have a turbine at one end of a critical machine and an electric motor at the other so that if power fails when the unit is electrically driven, the steam turbine automatically drives the machine. The turbine is often used for heat-balance purposes, such as in winter when all the exhaust steam can be used efficiently and the turbine is used as a reducing valve.

Condensing turbines expand the steam to low pressure; therefore, there is more heat energy used up in making power. The condensing medium is usually water from a lake or river (or salt water along the coast) or deep wells, and is recirculated through cooling towers. Several utilities in colder climates have installed air-cooled condensers. These require large heat-transfer surfaces. Condensing turbines are economically justified in larger sizes, primarily to drive generators. Condenser pressure can reach very low vacuum values. Provisions must be made to release the steam to the atmosphere such as relief valves in the event the cooling water is lost unless the condenser shell was built for the potential steam pressure that can be imposed on the shell from a loss of cooling water. Practically all condensing turbines are *multistage* machines with many rows of blades, and they generally operate over 600-psi inlet steam pressure.

Bleeder, or extraction, steam turbines are used to extract steam at a lower pressure than admission steam for process use, low-pressure power drives, or feedwater heating. Figure 10.14b shows a rotor with one extraction opening and the exhaust going to a condenser. All extraction turbines have a valve gear to maintain the required extraction pressure automatically.

In plants where new boilers with higher pressures have been installed, *topping* steam turbines may also be installed that use the higher-pressure steam but the exhaust from the new turbine is at the existing old boiler pressure, for example, 1250- to 600-psi old boiler plant header. The new turbine is called a *topping turbine* because the new pressure operation tops an existing pressure operation. Many multistage turbines are equipped to extract steam into a lower pressure, but at certain periods of plant operation, there may be excessive steam at this lower pressure. This excess steam is then used by back flowing it into the turbine's extraction line and then doing useful work in the remaining stages of the turbine. Such a turbine is called a *mixed-pressure* unit. The valve gear is arranged to admit any surplus steam from the lower-pressure steam line and then expand it to the exhaust end of the turbine. When plant conditions change, and more low pressure steam is needed, the valve gear automatically goes to the extraction mode of operation. Usually the steam pressure variations in the line determine whether extraction or back flow through the extraction line will be performed in these mixed-pressure units. It is important to have an emergency trip valve on the extraction line in mixed pressure units, and these require periodic checking.

All extraction lines from a turbine need *check valves* to prevent backflow of steam that could overspeed a turbine as it expands to lower pressure. In the *mixed-pressure* unit, a separate trip valve is required in the extraction line that cuts off the steam at the same time the high-pressure steam is cut off at the inlet to the turbine when an overspeed condition develops. It is important for service operators to make sure that both trip

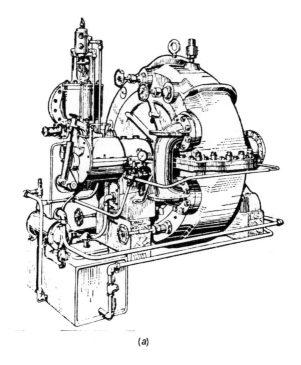

(a)

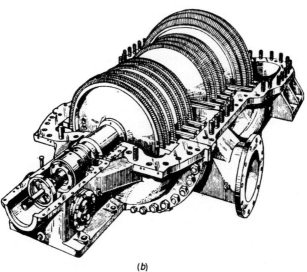

(b)

FIGURE 10.14 (a) Single-stage, back-pressure steam turbine with forced lubrication, oil-relay governor, and hand-operated nozzles for admitting additional steam; (b) 12-stage turbine with extraction after the fifth stage and condenser exhaust.

valves function properly during any overspeed tests. Figure 10.15 shows a double extraction, condensing turbine.

Multicylinder steam turbines achieve high capacity and are mostly used to generate electric power. In these units, arranged either in tandem or cross-compound, steam passes in sequence through more than one cylinder, partially expanding in each. Large units may have up to six cylinders as shown in the sketch in Fig. 10.12. In this arrangement, steam is reheated to 1000°F after passing through the high-pressure turbine, is further expanded in a reheat turbine, and then flows to four low-pressure cylinders, of which there two each on the high-pressure turbine train and the intermediate-pressure turbine train.

Figure 10.16 shows a tandem-compound extraction turbine with two cylinders, rated at about 30,000 kW. The high-pressure cylinder on the right has 2 velocity impulse gages, followed by 22 stages of reaction blades. A crossover pipe directs the steam to a double-flow low-pressure cylinder with nine stages on each side of the double flow.

Steam turbine governors vary considerably from constant-speed units to variable-speed operation; however, most governors can be classified as follows:

1. *Direct-acting centrifugal governor.* This is a flyball type, usually used for constant speed, with the biggest application being on mechanical-drive turbines. A governor weight is mounted on the shaft with a spring opposing the force due to centrifugal force on the governor weight. If the speed starts to rise above the governor setting, the weights will move further apart, and their movement causes the governor valve to start closing through the governor lever. This type of governor is extensively used on smaller turbines.

2. *Oil-relay centrifugal governor.* Here the governor weight actuates an oil pilot valve which puts oil pressure on a piston to open or close the governor valve. The pilot valve returns to its neutral position after each speed correction and is moved again only after there is another impulse from the governor weights. Oil pressure for operating this type of governor is supplied by a turbine-shaft-driven pump. This type of governor is used for steam pressures above 600 psi.

3. *Direct-acting variable-speed oil-relay governor.* This employs a turbine-shaft-driven oil pump that supplies oil pressure to an operating piston and discharges the oil back to a reservoir through an adjustable orifice valve. The operator can set the orifice valve for the desired speed, and the oil pump will maintain just enough pressure to keep the governor valve sufficiently open to carry the turbine load at the desired speed. If the speed increases above the setting, there is an increase in oil pressure which causes the governor to start closing until the desired speed is reached again. Speed is changed by the operator through an adjusting orifice valve. This type of governor may have a speed range change of 5 to 1 provided the turbine shaft assembly is designed for the lower speed operation.

4. *Oil-relay variable-speed governor.* This operates on the same principle as the direct-acting variable-speed oil governor, but instead of moving the governor valve directly, the piston operates the pilot valve through which oil is supplied to the piston. Two turbine-shaft-driven oil pumps are required for this type of governor. The operator can set the desired speed in the same way as for the previously described direct-acting variable-speed oil-relay governor.

5. *Emergency overspeed governor.* These are required on all steam turbines in order to cut off the steam when the regular governor cannot control the turbine speed within its designated rating. The overspeed trip is usually set for operation at 10 percent over rated speed. It should be tested for proper operation per the manufacturer's instruction, usually at least once per year. Most turbines also have provisions for hand tripping to make sure that the emergency linkages operate to shut off the steam admission valve(s). This test is important at each start-up since it makes sure there is no binding on the linkages and trip valve(s).

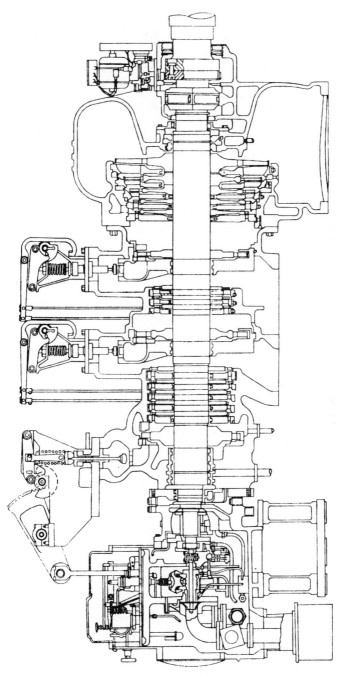

FIGURE 10.15 Double-extraction, condensing turbine showing constant pressure extraction control valves.

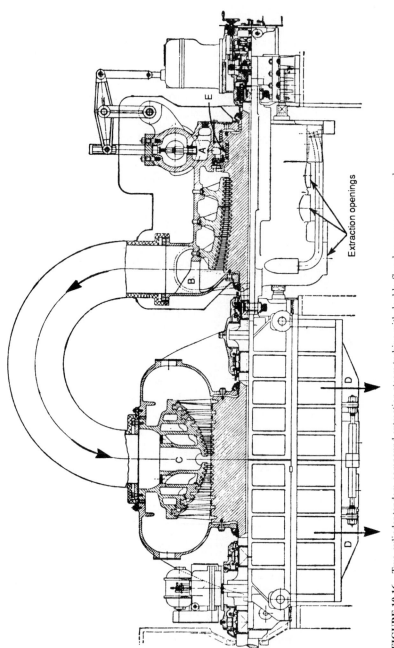

Extraction openings

FIGURE 10.16 Two-cylinder tandem-compound extraction steam turbine with double-flow low-pressure expanding steam to condenser.

TURBINE COMPONENTS

There are many parts that operators may have to inspect and maintain on steam turbines. Figure 10.17 shows some of the parts of turbines. There are many variations of blade designs, with the *fir tree root* attachment shown in Fig. 10.17 employed extensively on larger machines. The blades may be placed on the rotor stages or discs in a radial sliding manner, or side-entry design. Rotors may have *shrunk-on discs* or *solid forgings* with grooves machined into the forging for blade attachment. *Diaphragms* are the stationary blade discs as shown in Fig. 10.17.

Seals are extensively used on turbines to prevent steam from leaking out of the casing where the shaft emerges, to prevent air from entering the casing, and to separate the stages so that designed pressure drops take place between stages. Figure 10.18 shows some of the seals used on turbines. They are:

1. *Carbon rings* are used on smaller machines and are segments of carbon fitted around the shaft and held together by garter or retainer springs. The carbon rings are installed into grooves of a split box so that the rings can be removed and replaced when they become worn excessively.

2. *Labyrinth packings* are extensively used as seals on high-pressure turbines. These consist of a series of soft metal strips riding inside V grooves held in place by a housing. The close clearance between the rotating strips and the grooves forms resistance paths that throttle the steam so that each step causes the steam pressure to drop. The number of strips to prevent leakage is determined by the steam pressure that must be confined by the labyrinths.

3. *Water seals* employ a shaft-mounted runner that acts as a pump to create a ring of water under pressure that prevents the steam from escaping; instead it is directed to seal drains. It is essential to use treated water equivalent to water fed to the boiler in order to prevent pitting on the shaft underneath the water seal. Stress-corrosion cracks can start here on the shaft if treated water is not used for this type of seal application.

4. The *stuffing box* uses woven or soft packing material around the shaft, which is then compressed with a gland to offer resistance to steam flow from the turbine casing. This type of seal has the highest wear rate and thus requires frequent replacement. It is used on smaller turbines.

There are many *seal strips* used on the tip of blades on reaction turbines to prevent interstage leakage. These may ride between corresponding strips in the casing at that stage (see Fig. 10.17).

Dummy pistons are used on reaction turbines to act as an axial force counter to that created by the axial reaction force of the steam going through the blades. The term *dummy piston* dates from the time when a true piston was used for this purpose; present practice is to use strips on the shaft riding in grooves on the stationary part of the machine. The pressure drop toward the end of the strips produces the counter axial force that a piston once provided.

Bearings on turbines may be of the sleeve type with or without pressure lubrication, with only very small machines using oil ring lubrication. *Thrust bearings* are classified as follows: (1) babbitt-faced collar bearing, (2) tilting pivotal pad type, (3) tapered land bearing, and (4) roller contact ball or roller bearings. Figure 10.19*a* shows a Kingsbury thrust bearing, which is classified as a pivotal pad unit. In this bearing a collar on the shaft faces a number of thrust shoes or pads that are supported on pivots or buttons. This

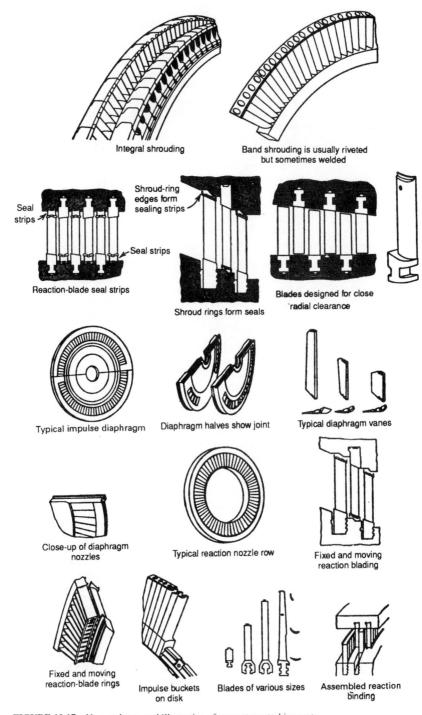

Integral shrouding

Band shrouding is usually riveted but sometimes welded

Seal strips

Shroud-ring edges form sealing strips

Seal strips

Reaction-blade seal strips

Shroud rings form seals

Blades designed for close radial clearance

Typical impulse diaphragm

Diaphragm halves show joint

Typical diaphragm vanes

Close-up of diaphragm nozzles

Typical reaction nozzle row

Fixed and moving reaction blading

Fixed and moving reaction-blade rings

Impulse buckets on disk

Blades of various sizes

Assembled reaction binding

FIGURE 10.17 Nomenclature and illustration of some steam turbine parts.

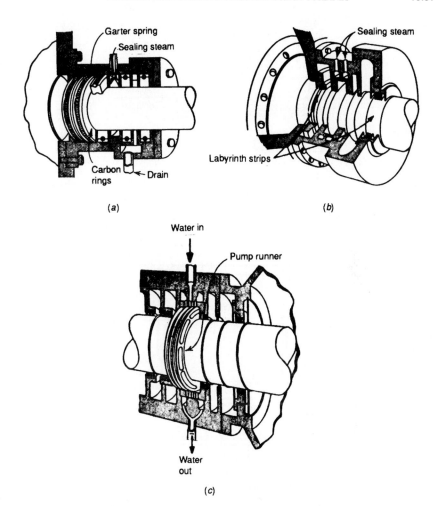

FIGURE 10.18 Seals are used in turbines to prevent leakage of steam along the shaft between stages and at the end of rotors and also to prevent air from entering the turbine at the low-pressure end or during idle periods. (*a*) Carbon seals are in segments, held by springs around rings; (*b*) labyrinth seals are circular strips or serrations almost touching shaft; (*c*) water seals have pump runners on shaft to hold water at periphery.

allows the shoes to tilt and form an oil wedge between the collar on the shaft and the shoes. The oil wedge carries any end thrust.

Any extra heavy axial thrust will show evidence of this by wiping on the shoes or babbit caused by being overheated from a loss of the oil wedge. If the oil wedge is lost, such as from heavy axial thrusts or loss of oil pressure, severe bearing damage can result (wiping and melted babbitt).

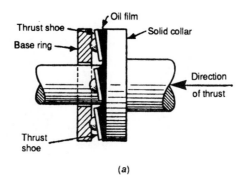

(a)

(b)

FIGURE 10.19   (a) Kingsbury thrust bearings feature tilted thrust shoes; (b) lubrication system for a variable-speed turbine has separate governor oil pump.

LUBRICATION SYSTEMS

Figure 10.19b shows a sketch of a lubricating oil system for a variable-speed turbine. Note that the governor has a separate oil pump. On constant-speed machines, such as for a generator drive, a single high-pressure oil pump may supply the governor oil, and the same pump will also supply the oil for the bearings at a lower pressure by passing the oil through a reducing valve. The shaft-driven oil pump will be backed by a starting oil pump and an emergency dc motor-driven oil pump.

Figure 10.20 shows a lubrication system for a turbo-generator. Note that in this system, the main oil pump is driven by the turbine shaft. The pump draws oil from the oil reservoir and discharges it through a strainer to (1) the primary and secondary governor

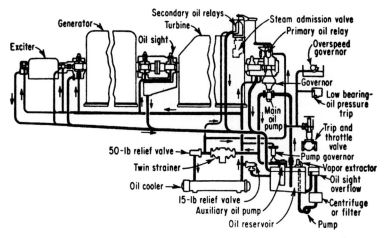

FIGURE 10.20 Circulating lube-oil system for a moderate-size steam turbo-generator. Larger units also have dc emergency oil pump backups.

relays, (2) the power cylinders, (3) the oil-operated trip, (4) the throttle valve, (5) the overspeed governor, and (6) the low-bearing oil-pressure trip. A relief valve maintains the oil-pump discharge at 50 psi and discharges the oil through an oil cooler to all bearings and gears needing lubrication. The bearing-oil pressure is kept at 15 psi by a relief valve which discharges to the oil reservoir. Oil from the bearings and power cylinders is returned to the oil reservoir. An auxiliary oil pump, used during starting and stopping, discharges oil through a check valve into the main oil-pump discharge piping. An overflow line from the bottom of the oil reservoir discharges to a centrifuge or filter. The clean oil is returned for reuse by a separate pump. A vapor extractor is provided in order to ventilate the oil reservoir.

OIL RECONDITIONING

Contamination and decomposition products are removed from oil in various ways. Three common methods are batch handling, continuous-total or full-flow handling, and continuous-bypass treatment.

In the batch-handling method, the entire oil charge is withdrawn and treated at one time. The most common method is to bypass a portion of the circulating oil flow to the treating equipment. This keeps the total impurities below the danger level and avoids large treatment facilities.

Use rags—never waste—to wipe the inside of tanks, bearings, or the other parts of a lube-oil system.

Operation and Maintenance

The general operating procedures given for steam engines also apply to steam turbines, with the turbine manufacturer's instruction to be followed in all cases. The rules given here are for a condensing turbo-generator, but most of them also apply to noncondensing machines.

1. Make sure all condensate is drained from the casing and the drain ahead of the throttle valve and steam chest before starting.

2. Check the oil level in the reservoir before starting the auxiliary oil pump, and then make sure the oil is flowing to all the bearings.

3. Establish water flow through the condenser, and start the air ejector and condensate pump.

4. When drains ahead of the throttle valve blow clear, open the throttle valve slowly but just enough to start the turbine rolling and to make sure there is no binding or unusual noises such as may occur from internal rubbing.

5. Trip the hand-operated emergency trip to see if the valves close per design. Reestablish steam flow, bring the speed up to 10 percent of the rating, close all casing drains, and start to establish vacuum conditions in the condenser.

6. Open seal drains and admit sealing steam, and, if so equipped, treated water to water seals. Start the drain condensers and vacuum pumps.

7. Continue to warm up the casing and rotor at low speed per the manufacturer's *ramp rate* instructions.

8. Slowly bring the speed up to the rated speed but do not dwell for any length of time at or near the critical speed. Note at what speed the governor establishes control and if this is per design. Open the throttle valve, trip the emergency trip and observe if the valves close correctly, and then reestablish speed.

9. With conditions normal on the turbine, proceed to excite the generator so that the voltage reaches line voltage and frequencies match before synchronizing the generator to the system.

Shutdown procedures follow the reverse of the above, starting with gradually diminishing the load to zero, disconnecting the generator from the line, and removing field excitation.

Many modern units are now monitored by computers during start-up and shutdown, with the operator making sure everything is performed in proper sequence from a central control room.

Service Conditions

There are many service conditions that may affect larger turbo-generators and large mechanical drive units. A few are listed here, concentrating on turbo-generators:

1. *Peaking and cycling service* on previously base-loaded machines can create repetitive thermal-induced stresses on shells, rotor bores, and steam chests that are integral with the high-pressure casing. These stresses can cause thermal-fatigue cracks to develop. It is important to follow the manufacturer's instruction on warm-up times and slow cooling in order to avoid low-cycle fatigue cracks, especially on abrupt sections of shells and rotor. Sudden and repeated load changes may also produce repetitive thermal stresses.

2. Wet steam *carryover* and *water induction* are two problems that operators must pay attention to so that no impact damage to blades occurs. When steam pressure drops, the load should be reduced so that steam does not start to condense in the lower stages and cause severe blade erosion. The last row of blades on condensing units is especially vulnerable to wet steam erosion. This row should be inspected visibly once a year through the rear manhole usually provided for condensing units.

Water induction may be experienced when a feedwater heater tube at boiler pressure develops a leak that flows to the lower-pressure steam side and could potentially flow into a hot turbine through extraction lines.

The ASME recommended guidelines to prevent water induction include:

a. Have drains on cold reheat piping on utility-size units.

b. Use thermocouples or other means to show when spray attemperators are putting excessive water into superheater lines.

c. Feedwater heaters should have high-low level alarms and also be equipped with automatic drains when the high-level alarm is activated.

d. Feedwater heaters should have check or nonreturn valves on the extraction lines supplying the heater.

e. Have automatic shutoff valves on all sources of water entering the feedwater heater so that the flow to the turbine extraction line is stopped in the event of a high-water-level alarm on the heater.

f. The ASME guide on the prevention of water induction also includes testing requirements; test heater-level controls and alarms, drain valves, and power-assisted check valves in extraction lines monthly. All valves essential to water induction prevention, such as attemperator spray valves, isolation valves for heaters, and extraction lines, should be subject to internal valve inspection for wear during planned turbine outages.

3. *Steam path deposits* also require operator attention. Two types of deposits are of concern: (*a*) Plugging deposits where the scale on turbine blades cause a reduction in efficiency and possibly sticking valves and control problems from rust and phosphates and sulfite used in water treatment and (*b*) contaminants that have a long-term effect and may cause caustic embrittlement and stress corrosion cracking. The contaminants of concern include sodium hydroxide, sodium sulfite, and chlorides, which can cause cracking. The best procedure is to maintain steam chemistry limits for contaminants that the turbine manufacturer specifies for the unit. Typical limits specified are pH, 8.55 to 9.26; oxygen, 11.6 ppb; sodium, 6.5 ppb; free OH, 0.1 ppb; silica, 17.4 ppb; copper, 5.0 ppb; iron 17.4 ppb; and phosphate, 7.0 ppb.

4. *Wear and tear* occur on such parts as nozzles, blades, seals, and bearings and will manifest themselves in operation by excessive shaft steam leakage and possibly vibration. Therefore, periodic internal inspections are required so that worn parts can be replaced and blades and other components can be inspected for cracks by visual and nondestructive testing methods.

5. *Vibration monitoring* is an excellent way to track the mechanical performance of steam turbines. A running record of vibration will reveal changes that may be occurring that are affecting machine performance. Most turbine manufacturers and some insurance companies involved with turbine breakdown insurance can offer guides on acceptable levels of vibration, usually based on the revolutions per minute of the machine. The term *signature analysis* is used when vibration data is analyzed to identify trouble spots on the machine for maintenance forecasting (see also Chap. 15).

6. *Performance monitoring* is another method that operators can use to keep track of a critical turbine's operation and its need for corrective action. The pounds of steam required to produce a certain output is established when the machine is new, or after a thorough overhaul, and this benchmark is then used to determine if the output performance (e.g., pounds of steam per kilowatt) of the machine is declining. A similar condition may be used on oil consumption to determine bearing wear that may be excessive, and changes in first-stage pressure for nozzle wear.

7. Daily routine checks and logging readings is an old operator's method of keeping track of machine condition. Inlet and exhaust steam pressures must be maintained

as close as possible to design. Investigate all changes from normal. Stage shell pressure checks, especially at the front end of the machine, can determine if the inlet nozzles are plugged with scale, eroded from wet steam cutting, or even broken from carryover. Bearing temperature checks are required to make sure proper lubrication is maintained.

8. *Planned overhauls* are scheduled during low production periods and are mainly used to correct conditions noted during operation such as seal leakage, vibration, high steam rate, and erratic control, to name a few. They are also used to inspect internal conditions, clean internal parts, and replace worn parts per the manufacturer's guidelines. It is also standard practice to perform nondestructive inspections on critical sections of the machine for possible cracks. These may include magnetic particle inspections, in some cases dye-penetrant and even ultrasonic tests on such parts as rotor bores. The overhaul also includes inspecting and correcting conditions, such as cleaning of condensers and oil coolers on the auxiliary equipment associated with turbine operation.

Condensers

The most important job of a condenser is to turn steam back into water and reduce the back pressure on the turbines so that maximum heat energy can be extracted from the steam. This is because the power produced in the expanding steam from atmospheric pressure to 29 in vacuum is about the same as if the steam were expanded from 200 psig, 600°F to atmospheric.

The second function of the condenser is to condense the steam so that the low-oxygen-content condensate can be used again for the boiler feedwater at full temperature.

In a condenser the steam is reduced in volume to that which it occupied as water, from which it was generated. Thus, condensing steam in a closed vessel produces a partial vacuum by reducing the volume of vapor.

Air ejectors are extensively used with condensers. An *air ejector* is a steam nozzle that discharges a high-velocity jet of steam at about 3500 ft/s (see Fig. 10.21). The steam flows across a suction chamber and through a venturi-shaped compression tube. The air or gases to be evacuated enter the ejector suction where they are entrained by a jet of steam and then discharged through the throat of the ejector. The velocity of the kinetic energy is converted into pressure in the throat of the ejector and compresses the mixture to a lower vacuum or a higher absolute pressure. Two-stage ejectors have a compression ratio of about 8:1 (the ratio between the discharge pressure and the suction pressure). The ejector discharges to either a small condenser or a feedwater heater where the steam is condensed, and the air and gases are vented to the atmosphere. The two jets, shown in Fig. 10.21*a*, are in series with intercondensers between the stages. These intercondensers condense the steam and cool the air-vapor mixture. Aftercondensers are used for the same purpose.

There are many types of barometric condensers; the ejector-jet, parallel-flow-contact type, shown in Fig. 10.21*b* is one. These exchangers condense the exhaust steam by bringing it into direct contact with the cooling water. They are simple and compact, but condensate mixes with the raw water.

A countercurrent barometric condenser uses disks to break up the flow of water (see Fig. 10.21*c*). It has a two-stage air ejector with an intercondenser. The condensables must be removed by a pump. Steam, with its noncondensable gases, flows in the opposite direction to the cooling water. Also, the cooling water breaks up into a spray to present the largest possible surface for steam absorption.

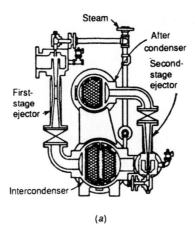

(a)

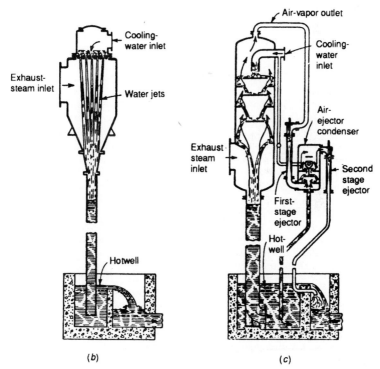

(b)   (c)

FIGURE 10.21   Ejectors are used in condensers. (a) Two-stage air ejector; (b) ejector-jet contact barometric condenser; (c) countercurrent barometric condenser.

QUESTIONS AND ANSWERS

Steam Engines

1. What is the difference between a conventional counterflow engine and the uniflow engine?

Answer. Counterflow engines, such as the D-slide valve and Corliss engines, have the steam flow into the end of the cylinder and then reverse direction to flow out at the same end of the cylinder. The exhaust at the inlet cools the cylinder walls and wastes thermal energy. In the uniflow engine, steam flows into the end of the cylinder, but the exhaust leaves in the middle of the cylinder as shown in Fig. 10.6, thus reducing heat loss and also reducing condensation of the steam inside the engine.

2. What is the sequence of events during the stroke of a steam engine?

Answer. The stroke sequence of a steam engine is as follows: (1) Admission—the steam port is uncovered, allowing steam to enter the cylinder. (2) Cutoff—steam valve closes, stopping the flow of steam. Steam expands and drives the piston. (3) Release—exhaust port opens and exhaust steam leaves the cylinder. (4) Compression—both the exhaust port and the steam port are closed. Some exhaust steam is trapped and compressed in the cylinder to cushion the end of the stroke.

3. What advantages do compound engines have over simple engines?

Answer. See Fig. 10.7. Compound engines have (1) greater steam economy, (2) smoother operation because the work is divided between two cylinders, and (3) smaller parts with less wear. A *cross-compound* engine has one high- and one low-pressure cylinder set side by side, driving the same shaft. Cranks are usually set 90° apart for smoother operation. A *tandem-compound* engine has two cylinders centered on the same piston rod. The low-pressure cylinder is placed nearest the crankshaft. An *angle-compound* engine has two cylinders set at 90°, usually to save floor space.

4. What is cutoff and how is it expressed?

Answer. Cutoff is the point at which the steam valve closes the steam port. It is expressed as a fraction of the stroke, thus one-half, three-quarters, etc. It can also be expressed as the apparent cutoff.

5. What is meant by the term *angle of advance* on a plain D-slide valve?

Answer. The cam driving the valve is set 90° ahead of the crank in the direction of rotation. But this is the zero position, and allowance must be made for the valve to be open when the piston is at the top of dead center. The lap of the valve plus the valve lead causes the cam to move forward to give this effect. The amount of cam movement is measured in degrees and is called "the angle of advance."

6. What is steam lap? How is it altered?

Answer. Steam lap is the fraction of an inch that the edge of the valve closes off the port opening when it is in the center of its travel. It can be altered by filing metal from the end of the valve to decrease the lap. To increase the lap, you must weld metal onto the edge of the valve and machine it straight, or use a larger valve.

7. What is valve lead? How is it altered?

Answer. Valve lead is the amount of opening in the steam port when the piston is exactly at the end of the stroke. It can be altered by adjusting the valve gear, moving the cam, or changing the length of the valve rod.

8. Why are indicator cards taken?

Answer. Indicator cards are taken to (1) prove that the steam and exhaust valves open and close at the proper time in relation to the position of the piston, (2) calculate the power developed by the piston on each side of the cylinder, and (3) calculate the amount of steam consumed by the engine. *Note:* The same analysis methods and calculations are applied to diesel engines as are applied to similar problems for reciprocating compressors.

9. What is a planimeter? How is it used by operating engineers?

Answer. A planimeter is used to measure the area of an irregular figure. It usually has a measuring wheel fixed to a set of arms. A stylus traces the shape of the figure, which is recorded by a calibrated vernier wheel. The vernier is calibrated to be read in square inches. These instruments are used to measure the work area of indicator diagrams. There are several types; follow manufacturers' instructions when using them.

10. What is the ihp of an 18- $\times$ 20-in double-acting engine, running at 200 r/min with a mean effective pressure (mep) of 47 psi?

Answer. It is:

$$\text{ihp} = \frac{2 \times PLAN}{33,000}$$

$$= \frac{2 \times {}^{20}/_{12} \times 0.7854 \times 18 \times 18 \times 200 \times 47}{33,000}$$

$$= 241.61$$

11. What is the mechanical efficiency of an engine which has an ihp of 275 and a brake horsepower (bhp) of 210?

Answer. It is:

$$\text{Mechanical percent efficiency} = \frac{\text{hp developed}}{\text{hp indicated}} \times 100$$

$$\text{Percent E} = \frac{\text{bhp}}{\text{ihp}} \times 100$$

$$= \frac{210}{275} \times 100$$

$$= 76.36$$

12. (1) What is the water rate of a reciprocating engine? (2) What is a normal or good-average water rate?

Answer. (1) The water rate is the actual measured pounds of water used by the engine to produce 1 ihp/h. To determine the water rate, you must know the ihp and the amount of water used in a given time. (2) A normal or good-average water rate depends on the type of the engine and the exhaust pressure. Here are some values in pounds per hour water used per indicated horsepower per hour:

Type of engine	Noncondensing engines	Condensing to 26-in vacuum
Uniflow	$20\frac{1}{2}$ lb	$12\frac{1}{2}$ lb
Corliss, 4 valve	22 lb	$20\frac{1}{2}$ lb
Slide valve	$26\frac{1}{2}$ lb	23 lb

13. What basic governor troubles are apt to occur?

Answer. (1) Hunting—alternate speeding and slowing of the engine—which means that the governor is too sensitive to load changes. (2) Sticking—failure to control speed, allowing the engine to run away or slow down—which means that the governor is not sensitive to load changes or parts are binding or worn.

14. What is a governor safety stop?

Answer. On throttling-type governors, the safety stop is a weighted arm that needs the support of a governor belt. If the belt breaks, the idler arm drops and shuts the steam supply valve to the engine. On Corliss units, the flyballs fall to the lowest position and knock off the safety cams; the cams disengage the catch blocks on the steam intake valves so that no steam is admitted to the engine.

15. Why is condensation or excessive carryover dangerous to reciprocating engines?

Answer. Because water is noncompressible. If an excessive amount of water gets into the cylinder, it will wreck the engine.

Steam Turbines

16. Why should a steam or moisture separator be installed in the steam line next to a steam turbine?

Answer. All multistage turbines, low-pressure turbines, and turbines operating at high pressure with saturated steam should have a moisture separator in order to prevent rapid blade wear from water erosion.

17. Under what conditions may a relief valve not be required on the exhaust end of a turbine?

Answer. If the manufacturer has provided information that the turbine shell was constructed for full-inlet steam pressure for the entire length of the shell. It is absolutely essential to have a relief valve to protect the shell in the event an exhaust valve is closed and high-pressure steam is admitted to the shell on the front end of the machine. Explosions have occurred when this happened.

18. What are some conditions that may prevent a turbine from developing full power?

Answer. The following may contribute to low power output: The machine is overloaded, the initial steam pressure and temperature are not up to design conditions, the exhaust pressure is too high, the governor is set too low, the steam strainer is clogged, turbine nozzles are clogged with deposits, and there is internal wear on nozzles and blades.

19. Why is it necessary to open casing drains and drains on the steam line going to the turbine when a turbine is to be started?

Answer.   To avoid slugging nozzles and blades inside the turbine with condensate on start-up; this can break these components from impact. The blades were designed to handle steam, not water.

20. What three methods are used to restore casing surfaces that are excessively eroded?

Answer.   They are (1) metal-spraying, (2) welding, and (3) insertion of filler strips or patch plates. The manufacturer should be consulted on the metallurgy involved so that the best method can be selected.

21. What is steam rate as applied to turbo-generators?

Answer.   The steam rate is the pounds of steam that must be supplied per kilowatt hour of generator output at the steam turbine inlet.

22. What is the most prevalent source of water induction into a steam turbo-generator?

Answer.   Leaking water tubes in feedwater heaters, which have steam on the shell side supplied from turbine extraction lines. The water at higher pressure can flow back into the turbine because the extraction steam is at a lower pressure. Check valves are needed on the steam extraction line to prevent the backflow of water into the turbine.

23. What is a regenerative cycle?

Answer.   In the regenerative cycle, feedwater is passed through a series of feed-water heaters and is heated by steam extracted from stages of a steam turbine. This raises the feedwater to near the temperature of boiler water, thus increasing the thermal efficiency of the cycle.

24. What is the reheating cycle?

Answer.   In the reheating cycle, superheated steam is expanded in a high-pressure turbine and then returned to the boiler's reheater to raise the temperature of the steam to the inlet temperature, usually to around 1000°F; it is then returned to the turbine to be expanded through intermediate-pressure turbines. In some cases, the steam is again returned for reheating in the boiler and then expanded in the lower-pressure sections of the turbine. The main purpose of reheating the steam on large turbo-gen-erators is to avoid condensation in the lower-pressure sections of the turbine, which can rapidly cause blade erosion problems from wet steam.

25. What does the Willans line show?

Answer.   The Willans line is a plot of throttle flow versus load, usually expressed in kilowatts; generally it is a straight line except for low and high loads. The Willans line is used to show steam rates at different loads on the turbine.

26. What are the two basic types of turbines?

Answer.   The two basic turbine types are the impulse and the reaction (see Fig. 10.9*b*).

27. What is the operating principle of an impulse turbine?

Answer.   The basic idea of an impulse turbine is that a jet of steam from a fixed nozzle pushes against the rotor blades and impels them forward. The velocity of the steam is about twice as fast as the velocity of the blades. Only turbines utilizing fixed nozzles are classified as impulse turbines.

28. What is the operating principle of a reaction turbine?

Answer.   A reaction turbine utilizes a jet of steam that flows from a nozzle on the rotor. Actually, the steam is directed into the moving blades by fixed blades designed to expand the steam. The result is a small increase in velocity over that of the moving blades. These blades form a wall of moving nozzles that further expand the steam. The steam flow is partially reversed by the moving blades, producing a reaction on the blades. Since the pressure drop is small across each row of nozzles (blades), the speed is comparatively low. Therefore, more rows of moving blades are needed than in an impulse turbine.

29. What are topping and superposed turbines?

Answer.   Topping and superposed turbines are high-pressure, noncondensing units that can be added to an older, moderate-pressure plant. Topping turbines receive high-pressure steam from new high-pressure boilers. The exhaust steam of the new turbine has the same pressure as the old boilers and is used to supply the old turbines.

30. What is an extraction turbine?

Answer.   In an extraction turbine, steam is withdrawn from one or more stages, at one or more pressures, for heating, plant process, or feedwater heater needs. They are often called "bleeder turbines."

31. What is a radial-flow turbine?

Answer.   In a radial-flow turbine, steam flows outward from the shaft to the casing. The unit is usually a reaction unit, having both fixed and moving blades. They are used for special jobs and are more common to European manufacturers, such as Sta-Laval (now ABB).

32. What is a stage in a steam turbine?

Answer.   In an impulse turbine, the stage is a set of moving blades behind the nozzle. In a reaction turbine, each row of blades is called a "stage." A single Curtis stage may consist of two or more rows of moving blades.

33. What is a diaphragm?

Answer.   See Fig. 10.17. Partitions between pressure stages in a turbine's casing are called *diaphragms.* They hold the vane-shaped nozzles and seals between the stages. Usually labyrinth-type seals are used. One-half of the diaphragm is fitted into the top of the casing, the other half into the bottom.

34. What are four types of turbine seals?

Answer.   (1) *Carbon rings* fitted in segments around the shaft and held together by garter or retainer springs. (2) *Labyrinths* mated with shaft serrations or shaft seal strips. (3) *Water seals* where a shaft runner acts as a pump to create a ring of water around the shaft. Use only treated water to avoid shaft pitting. (4) *Stuffing box* using woven or soft packing rings that are compressed with a gland to prevent leakage along the shaft.

35. In which turbine is tip leakage a problem?

Answer.   Tip leakage is a problem in reaction turbines. Here, each vane forms a nozzle; steam must flow through the moving nozzle to the fixed nozzle. Steam escaping across the tips of the blades represents a loss of work. Therefore, tip seals are used to prevent this.

36. What are two types of clearance in a turbine?

Answer. Turbine clearances are designated as (1) radial, the clearance at the tips of the rotor and casing, and (2) axial, the fore-and-aft clearance, at the sides of the rotor and the casing.

Turbine Bearings

37. What are four types of thrust bearings?

Answer. Thrust bearings are classified as (1) babbitt-faced collar bearings, (2) tilting pivotal pads, (3) tapered land bearings, and (4) rolling-contact (roller or ball) bearings.

38. What is the function of a thrust bearing?

Answer. Thrust bearings keep the rotor in its correct axial position.

39. What is a balance piston?

Answer. Reaction turbines have axial thrust because pressure on the entering side is greater than pressure on the leaving side of each stage. To counteract this force, steam is admitted to a dummy (balance) piston chamber at the low-pressure end of the rotor. Some designers also use a balance piston on impulse turbines that have a high thrust. Instead of pistons, seal strips are also used to duplicate a piston's counterforce.

40. What is a combination thrust and radial bearing?

Answer. This unit has the ends of the babbitt bearing extended radially over the end of the shell. Collars on the rotor face these thrust pads, and the journal is supported in the bearing between the thrust collars.

41. What is a tapered-land thrust bearing?

Answer. The babbitt face of a tapered-land thrust bearing has a series of fixed pads divided by radial slots. The leading edge of each sector is tapered, allowing an oil wedge to build up and carry the thrust between the collar and pad.

42. What is important to remember about radial bearings?

Answer. A turbine rotor is supported by two radial bearings, one on each end of the steam cylinder. These bearings must be accurately aligned to maintain the close clearances between the shaft and the shaft seals, and between the rotor and the casing. If excessive bearing wear lowers the rotor, great harm can be done to the turbine.

Turbine Seals

43. What is gland-sealing steam?

Answer. It is the low-pressure steam that is led to a sealing gland. The steam seals the gland, which may be either a carbon ring or labyrinth type, against air at the vacuum end of the shaft.

44. What is the function of a gland drain?

Answer. The function of a gland drain is to draw off water from sealing-gland cavities created by the condensation of the sealing steam. Drains are led to either the condenser air-ejector tube nest or the feedwater heaters. Often, gland drains are led to a low-pressure stage of the turbine to extract more work from the gland-sealing steam.

45. What is an air ejector?

Answer.   An air ejector is a steam siphon that removes noncondensable gases from the condenser.

Turbine Governors

46. How many governors are needed for safe turbine operation? Why?

Answer.   Two independent governors are needed for safe turbine operation. One is an overspeed or emergency trip that shuts off the steam at 10 percent above running speed (maximum speed). The second, or main governor, usually controls speed at a constant rate; however, many applications have variable speed control.

47. How is a flyball governor used with a hydraulic control?

Answer.   As the turbine speeds up, the weights are moved outward by centrifugal force, causing linkage to open a pilot valve that admits and releases oil on either side of a piston or on one side of a spring-loaded piston. The movement of the piston controls the steam valves.

48. What is a multiport governor valve? Why is it used?

Answer.   In large turbines, a valve controls steam flow to groups of nozzles. The number of open valves controls the number of nozzles in use according to the load. A bar-lift or cam arrangement operated by the governor opens and closes these valves in sequence. Such a device is a multiport valve. Using nozzles at full steam pressure is more efficient than throttling the steam.

49. What is the safe maximum tripping speed of a turbine operating at 2500 r/min?

Answer.   The rule is to trip at 10 percent overspeed. Therefore, $2500 \times 1.10 = 2750$ r/min.

50. What is meant by critical speed?

Answer.   It is the speed at which the machine vibrates most violently. It is due to many causes, such as imbalance or harmonic vibrations set up by the entire machine. To minimize damage, the turbine should be hurried through the known critical speed as rapidly as possible. *Caution:* Be sure the vibration is caused by critical speed and not by some other trouble.

Turbine Lubrication

51. How is oil pressure maintained when starting or stopping a medium-sized turbine?

Answer.   An auxiliary pump is provided to maintain oil pressure. Some auxiliary pumps are turned by a hand crank; others are motor-driven. This pump is used when the integral pump is running too slowly to provide pressure, as when starting or securing a medium-sized turbine.

52. Why is it poor practice to allow turbine oil to become too cool?

Answer.   If a turbine oil is allowed to become too cool, condensation of atmospheric moisture takes place in the oil and starts rust on the polished surfaces of the journals and bearings. Condensed moisture may also interfere with lubrication.

53. Steam blowing from a turbine gland is wasteful. Why else should it be avoided?

Answer.   It should be avoided because the steam usually blows into the bearing, destroying the lube oil in the main bearing. Steam blowing from a turbine gland also creates condensate, causing undue moisture in plant equipment.

54. Besides lubrication, what are two functions of lubricating oil in some turbines?

Answer.   In larger units, lube oil cools the bearings by carrying off heat to the oil coolers. Lube oil in some turbines also acts as a hydraulic fluid to operate the governor speed-control system.

Turbine Operation and Maintenance

55. What is meant by the water rate of a turbine?

Answer.   It is the amount of water (steam) used by the turbine in pounds per horsepower per hour or kilowatts per hour.

56. What are three types of condensers?

Answer.   (1) Surface (shell-and-tube), (2) jet, and (3) barometric.

57. Why is there a relief valve on a turbine casing?

Answer.   The turbine casing is fitted with spring-loaded relief valves to prevent damage by excessive steam pressure at the low-pressure end if the exhaust valve is closed accidentally. Some casings on smaller turbines are fitted with a sentinel valve which serves only to warn the operator of overpressure on the exhaust end. A spring-loaded relief valve is needed to relieve high pressure.

58. Why must steam turbines be warmed up gradually?

Answer.   Although it is probable that a turbine can, if its shaft is straight, be started from a cold condition without warming up, such operation does not contribute to continued successful operation of the unit. The temperature strains set up in the casings and rotors by such rapid heating have a harmful effect. The turbine, in larger units especially, should be warmed slowly by recommended warm-up ramp rates because of close clearances.

59. What should you do if you lost vacuum while operating a condensing turbine plant?

Answer.   If vacuum is lost, shut down immediately. The condenser cannot stand steam pressure; the condenser tubes may leak from excessive temperature. Excessive pressure will also damage the shell, the exhaust, and the low-pressure parts of the turbine.

60. What are the main causes of turbine vibration?

Answer.   Turbine vibration is caused by (1) unbalanced parts, (2) poor alignment of parts, (3) loose parts, (4) rubbing parts, (5) lubrication troubles, (6) steam troubles, (7) foundation troubles, and (8) cracked or excessively worn parts.

61. What is the purpose of a turning gear?

Answer.   Heat must be prevented from warping the rotors of large turbines or high-temperature turbines of 750°F or more. When the turbine is being shut down, a motor-driven turning gear is engaged to the turbine to rotate the spindle and allow uniform cooling.

62. What does the term *ramp rate* mean?

Answer. Ramp rate is used in bringing a turbine up to operating temperature and is the degrees Fahrenheit rise per hour that metal surfaces are exposed to when bringing a machine to rated conditions. Manufacturers specify ramp rates for their machines in order to avoid thermal stresses. Thermocouples are used in measuring metal temperatures.

63. What is the difference between *partial* and *full arc* admission?

Answer. In multivalve turbine inlets, partial arc admission allows the steam to enter per valve opening in a sequential manner, so as load is increased, more valves open to admit steam. This can cause uneven heating on the high-pressure annulus as the valves are individually opened with load increase. In full-arc admission, all regulating valves open but only at a percentage of their full opening. With load increase, they all open more fully. This provides more uniform heating around the high-pressure part of the turbine. Most modern controls start with full-arc and switch to partial arc to reduce throttling losses through the valves.

64. What is the steam rate in horsepower per hour and kilowatts per hour, and what is the heat rate for a turbo-generator carrying 74,500 kW with a steam flow of 875,375 lb/h, at a steam pressure of 750 psia and superheat temperature of 900°F?

Answer. The steam rate in kilowatts per hour is:

$$kW \cdot h = \frac{875,375}{74,500} = 11.75 \text{ lb/kW} \cdot h$$

$$\frac{74500 \times 3413}{2545} = 99,909 \text{ hp}$$

$$\text{steam rate} = \frac{875,375}{99,909} = 8.76 \text{ lb/hp} \cdot h$$

Enthalpy of steam at 750 psia and 900°F superheat = 1457.2 Btu/lb; therefore, heat rate = $1457.2 \times 11.75 = 17,122.1$ Btu/kW · h.

65. What are some common troubles in surface-condenser operation?

Answer. The greatest headache to the operator is loss of vacuum caused by air leaking into the surface condenser through the joints or packing glands. Another trouble spot is cooling water leaking into the steam space through the ends of the tubes or through tiny holes in the tubes. The tubes may also become plugged with mud, shells, debris, slime, or algae, thus cutting down on the cooling-water supply, or the tubes may get coated with lube oil from the reciprocating machinery. Corrosion and dezincification of the tube metal are common surface-condenser troubles. Corrosion may be uniform, or it may occur in small holes or pits. Dezincification changes the nature of the metal and causes it to become brittle and weak.

66. Where would you look for a fault if the air ejector didn't raise enough vacuum?

Answer. In this case, the trouble is usually in the nozzle. You will probably find that (1) the nozzle is eroded, (2) the strainer protecting the nozzle is clogged, or (3) the steam pressure to the nozzle is too low.

67. How would you stop air from leaking into a condenser?

Answer. First, find the leak by passing a flame over the suspected part while the condenser is under vacuum. Leaks in the flange joints or porous castings can be stopped with asphalt paint or shellac. Tallow or heavy grease will stop leaks around the valve stems. Small leaks around the porous castings, flange nuts, or valve stems can't always be found by the flame test. So, you might have to put the condenser under a few pounds of air pressure and apply soapsuds to the suspected trouble parts.

68. Do you stop cooling-water flow through a steam condenser as soon as the turbine is stopped?

Answer. You should keep the cooling water circulating for about 15 min or more so that the condenser has a chance to cool down gradually and evenly. Be sure to have cooling water flowing through the condenser before starting up in order to prevent live steam from entering the condenser unless it is cooled. Overheating can cause severe leaks and other headaches.

69. How would you stop a leaky tube in a condenser that was contaminating the feed-water?

Answer. To stop a leaky tube from contaminating the feedwater, shut down, remove the water-box covers, and fill the steam space with water. By observing the tube ends, you can find the leaky tube. An alternate method is to put a few pounds of air pressure in the steam space, flood the water boxes to the top inspection plate, and observe any air bubbles. Once you have found the leaky tube, drive a tapered bronze plug (well-coated with white lead) into each end of the tube to cut it out of service. This allows you to use the condenser since the tubes need not be renewed until about 10 percent of the tubes are plugged.

70. Why must condensate be subjected to salinity tests where brackish cooling water is used?

Answer. Condensate may leak from the cooling-water side to the steam side of the condenser and contaminate the feedwater, thus causing scale to form in the boilers. Or, brackish cooling water may leak into the steam space from cracked or porous tubes or from around the joints at the end of the tube ends, etc. By taking salinity readings of the condensate, leaks may be found before they can do any harm.

CHAPTER 11

INTERNAL COMBUSTION ENGINES AND TURBINES

This chapter addresses prime movers that use the products of combustion to generate power, not steam. In the spark-ignition gasoline and diesel engines, fuel is burned directly in the engine and not in a separate furnace that converts water into steam for use in steam engine or turbines. Gas turbines have combusters that heat air from compressors and then expand this hot product of combustion in turbine wheels. Since combusters in most gas turbines, especially those used for aircraft, are an integral part of the turbine arrangement, some specialists classify the gas turbine as an internal combustion engine. Larger units have their combuster separate from the compressor-turbine tandem arrangement that is a trademark of Aeroderivative jet engines.

There are many internal combustion engines installed for emergency power, such as in hospitals, large office buildings, power plants, and many manufacturing plants where a loss of power could cause consequential losses. Diesel generators are used where the power requirements for emergency service are much greater than a gasoline engine could provide. The diesel generator is also used extensively in municipal-owned power plants, on islands, at remote locations, and on ships. Many diesels are used in cogeneration plants where the cooling jacket water is used to heat malls, and in summer the hot water from the jackets is used in absorption air conditioning systems.

The gas turbine has shown remarkable growth to meet larger power requirements and the development of the combined-cycle power plant. Cogeneration, the simultaneous generation of power and process steam, has been encouraged by federal legislation. Any excess power generated at a cogeneration facility can be sold to the connected utility system. This has resulted in a rise in independent power producers, operating as a separate entity but selling power and steam to a nearby facility, with the excess power being sold to the utility.

Reciprocating spark-ignition engines are primarily used as a motor vehicle power source; however, until the advent of the jet engine at the end of World War II, this type of engine was also extensively used on aircraft, farm machinery, excavating machinery, and trucks. In the spark-ignition engine, a charge of air is drawn into the cylinder in the suction stroke shown in Fig. 11.1a with combustible gas or vaporized liquid fuel supplied by a carburetor. In the compression stroke, Fig. 11.1b, the mixture is compressed into a clearance space when it is ignited by spark plugs so that combustion takes place with a rise in pressure. This produces a downward power stroke on the piston as shown in Fig. 11.1c. On the return upward stroke, the burnt products are exhausted to the atmosphere, Fig. 11.1d, and the engine is ready to start the cycle again on the downward

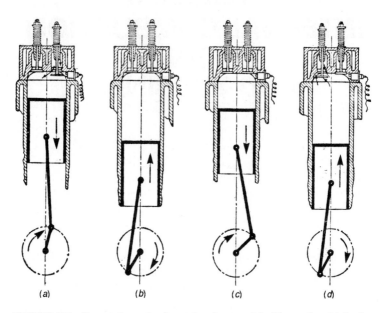

FIGURE 11.1 Four-stroke cycle of operation for a spark-ignition engine. (*a*) Suction stroke; (*b*) compression stroke; (*c*) power stroke; (*d*) exhaust stroke.

stroke by admitting filtered air and fuel to be compressed as shown in the figure. This cycle of operation is called a *four-stroke cycle* because it takes four strokes of the piston and two revolutions of the crankshaft to complete the cycle.

DIESEL-ENGINE CYCLES

The four-stroke cycle is also applied to *diesel engines,* but only the heat from compressing air is used to ignite injected fuel after the air is compressed to fairly high pressure and temperature in comparison to spark-ignition engines.

Figure 11.2*a* shows a spark-ignition *two-stroke cycle* engine in which the cycle is completed in two strokes and one revolution of the crankshaft. Figure 11.2*b* shows a *diesel two-stroke* cycle engine where at A the *scavenging* valves in the cylinder head open to admit a charge of compressed air from an air compressor. This blows *scavenging air* into the cylinder, forcing out the exhaust gases. The scavenging valve then closes in the cylinder head as shown in Fig. 11.2*b,* and as the piston moves upward, the exhaust ports are also closed. As the piston now moves upward, the air is compressed to about 500 psi and a temperature near 1000°F. At about top dead center, fuel is injected to provide a power stroke as shown in the illustration under B, and as the piston moves downward, the exhaust ports are opened to permit the exhaust of the spent fuel mixture so that the cycle can start again. Thus, there is one expansion per one revolution of the crankshaft, or a two-stroke cycle per cylinder.

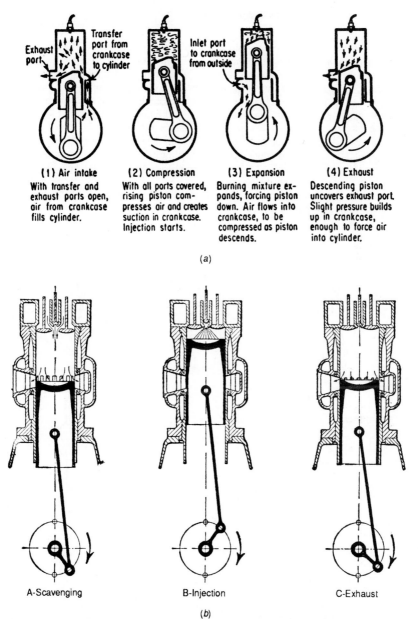

(a)

(b)

FIGURE 11.2 Two-stroke cycle engines are used in spark-ignition and diesel engines. (*a*) Two-stroke cycle of operation for a spark-ignition engine; (*b*) two-stroke cycle sequence of operation for a diesel engine using scavenging air supplied by an air compressor.

ENGINE CYLINDER ARRANGEMENTS

Figure 11.3 shows some of the possible engine cylinder arrangements. The radial engine was extensively used in aircraft applications until the jet engine was developed. There are still many smaller planes using positive displacement piston engines. Most automobile and truck engines are in-line or the V type. The horizontally opposed engine is usually applied to stationary diesel power drive, such as compressors or generators. Figure 11.4*a* lists some of the parts in a four-stroke diesel engine, while Fig. 11.4*b* shows a General Motors two-stroke cycle supercharged diesel engine.

In *supercharging,* air is forced by means of a compressor into the cylinder during the compression stroke so that more fuel can be burned, and thus the output power increased. The design requires a balance between increasing output and engine cooling; otherwise engine overheating can occur.

Ignition timing is an important feature in spark-ignition engines that use spark plugs, as is *valve timing.* Figure 11.5 shows a valve-timing diagram for one spark-ignition

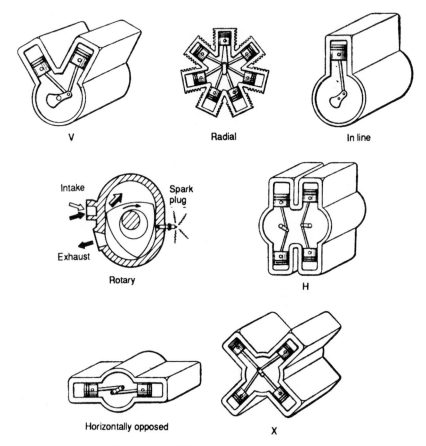

FIGURE 11.3 Types of engines by cylinder arrangement.

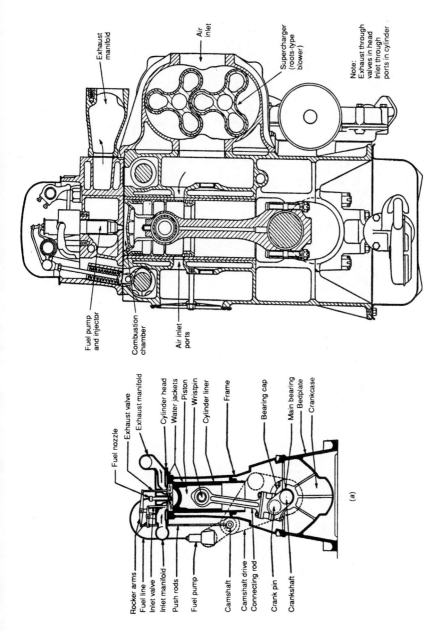

FIGURE 11.4 (*a*) Parts nomenclature of a four-stroke cycle engine; (*b*) two-stroke cycle engine features supercharging to increase output.

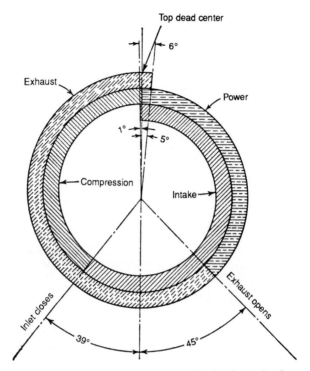

FIGURE 11.5   A valve-timing diagram and cycle of operation for a spark-ignition engine used on automobiles.

engine. The inlet valve opens 1° before the piston reaches top dead center on the suction stroke, and as the piston moves downward, the air-fuel mixture enters from the carburetor. The inlet valve closes 39° after the piston reaches bottom dead center, and then as the piston rises, the mixture is compressed with all valves closed. Just before the beginning of the next downward stroke, the spark plugs ignite the mixture for the power stroke. The ignition can raise cylinder pressures to over 400 psi. Note that the exhaust valve opens 45° before the piston reaches bottom dead center and closes 5° after the piston reaches top dead center.

INDICATOR DIAGRAMS

A *comparison* of the ideal spark-ignition and diesel engine, where little heat is lost, is possible by looking at the indicator diagrams in Fig. 11.6. Figure 11.6a shows a spark-ignition, four-stroke cycle engine, where V_2 is the clearance volume, and 0-1 is the piston displacement, or volume, swept by the piston. The *line* 0-1 represents admission at atmospheric pressure; 1-2 is compression; 2-3 is the rise in pressure caused by ignition; and 3-4 is expansion, or the working stroke. At point 4 the exhaust valve opens with the pressure falling to atmospheric. Line 1-0 represents the exhaust stroke. If the engine were operating on the two-stroke cycle, lines 0-1 and 1-0 would be eliminated, and the

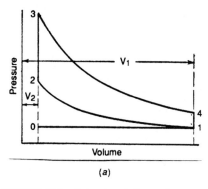

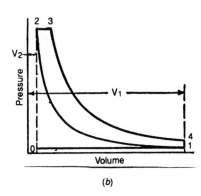

(a)   (b)

FIGURE 11.6 Comparison indicator diagrams of ideal four-stroke spark-ignition (*a*) and four-stroke diesel engine (*b*).

diagram would be represented by 1-2-3-4-1. Note the increase in pressure from 2-3 is at constant volume. The *air standard* is used in studying the spark-ignition engine, with the working substance assumed to act as air in compression instead of as a mixture. Other assumptions in the air standard are that the working substance has a constant specific heat, there is no transfer of heat between the gas and engine metal, there is adiabatic compression and expansion, and heat is added and rejected at constant volume as shown by lines 2-3 and 4-1 in the ideal indicator diagram for a spark-ignition engine.

In the ideal diesel cycle shown in the indicator diagram in Fig. 11.6*b,* air enters the cylinder along line 0-1, then is compressed to about 500 psi along the curve 1-2, and at 2, fuel is *forced* into the cylinder. The high temperature of the compressed air then ignites the air-fuel mixture; however, the supply of fuel continues as the piston moves back, causing combustion at constant pressure, line 2-3. The products of combustion then expand through the curve 3-4, providing the power stroke. At 4, the exhaust valve opens, and the pressure drops to point 1. On the fourth stroke, along line 1-0, the products of combustion are exhausted from the cylinder. Some products may remain in the clearance space and are mixed with the incoming air on the beginning first stroke of the cycle. Note that in this ideal indicator diagram of a diesel engine, all heat is supplied to the engine along line 2-3, and all heat is rejected along line 4-1.

DEFINITIONS

Definitions of interest to service operators include the following:

The *diesel engine* is an internal combustion engine in which the fuel is ignited by the heat of compression, whereas in a *spark-ignition engine* the ignition is by spark plugs timed to ignite when the compression of air and fuel is completed.

The *dual-fuel diesel engine* is one which may be operated by liquid oil fuel or gas and is equipped with controls to permit operating with either fuel.

The *single-acting engine* is one that has the working medium on one side of the piston, while in a *double-acting* engine it is on both sides of the piston.

The *opposed piston engine* uses the working medium simultaneously on two pistons in the same cylinder.

The *trunk piston engine* has the connecting rod connected directly to the wristpin in the piston, while a *crosshead engine* has the connecting rod connected to a crosshead traveling in guides, and the crosshead in turn is connected to the corresponding piston.

Air injection uses a quantity of high-pressure air which forces its way through a mechanically operated valve into the power cylinder and carries the fuel charge with it.

Mechanical injection refers to the method of injecting liquid fuel into the power cylinder by using a fuel pump to inject it. Subdivisions of mechanical fuel injection include:

1. Pump-timed injection by the action of a fuel injection pump plunger
2. Pump-timed injection using a distributor, in which the fuel pump both times the injection and also meters the fuel charge
3. Common-rail system in which a fuel pump supplies fuel to a header with the fuel being passed from this header to each cylinder at the proper time through mechanically operated valves

Scavenging air is air at low pressure that is introduced during the exhaust cycle into the cylinder to force out the products of combustion in the cylinder. Scavenging air may be compressed by (1) a separate compressor, (2) a direct engine-driven blower or compressor, (3) an under-piston scavenger by which the air is compressed on the underside of a power piston in a chamber which is separated from the crankcase of the engine, and (4) crankcase scavenging in which air is drawn from the atmosphere into individual crank chambers per cylinder by the upstroke of the power piston and then is compressed by the downstroke of the piston, with suitable air valves provided.

Supercharging is supplying high-pressure air into the intake of an engine to supply more combustion air for increased horsepower output. Methods of supercharging include (1) exhaust gas from the engine driving a turbine-driven blower and (2) mechanical-driven blower off the engine shaft, or by a separately driven electric motor.

Force-feed lubrication is lubricating engine parts with oil under pressure produced directly by a metering pump, *while gravity-feed lubrication* depends on lubricating engine parts under the action of gravity from an elevated supply.

A *starting air compressor* is used to supply high-pressure air to a diesel engine at a pressure sufficient to start the diesel. An *injection air compressor* furnishes air at a pressure sufficient to inject fuel into the engine.

Displacement.   *Engine displacement* is the volume in *cubic feet* per minute swept by all the pistons during the power stroke and is equal to the number of cylinders times the area of piston in square feet times the stroke in feet times number of power strokes per cylinder per minute. *Piston displacement* of an engine is the cylinder volume in *cubic inches* swept by the piston and is equal to the number of cylinders times the area of each piston in square inches times the stroke in inches.

Indicated horsepower (ihp) is the horsepower developed in the cylinder and is determined from the mean effective pressure shown by indicator cards, the engine speed, and the cylinder bore area, and stroke:

$$\text{ihp} = \frac{PLAN}{33,000}$$

where P = mean indicated pressure, psi
L = stroke of piston, ft
A = net bore or piston area, in^2
N = number of power strokes per minute

Piston speed is the total travel in feet made by each piston in 1 min, or piston speed = stroke in feet times r/min times 2.

Indicated thermal efficiency is equal to the ratio of the heat equivalent of 1 hp (2545 Btu) per hour to the number of Btu's suppled per indicated horsepower · hour.

Examples.   Here are some examples:

1. What is the engine displacement of a two-cycle engine with four 12-in diameter cylinders, a 24-in stroke, operating at 200 r/min?

$$\text{Engine displacement} = \frac{12 \times 12 \times 0.7854 \times 24 \times 200 \times 4}{144 \times 12} = 1256.6 \text{ ft}^3$$

2. What is the piston speed for above engine?

$$\text{Piston speed} = \frac{24}{12} \times 200 \times 2 = 800 \text{ ft/min}$$

3. What is the piston displacement for the engine in example no. 1?

$$\text{Piston displacement} = 4 \times 12 \times 12 \times 0.7854 \times 24 = 10{,}857.3 \text{ in}^3$$

4. What would the engine displacement be for the engine in example no. 1 if it were a four-cycle engine? Since there is one power stroke every two revolutions, the engine displacement is one-half that shown in example no. 1, or 628.3 ft^3.

5. What is the indicated horsepower for the engine in example no. 1 if the mean indicated pressure is 400 psi?

$$\text{ihp} = \frac{400 \times 2 \times 12 \times 12 \times 0.7854 \times 200 \times 4}{33{,}000} = 2193.4 \text{ hp}$$

Note that the stroke is 24/12 = 2 ft.

6. The engine in example no. 1 consumes 895 lb of fuel oil per hour with a heating value of 19,100 Btu/lb. Calculate the indicated thermal efficiency of the engine in delivering the indicated horsepower shown in example no. 5.

$$\text{Indicated thermal efficiency} = \frac{\text{ihp} \times 2545}{W_f \times H_f}$$

where W_f = fuel supplied per hour, per pound for liquid fuels
H = higher heating value or specified lower heating value, Btu/lb

Substituting,

$$\text{Indicated thermal efficiency} = \frac{2193.4 \times 2545}{895 \times 19{,}100} = 32.7\%$$

The *brake horsepower* is the horsepower delivered at the output end of the shaft of the engine. The ratio of brake horsepower (bhp) to indicated horsepower is the *mechanical efficiency.*

Example. The engine in example no. 1 drives a generator delivering 1200 kW with the generator having an efficiency of 93 percent. What is the mechanical efficiency of the engine if the indicated horsepower is as determined in no. 5 above?

$$\text{Engine shaft output} = \frac{1200}{0.93} \times \frac{3413}{2545} = 1749 \text{ hp}$$

$$\text{Mechanical efficiency} = \frac{\text{bhp}}{\text{ihp}} = \frac{1749}{2193} = 80\%$$

Diesel ratings. Diesels are rated at sea level by manufacturers. This is the net brake horsepower that the engine will deliver continuously when located at an altitude not over 1500 ft, with atmospheric temperature not over 90°F and a barometric pressure not less than 28.25 in of mercury. Overload guarantees normally stipulate 10 percent in excess of the full load rating for 2 continuous hours but not to exceed a total of 2 h in any 24-h period of operation. The power which an internal combustion engine delivers decreases as the altitude increases because of less dense air being supplied to the cylinder. Engine builders can supercharge an engine to overcome the altitude effect.

Higher intake temperatures also decrease output because hotter air is less dense, resulting in less oxygen for burning in the cylinder.

FUEL CHARACTERISTICS

Gasoline is a mixture of the various hydrocarbons, depending on the method of refining. The important properties for operating engineers are:

1. *Detonation characteristics* with the present measurement being expressed by the *octane rating.* The octane rating was derived by using a "standard" engine with standard fuels because poor detonation can be caused by many factors such as poor timing and worn spark plugs, to name a few. The standard fuels used are Iso-octane (2,2,4 trimethyl pentane), with a rating of "100 octane" as the best fuel. On the low side of the scale is normal heptane, which has very poor detonation qualities, and this was assigned a 0 octane number. Thus, the octane number is a measure of how well a mixture of fuel such as gasoline will avoid knocking because of poor detonation in the cylinder. As can be noted, gasoline engine manufacturers specify the octane rating that should be used in their engines in order to prevent objectionable knocks in operation.

2. *Volatility* of a gasoline is a measure of its starting capability and indicates if a portion of the fuel has a low enough boiling point to form a combustible mixture at existing air temperatures. *Vapor-lock* is caused by a fuel vaporizing in the fuel lines due to low vapor pressure and is also related to volatility. Best operating performance is obtained with a gasoline with the lowest distillation temperature. Crankcase dilution is avoided by having a fuel that does not condense or fails to vaporize in the engine.

3. *Gum and varnish deposits.* The fuel should not deposit these undesirable products on cylinders or pistons.

4. *Sulphur* in gasoline fuel reduces the octane rating and produces corrosion products as well as environmental harmful discharges.

DIESEL FUELS

The *diesel fuel knock rating* is also determined by testing fuels in a standard diesel engine, which is comparable to the spark-ignition engine. Two reference fuels are used: (1) *cetane* is assigned a rating of 100 cetane because of its ability to burn smoothly and ignite quickly, thus avoiding knocking. (2) On the low-scale side of poor detonation characteristics for diesels is alpha-methyl naphthalene with a rating of 0 cetane. Thus, a fuel having equivalent ignition properties to a mixture of 40 percent of cetane and 60 percent of alpha-methyl naphthalene, by volume, is classified as having a 40 cetane rating.

Other diesel fuel characteristics that are important in diesel operation are:

1. *Specific or API gravity* is the weight per unit volume, expressed either in specific gravity or by API degrees, with the conversion being

$$\text{API gravity in degrees} = \frac{141.5}{\text{sp.gr. at } 60°F} - 131.5$$

Gravity is related to heat content. See Table 11.1.

2. *Heat content* is expressed in high- and low-heat content value. The high-heat value includes the heat liberated by condensation of the water formed by fuel combustion, which the low-heat value does not. Most fuel oil consumption guarantees are based on the high-heat value.

TABLE 11.1 Some Typical High and Low Heating Values for Diesel Fuels

Gravity, API	Sp gravity, at 60°F	Weight fuel, lb/gal	High heating value		Low heating value	
			Btu/lb	Btu/gal	Btu/lb	Btu/gal
44	0.8063	6.713	19,860	133,500	18,600	125,000
42	0.8155	6.790	19,810	134,700	18,560	126,200
40	0.8251	6.870	19,750	135,800	18,510	127,300
38	0.8348	6.951	19,680	137,000	18,460	128,500
36	0.8448	7.034	19,620	138,200	18,410	129,700
34	0.8550	7.119	19,560	139,400	18,360	130,900
32	0.8654	7.206	19,490	140,600	18,310	132,100
30	0.8762	7.296	19,420	141,800	18,250	133,300
28	0.8871	7.387	19,350	143,100	18,190	134,600
26	0.8984	7.481	19,270	144,300	18,130	135,800
24	0.9100	7.578	19,190	145,600	18,070	137,100
22	0.9218	7.676	19,110	146,800	18,000	138,300
20	0.9340	7.778	19,020	148,100	17,930	139,600
18	0.9465	7.882	18,930	149,400	17,860	140,900
16	0.9593	7.989	18,840	150,700	17,790	142,300
14	0.9725	8.099	18,740	152,000	17,710	143,600
12	0.9861	8.212	18,640	153,300	17,620	144,900
10	1.0000	8.328	18,540	154,600	17,540	146,200

3. *Viscosity* is a measure of the resistance of a fluid to flow and is expressed by the Saybolt viscometer. For injection conditions, fuel oil must have a low viscosity in order to flow freely and atomize easily. The Saybolt viscosity required for most engines is between 35 and 70 s as reported at 100°F.

4. *Flash point* is that temperature where an open flame above the fuel surface will cause ignition of the fuel vapors. It is important in storing and handling of diesel fuels in transportation and fire protection.

5. *Pour point* is the temperature at which a fuel will cease to flow, such as in cold weather. It determines the heating equipment required in the fuel-handling system.

6. *Volatility and distillation range.* Volatility determines if the oil or fuel will atomize and ignite, while distillation is expressed in a range to be as low as possible without affecting the flash point or the viscosity. A low 10 percent point may cause trouble in the fuel pump injector because of vapor formation. The temperature for the first 10 percent distilled for medium diesel fuels will range from 400 to 480°F. The 90 percent points range up to 725°F and are used in low-speed engines.

7. *Ignition quality* is defined as the relative rapidity with which ignition starts following the beginning of the injection of the fuel oil. The accepted method for comparing ignition quality is the cetane number, with 100 cetane being a high-quality fuel.

8. *Water and sediment, or cleanliness,* of the fuel oil is important because water can reach the nozzles and cause misfiring as well as corrosion. Sediment can cause erratic governing and abrasive wear. Specifications usually limit water and sediment to 0.1 percent maximum. Water and impurities may enter the oil during transportation and storage. The cleanliness of a fuel oil is determined by centrifuging.

9. *Ash* represents the ash content in the products of combustion and is objectionable due to its abrasive action on engine internal parts.

10. *Sulfur* in fuel oil forms sulphur dioxide, which when combined with moisture, forms sulphurous acid which attacks cylinder walls, exhaust valves, and piping. The usual maximum allowed is 0.5 percent.

Fuel additives are used in hydrocarbon fuels to improve the detonation characteristics, with one of the first being tetra-ethyl lead. Refinery research has developed other additives, but the basis of octane and cetane rating remains the same, with the higher numbers indicating a fuel that will burn smoothly and avoid knocks from poor flame propagation in the cylinder.

SPEED AND LOAD CONTROL—GOVERNING

In the *spark-ignition engine,* a spark must ignite a combustible mixture, and a throttle valve in the carburetor is used to open and close the passageway into the intake manifold. The throttle regulates the amount of charge, which is of a fairly definite air-to-fuel ratio from idling to full load, either manually, such as in automobiles, or by speed governors responding to increases or decreases in load.

In most *diesel engine governing,* throttling is not used on the intake stroke, but a constant amount of air is inducted into the engine, and variations in load or speed are achieved by varying the amount of fuel injected into the constant amount of air. Two types of governors are used for stationary diesel engines: (1) The *isochronous* governor which maintain constant speed, or zero speed drop. It is usually applied to electric generator drives. (2) The *nonisochronous* governor, where speed regulation may not be as critical as for a generator drive, with speed variation capability being 5 to 8 percent of the rated speed.

Centrifugal-powered governors are used only for the nonisochronous governor. This is a standard mechanical governor in which power to operate the fuel control linkage is supplied by flyweights. Relay-powered governors are always used for isochronous governors and may be also used for nonisochronous types. Relay-powered governors are those in which power for operating the fuel control linkage is supplied by a suitable relay triggered by hydraulic, electric, or electronic means.

The *sensitivity* of a governor is the percentage of speed change required to produce a corrective movement of the fuel control mechanism. *Speed regulation* is the same as for most prime movers and expresses, as a percentage, how well the governor regulates the speed from no load to full load, or

$$\% \text{ regulation} = \frac{\text{no-load speed} - \text{full-load speed}}{\text{full load speed}} \times 100$$

Overspeed stops are required on all engines where a sudden drop in the load could cause the engine to overspeed well beyond its rating and do extensive damage. This overspeed stop must be independent from the regular governing system.

Uniform governing characteristics are desirable for diesels in a plant that drive ac generators since this permits changing the loading of individual units as a percentage of the total plant-load change.

Large diesels have *air starting* systems in which compressed air is admitted to the engine by air starting valves that admit compressed air to the power cylinders in a predetermined sequence (see Fig. 11.7). Air pressures required for starting can vary from 250 to 1000 psi, with the most prevalent being 250 to 350 psi. Other starting methods used are electric motor, air motor, gasoline engine, and generator operated as a synchronous motor for starting purposes.

Diesel *jacket water cooling* may be of the open and closed type. In the open type, water is passed through the jacket and is discharged or recirculated through cooling towers with makeup water added. In the closed system, water is passed through a heat exchanger, radiator, or a cooling tower with a heat exchanger for the jacket water. In short, in the closed system, water from the jacket is returned, while the cooling medium is recirculated in its own flow and not wasted. The closed systems prevent buildup of scale in the jacket in comparison to raw water flow with no recirculation.

An emergency water connection from a municipal water supply to the jacket cooling system is most desirable in the event of a failure in the regular jacket cooling system; however, the water should be tempered before being introduced into the jacket to prevent thermal shock.

Intake air for diesels should be filtered and a muffler installed to reduce the noise level as shown in Fig. 11.8. Mufflers are also needed on the exhaust to prevent excessive neighborhood noise levels as shown in Fig. 11.8

Shutdown alarms and trips are an essential safety feature of diesel operation used for stationary service. As a minimum, the following should be installed: overspeed trip, loss of jacket cooling water alarm and trip, low oil pressure alarm and trip, and high oil temperature alarm and trip.

Diesel operation and maintenance practices usually start with plant service operators becoming familiar with the plant layout and then reviewing plant and manufacturer's instructions for the diesels under their control. An operator should become familiar with all piping and valves for engine fuel, cooling water, lubrication, air starting, air intake, and exhaust systems as well as mechanical and electrical auxiliaries. All well-managed power plants have clean engines, auxiliaries, and engine room surroundings. Therefore, operators should not neglect the "housekeeping" functions since they indicate dedication to their responsibilities in assuring a safe and reliable plant. It is also good practice to periodically review the manufacturer's instruction book(s) with respect to such items as starting, stopping, operation, capacity limits, allowable pres-

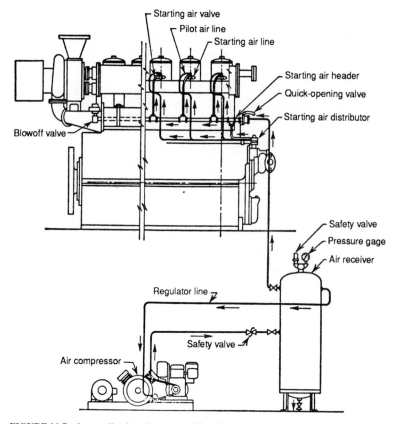

FIGURE 11.7 Larger diesel engines use auxiliary high-pressure air system to start the engine.

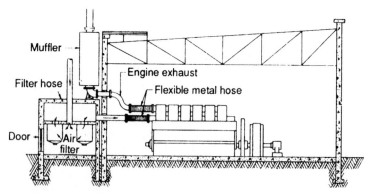

FIGURE 11.8 Diesel engines require outside intake air with filters and mufflers on the intake and exhaust for noise abatement.

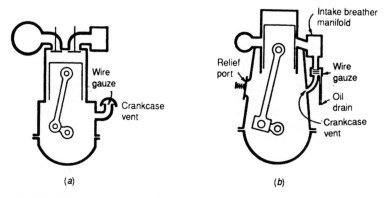

(a) (b)

FIGURE 11.9 Crankcase explosion prevention is important on diesel engines. (*a*) Crankcase vent requires screen to prevent an outside source from igniting crankcase oil vapors; (*b*) relief port acts as relief valve when crankcase pressure rises. Do not stand near this port.

sures and temperatures, clearances or wear and tear that can be tolerated, adjustments, and recommended inspection and overhaul schedules for the different parts of the engine.

A *log book* is an essential feature of good operating practices. The log should show starting and stopping times for each engine, load readings, engine operating data such as fuel pressure, exhaust temperatures, and similar data for future reference as to engine performance. Space should be available on the log to record unusual events or conditions on a shift, including notations that some item on the engine needs maintenance attention.

Crankcase explosions are caused by oil vapors in the crankcase being ignited by a spark, a red-hot metal such as a bearing, piston blowby, or outside flame, such as a burning cigarette. It is good practice not to remove crankcase doors or access covers when the engine is hot to avoid hot crankcase oil vapors mixing with incoming air. The engine should be cooled for at least 10 min under normal operating conditions before inspecting the crankcase area. No open lights should be placed in the crankcase until the engine is cool.

Lubricating oil is vaporized in the crankcase by the motion of the crank, connecting rod, and crankpin spraying the oil through the air in the crankcase. Oil vapor needs a source of ignition and the correct air-fuel ratio to become explosive. A lubrication failure on a bearing could set off a crankcase explosion; operators must be especially careful to avoid this since a plant fire could result.

A spark can result from bare metal striking against bare metal, as when an indicator gear comes adrift or when the telescope piping or lube-oil lines strike another metal part. Piston blowby is common but not dangerous unless red-hot gases blow past broken piston rings into the vaporized lube oil. Blowby is dangerous when it burns oil off the cylinder walls, causing a hot spot from the galled metal. Failure of the lube oil in the wrist-pin, main, or crankpin bearings, also causes metal-to-metal contact; dry metal parts seize, causing sparks. If the vent pipe is unscreened (see Fig. 11.9*a*), an outside source may ignite the crankcase oil vapors. A vent connection from the intake manifold to the crankcase must have wire gauze and a drain, as shown in Fig. 11.9*b*. If not, sparks from the leaky valves will work in reverse and find their way into the crankcase.

The damaging crankcase explosion is always a double one. The first explosion is mild; it may rupture the crankcase or blow off a crankcase door. Then, air rushes into the crankcase through this opening and combines with the hot oil vapor, resulting in a big explosion.

To guard against explosion, some vapors are removed by a vent from the crankcase and intake manifold (see Fig. 11.9b). Other engines carry a slight vacuum in the crankcase. This vacuum is set up by suction to the scavenging blower, which removes vapors and air needed to support an explosion. Another precaution places spring-loaded relief ports on the crankcase. After the first explosion, the spring-loaded ports open to relieve the pressure and then snap shut, closing out the air needed to support the violent second explosion. Make sure that your relief ports work freely. Let the engine cool as much as practical before opening the crankcase for maintenance.

Lubricating oil filters and strainers should be inspected and cleaned at regular intervals. Oil can be reclaimed by the use of fuller's earth-type filters with provisions for heating the oil so that the action between the oil and the fuller's earth is accelerated and intensified. Heating the oil to a higher temperature will help drive off any moisture and any fuel dilution. Vacuum pumps are also used to remove vapors without overheating the oil, which may cause additive loss.

Centrifuging oil removes moisture but may not remove oxidation products since they may have the same specific gravity as the oil. *Oil coolers* must be kept clean both on the water- and oilside in order to avoid losing heat-transfer capacity. Good instrumentation on the oil system will help to detect dirty filters and poor heat transfer, which cause changes from normal pressure and temperature readings.

Cooling-water systems must be maintained free of scale and sedimentation in order to avoid engine overheating. It is good practice to use treated water for the jackets in order to avoid corrosion inside of them. Be careful in changing pumps or filters on water cooling systems. Engines can overheat if the proper valves are not opened during these changes.

Starting air systems must be kept free of condensed moisture in the air lines or air storage tank, and this requires draining the condensate on a regular basis, including prior to placing a diesel engine into service.

Crankshaft alignment and strain gage testing are important procedures during engine inspections. This will assist in detecting cracks that may be developing and progressing in the corner radius of the crank webs. This area of the crankshaft should be checked for fatigue cracks by nondestructive testing methods, such as a dye penetrant. Make sure the dye is cleaned off after the inspection.

Diesel overhaul inspections require piston, cylinder, and piston ring inspections, usually after so many hours of operation. Carbon buildup on the rings may prevent them from performing per design and thus may cause excessive piston and cylinder wear. The manufacturer's instructions should be used as guide in scheduling cylinder, piston, and ring wear inspections, unless signs of internal problems in this area of the piston manifest themselves during operation. These signs could include excessive blowby, knocking, loss of output, and similar poor performance criteria that will require shutting the engine down for inspection and repairs before more damage may result.

Balancing the load on each cylinder of a diesel engine is important to prevent unbalanced torques on the crankshaft that can lead to cracking in highly stressed areas of the crankshaft. The recommended procedure for balancing cylinder loads is usually provided by the engine manufacturers but they generally are: (1) Record the exhaust temperature of each cylinder. (2) Adjust the fuel-pump metering control rod so that the temperature is the same from all cylinders. (3) Check the cylinder temperatures again after adjusting the fuel. (4) With all exhaust temperatures the same, take indicator cards at full load *only*. These cards show whether or not each cylinder is doing its share of the work. Remember that cylinder temperatures may balance perfectly, yet combustion pressures may be out of balance. This is the most important step in the balancing operation. (5) Study the pressure cards for unbalance by checking them against the engine maker's instruction book. (6) Make adjustments to bring all the cylinder firing pressures to the same level.

Smoky exhaust can result from poor compression, an overloaded engine, water in the fuel, the wrong type of fuel, not enough fuel-valve lift, etc. The fault might lie in one cylinder or in a number of cylinders. A quick way to find the faulty cylinder is to shut off the fuel to one cylinder. Then go outside and look at the exhaust. If the exhaust clears, it is evident that the trouble is in that cylinder. If there is no difference, keep cutting out one cylinder at a time while checking the exhaust. If there is no change in the exhaust when all cylinders have been tested, the trouble is common to more than one cylinder. An exhaust pyrometer also gives you this information.

Always *analyze operating data* as part of your procedure in keeping a diesel plant operating efficiently and safely. Fuel consumed per horsepower · hour or kilowatthour is a good index to follow in planning maintenance or adjustments. Radical changes in pressures and temperatures should require immediate attention and correction before more serious damage may result. By being alert for changes in noise and similar manifestations of engine operation, problems that could become serious are avoided by the attentive operator.

For those diesels kept in reserve for *emergency duty,* it is important to test-run them on a periodic basis to make sure all engine performances are normal. Be especially alert to lubricating oil degradation by the entrance of moisture from leaking oil coolers or water from cooling jackets. Check the oil level periodically, and watch for a *rising level* since this will indicate water is entering the lube system. Drain water and condensate from all air-starting equipment prior to test runs. For those engines kept in separate rooms, and in northern climates, it is good practice to keep the room heated to reduce the risk of condensation intruding into the lube systems and also to avoid *jacket water freezing.*

GAS TURBINES

The gas turbine is one of the fastest growing prime movers (1) for industrial power and process application, (2) for utility peaking and life extension of older steam plants with the addition of a gas turbo-generator, and (3) with the rapid expansion of independent power producers (IPP) that use the combined cycle consisting of gas turbo-generator, waste heat to waste heat boiler, and then steam to steam turbo-generator to generate power and sell steam to an adjoining property. Manufacturers continue to develop gas turbines that have higher efficiencies, lower emissions, and increased reliability and maintainability. Facility service operators may become involved with this rapidly developing prime mover.

A *simple cycle of operation* is illustrated in Fig. 11.10a: (1) An axial flow, multistaged air compressor compresses air from the atmosphere. The compressor is driven by the reaction-type gas turbine once the system is started. (2) Liquid or gaseous fuel is injected into the combuster and is burned with part of the air discharged from the compressor; however, the remaining and greater part of the compressed air flows through the annular space around the combuster and, when emerging from the annular space, mixes with and cools the products of combustion to the design inlet temperature for the gas turbine. (3) The gas flow then enters the multistaged gas turbine to be expanded to atmospheric discharge. The expansion of this gas through the gas turbine produces enough power to drive the axial flow compressor, and the remaining power is used to drive a generator in the illustration. Figure 11.10b shows a typical layout for a simple gas turbine cycle arranged in tandem.

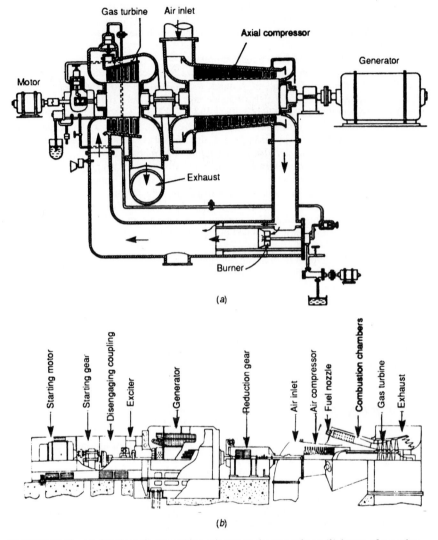

FIGURE 11.10 (*a*) Simple-cycle gas turbine exhausts to the atmosphere; (*b*) layout of a tandem-arranged gas turbine driving a generator.

There are many variations of the simple cycle. For example:

1. *Aeroderivative gas turbines* as shown in Fig. 11.11 have the exhaust gas go through a venturi nozzle to produce the thrust for airplanes. Many of these aeroderivative turbines have been converted for power drive by installing an *expander turbine,* with the jet engine being the gas producer. The expander turbine may be on a separate shaft.

2. Aeroderivative gas turbines use many combusters, arranged in a ring around the turbine.

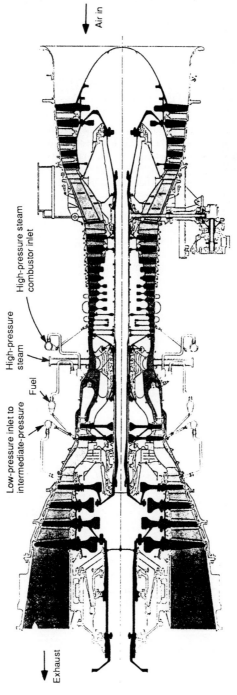

FIGURE 11.11   Aeroderivative gas turbine converted to power production uses steam injection for NO$_x$ control and power augmentation. (*Courtesy,* Power Magazine.)

3. *Industrial gas turbines* are increasingly using steam injection for NO_x control and also in some cases to boost output.

4. The turbine shown in Fig. 11.11 has a high- and low-pressure expander turbine on the same shaft with the compressor. Many units use one gas turbine to drive the compressor and the second to drive a generator, but on a different shaft.

5. In the regenerative cycle, Fig. 11.12a, the exhaust from the gas turbine preheats the air going to the combuster, thus saving fuel.

6. Figure 11.12b shows a *combined-cycle* where a heat-recovery boiler uses the exhaust heat from the gas turbine to generate steam to drive a steam turbo-generator. Figure 11.12c shows a two-pressure heat-recovery boiler supplying a high- and low-pressure section of a steam turbo-generator. *Duct auxiliary* firing is used to augment the energy input to the heat-recovery boiler, usually for peak loading purposes.

7. Figure 11.12d shows a combined cycle, using two steam pressures to supply a steam turbo-generator but with extraction steam from the turbines being used for feedwater heating. Many IPPs use the different pressures, or extraction steam, to supply a nearby plant process with steam and thus the facility becomes a *cogeneration* plant, which the federal government encourages under the PURPA Act because it is considered a fuel-saving method.

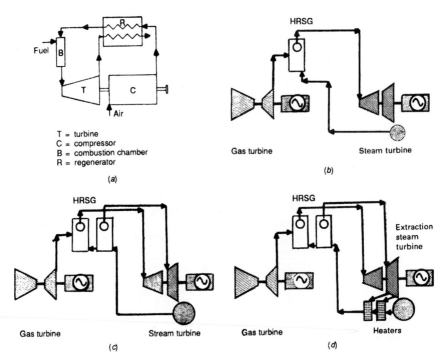

FIGURE 11.12   Some gas turbine cycles. (*a*) Regenerative cycle; (*b*) single-pressure combined cycle; (*c*) two-pressure combined cycle; (*d*) two-pressure combined-cycle cogeneration with steam extracted for feedwater heating or process use.

High-Temperature Materials

Gas turbine development concentrated on developing materials that could survive the high temperatures that blades and nozzles may be exposed to, but even more important, the *thermal efficiency* could be improved by raising the turbine inlet temperature and the pressure ratio as is shown in Fig. 11.13a. Figure 11.13b shows the theoretical pressure-volume diagram for the simple gas turbine. Note its similarity to the diesel diagram with respect to combustion taking place at constant pressure from B to C. This is called the *Brayton cycle*. As the curves in Fig. 11.13a show, for any specified turbine inlet temperature, there will be a corresponding best pressure ratio which will result in maximum thermal efficiency. The optimum value of pressure ratio depends on the turbine and compressor efficiencies. Gas turbines are being built with pressure ratios of 13.5 to 1 and with gas turbine inlet temperatures of 2190°F; this is possible because of (1) gas turbine material development, (2) blade and nozzle cooling methods, and (3) coating buckets or blades with alloys of cobalt, chromium, aluminum, and ytrium in order to increase the resistance of these critical components to hot corrosion and oxidation.

Factors affecting the materials used for gas turbines differ somewhat from those used in steam turbines because of the high-temperature exposure and the fact that the material is exposed under high stress to products of combustion and, with steam injection, to the reaction of steam with the products of combustion. Practically all gas turbine material is of the chromium-nickel alloy type. Chromium offers resistance to hot oxidation and corrosion. Nickel improves ductility. Check your machine's specifications for the material used for blades, nozzles, and combusters that are all exposed to high-temperature gas in operation. The following lists some items that affect gas turbine material life:

1. *Stress-rupture tests* at different temperatures shown in Fig. 11.14a show that the materials lose strength at a more rapid rate at elevated temperatures.

2. Figure 11.14b shows *creep rates* for some alloy material used in gas turbines. *Creep* is the loss of ductility of high-temperature alloys with time, making them brittle near the end. Any shock load at this point will cause rapid crack failures. Creep must be considered during material lifetime evaluation.

3. *High-temperature oxidation* of high-temperature alloys in gas turbine service is an important consideration, both in material selection and during inspections. Poor oxidation and corrosion resistance of material reduces the cross-sectional area of the material resisting the loads, and as a result the material has higher stress imposed on it, and cracks may develop or even rupture. Oxidation tests are usually conducted by the gain in weight method because the material is combining with oxygen. Some tests show an increase in weight of about 1 mg/cm^2 at 1500°F in 1000 h of exposure. Metal-spray coatings have been developed to resist hot oxidation, but these should be checked during inspections to make sure they have not been eroded.

4. *Starts and stops* impose thermal stresses on vital gas turbine parts, and this is especially applicable to gas turbines in peaking service. Gas turbine manufacturers assign a "used life factor" to turbine blades and buckets that are exposed to so many starts and stops, and these hours are *added* to the number of hours that the machine was in service. Generally, gas turbines are inspected internally on the "hot gas path" about once per operating year, meaning combusters and the gas turbine that is exposed to high temperatures are inspected. The purpose of the inspection is to note wear and tear, signs of cracking, creep degradation, and the effects of hot oxidation as well as steam injection for those units so equipped. The air compressor and auxiliaries are subjected to inspection at about the 25,000 equivalent hours of operation.

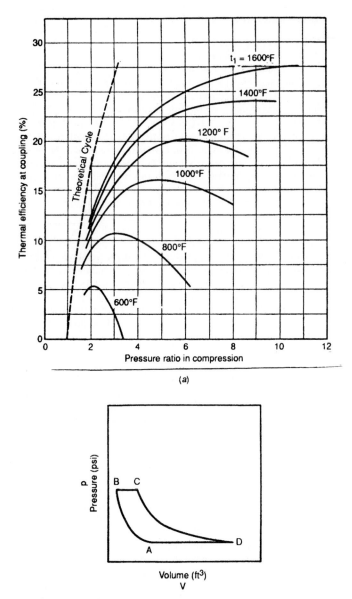

FIGURE 11.13 (*a*) Thermal efficiency of gas turbine is dependent on high gas inlet temperature and compressor ratio; (*b*) Brayton or theoretical combustion cycle for simple gas turbine arrangement.

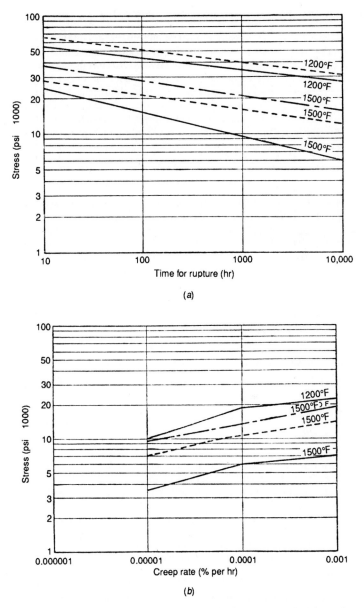

FIGURE 11.14 (*a*) Stress-rupture tests show high-alloy materials lose strength at higher temperatures with time; (*b*) creep rate tests show high-alloy material's ductility declines at high temperature with time.

Expected Blade Life

It is important for operators to request information from the gas turbine manufacturer on the question of expected blade life as a result of material degradation from creep and oxidation. Some manufacturers of older aeroderivative gas turbines that have been converted to stationary service recommend first-stage blades be renewed after 5 years of equivalent operating hours, or after about 25,000 h. Other manufacturers will assign another 5 years life *if* conditions show no heavy wear and tear after each hot gas path inspection. Be aware of the possibility of blade renewal as a potential maintenance budget item. When purchasing a gas turbine, it is important to request information on the expected life of the components.

Fuels

Firing oil for gas turbines in the beginning of their use for stationary service was found to have rapid deterioration, especially the first-stage nozzles as a result of using fuel oils high in vanadium and sodium. As a result, permissible limits of contamination in fuel oil and gas fuel were established, and this has also resulted in fuel oil treatment before it is introduced into the combuster. One manufacturer recommends the contaminant limits for natural gas firing, 5-ppm alkali-sulfate. The recommendation for oil-type fuel is:

Lead	5 ppm maximum
Calcium	10 ppm maximum
Sodium plus potassium	5 to 10 ppm maximum
Vanadium if no additive is used	2 ppm maximum
Vanadium if additive is used	500 ppm maximum
Sulfur if waste heat boiler is used	0.5 percent or boiler/steam turbine manufacturing specifications
Suspended solids filtering	5 μm mechanical or low-pressure air atomizers; 25 μm for high-pressure air atomizers
Total ash in the oil	Not over 2000 ppm

Methods for treating oil include removing the sodium by washing and centrifuging; the corrosive effect of vanadium is reduced by the introduction of a suitable additive, such as a water solution with magnesium sulfate thoroughly mixed with the oil to form an emulsion.

Operators should become familiar with any fuel treatment that may be used on their gas turbine so that acceptable contamination limits are maintained in operation.

NO$_x$ control on gas turbines burning natural gas is another operating consideration. The higher the temperature at which natural gas is burned, the more NO$_x$ is formed. *Steam injection* is used to lower the temperature and thus reduce the NO$_x$ level to acceptable jurisdictional criteria. For example, in the Los Angeles area, the maximum NO$_x$ emission level permitted is 0.15 lb/MW · h from any industrial facilities.

Selective catalytic reduction (SCR) is also used to reduce NO$_x$ emissions. The SCR process involves injecting air-diluted ammonia vapor into the flue gas of a combustion stream through a series of injector manifolds upstream of the catalyst. The SCR injection system automatically meters the proper amount of ammonia vapor needed for the catalytic process. Catalyst units for the SCR system consists of honeycomb-shaped cells of ceramic material that contain metallic oxides. As the combustion gas flows through the catalyst cells, the nitrogen oxides are converted to nitrogen and water vapor.

Low NO$_x$ burners are under development and in use that permit burning at lower temperature, using methods such as premixing fuel and air and using swirlers for better fuel-air combustion at lower temperatures.

Retrofitting or repowering is also boosting gas turbine use. Typically, a utility may replace an older boiler with a gas turbine and waste heat boiler that then supplies steam to existing steam turbo-generators and feedwater heaters. This boosts the overall thermal efficiency in comparison to installing a new boiler from 35 percent to between 40 and 50 percent. Existing turbo-generators are usually also retrofitted with any new gas turbine-waste heat boiler installation by replacing, if necessary, parts that have outlived their useful life, such as worn blades or windings on the generator. This type of repowering is faster and costs less in terms of dollars per kilowatthour than it would to install a complete new facility.

Controls

Combined-cycle plants have been heavily automated to maximize output and minimize downtime. Controls have been developed to track more than one gas turbine combination, with future developments having the capability of diagnosing readings and supplying data that need the operator's attention. Also in development are expert systems to tell what needs correction. An important feature incorporated in these systems is tracking performance and operating data to schedule future maintenance. Sensors and transmitters distributed through the combined-cycle plant supply this information to a computer. The operator in the control room is linked to the plant by video displays that show the status of each subprocess and is provided with function keyboards that permit manual intervention if necessary. Each specific plant subprocess, such as the gas turbine plant, heat recovery boiler, or the steam turbine set, has its own control system. All subprocess controls have start-up and shutdown sequences preprogrammed into the computer. Many possess intelligent logic that can interrupt the starting sequence if deviations from set starting procedures occur.

Vibration monitoring is a standard feature of these automated systems, because of the recognized value of tracking vibration to detect imbalance, looseness, misalignment, rubbing, and bearing degradation.

Most systems installed on newer combined-cycle plants are programmed to generate daily reports, similar to an operator log summary. Weekly and monthly production records on kilowatthours produced, steam generated, and fuel used are among some of the summaries that are printed out for plant use and future reference.

Typical *alarm and shutdown* functions per ASME, ANSI B 133.4 standard are the following:

	Recommended Function	
Item	Alarm	Trip
Low lube oil pressure	x	x
Overspeed	x	x
Turbine exhaust temperature	x	x
Excess vibration in train	x	x
Inlet air filter, high differential pressure	x	x
Flame failure	x	x
Standby lube oil pump operation	x	
Low fuel supply pressure	x	
High and low oil reservoir level	x	
High lube oil cooler outlet temperature	x	

	Recommended Function	
Item	Alarm	Trip
High bearing temperature	x	
High differential pressure lube oil filter	x	
Fire protection initiation	x	x
Hazardous atmosphere detection	x	x
Protective devices on auxiliaries or process operating	x	x

Additional trip functions may be selected by the user or manufacturer. In starting a gas turbine, it is important to provide automatic *purging* of sufficient duration so that any residue combustibles are displaced from the inlet through the entire exhaust system, before the unit is fired.

Fire protection is required in gas turbine systems. Detectors with the ability to sense a fire and then initiating automatic extinguishing systems should be installed at points where fuel or fire may spread. The gas turbine must have alarms and trips that go off when a fire is sensed. *Control failures* on a gas turbine should have provisions that do not endanger the machine but provide an alternate safe operation path or safety shutdown. Many systems have redundant controls.

Water treatment is essential in combined-cycle plants for the heat-recovery steam generator and where steam or water injection is used. One gas turbine manufacturer requires that no more than 30 ppb of sodium, potassium, vanadium, or lithium should be contained in the injected steam to prevent hot gas path chemical attacks on metals, such as stress corrosion.

Low ambient air temperatures can cause icing in the air-inlet plenum of the compressor. One means of protecting against icing is a thermostatically controlled valve that supplies exhaust gas to the inlet plenum to prevent freezing temperatures.

Axial air compressors usually come equipped with a *blowoff* system to prevent *surging*. These compressors have steep pressure-volume characteristics, which may cause harmful pulsations when pressure conditions are beyond the normal operating range. Violent surging can result in high vibration and equipment damage. Manufacturers establish the discharge pressure on their units; if they are exceeded, surging may occur. Blowoff valves are provided that open at set pressures and also open and release the air from the compressor on shutdown. When the machine is started, these valves close automatically. Usually an external source of air is used to keep the valves closed during start-up.

Because of the rapid development in gas turbine technology, and its application to combined cogeneration systems, operators will find new opportunities as this use of gas turbines expands in utilities and industrial plants and with independent power producers. It is important, however, to carefully review all operation, testing, and maintenance instructions for the machine under your custody and control since manufacturers differ in their designs.

QUESTIONS AND ANSWERS

1. By what other name is the spark-ignition cycle known, and what are the events in this cycle?

   Answer.   It is known as the Otto cycle after the German developer. The events in this cycle are intake stroke; compression stroke; ignition, combustion, and expansion stroke; and exhaust stroke.

2. What is the other name for a compression-ignition engine?

Answer. Another name is diesel, after its inventor, Rudolph Diesel.

3. What is the piston displacement for a six-cylinder engine, $3\frac{1}{2} \times 3\frac{1}{4}$ in?

Answer. $0.7854 \times (3.5)^2 \times 3.25 \times 6 = 187.6$ in^3

4. A four-stroke-cycle diesel was given an indicator diagram test that showed an area of 0.458 in^2 with a diameter length of 2.41 in. The indicator used had a spring constant of 400 psi/in. The one cylinder tested was 22×30 in, operating at 180 r/min. What is the imep and ihp for the one cylinder?

Answer. They are:

$$\text{Mean height of diagram} = \frac{0.458}{2.41} = 0.19 \text{ in}$$

$$\text{imep} = \text{mean diagram height} \times \text{spring constant} = 0.19 \times 400 = 76.0 \text{ psi}$$

$$\text{ihp} = \frac{PLAN}{33,000} \times \text{power stroke per revolution}$$

$$= \frac{76 \times 30/12 \times 0.7854(22)^2 \times 180}{33,000} \times 1/2 = 197 \text{ hp/cylinder}$$

5. A diesel engine operating at 800 r/min uses 0.25 lb of fuel in 4 min. The torque developed in a test machine indicates 56 lb · ft. What is the specific fuel consumption for the engine?

Answer. Specific fuel consumption is defined as the fuel used per brake horsepower; therefore it is necessary to first find the brake horsepower. Use hp $= TN/5252$ where $T =$ torque in lb · ft, and $N =$ r/min. Substituting, bhp $= (56 \times 800)/5252 = 8.53$ hp. Fuel consumption is per bhp · h $= 0.25(60)/4(8.53) = 0.44$ lb/bhp · h.

6. What engine performance ratings are used to compare engine performance?

Answer. (1) Specific fuel consumption, lb/bhp · h, (2) brake mean effective pressure, psi, (3) specific weight, weight of engine per hp, or lb/bhp, (4) output per unit displacement, bhp/in^3.

7. Why is the octane number significant for spark-ignition engines?

Answer. Detonation problems can exist in a spark-ignition engine that produces knocking and rough operation. The octane number classifies the antiknock characteristic of a fuel. A high octane number means smoother combustion in the cylinder.

8. What does a high cetane number indicate in fuels for diesel engines?

Answer. A high cetane number indicates the fuel will burn smoothly and ignite quickly to prevent detonation and knocking in the cylinder during combustion.

9. What is the function of a carburetor in spark-ignition engines?

Answer. The carburetor meters the fuel in comparison to supply air being drawn into the cylinder so that a combustible mixture is formed in the cylinder to be ignited by a spark.

10. What does the term *scavenging air* mean?

Answer. In a two-cycle engine, usually separate low-pressure air from a blower is supplied or blown into the power cylinder during the exhaust period to help clean the cylinder of exhaust gas. This displacement furnishes fresh air for the next cycle. Scavenging air is always used for two-cycle engines and sometimes for four-cycle engines.

11. What are four methods of air scavenging?

Answer. Four methods of air scavenging are:

1. *Separate scavenging.* Air is compressed in a compressor or blower that is driven by an independent source.
2. *Integral scavenging* (see Fig. 11.15a). Air is compressed in a compressor or blower that is directly connected, geared, or belted to the engine scavenged.
3. *Under-piston scavenging.* Air is compressed on the underside of the power piston in a chamber separate from the crankcase. The upstrokes of the power piston draw air into this chamber from the atmosphere. The downstrokes of the power piston compress the air.
4. *Crankcase scavenging.* Here, the upstrokes of the power piston draw air from the atmosphere into individual crank chambers, one on each cylinder. Then, the air is compressed in the crank chambers by the downstrokes of the power piston.

12. What does the term *supercharging* mean?

Answer. Supercharging is the process of supplying the intake of an engine with air at a density greater than the density of the surrounding atmosphere. This increased density is retained in the cylinders at the start of the compression stroke. *Result:* The same engine can burn more fuel; thus the engine's horsepower is increased.

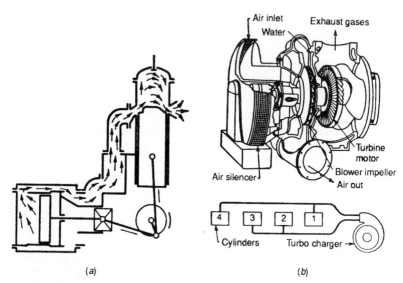

FIGURE 11.15 (*a*) Scavenging is used to help clean a cylinder of exhaust gas; (*b*) supercharging is used to force more air into the cylinder so that more fuel can be burned and thus boost the power output of an engine. In (*a*) a scavenger compressor is driven by the crankshaft (integral type). In (*b*) an exhaust gas turbine is used to drive a supercharging air compressor.

13. How are engines supercharged?

Answer. See Fig. 11.15*b*. There are two general ways to supercharge an engine. (1) Exhaust gases from the engine cylinder are used to drive an exhaust-gas turbine which is directly connected to a blower that supplies air to the engine's intake manifold. There are no mechanical connections between the engine and the turbine. (2) A blower that supplies air to the intake manifold is either driven directly from the engine or separately by an electric motor or other prime mover. When separately driven, the extra power needed to drive the blower is added when calculating the net bhp rating of the engine.

14. How does a unit injector work?

Answer. A unit injector combines a pump, fuel valve, and nozzle in a single housing, thus eliminating discharge tubing. This permits very high injection pressures, up to 20,000 psi. A plunger has a helical groove in its side that connects to a space under the plunger. Two ports are on one side and the inlet and bypass ports are above the plunger. With the plunger in its highest position, the fuel enters the inlet port until the lower edge of the plunger closes it and the pressure rise begins. When the pressure reaches its proper level, oil discharges through a valve on the nozzle. As the downstroke begins, the lower edge of the helix passes the bypass port, releasing pressure. Then a spring closes the nozzle valve. The bypass continues to the end of the stroke. By turning the plunger, you can change the effective stroke.

15. What is the ignition quality of a diesel fuel?

Answer. Ignition quality is expressed by an index called the cetane number; it is one of the most important properties of a diesel-fuel oil. High-speed diesel engines need a cetane number of about 50. Its value as a diesel-fuel characteristic is similar to the octane number of gasoline. Ignition quality not only determines the ease of ignition and cold starting but also the kind of combustion obtained from the fuel. A fuel with a better ignition quality (a higher cetane number) gives easier starting, even at low temperatures; quicker warm-up; smoother and quieter operation; lower maximum cylinder pressures; and more efficient combustion, hence lower fuel consumption.

16. What is speed droop?

Answer. Speed droop is the decrease in engine speed from no load to full load. It is expressed in revolutions per minute or as a percentage of the normal or average engine speed. For example, to find the speed droop of a governor if an engine has a normal speed of 1400 r/min and a no-load speed of 1450 r/min:

$$\text{Speed droop} = \frac{100\,(n_1 - n_2)}{n_2}$$

$$= \frac{100\,(1450 - 1400)}{1400} = 3.57 \text{ percent}$$

where n_1 = no-load speed
n_2 = normal speed

17. What is isochronous governing?

Answer. Isochronous governing maintains engine speed at a truly constant rate, regardless of the load; thus, there is perfect speed regulation and zero speed droop.

18. What is hunting, and what causes it?

Answer. A continuous fluctuation of engine speed, slowing down and speeding up from the desired speed, is called hunting. It is due to undercontrol or too high a sensitivity range of the governor. Governor hunting is due to the lag in action of the control mechanism, caused by poor sensitivity. This results in a large speed change before any governor action takes place. The engine slows down or speeds up too much. Then, once the controls begin to move, they continue to move, even after the correct speed is obtained. This results in an overcorrection of engine speed in the opposite direction. Hunting may also be caused by a governor that is too small. The lag in action permits too great a change in engine speed to occur during the change in engine load before correction, even if the governor is very sensitive. Reducing the friction in the operating mechanism of the governor will tend to increase its stability.

19. What is sensitivity of a governor?

Answer. A change in engine speed required before the governor makes a corrective movement of fuel control is known as sensitivity. It is generally expressed as a percentage of the normal or average engine speed. For example, to find the sensitivity of a governor which at one-half load begins to act with an increasing load at 1417 r/min and with a decreasing load at 1429 r/min:

$$\text{Sensitivity} = \frac{(n_{max} - n_{min})}{n_{max} + n_{min})/2} \times 100$$

$$= \frac{1429 - 1417}{1417 + 1429} \times 200$$

$$= 0.84 \text{ percent}$$

20. Why do diesel engines overheat?

Answer. Overheating may be caused by (1) not enough cooling water, (2) deposits of scale on the cylinder and in the cylinder-head water jackets, (3) a bypass valve in the cooling tower being open, (4) not enough piston lubrication, (5) faulty lube oil, (6) lube-oil filters that need cleaning, (7) a worn lube-oil pump, (8) incorrect fuel-injection timing, (9) a fuel nozzle that is carbonized, or (10) an after-dribble condition.

21. What are some causes of a noisy diesel engine?

Answer. Mechanical and fuel knocks are the two basic causes of noise in a diesel engine. Mechanical knocks may be from a worn piston pin or bearing, too much clearance in the crankpin bearing, a loose flywheel on the shaft, or worn pistons or liners that cause piston slap. Fuel knocks may result from early fuel injection, an injection system that is out of order, improper fuel, or too high an injection-air pressure.

22. What is the brake horsepower of a diesel engine?

Answer. The bhp is the horsepower delivered by the shaft at the output end. The name is derived from the fact that it is determined by a brake in engine testing.

23. What is the brake mean effective pressure (bmep) of a diesel engine?

Answer. The bmep is figured by the following formula:

$$bmep = \frac{bhp \times 33,000}{LAN}$$

where bhp = brake horsepower per cylinder
L = stroke of piston, ft
A = net piston area, in^2
N = number of power strokes per min

Gas Turbines

24. What are the main components of a gas turbine?

Answer. Air compressor, combuster, and flue gas expander turbine.

25. What is the effect of ambient temperature on gas turbine performance?

Answer. Most designs are based on an air temperature of 60°F entering the air compressor. In general, unless preheating is used, the thermal efficiency will decrease about 1 percentage point for each 15° rise in air inlet temperature.

26. What is the effect of creep on high-alloy material used in gas turbines?

Answer. The high-alloy materials may show a loss of ductility at high temperature over a period of time and thus become brittle and prone to cracking. The higher the temperature, the more rapidly the material loses its strength and ductility.

27. To avoid condensation, what precautions are necessary for natural gas fuel that supplies a gas turbine?

Answer. Most gas turbines for industrial use are hooked up to pipeline-quality natural gas with liquid fuel, usually no. 2, as backup. The natural gas should go through water knock-out vessels before entering the gas turbine room. These vessels normally contain coalescing and filtering elements. To avoid condensation, the natural gas temperature should be maintained at least 50°F above its dewpoint. Pockets must be avoided in pipe runs so that condensate will not collect in low spots. Piping is usually stainless steel.

28. Where may you expect to find thermal stress cracks in a gas turbine when making internal inspections?

Answer. Thermal stress cracks appear in those components which are subjected to high temperatures, such as combustors, entrance nozzle segments, and first-stage blades, and which have abrupt changes in stiffness or configuration that produce stress concentration.

29. What are some of the factors that may determine the service life of gas turbine components?

Answer. The manufacturers express this as used-up service life based on their methods of calculating remaining life. Factors to consider include starts and stops, load and temperature swings, running hours, whether components have protective coatings, material creep strength, endurance limit for fatigue strength evaluation, method of blade cooling, effect of steam injection, and erosion wear noted during inspections. This also includes hot gas metal oxidation. The trend is to convert many of these items to equal a certain amount of actual running hours, adding them to the running hours, and then comparing them to the manufacturer's predicted life. In this way, a decision is made for a planned replacement before a failure can be expected. It is important for operators to become familiar with the manufacturer's method of assigning remaining life because many of these are influenced by operating practices.

30. What is combined-cycle cogeneration?

Answer. A combined cycle uses a gas turbine or diesel, usually driving a generator in which the exhaust gases are directed to a waste heat-recovery boiler or heat-recovery steam generator (HRSG), and the steam from the HRSG is directed to a steam turbo-generator for additional electric power production. The use of the exhaust heat from a gas turbine improves the overall thermal efficiency. In cogeneration, electric power is produced, but part of the steam from the HRSG or from extraction from the steam turbine is used for process heat, hence the term *cogeneration*—the simultaneous production of electric power and process heat steam.

31. Why else is steam from an HRSG used?

Answer. For steam injection into the gas turbine for NO_x control.

32. What gas is used in the SCR method of controlling NO_x?

Answer. Air-diluted ammonia vapor is injected into the flue gas stream before it enters the catalyst units consisting of honeycomb-shaped ceramic material. These cells, with the ammonia vapor, convert nitrogen oxides to nitrogen and water vapor for discharge into the atmosphere.

33. What are some conditions that require a gas turbine train trip?

Answer. Trips are required for overspeed, low lube oil pressure, high turbine exhaust temperature, excess vibration, flame failure in combustor, inlet air filter having high differential pressure, and any initiation of fire protection around the unit.

34. What is the Brayton cycle?

Answer. This is the basic gas turbine cycle of adiabatic compression of air in a compressor, constant pressure burning in a combustor, followed by adiabatic expansion of the working fluid in a turbine to produce mechanical power.

35. Why are fuel considerations important in gas turbines?

Answer. The products of combustion come in direct contact with turbine blading. This dictates using fuels that produce combustion products that do not cause high-temperature corrosion or oxidation, erosion, and deposition of ash on blades. Preferred are natural gas, properly prepared syngas, and distillate oils. Most other liquid fuels require external treatment to remove harmful vanadium and sodium. Vanadium pentoxide and sodium sulfate are the principal ash components formed at higher temperatures. This ash adheres to blades and causes corrosion on the blades.

36. What provisions in the layout of a combined-cycle should be considered?

Answer. It is important to consider the use of a bypass stack that will permit operating the gas turbo-generator in the event of a forced outage on the HRSG or steam turbo-generator. However, in certain states, such as California, also to be considered are NO_x limits that require steam injection and loading to limit the exhaust temperature coming out of the simple-cycle gas turbine so that they do not exceed jurisdictional limits.

37. How does steam injection lower the NO_x discharged to the atmosphere by a gas turbine?

Answer. Adding steam in the combustion zone of the turbine lowers both the flame and gas temperature, preventing up to 80 percent of uncontrolled NO_x formation.

CHAPTER 12
BASIC ELECTRIC CIRCUITS, GENERATORS, AND MOTORS

All plant service operators come in contact with electric circuits and machinery in operating and maintaining the services of a plant. Most electric power to the facility probably comes from a utility source; however, this chapter will review generators because some plants do generate their own power, and many have emergency generators. It is also prudent to review some basics for the benefit of those readers who may need a refresher on the many characteristics that electric power can take. Electric energy is converted into many forms—power to move machines, lights to illuminate occupancies, and heat to keep the occupancy warm or to melt iron ore such as in an electric arc or induction furnace, and it is used to make chemical changes in processes and to control and calculate the many different applications of electric relays.

DC CIRCUITS

DC circuits were the first to be developed as electric energy came into use. The typical dc circuit has an electromotive force, or voltage, to make electrons flow in a completed circuit. The rate of flow of electrons is measured in amperes. There is resistance to flow, and this is expressed in ohms. Ohm's law for a simple dc circuit is

$$I = \frac{V}{R}$$

where I = current, A
V = voltage V
R = resistance, Ω

In a series circuit (see Fig. 12.1a), resistances are added before applying Ohm's law. In a circuit in which the resistances are parallel to the voltage source (see Fig. 12.1b), the equivalent resistance of the parallel circuit resistance can be found by using the well-known equation

$$\frac{1}{R} = \frac{1}{R_1} + \frac{1}{R_2} + \frac{1}{R_3} + \frac{1}{R_4}$$

After R is solved for, the current flowing from a voltage source, such as a generator, can be calculated.

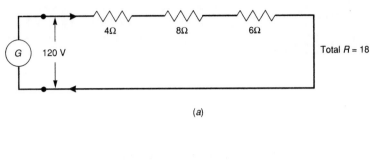

(a)

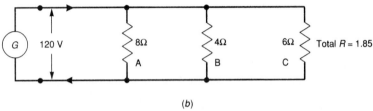

(b)

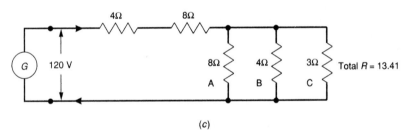

(c)

FIGURE 12.1   (a) Electric circuit with resistance in series; (b) electric circuit with resistance in parallel; (c) electric circuit that combines series and parallel resistances.

Figure 12.1c shows a combined series and parallel arrangement of resistance supplied by a generator whose terminal shows 120 V. The generator produces the current and voltage to make the current flow through the entire system. Each resistor causes a volt drop, equal to IR, and the IR drops must total the generator's terminal voltage for equilibrium to exist. Let us illustrate this by calculating the currents in the circuits shown in Fig. 12.1:

1. *Series resistance* as shown in Fig. 12.1a:

$$R = 4 + 8 + 6 = 18 \ \Omega$$

$$I = \frac{120}{18} = 6.7 \ \text{A}$$

2. *Parallel resistance* as shown in Fig. 12.1*b:*

$$\frac{1}{R} = \frac{1}{8} + \frac{1}{4} + \frac{1}{6}$$

$$= \frac{3 + 6 + 4}{24} \qquad \text{then} \qquad R = \frac{24}{13} = 1.85 \ \Omega$$

$$I = \frac{120}{1.85} = 64.9 \ \text{A}$$

3. *Series-parallel* resistance as shown in Fig. 12.1*c:* It is necessary to first calculate the equivalent resistance of the parallel circuits:

$$\frac{1}{R} = \frac{1}{8} + \frac{1}{4} + \frac{1}{3}$$

$$= \frac{3 + 6 + 8}{24} \qquad \text{then} \qquad R = \frac{24}{17} = 1.41 \ \Omega$$

This resistance can now be added to the resistances in series:

Total circuit resistance $= 1.41 + 4 + 8 = 13.41 \ \Omega$

$$I = \frac{120}{13.41} = 8.95 \ \text{A}$$

To calculate the current flowing in the three parallel circuits designated *A*, *B*, and *C* in Fig. 12.1*c*, the current flow in each circuit is inversely as the resistance, or

$$\frac{I_A}{I_B} = \frac{R_B}{R_A} \qquad \frac{I_B}{I_C} = \frac{R_C}{R_B} \qquad \frac{I_C}{I_A} = \frac{R_A}{R_C}$$

Let us calculate the currents flowing in *A*, *B*, and *C* of Fig. 12.1*c:*

$$I_A + I_B + I_C = 8.95 \ \text{A}$$

$$\frac{I_A}{I_B} = \frac{4}{8} = 0.5 \qquad \text{or} \qquad I_B = 2 \ (I_A)$$

$$\frac{I_A}{I_C} = \frac{3}{8} = 0.375 \qquad \text{or} \qquad I_C = 2.67 \ (I_A)$$

Substituting the I_B and I_C value in terms of I_A in the total ampere equation,

$$I_A + 2(I_A) + 2.67(I_A) = 8.95$$

and solving,

$$I_A = 1.58 \ \text{A}$$

$$I_B = 2 \times 1.58 = 3.16 \ \text{A}$$

$$I_C = 2.67 \times 1.58 = 4.21 \ \text{A}$$

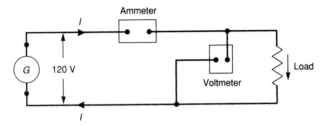

FIGURE 12.2   Ammeter and voltmeter connections. The ammeter measures the quantity of electricity flowing per second in the circuit; therefore, it is connected so that the current to be measured passes *through* it. The voltmeter measures potential difference between two points; therefore, it is connected *across* those points to be measured.

Ammeters

Figure 12.2 illustrates that current flowing in a line is measured with an ammeter that must be connected in series with that line to obtain the current. Never connect an ammeter across the lines because then the low resistance of an ammeter will take a current several times its rating and may burn up internally or the pointer may become severely bent.

Voltmeters

Voltage is measured by a voltmeter that measures the *potential difference* between two points. The voltmeter, therefore, is connected *across* or *between* the lines in order to measure the potential difference as is illustrated in Fig. 12.2.

Another relationship from Ohm's law is that $V = IR$, or a volt drop occurs in a circuit where a current flows through a resistance.

Example.   See Fig. 12.1c. What is the voltage across the parallel circuits A, B, and C if conductor resistance is neglected? Calculate the current flowing through the parallel circuits A, B, and C using this line voltage.

The voltage for the *across the line* parallel circuits A, B, and C in Fig. 12.1c is 120 − 8.95(4 + 8) = 12.6 V with 8.95 being the current flowing through the two resistors in series. Then,

$$I_A = \frac{12.6}{8} = 1.58 \text{ A} \qquad I_B = \frac{12.6}{4} = 3.15 \text{ A} \qquad I_c = \frac{12.6}{3} = 4.20 \text{ A}$$

Note that these values agree with the previously calculated amperes using the total resistance for the series-parallel circuit of Fig. 12.1c.

In general, electric conductors offer series resistance to current flow, while appliances, motors, and lights are connected in parallel, or across the line voltage that is being supplied in that particular circuit. All circuits in a plant can be broken down into series and parallel resistances. The parallel circuits are combined and then added to the series resistance so that the amperes to be carried per circuit is determined by treating the total resistance as being in series.

Another electrical term is the *coulomb* (C), which is the *quantity* of electricity conveyed by 1 A in 1 s. Amperes are the *rate* of flow of electricity.

Electric power is the rate of doing work. The term *power* was defined in the early days of the steam engine by determining that 33,000 ft · lb/min is equal to 1 hp. As can be noted, power is the *rate* of doing work. In the same way, *electric power* is defined as 1 C/s being carried through a potential difference of 1 V, which is equal to 1 J/s, or 1 W. Since 1 C/s equals 1 A, 1 W = volts × amps, or

$$\text{Power, } P = VI \text{ W} \qquad \text{with kilowatts being} = 1000 \text{ W}$$

The above equation is very useful in many electrical problems. From this equation is derived the I^2R heat losses that may occur from resistance. It was established that amperes × resistance = volt drop across the resistance, or $V = IR$. If we substitute this in the power equation, the power lost across the resistor is $(IR)I = I^2R$. Anytime there is resistance in an electric circuit, there is an I^2R loss.

Example. A field rheostat on a dc motor carries 10 A. It has resistance of 12.8 Ω. What would be the heat loss in watts and Btu? Use $P = I^2R$ or $P = 10^2 \times 12.8 = 1280$ W = 1.28 kW = 4369 Btu.

The power equation can also be converted to the following for some problems. Ohm's law is $I = V/R$ and substituting this value for I in the power equation, we obtain

$$P = V\frac{V}{R} = \frac{V^2}{R}$$

Example. Determine the resistance of a 100-W lamp connected to a 120-V circuit. Use

$$P = \frac{V^2}{R} \qquad \text{or} \qquad R = \frac{V^2}{P} \qquad \text{then} \qquad R = \frac{120^2}{100} = 144 \text{ Ω}$$

Example. An electric toaster is rated at 1260 W and 10.5 A. What voltage is needed across the resistor? Use

$$V = \frac{P}{I} \qquad \text{or} \qquad \frac{1260}{10.5} = 120 \text{ V}$$

Figure 12.3 shows the connections to be made to connect a *wattmeter,* which measures the power going to the load by measuring both current and volts with the product being watts. Two terminals for the ampere flow are connected in series with the load, while the other two terminals for the volts are connected across the load.

A *watthour* meter has an internal mechanism that correlates the power of the load with time used, to provide a reading in kilowatthours of *energy* consumed. *Energy* is defined as the product of electric power and time, or Energy = VIt. Utility bills always show the energy consumed by a power customer, and this is expressed in kilowatthours.

Example. A generator in a plant is used for cyclic service. From 8 a.m. to 5 p.m. it has an average load of 11,500 kW. From 5 p.m. to midnight, the average load is 7500 kW. From midnight to 8 a.m. the average load is 6500 kW. What is the energy delivered in the 24-h period? The energy delivered is $9 \times 11,500 + 7 \times 7500 + 8 \times 6500 = 214,000$ kWh.

Resistance and Conductance

Resistance to current flow is measured in ohms. Insulators which are considered to have infinite resistance are measured for resistance to electric current flow, but it is in terms of millions of ohms, or *megohms*. Ohms of resistance are measured by ohmmeters or by using a voltmeter and ammeter and then applying Ohm's law.

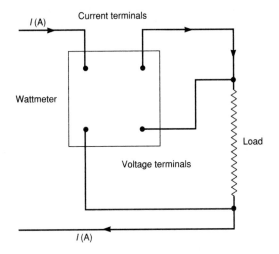

FIGURE 12.3 A wattmeter provides the power in watts consumed by the load by combining the readings from an ammeter connected in series with the load and a voltmeter that is connected across the load.

For very low resistance, such as bus bars, the ohm is too large a unit, and measurements are made in microhms, which are equal to 1/1,000,000 Ω.

The resistance of a conductor varies directly with its length and inversely with the conductor's cross-sectional area, or

$$R = p\,\frac{L}{A}$$

where p = specific resistance per cubic centimeter or cubic inch depending on units used. Resistance increases with length but decreases with an increase in the conductor's cross-sectional area.

Conductance is that property of a material which tends to permit the flow of electric current, and, mathematically, is the reciprocal of resistance. Copper, aluminum, silver, and tungsten are good conductors, while insulators are poor ones.

Cables and Wire Sizes

Electric wires and cables are designated by their cross-sectional areas as an indication of their ability to carry up to a certain value of current. The diameter size of wires is expressed in *mils* with 1 mil = one-thousands of an inch (0.001). Thus, a wire ⅜ in in diameter is 375 mils in diameter. However, industry practice is to express the ability of a wire to carry current by its cross-sectional area as derived from the mil diameter size, and this is called *circular mils*. A circular mil is the area of a circle having a diameter of 1 mil and in equation form is circular mil = D^2 where D = diameter of wire in mils. Note the area of this circle is 0.7854 D^2 mils².

Example.   A wire is 0.3648 in in diameter. What is its mil diameter and circular mil size? The mil diameter is 364.8 mils. The circular mil is $(364.8)^2 = 133,100$ circular mils. The circle area is 0.7854 × 133,000 = 104,536 mils².

The *circular-mil-foot* is also used in cable work. This is the resistance of a wire having a crosssection of 1 circular mil and a length of 1 ft. For copper at 68°F, the resistance of a circular-mil-foot is 10.37 Ω. This makes it possible to determine the resistance of a length of cable with a certain size.

Example. A 750,000-circular mil cable is 2500 ft long. Determine its resistance. For 1 circular-mil-foot, the resistance = 10.37 × 2500 = 25,900 Ω. The larger the circular mils, the smaller the resistance, therefore,

$$R = \frac{25,900}{750,000} = 0.0346 \ \Omega \quad \text{or} \quad R = 10.37 \times \frac{2500}{750,000} = 0.0346 \ \Omega$$

AWG Designation

The most common reference to wire size is the American Wire Gage (AWG), previously called the Brown & Sharpe gage. Wire sizes are graded by numbers, with the smallest size designated as no. 40 (see Table 12.1). The largest size for the number designation is no. 0000, called 4/0. Above this size, about 250,000 circular mils, cable designation in circular mils comes into effect. As Table 12.1 indicates, the higher the gage number for a wire, the smaller is its diameter and current carrying capacity. Thus, a no. 14 is smaller than a no. 8.

Temperature has an effect on conductor resistance, with the resistance *increasing* with a rise in temperatures of the conductor, or

$$R_t = R_o(1 + \alpha t)$$

where R_t is the resistance at temperature, t and R_o are the resistance of the conductor *at 0°C,* and α is the temperature coefficient of resistance at 0°C. For copper, $\alpha = 0.00427$.

Example. A winding has a resistance of 30 Ω at 20°C (68°F). Calculate the resistance of the winding at 60°C (140°F). It is necessary to first find the resistance at 0°C, or R_o:

$$R_o = \frac{30}{1 + (0.00427 \times 20)} = \frac{30}{1.085} = 27.65 \ \Omega \text{ at } 0°C$$

For 60°C,

$$R_{60} = 27.65(1 + 0.00427 \times 60) = 34.73 \ \Omega$$

NEC *Conductor ampere capacity* involves not only the conductor's cross-sectional area but also the insulation rating of the conductor, the grouping together of conductors, and the environment or service it will be subjected to. Current limits are set by conductor manufacturers and the *National Electrical Code* (NEC) and are based on safe operating temperatures for the current-carrying parts and the insulation around the conductor. Current limits are also established by permissible I^2R losses and volt drops in the circuits.

The NEC is a standard all service operators should have at their disposal because it contains good safety requirements for most electrical equipment found in a facility. It is available from the National Fire Protection Association, Batterymarch Park, Quincy, Mass 02269. *Insulation* per NEC on conductors is classified by suitability of possible applications, such as (1) dry locations, (2) dry and damp locations, (3) dry and wet locations, (4) wet locations, or (5) switchboard wiring only. The NEC also lists temperatures permissible for conductors made of the many different insulation material available today.

The NEC includes requirements for *service-entrance* conductors, *feeder circuits,* and *feeder branch circuits* (see Fig. 12.4).

TABLE 12.1 American Wire Gage Sizes in Gage Number and Circular Mils for Standard Annealed Solid Copper Wire in English Units

Gage number	Diameter, mils	Cross section Circular mils	Cross section Square inches	Ohms per 1,000 ft 25°C (77°F)	Ohms per 1,000 ft 65°C (140°F)	Ohms per mile 25°C (77°F)	Pounds per 1000 ft
0000	460.0	211,600.0	0.166	0.0500	0.0577	0.264	641.0
000	410.0	168,100.0	0.132	0.0630	0.0727	0.333	508.0
00	365.0	133,225.0	0.105	0.0795	0.0917	0.420	403.0
0	325.0	105,625.0	0.0829	0.100	0.116	0.528	319.0
1	289.0	83,521.0	0.0657	0.126	0.146	0.665	253.0
2	258.0	66,564.0	0.0521	0.159	0.184	0.839	201.0
3	229.0	52,441.0	0.0413	0.201	0.232	1.061	159.0
4	204.0	41,616.0	0.0328	0.253	0.292	1.335	126.0
5	182.0	33,124.0	0.0260	0.319	0.369	1.685	100.0
6	162.0	26,244.0	0.0206	0.403	0.465	2.13	79.5
7	144.0	20,736.0	0.0164	0.508	0.586	2.68	63.0
8	128.0	16,384.0	0.0130	0.641	0.739	3.38	50.0
9	114.0	12,996.0	0.0103	0.808	0.932	4.27	39.6
10	102.0	10,404.0	0.00815	1.02	1.18	5.38	31.4
11	91.0	8,281.0	0.00647	1.28	1.48	6.75	24.9
12	81.0	6,561.0	0.00513	1.62	1.87	8.55	19.8
13	72.0	5,184.0	0.00407	2.04	2.36	10.77	15.7
14	64.0	4,096.0	0.00323	2.58	2.97	13.62	12.4
15	57.0	3,249.0	0.00256	3.25	3.75	17.16	9.86
16	51.0	2,601.0	0.00203	4.09	4.73	21.6	7.82
17	45.0	2,025.0	0.00161	5.16	5.96	27.2	6.20
18	40.0	1,600.0	0.00128	6.51	7.51	34.4	4.92
19	36.0	1,296.0	0.00101	8.21	9.48	43.3	3.90
20	32.0	1,024.0	0.000802	10.4	11.9	54.9	3.09
21	28.5	812.3	0.000636	13.1	15.1	69.1	2.45
22	25.3	640.1	0.000505	16.5	19.0	87.1	1.94
23	22.6	510.8	0.000400	20.8	24.0	109.8	1.54
24	20.1	404.0	0.000317	26.2	30.2	138.3	1.22
25	17.9	320.4	0.000252	33.0	28.1	174.1	0.970
26	15.9	252.8	0.000200	41.6	48.0	220.0	0.769
27	14.2	201.6	0.000158	52.5	60.6	277.0	0.610
28	12.6	158.8	0.000126	66.2	76.4	350.0	0.484
29	11.3	127.7	0.0000995	83.4	96.3	440.0	0.384
30	10.0	100.0	0.0000789	105.0	121.0	554.0	0.304
31	8.9	79.2	0.0000626	133.0	153.0	702.0	0.241
32	8.0	64.0	0.0000496	167.0	193.0	882.0	0.191
33	7.1	50.4	0.0000394	211.0	243.0	1,114.0	0.152
34	6.3	39.7	0.0000312	266.0	307.0	1,404.0	0.120
35	5.6	31.4	0.0000248	335.0	387.0	1,769.0	0.0954
36	5.0	25.0	0.0000196	423.0	488.0	2,230.0	0.0757
37	4.5	20.3	0.0000156	533.0	616.0	2,810.0	0.0600
38	4.0	16.0	0.0000123	673.0	776.0	3,550.0	0.0476
39	3.5	12.3	0.0000098	848.0	979.0	4,480.0	0.0377
40	3.1	9.6	0.0000078	1,070.0	1,230.0	5,650.0	0.0299

NOTE 1.—The *fundamental resistivity* used in calculating the tables is the International Annealed Copper Standard, *viz.*, 0.15328 Ω (meter, gram) at 20°C. The *temperature coefficient* for this particular resistivity is $a_{20} = 0.00393$, or $a_0 = 0.00427$. The *density* is 8.89 g. per cubic centimeter.

NOTE 2.—The values given in the table are only for annealed copper of the standard resistivity. The user of the table must apply the proper correction for copper of any other resistivity. Hard-drawn copper may be taken as about 2.7 percent higher resistivity than annealed copper.

NOTE 3.—Pounds per mile may be obtained by multiplying the respective values above by 5.28.

Source: From Circ. 31, U.S. Bureau of Standards.

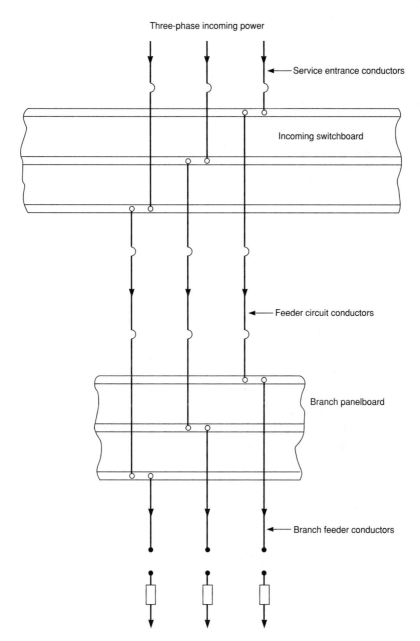

FIGURE 12.4   NEC classification of facility conductors.

There must be a way to disconnect all conductors in a facility from the service-entrance conductors. A maximum of six switches or circuit breakers is permitted to disconnect the facility conductors from the service entrance conductors.

For service disconnection methods rated 1000 A or more, and with grounded wye connections (ac current is covered later) of more than 150 V to ground, but not exceeding 600 V phase to phase, *ground-fault protection* of equipment is required. However, this ground-fault protection requirement (1) does not apply to fire pumps and (2) does not apply to a continuous industrial process where a sudden ground-fault trip may introduce additional or increased hazards.

Most feeder and branch-circuit conductors are required to be protected by over-current-protective devices connected at the point the conductors receive their supply. There are exceptions for conductors with short-length taps. In general, the feeder and branch-circuit protection should operate first, in a cascading-type facility protective scheme, and the service-entrance conductors should only operate after the feeder and branch-circuit protectors have tried to clear a fault on their circuit. Short-circuit studies are used to set fault protection in a cascade style and also to develop the available short-circuit currents per circuit.

AC CIRCUITS

Most electric power used today is alternating current with the chief advantage being the ability to raise and lower voltages for transmission and distribution by means of transformers. AC generators can be built in larger sizes in comparison to dc generators since they do not have a commutator.

Sine waves are generated when a coil rotates at constant speed in a dc magnetic field. A voltage is generated and, if applied to a circuit, an ac current flows due to the potential difference (see Fig. 12.5a). AC *cycles* are related to the passing of a conductor across the magnetic field of a north and south pole as shown in Fig. 12.5b. Thus, there is a repeat of the wave that is expressed as frequency, or cycles per second, when the conductor passes a *pair* of poles as shown in Fig. 12.5b. In equation form,

$$\text{Frequency} = \frac{PS}{120} \text{ cycles per second}$$

where P = number of dc poles and S = r/min of the rotor.

Example. How many poles would a 60-cycle alternator have with a speed of 120 r/min?

$$P = \frac{120(\text{frequency})}{120} = \frac{120 \times 60}{120} = 60 \text{ poles}$$

The generation of an emf or voltage in a coil passing through a magnetic field involves *magnetic induction,* which is treated more thoroughly in electrical engineering texts. The alternating current generated also involves *phase angles* between the emf generated and the induced current, as well as the *effective value* of current since it varies from zero to a maximum value in a cycle. The effective value of the sine-wave current is the root-mean-square (rms) value, which is equal to the ampere that will produce the same I^2R heat effect as 1 A of dc current. In a sine wave, the effective, or rms, value of current is equal to 0.707 times the maximum value, or peak, of the current wave.

It is the effective value of the sine wave, and of the voltage, that is shown on equipment nameplates. Electrical instrument readings of amperes and voltages also show the rms values.

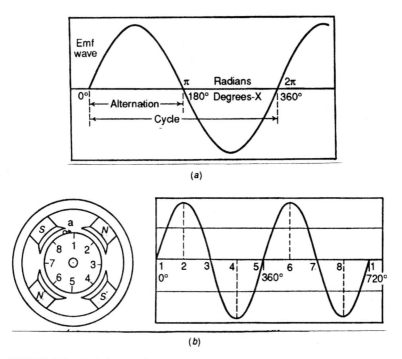

FIGURE 12.5 (*a*) A sine wave is generated when a coil rotates at constant speed in a north-south pole magnetic field; (*b*) cycles of an alternating sine wave are determined by speed and number of poles. A four-pole arrangement produces two cycles per revolution.

By using rms values on ac currents and voltages, Ohm's law can be applied to single-phase circuits with only resistance because the rms value is equal to an equivalent dc current. The power equation of $P = VI$ also applies to single-phase circuits with only resistance using rms values. The *phase angle* effect in ac circuits is caused by the fact that ac circuits have (1) *resistance,* (2) *inductance,* and (3) *capacitance.* If a circuit has only resistance, the current and voltage waves pass through their zero and maximum values at the same time and in the same direction. In such a case of pure resistance, voltages and currents are *in phase.* Out-of-phase conditions are caused by circuits with inductance and capacitance. If a current reaches its zero and maximum value at a *later period* than the voltage wave, the current is *lagging* the voltage. If the voltage wave follows the current wave, the current is classified as *lead* current.

Inductance in a circuit is caused by coils or other electromagnetic devices which produce self-induced emf due to the varying flux produced by an alternating current. This induced emf opposes any change in the current flowing in the circuit. While inductance is measured in units called henrys (H), the opposition that inductance presents to the flow of alternating current is called inductive reactance, X_L and is equal to $2\pi fL$ where f = frequency cycles per second, and L = value of inductance, H. In a circuit containing only inductance, the current *lags* the applied voltage by 90° out of phase. Electrical equipment often has both inductance and resistance to current flow, as is the case with motors, generators, and transformers. The resistance to current flow that combines resistance, inductance, and capacitance is called *impedance* and is expressed in ohms.

Capacitors are devices used in electric circuits based on their ability to hold electric charge. They are extensively used for power factor correction and in electronic circuits.

A capacitor connected in an ac circuit offers opposition to current flow, and this is based on the frequency and applied voltage, and has an ohm value of

$$X_C = \frac{1}{2\pi f C}$$

where f = frequency of the voltage, cycles per second, and C = value of capacitance, farads (F). In a purely capacitive circuit, the current *leads* the voltage by 90° out of phase.

Impedance

AC circuit resistance can now be combined to include resistance, inductance, and capacitance with the following equation for impedance Z:

$$Z = \sqrt{R^2 + (X_L - X_c)^2}$$

where X_L = inductive reactance, Ω
      X_c = capacitive reactance, Ω
      R = resistance, Ω

Note that $X = X_L - X_c$, and the negative sign is due to the fact that one lags the applied voltage and the other leads the applied voltage. Impedance is a vectorial addition of resistance, inductance, and capacitance, as is illustrated in Fig. 12.6a. With pure resistance in the circuit, $Z = R$, and there is no angle Q. With inductance and resistance in the circuit, Q will be between 0° and 90°. Inductance current is used to magnetize coils and therefore does not contribute to power output. Since power is defined as VI or I^2R, the cosine value of the angle Q is called power factor because "apparent power" IZ has magnetizing and "true power," which is I^2R. Therefore, to obtain true power, for single-phase alternating circuits use:

$$P = I^2Z(\cos Q) \qquad \text{or} \qquad P = I^2Z \times \text{power factor, P.F.}$$

Power factors are expressed in decimals equal to the numerical value of the cosine of the angle Q. Also useful is P.F. = kW/kVA.

Example. A generator delivers 10,000 kVA at 0.8 P.F. Determine the power output. The power output is in kilowatts, while the kVA value is the apparent power, which includes magnetizing current. P.F. = kW/kVA; therefore, kilowatt output = 0.8 × 10,000 = 8000-kW output.

A *series ac circuit* is shown in Fig. 12.6b with the accompanying vector diagram. The vector diagram is explained as follows. The current is the same in all parts of the circuit. The current I is laid off horizontally, and since IR is in phase with the current, this is also laid off on the horizontal line. The voltage IX_L across the inductance is laid off at right angles to the current and leads the current by 90°. IX_c is also laid off at 90° from the current, but it lags the current. The difference between these two is IX, and this is vectorially added to IR to give IZ, or the voltage drop across the series circuit.

Example. In Fig. 12.6b, $R = 50$ Ω, $X_L = 56.6$ Ω, $X_C = 106$ Ω, and V = 120 V at 60 cycles. Find (1) impedance of the circuit, (2) current, I, (3) voltage across the resistor, inductance, and capacitor, (4) cos Q or P.F., and (5) power supplied to the circuit:

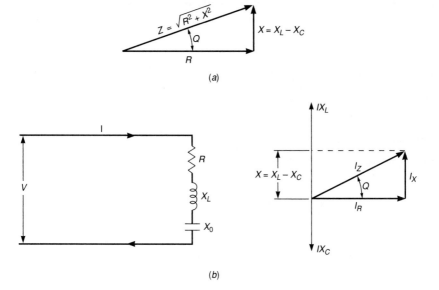

(a)

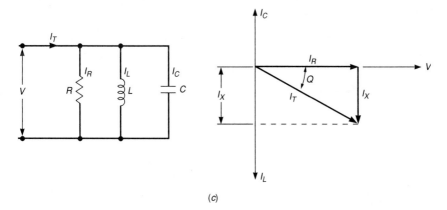

(b)

(c)

FIGURE 12.6   AC circuits require vector additions. (a) Vector additions of R and X to obtain impedance Z; (b) ac circuits with R, X_L, and X_C in series with vector addition diagram; (c) ac circuit with R, X_L,
and X_C in parallel with vector addition diagram.

1. $Z = \sqrt{50^2 + (56.6 - 106)^2} = 70.3 \; \Omega$

2. $I = \dfrac{V}{Z} = \dfrac{120}{70.3} = 1.71 \; A$

3. $V_R = IR = 1.71 \times 50 = 85.5 \; V$

   $V_L = IX_L = 1.71 \times 56.6 = 96.8 \; V$

   $V_C = IX_C = 1.71 \times 106 = 181.1 \; V$

4. $\text{Cos } Q = \dfrac{R}{Z} = \dfrac{50}{70.3} = 0.711 = \text{P.F.}$

   $P = VI \cos Q$

5. $P = IZ \times I \times \cos Q = 1.71 \times 70.2 \times 1.71 \times 0.711 = 146 \; W$

also

$P = I^2R = 1.71^2 \times 50 = 146 \; W \text{ (checks)}$

AC Parallel Circuits

See Fig. 12.6c. The voltage is common to the three parallel circuits with the resistance current I_R in phase with the voltage. The inductance current I_L lags the voltage by 90°. The capacitive current I_C leads the voltage by 90°. Current I_X is equal to the difference between the inductance and capacitive currents but still lags the voltage by 90°.

Example. In Fig. 12.6c, resistance is 10 Ω, inductive reactance is 8 Ω, and the capacitive reactance is 15 Ω. Determine total current, current in each branch, power factor, and power to the circuit if the voltage is 120 V, 60 cycle:

$$I_R = \frac{120}{10} = 12 \; A$$

$$I_L = \frac{120}{8} = 15 \; A$$

$$I_C = \frac{120}{15} = 8 \; A$$

The vector diagram shows

$$I_T = \sqrt{12^2 + (15 - 8)^2} = 13.9 \; A \text{ lagging}$$

$$\text{Power factor} = \frac{I_R}{I_T} = 12/13.9 = 0.864$$

$$\text{Power} = VI_T \times 0.864 = 120 \times 13.9 \times 0.864 = 1440 \; W = 1.44 \; kW$$

$$= VI_R = 120 \times 12 = 1440 \; W = 1.44 \; kW$$

THREE-PHASE CIRCUITS

Generation and distribution of electric energy is most economical with three-phase electric circuits. Practically all ac power is generated in three-phase generators by the utilities and industrial plants. The generators producing three-phase power have three separate windings, set apart 120°; thus each turn of one winding is physically displaced 120° from the same turn on either of the other two windings. A sinusoidal voltage is produced in each of the three windings as they cut through the magnetic lines of flux that are produced by a dc field of the rotor. The voltage wave generated by each winding is also 120° apart in time with respect to wave generation. This difference in time of wave generation is called *phase difference*. The phases are usually labeled A, B, and C. The most common methods of connecting the three phases are shown in Fig. 12.7, and these are called wye, or Y, and the delta connections. Note the Y connection has a common connection for all three phase windings, called the neutral. If this point is connected to a fourth outlet, usually grounded, it is called a four-wire Y connection.

Y-Connection Voltages

In a three-phase Y system that is balanced, the three line voltages shown in Fig. 12.7*a* are equal and differ in phase by 120°. Each line voltage differs in phase by 30° from one

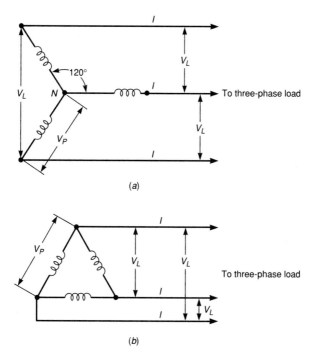

(a)

(b)

FIGURE 12.7   (*a*) Three-phase Y connection; (*b*) three-phase delta connection.

of its phase voltages. The three line voltages are each equal to each other in magnitude but are equal to $\sqrt{3}$, or 1.732, times the *phase voltage.*

Example. A generator delivers a three-phase line voltage to a facility at 4160 V. It is Y connected. What would be the phase voltage or the voltage across one of the three windings with a balanced load?

$$\text{Phase voltage} = \frac{4160}{\sqrt{3}} = 2402 \text{ V}$$

Y-Connection Currents

Figure 12.7a illustrates that the line currents are the same as the winding, or phase currents in a Y connection.

Y-Connection Power

The power delivered in a three-phase Y connection is

$$kW = \frac{1.732 \, V_L \times I_L \times \text{P.F.}}{1000}$$

Example. A generator is rated at 13,800 V, three-phase, and 418 A at 0.85 P.F. What is the kilowatt rating if it is Y connected?

$$kW = \frac{1.732 \times 13,800 \times 418 \times 0.85}{1000} = 8492 \text{ kW}$$

Example. A three-phase load that is Y connected and operating at 2400 V shows 500 kW. What is the line current supplying this load if the power factor is 90 percent?

$$I_L = \frac{1000 \times kW}{1.732 \times V \times \text{P.F.}} = \frac{1000 \times 500}{1.732 \times 2400 \times 0.9} = 133.6 \text{ A}$$

Delta-Connection Voltages

See Figure 12.7b. The line voltages in a balanced three-phase system *equal* the phase voltages.

Delta-Connected Currents

In a balanced three-phase delta system, the line current is $\sqrt{3}$ times the *phase current.*

Example. A transformer has a delta connection where an ammeter shows 560 A at a voltage of 460. What is the coil or phase current, and what is the kilovoltampere output if a balanced three-phase system exists?

$$kVA = \frac{1.732 \times V \times A}{1000} = \frac{1.732 \times 460 \times 560}{1000} = 446 \text{ kVA}$$

$$\text{Phase current} = \frac{\text{line current}}{1.732} = 323 \text{ A}$$

$$\text{Phase voltage} = 460 \text{ V}$$

Power in Delta System

The equation is the same as for the Y-connected system, or

$$kW = \frac{1.732 \times V_L \times I_L \times \text{P.F.}}{1000}$$

Example. A three-phase delta-connected induction motor operating at 60 cycles is rated 2300 V, 500 hp, and has an efficiency of 94 percent and a power factor of 85 percent. Determine at rated load (1) kilovoltampere input, (2) kilowatt input, (3) line current, and (4) phase or coil current.

1. $kVA \text{ input} = \dfrac{kW}{0.94 \times \text{P.F.}} = \dfrac{500 \times 0.746}{0.94 \times 0.85} = 467 \text{ kVA input}$

2. $kW = \dfrac{hp \times 746}{1000 \times 0.94} = \dfrac{500 \times 746}{1000 \times 0.94} = 397 \text{ kW input}$

3. $\text{Line current} = \dfrac{kVA \times 1000}{1.732 \times 2300} = \dfrac{467 \times 1000}{1.732 \times 2300} = 117.2 \text{ A}$

4. $\text{Phase current} = \dfrac{117.2}{1.732} = 67.7 \text{ A}$

The Y connection with the neutral brought out in a fourth wire is extensively used where a facility has a motor load and lightning load, such as a 480Y/277-V system. The 480-V system is the line voltage and is used for power loads, while the 277-V connection to the neutral is the same as the phase voltage, or $480/1.732 = 277$ V. A similar system is the 208Y/120-V system.

When the neutral on a Y connection is grounded, it is called a *system ground,* which protects the system in the event a ground develops on any conducting part. With the neutral grounded, another ground causes a short circuit between two ground points, thus causing protective devices to operate immediately to clear the fault in the circuit where the ground occurred.

Equipment grounding is for protection against shock. For example, if a motor has a grounded winding, the frame will have a potential to ground, and if the frame is not grounded, a person touching the frame could receive a fatal shock as current from the motor frame flows through the person's body to ground. With the motor frame grounded, there is no potential difference for this to occur. Always make sure that the equipment under your care is properly frame grounded. See the NEC Article 250 on grounding requirements.

Kvar

Another useful relationship in three-phase circuits is kilovoltamperes with kilowatts and kilovars, or kiloreactive voltamperes. In equation form,

$$kVA = \sqrt{kW^2 + kvar^2}$$

When a circuit has unity power factor, the kilovar is zero. This equation is useful in calculating the kilovars needed to correct a power factor. Poor or low power factor requires a utility or generating plant to deliver more kilovoltamperes, or more current, which in turn requires larger copper lines, and since larger currents produce more I^2R heat, there is a loss of efficiency. Therefore, utilities have a power factor penalty, and most plant engineers try to establish a high power factor to reduce this penalty. Synchronous motors and capacitors are used to improve a plant's power factor by supplying leading kilovars to offset the lagging kilovars that an induction motor load produces on a plant's electrical system.

Example. A facility has a heavy induction motor load of 2000 kW at 0.65 P.F. Calculate the leading kilovars required to obtain a power factor of 0.9 (see Fig. 12.8). The problem is solved by first calculating the existing kilovars with 0.65 P.F. and then calculating the kilovars that would exist with 0.90 P.F. The power load of 2000 kW remains the same.

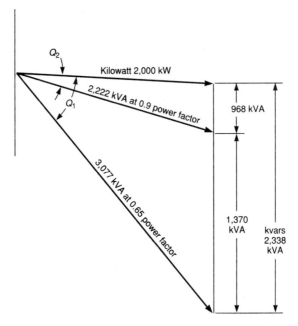

FIGURE 12.8 Illustration of kilowatts, kilovoltamperes, kilovars, and power factor, or cos Q.

$$\text{kVA at 0.65 P.F.} = \frac{2000}{0.65} = 3077 \text{ kVA}$$

$$\text{kVA} = \sqrt{\text{kW}^2 + \text{kvar}^2}$$

$$\text{kvar} = \sqrt{\text{kVA}^2 - \text{kW}^2}$$

$$\text{kvar} = \sqrt{3077^2 - 2000^2}$$

$$\text{kvar} = 2338 \text{ kVA lagging at 0.65 P.F.}$$

$$\text{kVA at 0.9 P.F.} = \frac{2000}{0.9} = 2222 \text{ kVA}$$

$$\text{kvar} = \sqrt{2222^2 - 2000^2}$$

$$\text{kvar} = 968 \text{ kVA at 0.9 P.F.}$$

Required kilovars to make the facility have this power factor are 2338 − 968 = 1370 kvar leading to offset the lagging kilovars at 0.65 P.F. Figure 12.8 illustrates this calculation.

The $\sqrt{3}$, or 1.732, is used to calculate the kilovoltamperes of a circuit with volts and amperes known if the circuit is a three-wire three-phase system. There are some two-phase systems used where two phases are connected and insulated from each other, and this connection can vary from a four-wire to a three-wire system, where $\sqrt{2}$, or 1.442, instead of the $\sqrt{3}$ is used to calculate the kilovoltamperes of a circuit. However, if there are two independent single-phase circuits, the kilovoltamperes are calculated per phase, and they are added to each other to obtain the total kilovoltamperes for the circuit.

ELECTRICAL SAFETY PRACTICES

When working around electric circuits or electrical equipment, extreme care is needed because electric energy does not give a warning of immediate failure with knocking or other noise as mechanical equipment may do. The threat of injury is always there and is especially pronounced when working around energized electrical equipment. Proper safety rules must be followed in order to avoid the risk of injury. It is important for facility service operators to be familiar with electrical equipment under their maintenance or operational control. However, if any testing, adjustments, or repairs are needed, it is strongly recommended that such work be performed by experienced electrical technicians or electricians.

General Safety

1. Make sure all live electrical parts are properly guarded and insulated.
2. Make sure mechanical rotating parts are guarded so that clothing is not caught in them.

3. Mark all incoming feeders and switches, branch circuit feeders, and switches so that the power can be cut off promptly in an emergency or if maintenance work is required on that circuit.

4. Always rope off or place guards around equipment that may be hazardous to others when maintenance work is required.

5. Tag and/or lock out switches for equipment that is to be out of service for inspection or maintenance before proceeding.

6. Consider all circuits to be alive unless electrical testing shows the circuit to be disconnected from all sources of energy.

7. Avoid using metal-cased flashlights and other conducting parts around live electrical equipment. Wear proper clothing and shoes. Be especially careful in wet and damp surroundings when inspecting operating equipment. Make sure all frames are properly grounded to avoid a possible shock when touching the frame of the electrical equipment.

8. Use approved safety rubber gloves, mats, and blankets on high-voltage inspection and maintenance work.

9. Wear cover-all or cup-type goggles with approved plastic lenses when working on or close to live circuits or equipment.

10. Treat all manhole work as a confined-space workplace and follow the safety procedures in OSHA rules for a confined space.

11. Treat dead circuits as though they were alive. Many circuits can be remotely energized. Make sure disconnect switches are completely open and all phases of the switch are open.

12. Before energizing equipment, make sure all tools, obstructions, driven equipment, valves, and similar items are in proper position and condition for operation and there are no other personnel working on the equipment to be energized.

13. Take extra precautions when working on damp or wet surfaces by using additional insulating material between you and the ground.

14. Keep oil cans, dusters, rags, and wiping cloths away from moving machinery or switchboards to prevent contact with live parts and to reduce the fire hazard if a spark is generated from the electrical equipment.

15. Clean electrical equipment only when it is completely deenergized. Make sure that adequate ventilation is provided when using solvents for cleaning. Wear personal protective equipment such as goggles, gloves, and suitable clothing when working with cleaning solvents.

16. Follow the safety rules provided by the manufacturer of the equipment during inspection and maintenance.

DC GENERATORS AND MOTORS

DC Generators

DC power is still used for many variable-speed applications such as elevator, paper machine, and similar machine drives where the speed has to be varied. DC generators are used to provide the dc current for the dc motor load since dc lighting has been replaced by ac circuits. Even dc generators are being phased out by solid-state rectifiers that convert ac current to dc current. However, facility service operators may have dc generators driven by steam engines or as part of motor-generator sets, so a brief review will be made of this emf producer.

An electric generator converts mechanical energy into electric energy. As a conductor moves through a magnetic field, an emf is induced in the conductor, and if this conductor is connected to an external circuit, current is delivered. The power required to rotate the conductor (called armature) at no load is very small, however, when the circuit is closed to an external load, the action of the ensuing current flow produces a force which opposes the rotation (armature reaction), and this requires more mechanical energy to be supplied to maintain constant speed.

See Figure 12.9*a.* The emf produced in a coil rotating in a dc field is a sine wave. A direct current must always flow into the external circuit in the same direction; therefore,

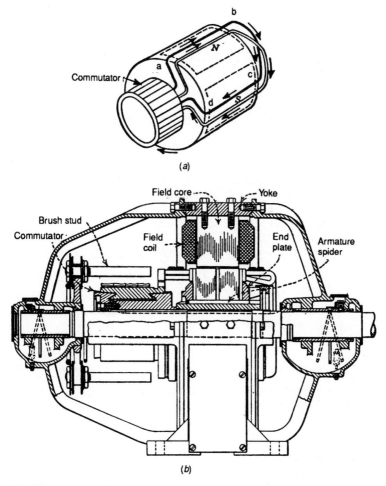

FIGURE 12.9   Features of a dc generator. (*a*) An armature produces a sine wave, which is rectified by commutators in a dc generator; (*b*) major dc generator components.

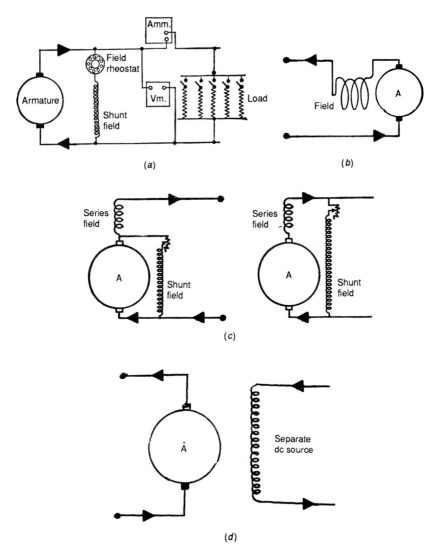

FIGURE 12.10 Types of generators classified by field connections. (*a*) In a shunt-generator, the field is connected across the armature conductors; (*b*) in a series field-generator, the field is connected in series with the armature conductors, (*c*) in a compound generator, the field consists of both series and shunt windings; (*d*) separate dc field excitation generator.

commutators are used to rectify the sine wave by having copper bars in each coil to reverse the connection to the external circuits as the current in the sine-wave reverse. Brushes are used to pass over the cuts in the bars when the coil is perpendicular to the magnetic field, or in the so-called *neutral plane*. Thus, the armature, commutator, and brush arrangement make the negative half of the sine wave become positive. The resultant emf is considered a direct, or dc, current. A relatively large number of commutator segments are necessary so that the generated emf ripples are not too pronounced on the dc waveform and the voltage between bar segments on the commutator is not too high so that flashovers are avoided.

Figure 12.9*b* shows a small dc generator. DC generators and dc motors are classified by the method of field windings supplying excitation current to the armature. Thus, there are *shunt, series, compound-wound,* and *separate excitation fields.*

In a *shunt-field generator* (see Fig. 12.10*a*), the field circuit is connected across the armature terminals, usually in series with a rheostat that varies the field current by introducing or lowering the resistance of the field circuit. This varies the strength of the magnetic field. Shunt fields must have high resistance to avoid excessive current flow from the armature current. Shunt generator voltage drops with increase in load. It is important to know the voltage at the terminals of a generator for each value of current it delivers. This relationship of voltage and current is the *characteristic* of the generator. The shunt generator characteristics depend on the degree of field saturation, or magnetic circuit, and on what part of the saturation curve the generator is operating. In general, a shunt generator has a drooping voltage and speed characteristic with load increases.

The *series-field generator* (see Fig. 12.10*b*), has the field winding connected in series with the armature and the external load. An increase in load current causes the field to become stronger and thus builds up generator voltage. However, increases in current flow also cause an increase in voltage drop across the armature and the field, or $I_a(R_a+R_s)$. The saturation curve of a series-field generator reaches a certain maximum beyond which armature reaction becomes so great as to cause the saturation curve to droop sharply and the generator's terminal voltage drops very rapidly.

The *compound-field* generator (see Fig. 12.10*c*) combines a shunt field with a series field in order to overcome the drop in terminal voltage that is characteristic of a shunt generator. If a series winding is placed with the armature, it aids the shunt turns by increasing the flux as the load increases. The effect is to increase the induced voltage on the generator. Usually the design of series ampere turns increases the voltage to balance the combined drop in voltage due to armature resistance and armature reaction, and a steady output voltage is obtained with increase in load.

The *separate-field* excitation generator (see Fig. 12.10*d*) receives dc current from a source other than the generator. This permits finer control of excitation current but usually requires a separately driven exciter or a connection to a dc bus supplied by other generators in the facility.

Voltage buildup on integral field windings with a generator depends initially on residual magnetism in the magnetic circuit. This magnetism produces an initial field current as a result of residual field voltage. Each value of field current produces a larger voltage than the previous value, which in turn increases the dc voltage so that the machine builds up to rated voltage until the field resistance line crosses the field saturation curve, at which point the machine cannot build up any more voltage.

A common problem in starting a dc generator is failure to build up voltage. The reason may be:

1. Shunt field is connected so that when the initial current starts, it "bucks" the residual magnetism current. To test for this, open the field circuit, and if the voltage rises, the field current is bucking the residual magnetism, and the field connections may have to be reversed. If opening the field circuit produces no effect on the voltmeter, there may already be an open in the field circuit.

2. The field resistance is too high. Reduce the field resistance and note if voltage builds up.

3. Imperfect brush contacts may cause high resistance from commutator to brush. Since the shunt field is connected across the armature, poor brush contact is the same as high resistance in the series field, as in no. 2.

4. There may be no residual magnetism due to long idleness or jarring of the machine. The field voltmeter should read zero. If the generator has a series field, a low-voltage source such as a battery may be connected across the series field, and this may produce enough magnetism to cause the machine to build up voltage. If there is no series field, the field terminals may have to be connected temporarily across a separate supply circuit in order to build up the residual magnetism, often called *flashing the field*.

Commutating poles or *interpoles* are used on dc machines in order to assist commutation by preventing sparking of the brushes at different loads. These coils are usually placed in the neutral position, or when the armature current sine wave is just being reversed, and the interpoles are placed midway between each pair of field poles. The interpoles carry armature current to overcome armature reaction and thus reduce commutator sparking.

Automatic *voltage regulators* are used to maintain the voltage of a dc generator fairly constant with load and speed changes. The regulators vary the field strength of the generator automatically with any changes in line voltages.

DC Motors

DC motors and generators are almost identical in mechanical design but differ in electrical design; therefore, it is better to use a dc motor for an application and a dc generator for generating dc power because they operate more efficiently and perhaps more safely for the service for which they were designed.

DC motors operate on the principle that a conductor carrying current in a magnetic field develops a force on the conductor which is proportional to (1) the strength of the magnetic field, (2) the magnitude of the current in the conductor, and (3) the length of the conductor lying in the field. Force times distance from a center to where the force is applied is called the torque that a motor develops to drive machinery. In dc motors, the torque developed is proportional to the armature current and to the strength of the magnetic field. When a motor is in operation, the armature current is determined not only by the resistance value of the armature but also by the counteremf, or back electromotive force. The counteremf must always be less than the terminal or impressed voltage if current is to flow *into* the motor's armature. The power input to a motor armature is

$$P_m = VI_a - I_a^2 R_a$$

where V = line voltage to the armature, I_a = armature current, and R_a = armature resistance. The above equation can be transferred to

$$P_m = I_a(V - I_a R_a)$$

The term $V - I_a R_a$ is the counteremf of the motor. The mechanical power developed within the armature of a dc motor is equal to the product of this counteremf and the armature current. Note that $I_a^2 R_a$ is the power lost in the armature from resistance.

Example. A dc motor has a terminal voltage of 110 V, and the armature records 90 A. The armature resistance is 0.05 Ω. Determine (1) the counteremf, and (2) the mechanical power developed in the armature.

The counteremf $= 110 - 90 \times 0.05 = 105.5$ V

$$P_m = 105.5 \times 90 = 9495 \text{ W} = 12.73 \text{ hp}$$

DC motor classification follows dc generator classification as shown in Fig. 12.10 except that current is flowing *into* the armature.

Shunt motors are considered constant-speed motors even though the speed drops slightly with load. The shunt motor also can be programmed for adjustable speed, where the speed is set at the desired value and then remains substantially constant at that speed as the load varies. It is thus suited for a drive where a substantially constant speed is required, such as machine tools, blowers, and spinning frames.

The *series motor* has a speed characteristic that varies greatly as the load or armature current changes. With a light load, the speed may increase 3 to 5 times as fast as at no load. A series motor has a large starting capability under heavy loads and is used on railroad cars where heavy starting capability is needed. The dc motor *can overspeed* to destructive speeds if there is no load on the motor. The effect is cumulative; as the field current rises with each increase in speed, more torque becomes available. For this reason, a no-belt-type drive should be used on a series motor because if the belt breaks, a no-load situation will occur on the motor, and it will "run away" like a steam turbine with a defective governor and overspeed trip. In fact, many manufacturers include overspeed trip devices on the control scheme of series motors.

The *compound-connected motor* combines the characteristics of a series and shunt motor since both have windings. The speed changes as the load changes but not as much as a series motor does and more than a shunt motor. If the series winding aids the shunt winding, the motor is termed *cumulative compounded,* and this type of motor develops a high torque with sudden increases in load.

To *reverse the direction* of a dc motor, either the armature alone or the field alone must be reversed. If both are reversed, the direction of rotation remains the same.

When *starting* a dc motor, the resistance, or rheostat, must be connected in series with the motor armature. This resistance must be gradually cut out as the armature comes up to speed. Automatic starters are used in order to cut out the starting resistance at a definite rate. This avoids the blowing of fuses or the opening of circuit breakers due to too rapid acceleration and armature current rise.

Dynamotors convert dc power from one voltage to another by having one winding act as a motor and the other winding delivers power as a generator. Both windings cut the same magnetic field at the same speed, and as a result their induced emf is proportional to their winding turns. The armature windings are on the same shaft but with two commutators. One application of the dynamotor is on automobiles where the dynamotor is used as both starter and generator for the automobile's electrical system.

AC GENERATORS AND MOTORS

AC Generators

In an ac generator, the armature, called a *stator,* is stationary, and the dc field rotates. Thus current is taken directly from stator connections, and these could be single-, three-, or two-phase stator arrangements. It was previously established that a conductor rotating in a magnetic field generates a sine wave of alternating current. There are many advantages of having a stationary armature where the sine wave is generated by a rotating field rather than by a dc generator with its commutator: Higher voltages are possible on a stator in comparison to a rotating armature, and the generator can be much larger in kilowatt output.

Rotating Field

The dc current to the rotating field is conducted to the field windings through slip rings. A direct-connected exciter to the prime mover supplies the dc current. More modern machines use rectifiers to supply the dc current. Many also employ brushless excitation by the use of diodes, silicon-controlled rectifiers, and field resistors. Figure 12.11 illustrates the cross section of a steam-engine driven *alternator,* as ac generators are described. The pole pieces are made of laminated sheet steel to avoid eddy currents being generated in the iron. The stator also has laminated steel, and slots are cut to house the coils. Two general types of slot construction are used: open slot and closed slot. Both require wedges to hold the coils so that centrifugal force does not throw them out of the slots. The field poles shown in Fig. 12.11 are called *salient poles,* and this construction is only used for low-speed alternators of relatively small kilowatt size. The cylindrical, or nonsalient-type, pole rotor is necessary for direct- and turbine-driven alternators. Figure 12.12 shows a cylindrical dc rotor for a utility-size turbo-generator. The rotor is wound with strip copper that is properly insulated for the voltage to be encountered. Cooling slots are provided in the forging. The end turns of the windings are supported by retaining rings of high-alloy material. Utilities are switching to 18Cr-18Ni alloyed steel on retaining rings on hydrogen-cooled units to prevent stress-corrosion cracking on this highly stressed shrunk-on ring. Moisture in the hydrogen gas emanating from hydrogen gas heat exchangers have caused lower alloy steel rings to develop cracks.

Stator and rotors may be air cooled on the smaller sizes and hydrogen cooled on larger sizes and may also employ hollow conductors in the stator winding, where water of high purity is circulated. This additional cooling permits higher output with smaller frame sizes.

Most alternators are Y connected in order to avoid harmonic voltages with delta connections.

Ratings. Alternators are rated in kilovoltamperes with the power factor specified in the nameplate. On larger machines, the rating may vary depending on the hydrogen

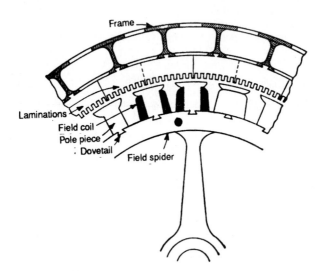

FIGURE 12.11   Salient pole rotors are used on engine-driven alternators of lower kilowatt rating.

FIGURE 12.12　Cylindrical, or nonsalient pole, rotor for large alternator. (*Courtesy, General Electric Co.*)

pressure; therefore, manufacturers list the ratings for the different hydrogen pressures at which the machine may be operated. It is the I^2R loss in the stator windings that limits the output of a machine and is independent of the power factor rating of the machine. It is thus essential to watch the stator amperes that the manufacturer specifies at the different hydrogen pressure ratings in order to make sure the machine operates within this ampere rating.

Example.　A three-phase generator shows the following data on the nameplate:

75,000 kVA
63,750 kW
0.85 P.F.
13,800 V
3138 A on stator
3600 r/min

The load noted is 74,000 kW and 3100 A. Is the generator overloaded because the kilowatt load exceeds the kilowatt rating? The ampere load is not exceeding the ampere rating; therefore, the I^2R loss is within rating (i.e., the machine should not be overheating). The kilovoltampere load is $13.8 \times 3100 \times 1.732 = 74,095$ kVA, which is within the machine's rating. The operating power factor is $74,000/74,095 = 99.9$ percent, and this is why the machine can carry a high kilowatt load; the power factor is above the rated power factor. The machine is not overloaded because the stator amperes are within rating as is the kilovoltampere load.

Voltage Regulators.   Voltage regulators are used to control the voltage of a generator within set limits. It operates automatically by comparing the output voltage of the generator with a voltage reference level within the regulator. Deviations in alternator output voltage result in adjustments up or down in the field current so that the generator voltage returns to the set points of voltage. The standard voltage regulation on engine-driven units is ±2 percent. On turbine-driven units of larger size, the regulation is much closer.

Most alternators have their own *switchboard panel,* which generally should have the following equipment to aid in operating the generator(s):

Bus voltmeter and generator voltmeter

AC and dc ammeter

Synchroscope for synchronizing a generator to the bus where outside power or another generator is paralleled

DC voltmeter for excitation

Two-pole field switch with discharge contact when opening the switch to drain off induced voltage

Line circuit breaker

Automatic voltage regulator

Alternator field rheostat

Current and voltage instrument transformers

Synchronizing switch

Indicating kilowatt meter

Possibly a governor motor control switch

Protective Devices.   Abnormal conditions can develop in an ac generator due to failure of insulation but also from (1) stator overheating, (2) loss of excitation, (3) loss of synchronism if paralleled, (4) motorizing, (5) phase-current unbalance, (6) grounded field, (7) overvoltage and undervoltage, and (8) loss of frequency. Protection can be provided for these abnormal conditions, but many are the option of the purchaser or system design engineer due to costs or system design considerations.

Failures in the stator windings involve higher voltages and the possibility for excessive currents. Protection required is against phase-to-phase shorts, short-circuited turns, open circuits, and grounds. Stator failures may start as a ground but quickly develop into one or more of the other type of faults. A brief description of protective device functions should assist the reader in checking their installation.

Percentage-differential relaying is used to protect the generator from stator short circuits. The relays must be arranged to trip the generator's breaker and field circuit breaker and must be connected to the prime movers trip scheme.

External-fault back-up relaying is used to disconnect the generator if the primary protection does not clear a short circuit on the high-voltage bus or on the circuits connected to the bus.

Stator overheating protection is provided in case of slow overheating, such as that due to loss of cooling air, and consists of resistance temperature detectors connected to relays that may alarm only and then trip the generator breaker.

Loss-of-excitation relays are required because of the danger of overheating the rotor. Some relays ring an alarm first, followed by a breaker trip.

Loss of synchronism relaying is applied primarily to unattended stations, where the generators are paralleled.

Reverse-power relays are used to prevent motorizing a generator. On large steam-turbine-driven units, blade overheating can occur if no steam is flowing and the generator is accidentally motorized by switching mistakes. Reverse-power relays generally have some time-delay feature so that the generator will not be tripped by momentary power disturbances.

Negative-phase-sequence relays protect the generator against *system faults* which cause unbalanced currents in the generator, such as phase-to-ground and phase-to-phase faults. When such faults occur on the system, it causes unbalanced phase currents in the generator or negative phase sequence current that can quickly overheat the rotor.

Grounded field relays are used to detect a ground in the field. It must be remembered that if a second ground were to develop, the field would suffer short-circuit damage. Most grounded field relays first ring an alarm to warn the operator. Tripping may follow, and this depends on what disturbance could occur to the system if the generator was suddenly taken out of service.

It is important for operators to become familiar with the protection provided to the machines under their control, and this includes knowing what is only alarmed, and what trips the generator and its prime mover.

Induction Motors

The most used ac electric motors are induction motors, consisting of a squirrel cage, the wound motor, and the synchronous motor. Single-phase motors include (1) the series motor, (2) the repulsion motor, and (3) the single-phase induction motor. Most single-phase motors are small, while service operators deal mostly with three-phase power drives.

Induction motors are the most widely used ac motors. The stator is stationary and receives alternating current, which induces a magnetic force in the rotor. Figure 12.13 shows four field poles inducing armature currents in the cylinder. As is illustrated, a revolving electric field causes the cylinder to rotate in the opposite direction. The induction motor operates on the principle of the rotating field, but the rotating field is produced by polyphase currents in the stator windings. Such rotating fields are produced electrically entirely by the stationary stator. The reader is referred to electrical engineering texts for a further explanation of induction motor theory.

The field or stator has a *synchronous speed* in its electric rotating field, the same as an alternator:

$$N = \frac{120(f)}{p}$$

where N = synchronous speed
   f = frequency, cycles per second
   P = number of poles

For example, a four-pole, 60-cycle motor has a synchronous speed of

$$N = \frac{120 \times 60}{4} = 1800 \text{ r/min}$$

The rotor of an induction motor can never reach the synchronous speed of the rotating stator field because if it did, the cutting of conductors by magnetic flux would cease

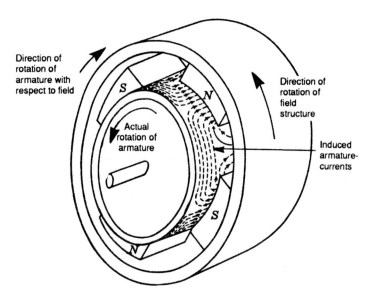

Direction of rotation of armature with respect to field

Direction of rotation of field structure

Actual rotation of armature

Induced armature-currents

FIGURE 12.13 The induction motor has a rotating electric field inducing currents in a rotor, and the revolving magnetic field pulls the rotor to turn with it.

producing induced currents in the rotor, and therefore there would be no torque to turn the rotor. Thus, the rotor revolutions per minute of an induction motor are always less than the synchronous speed of the rotating field by an amount called *slip*. For example, a four-pole machine with a synchronous speed of 1800 r/min may operate at 1750 r/min, producing a revolution slip of 50 r/min. Slip is usually expressed as a percentage of the synchronous speed. For the above motor it would be 50/1800 = 0.028 = 2.8 percent. It has also been established that the *rotor frequency = stator frequency × slip*.

Example. An induction motor operates at 3550 r/min. How many poles does the motor have, what is the slip, and what is the rotor frequency if the supply is 60 cycles? This is a two-pole machine. The slip is (3600 − 3550)/3600 = 0.014 = 1.4 percent. The rotor frequency = 0.014 × 60 = 0.84 cycles per second.

The induction motor can be used as a *frequency changer* if the rotor is driven mechanically at the proper speed. Current is taken from the rotor at the desired frequency by means of three slip rings, while the stator receives normal frequency power.

Variable-frequency induction motor drives can convert the basically constant speed induction motor to variable speed operation by electronically varying the frequency and voltage supplied to the induction motor. The growth of electronics has made it possible to change frequency on power supplies by means of electronic converters. If the frequency varies, so will the speed of an induction motor. However, magnetic field strength varies inversely from the frequency, and lowering the frequency too much would push the field strength unacceptably high, so the applied voltage is lowered with the frequency. This is performed in the converter of a variable-speed controller for the induction motor by a solid-state rectifier converting the ac line power to the required voltage in dc form, and then an inverter section transforms the dc to an ac output at the required voltage and frequency.

The *squirrel-cage motor* (see Fig. 12.14a) is the most widely used induction motor. The core of the rotor is made of slotted punchings. The windings in the rotor consist of either copper or aluminum bars placed in the slots. The bars are connected together by bar rings called end rings. The bars are welded to the end rings. Larger motors, 30 hp and over, have die-cast rotors with the bars, end rings, and cooling rotor fans cast integral with one another. The slots must be closed to encase the molten metal in fabrication. This type of construction produces a very rugged rotor.

Some characteristics of an induction motor are:

1. The breakdown torque is proportional to the square of the line voltage.

2. The breakdown torque is reduced by an increase in the stator resistance and by an increase in the stator and rotor reactance.

3. The breakdown torque is independent of the rotor resistance.

Wound-Rotor Motor.   The wound-rotor motor is an induction motor, but it has a rotor construction that permits resistance to be introduced into the rotor circuit by means of three slip rings. By introducing resistance, higher starting torques are possible. Some variable-speed operation can also be achieved. Thus, wound-rotor motors are used where considerable starting torque is required and where some speed adjustment is desirable. Applications include centrifugal compressor drives for air conditioning where refrigeration load varies with temperature, cranes and hoists, elevators, calendars, and variable pump-load drive.

Starting Induction Motors.   A squirrel-cage motor has low resistance so that the in-rush current with across-the-line starting is 4 to 5 times the rated current. This may not disturb the line voltage if the motors are relatively small, but on larger motors, the in-rush of current can trip plant circuit breakers; therefore, reduced voltage starters are used to limit the in-rush current, and when the motor is near rated speed, the compensator starter places the motor "across the line," or at full-rated voltage. *In the Y-delta* method, the stator windings in the starter are first connected in Y across the line, thus applying only $1/\sqrt{3}$, or 58 percent, of the normal voltage to each phase, and this makes the line current one-third the value that it would have in across-the-line starting. When the motor is up to a certain speed, the controller activates a switch to connect the motor delta across the line. Two windings are also used, where the windings are connected in series when starting and in parallel when running. Service operators dealing with larger motors should become familiar with the system used for starting their larger motors.

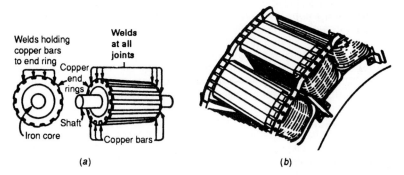

FIGURE 12.14   (*a*) Squirrel-cage induction rotor; (*b*) amortisseur cage in a synchronous motor, and in engine-driven alternators, has a short-circuited winding on top of dc pole pieces that help start a synchronous motor or dampens oscillations on engine-driven alternators.

Induction motors must have NEC letters on their nameplate to show motor input by kilovoltamperes per horsepower with locked rotor. This code letter, indicating motor input with locked rotor, is used to determine branch-circuit short circuits and ground fault protection for sizing conductors and protective equipment. Code letters range from A to V. See Article 430 of the NEC.

Motor Enclosures. The NEC has classifications of occupancies which determine the type of enclosure that electrical equipment must have to operate in that surrounding. For example, Class 1 is classified as a location in which flammable gases or vapors may be present in the air sufficient to produce explosive or ignitable mixtures. For such an environment, an *explosionproof* motor frame is required. *Totally enclosed fan-cooled* motors have internal windings that are isolated from the outside air. This type of enclosure is used in areas with high levels of contamination that can harm windings and bearings, such as outdoor, wet surroundings.

Protected enclosures are open to the outside environment, allowing outside air to cool the windings, and are further broken down to (1) dripproof, (2) weather-protected, type 1, and (3) weather-protected, type 2. Motor enclosure selection is an important consideration in specifying motors and is quite often ignored. For example, many cooling tower motors fail because they are only dripproof, whereas a totally enclosed fan-cooled motor would protect the motor from horizontally driven rains that such motors are exposed to.

Synchronous Motors

Synchronous motors have features in common with ac alternators and generators. Both have stationary stators and a dc rotor for excitation. The speed is fixed or constant, depending on the frequency of the incoming ac current. The rotating field poles receive direct current from a separate dc generator called an exciter. The dc field poles have, in addition to spool windings, a short-circuited grid of copper wires inserted in slots in the pole faces. These are called *amortisseur* or *damper windings.* This is a continuous squirrel-cage winding that helps the synchronous motor come up to speed before the dc field is cut in (see Fig. 12.14*b*). The purpose of the damper winding is also to damp the tendency of the rotor to oscillate while rotating.

When a three-phase current is applied to the stator, it also produces a magnetic field that revolves synchronously with the supply circuit. This acts on the amortisseur winding to produce a starting torque as in an induction motor. Near rated speed, excitation produces north and south poles which "lock into" position with the revolving magnetic field of the stator, and the rotor rotates at synchronous speed that is determined strictly by the frequency of the supply current and number of poles. There is no slip. This means variations in load do not cause any speed change.

By varying the field current, the power factor can be varied from unity to lagging with a weak field and to leading with a strong dc field. For this reason, synchronous motors are extensively used to drive machines at constant speed and as a means to improve a plant's power factor by supplying leading current where the motor acts as a capacitor.

Most ac motors have a maximum permissible load beyond which they rapidly lose speed and stop. This is called the *pull-out* or *breakdown torque.* Both the pull-in and pull-out torques, as well as the starting torque, are provided by the manufacturers as a percentage of the full-load torque, which is 100 percent. The guaranteed breakdown torque ranges between 175 and 200 percent of full-load torque.

Direction of rotation may be changed on motors as follows:

1. *DC motors.* Transpose either the field or armature connections *but not both.* However, check with the manufacturer's instructions for exact details.

2. *Polyphase ac motors.* On two-phase motors, the connections to one phase should be interchanged. On three-phase motors interchange two of the leads.

3. *Single-phase motors.* Follow the manufacturer's instructions since this varies by the type of single-phase motor and may include brush shifting instructions.

IMPORTANCE OF INSULATION

Rotating electrical equipment depends on continuous operation; therefore it is important to have the proper insulation for the application and then maintain this insulation. Some motors are subjected to environmental conditions, such as dirt, oil, and grease that shorten the life of some insulation material, and unexpected winding failures may result. It is important to keep track of the condition of the insulation by periodic visual inspection and by instrument tests, such as using a megohmeter, or *megger.* The insulation resistance should be recorded for future reference. This will help in noting trends of declining values, and corrective action can be taken before a failure occurs. This is especially important on rotating equipment with high replacement or repair costs or which can seriously affect a plant service or process flow.

There are many types of insulation material available based on thermal endurance since it is heat that determines the amount of output that a manufacturer will stamp on the nameplate. Insulation has been defined by the American Institute of Electrical Engineers (AIEE) into the following classes:

Class O. 90°C limit. Made from combinations of cotton, silk, and paper without impregnation.

Class A. 105°C limit. Made from combinations of cotton, silk, and paper but impregnated in a dielectric such as oil compounds.

Class B. 130°C limit. Made from combinations of materials such as mica and glass fiber with suitable bonding substances.

Class F. 155°C limit. Made from combinations of materials such as mica and glass fiber with bonding agents for the higher-temperature rating.

Class H. 180°C limit. Made from combinations of materials such as silicone elastomer, mica, and glass fiber with suitable bonding agents such as silicone resins.

Class C. 220°C limit. Made entirely from mica, porcelain, glass, quartz, and similar inorganic materials.

The above temperature limits are defined as applying to *"hottest spot"* in the machine or apparatus. The *limiting observable temperatures* are determined by subtracting a specified number of degrees from the above listed temperatures. With the three methods used to determine temperatures, the following are the limiting observable temperature that should not be exceeded;

Measurement method	Limiting observable temperature, °C				
	Class O	Class A	Class B	Class F	Class H
Thermometer	75	90	110	Still	150
Resistance or imbedded detector	80	100	120	Open	165

The Class F limit is under investigation for establishing limits. The AIEE has also established *limiting observable temperature rises,* which are used on nameplates to show insulation rating. These are obtained by *subtracting* 40°C from the limiting observable temperature, which is the base agreed to be the surrounding air temperature. All the above temperature limits are for air-cooled open machines.

INSULATING OIL

Insulating oil is used in circuit breakers, transformers, starting devices, and similar apparatus. The oil must have high dielectric strength and a high flash point, maintain its fluidity, and contain no sulphur, acid, or alkali. It degrades with service and requires periodic testing to make sure it is still of insulating quality. Since transformers are the greatest users of insulating oil, oil testing is covered in Chap. 13.

TESTING ROTATING ELECTRIC MACHINES

Critical rotating electric machines in a plant degrade with use from various combinations of progressive aging and drying out of the insulation, causing it to crack and produce resultant current leakage paths; moisture, oil, and grease accumulations; and in some cases harmful dusts, chemical contaminants affecting the windings, mechanical damage such as abrasion from gritty dust, and abnormal operating occurrences such as undervoltage conditions, single-phasing, and shock overloads.

Electrical testing can be classified as being performed (1) for preventive maintenance, (2) as a diagnostic test to determine where there may be a problem, and (3) for proof-testing to establish that the equipment is safe for further operation.

Most plants perform preventive testing as part of the maintenance function. The most common plant test is the insulation resistance test with a *megger.* The most common unit is a 500-V dc unit for spot checking. The readings obtained in megohms must be interpreted in relation to the condition of the windings. Thus, a machine in a paper mill that has been idle may have accumulated moisture that will affect the insulation reading; however, the reading can be used for deciding whether the machine needs drying out before returning it to service. Dirty windings may have a similar initial low reading. Insulation resistance instrument manufacturers provide good instructions on how to use the megger and offer guides on interpreting the readings. The chief value of insulation testing is to spot trends with respect to whether readings are the same and still acceptable or whether a declining trend is developing that requires further analysis as to cause. However, readings of low value, such as below, 1 MΩ, need immediate correction.

Dielectric absorption tests on larger equipment are often used to supplement spot megger readings. The test voltage applied should match the rating of the machine being tested. For example, a 2300-V stator winding can take a 2500-V test voltage, but if the machine has a synchronous motor, the field winding test voltage has to be lower because its rated voltage is usually 250 V or less. The test voltage is applied for 10 min with readings taken at 30 s and every minute thereafter. The readings are plotted, megohms versus time. A rising curve in which dielectric absorption is good is an indication of clean dry windings. A flat or declining curve is the result of leakage current through or over the surface of the windings and generally is an indicator of wet, dirty, or dried-out windings due to age. There are guiding ratios between the 30-s, 1-min, and 10-min readings provided in the literature that assist in interpreting the readings.

Other tests are made, but these are usually performed by specialists from the manufacturer of the machine. Some can be destructive, such as overvoltage or *hi-pot test.* This test is made to obtain assurances that the electric machine has the dielectric to withstand a voltage that stresses the dielectric. This allows for detecting and repairing an incipient insulation weakness during scheduled overhaul periods, thus trying to prevent unscheduled outages from insulation deterioration.

Power factor tests are made by special bridge circuits. Trending is important on power factor values because an increasing power factor on the same machine over a period of time is believed to show that a deteriorating condition is developing on the insulation.

Making *rotor winding impedance* tests of the total winding and individual coils and comparing these values to new conditions is a very effective way of detecting and locating turn-to-turn shorts or faults. Routine visual inspections and basic dielectric tests should be made annually on a plant's critical rotating electric machines.

MAINTENANCE PROGRAMS

A regular maintenance program on critical electric machines has economic advantages over operating the unit until failure occurs. Outages can be planned if a maintenance program is established. Here are some items that need attention:

1. *Visual inspection* to detect vibration, overheating, which may be due to plugged cooling passages, insufficient cooling air or other similar reasons, including overloading; cleanliness of windings, connections, and control equipment; worn or frayed contact points in controllers; loose connections on lead connectors and in control apparatus, usually evident by signs of heat; and loose wedges on windings and loose stator or field bars. Check for loose and hot bearings during lubrication checks.

2. *Overhaul inspections* to detect worn parts; repair those that show any of the above distresses.

3. *Testing* for insulation resistance, level of vibration, and its correction; making oil and similar tests recommended by manufacturers for their machines.

4. Inspection and *testing of safety relays* to make sure they are set properly, have not been bypassed, and are in good operating condition. This includes overload relays, circuit breakers, disconnect switches, and fuses.

5. It is important to *record findings,* readings, and test results on an equipment card when performing preventive maintenance work. This will assist in evaluating the future condition of the electric machine and will assist in future planning for repairs, tests to be made, and how soon another thorough inspection should be scheduled.

QUESTIONS AND ANSWERS

1. What is emf?

   Answer. The electromotive force, or emf, is the voltage that forces electrons to flow in a circuit. The term is usually applied to the voltage generated within a generator. An electric generator produces an emf through the relative motion between a magnetic field and a conductor passing through it. An electric current flows when a potential difference or voltage of sufficient strength is established in the generator.

2. What is Ohm's law?

   Answer. This establishes the relationship between voltage V and amperes I that flow in a circuit with the resistance R to that flow. For dc current $I = V/R$. For single-phase ac circuits $I = V/Z$, where Z = impedance in place of resistance in the circuit.

3. Resistances of 5, 6, and 7 Ω is in a series in a 110-V dc line. What is (1) the total resistance in the circuit, (2) the amperage in the circuit, and (3) the power in the line?

   Answer. They are:

$$1.\ R_t = R_1 + R_2 + R_3$$
$$= 5 + 6 + 7$$
$$= 18\ \Omega$$

$$2.\ I = \frac{V}{R} = \frac{110}{18} = 6.111\text{A}$$

$$3.\ P = IV = 6.111 \times 110 = 672.21\ \text{W}$$

4. In a parallel-circuit of a dc circuit, resistances of 5, 6, and 7 Ω are placed in parallel on a 110-V line. What is (1) the total resistance, (2) the total amperage, (3) the voltage through each resistance, and (4) the power in the line?

   Answer. They are:

$$1.\ \frac{1}{R_t} = \frac{1}{R_1} + \frac{1}{R_2} + \frac{1}{R_3}$$
$$= \frac{1}{5} + \frac{1}{6} + \frac{1}{7}$$
$$\frac{1}{R_t} = \frac{42 + 35 + 30}{210}$$
$$R_t = \frac{210}{107} = 1.962$$

$$2.\ I = \frac{V}{R}$$
$$= \frac{110}{1.962}$$
$$= 56.06\ \text{A}$$

  3. Voltage = 110 V (since the resistances are connected
               across the 110-V line)

  4. $P = IV$

   $P = 56.06 \times 110$

   $P = 6166.60 \text{ W} = 6.17 \text{ kW}$

5. What is the designation system of numbers to gage copper conductors?

Answer. The American wire gage in which 40 is the least size and 0000 the largest.

6. What is the approximate ratio between gage numbers?

Answer. The diameters double for each three gage numbers. This means that size 4 is twice as large as size 7; size 7 is twice as large as size 10; etc.

7. What is the circular mil system of sizing conductors?

Answer. All round conductors' diameters are first converted to one-thousand of an inch, or mils. Thus, a $\frac{1}{8}$-in diameter conductor is equal to 0.125 in, and this is equal to 125 mil. A circular mil is the diameter in mils of a round conductor squared. Thus, the $\frac{1}{8}$-in conductor would be $125^2 = 15,625$ circular mils.

8. A copper bus bar has a $\frac{1}{8}$- $\times$ $1\frac{1}{2}$-in rectangular shape. What are the equivalent circular mils of this copper bar?

Answer. First solve for the square mils, or $125 \times 1500 = 187,500$ square mils. With $D =$ the diameter of a round conductor in mils, then

$$\frac{\pi D^2}{4} = 187,500 \text{ square mils}$$

Now since circular mils $= D^2$, all that is necessary is to solve for D^2 to obtain the equivalent circular mils, or $D^2 = 187,500/0.7854 = 238,732$ circular mils.

9. How does the NEC classify facility service conductors?

Answer. By their service as follows: (1) service entrance, (2) feeder circuits, (3) feeder branch circuits.

10. What would be the speed of an alternator that had 12 poles and operated at 60-cycle current?

Answer. Speed $= 120 \times 60/12 = 600$ r/min.

11. What is the effective value of a sine-wave current?

Answer. In a sine wave, the effective, or rms, value, of current is equal to 0.707 times the maximum peak value of the sine wave. This value is equal to the dc amperes that will produce the same I^2R heat effect. All nameplates show effective value of current on ac equipment.

12. What three components comprise the impedance in an ac circuit?

Answer. Resistance, inductance, and capacitance.

13. See Fig. 12.6b. This is an ac series circuit with $R = 40$ Ω, $X_L = 30$ Ω, and $X_c = 20$ Ω. What is the impedance of the circuit, current, power factor, and power supplied to the circuit if the voltage is 440 V?

Answer. They are:

$$\text{Impedance } Z = \sqrt{40^2 + (30 - 20)^2} = 41.2 \ \Omega$$

$$\text{Current } I = \frac{V}{Z} = \frac{440}{41.2} = 10.7 \text{ A}$$

$$\text{P.F.} = \frac{R}{Z} = \frac{40}{41.2} = 0.97$$

$$\text{Power} = V \times I \times \text{P.F.} = 440 \times 10.7 \times 0.97 = 4.567 \text{ kW}$$

14. A three-phase generator shows 2300 V, 3000 A, and 9400 kW as a load. What is the kVA output and power factor? What are the kvars?

Answer. They are:

$$\text{kVA} = \frac{1.732 \times V \times I}{1000} = \frac{1.732 \times 2300 \times 3000}{1000} = 11{,}950 \text{ kVA}$$

$$\text{P.F.} = \frac{\text{kW}}{\text{kVA}} = \frac{9400}{11{,}950} = 0.79, \text{ or 79 percent}$$

$$\text{kvar} = \sqrt{11{,}950^2 - 9400^2} = 7378 \text{ kvar}$$

15. What are three types of dc generators? Name them by method of excitation.

Answer. Shunt field, series field, compound-wound, and separate excitation are the most common excitations for dc generators.

16. What is residual magnetism? Describe the need of residual magnetism in dc generators.

Answer. Residual magnetism is the permanent magnetism in the iron poles of a generator. When a generator is started, the revolving wire windings of the armature cut magnetic lines of force and generate electricity. The flow of current builds up the field magnetism to the proper value. If it were not for residual magnetism, the machine would not build up electricity, and outside excitation would be needed.

17. Why do brushes spark during commutation? How is this overcome?

Answer. The inductance of the coils discharges following the reversal of current. If this occurs at the moment of commutation, a spark results. This is overcome by one of two ways: (1) neutral plane shift or (2) interpoles.

In neutral plane shift, the brush rigging is on a frame that can be swung about the commutator. By shifting the brushes slightly ahead of the neutral plane, current is induced opposite to that of the inductance and counteracts the spark effect. The point varies with the load and the brushes are set by trial and error until right.

Interpoles, sometimes called *commutating or auxiliary poles,* are small auxiliary poles mounted between the main poles. Line current or a proportion of it passes through the interpole windings to produce a magnetic field of a little more than the armature produces. This counteracts the spark-producing armature reaction or induced current and minimizes the sparking. These poles require no adjustment since the manufacturer sets the brushes to the neutral plane of the machine. Interpoles have replaced the old neutral-plane-shift devices.

18. Why are armature cores, field pieces, and transformer cores made of laminated stock?

Answer. By making these pieces of laminated metal and by insulating them with a compound from each other, no path of any length is furnished. This minimizes the eddy currents and the heat generated by them. If they were made solid, eddy currents would set up and cause heating.

19. What are the dc motor types?

Answer. The dc motor types are (1) series, (2) shunt, (3) compound, and (4) universal (ac or dc types).

20. Why are interpole windings used on dc motors or generators?

Answer. Interpole windings are small windings between the poles. They hold the neutral on the armature to minimize sparking on the brushes. The coils of these poles are in series with the armature.

21. How are dc motors reversed?

Answer. These motors can be reversed in two ways: (1) reverse the leads to the armature or (2) reverse the leads to the field coils. (Use either method, but *not both* at the same time.)

22. What is the basic method of speed control in dc motors?

Answer. The speed of dc motors is controlled by changing the field strength. Weakening the field increases speed. Resistances are used to reduce the current to the field. A rheostat is commonly used for this purpose.

23. What happens if the voltage is low?

Answer. Low voltage will cause the motor to overheat; torque is reduced in proportion to the square of the voltage. The motor may not start, or it may stall under a load. Current amperage will rise and overheat the motor. A 10 percent temperature change is the limit a motor can stand.

24. A dc motor has an armature input voltage of 230 V and 110 A. The armature resistance is 0.25 Ω. What is the electric power developed within the armature?

Answer. Use $P_m = I_a(V - I_a R_a)$ and substituting,

$$P_m = 110\,[230 - (110 \times 0.25)] = 22{,}215 \text{ W} = 22.2 \text{ kW} = 29.9 \text{ hp.}$$

AC Generators and Motors

25. What kind of current does an exciter furnish? Where are exciters used?

Answer. Exciters furnish direct current. They are used on all types of machines that are not self-excited, whether alternating current or direct current. They are always used on synchronous motors and mostly on ac generators.

26. What additional pieces of equipment are needed on ac generators that dc generators don't require?

Answer. AC generators require (1) an excitation source, such as a small dc generator, (2) a rheostat to control the voltage of the exciter, and (3) synchronizing indicators, either synchroscope or lamps.

27. What is an alternator?

Answer. A machine that produces alternating current. In an alternator, the rotor is the dc field, and the stator (stationary winding) becomes the armature. The dc excitation is fed to the poles which are mounted on the rotating shaft of an alternator. Magnetic flux sweeps past the fixed windings of the stator coils mounted on the frame. Voltage and current are induced when the shaft is turned by an engine or other drive. Depending on the number of coils and their connections and the number of pairs of poles, alternating current having single or polyphase characteristics is produced at a frequency determined by the following equation:

$$f = \frac{PN}{120}$$

where f = cycles per second
P = number of poles
N = r/min

28. What are the two types of revolving fields used for alternators?

Answer. For lower speed and capacity, polar projections called salient pole construction are used. For higher speeds and larger capacities, a distributed winding embedded in the face of a revolving forged cylinder called a nonsalient pole field is used.

29. What is a brushless excited generator?

Answer. In a brushless ac generator, the power from the direct-connected exciter is supplied to the revolving field of the generator through a rotating rectifier assembly. The field of the exciter is supplied direct current from a silicon-controlled rectifier regulator. The regulator senses voltage changes on the ac output and then controls the dc input to the exciter field. This construction uses no brushes to supply direct current to the revolving field and thus eliminates a source of arcing across slip rings due to carbon dust buildup.

30. What type of alternator has damper or amortisseur windings?

Answer. Damper or amortisseur windings consist of copper bars embedded in salient pole faces and are connected together to make a squirrel-cage winding. The purpose is to smooth the operations of generators when the drive is an internal combustion engine, whose crank effort may produce oscillations or hunting effects, and this may produce cross- or synchronizing currents in the generator. Damper or amortisseur windings are also employed on synchronous motors for starting the motor as an induction motor (see Fig. 12.14*b*).

31. What is the difference between dc and ac generator paralleling for engine-driven units?

Answer. The requirements for parallel operation of dc generators are that the (1) polarities of the generators must be alike, (2) voltage of the incoming machine must be the same as the running machines, (3) field rheostats must be used to adjust the load between the machines, and (4) the machines must have compatible operating characteristics that allow load changes to divide between the machines. A second ac generator is paralleled with a generator on the line as follows for engine-driven units: The two generators must be synchronized. (1) Voltages must be equal, (2) they must be in phase, and (3) the frequency must be the same. To parallel, do the following: (1) Bring the second machine up to speed with switches open. (2) Cut in full resistance on the field rheostat. (3) Adjust exciter voltage to the normal excitation voltage. (4) Close the field switch. (5) Adjust the generator field resistance until the voltage is the same as on bus bars from the other machines. (6) Synchronize with the lamps or synchroscope. At the instant of synchronization, turn the main switch. (7) Adjust the rheostat until cross flow is at a minimum. (8) Adjust the governors of the engines so that load is equally divided.

32. If you use lamps to synchronize ac generators, should you cut in at "light" or "dark"?

Answer. In the United States, dark is preferred because judging maximum brightness is difficult. Besides, the light period is much longer than the dark period. The danger of synchronizing dark is that the filament of the synchronous lamp may be broken at the moment of synchronization and may cause an accident. To prevent this, two synchronizing circuits or two lamps in paralleled circuits are used.

33. How is an ac generator cut out when it is in parallel with others?

Answer. (1) Cut down on the governor until the no-load point is reached. (2) Adjust the resistance in the field circuit until the armature current is at its minimum. (3) Open the main switch. *Caution:* don't open the field switch before the main switch. If the field is disconnected, a heavy current will flow between the two armatures. (4) Small units may be disconnected by opening the main switch and allowing the other units to furnish the full load.

34. What is a synchroscope?

Answer. A synchroscope instantaneously indicates phase differences and is used for synchronizing polyphase generators. A hand moves on a dial to indicate whether the phase is ahead or behind the bus-bar load. When the machine speed and phase are right, the synchroscope hand indicates the moment to cut in.

35. What are the ac-motor types?

Answer. AC motors are classified as (1) universal, (2) single-phase, and (3) polyphase. Polyphase motors may be squirrel-cage, wound-rotor, or synchronous types.

36. What is a wound-rotor induction motor?

Answer. A wound-rotor induction motor operates on the same principle as does a squirrel-cage motor, except that wire windings are used on the rotor. The ends of these windings are led out to slip rings (one per phase) where they are further connected to external banks of resistance through rheostat control. The external resistance permits cutting in and out of resistance to the rotor for speed control and/or higher starting torque.

37. What is the difference between an induction motor and a synchronous motor?

Answer. An induction motor has a revolving electric field in the stator coils. This is produced by the alternating current, which is usually of 60 Hz. The revolving field induces a magnetic reaction with the rotor windings (or bars) and causes the rotor to spin.

A synchronous motor has outside dc excitation to produce a magnetic field. Coils are mounted on a revolving spider frame, and excitation is led to them through brushes and slip rings mounted on the rotor shaft. The stator windings receive current from an ac source in phase with the generator. The magnetic reaction of the alternating current flowing in the stator with the direct current in the rotor causes the rotor to spin.

38. How is a three-phase motor reversed?

Answer.   You can reverse a three-phase motor by interchanging any two stator leads.

39. What are reverse-phase relays? Where are they compulsory?

Answer.   They are relays that will trip out if the phase leads are reversed ahead of the control board of a multiphase (three-phase) motor. They are compulsory on electric elevators because the controls would act one way and the hoist motor would act in the other direction if the phase leads were accidentally changed.

40. What is part-winding full-voltage starting?

Answer.   In this type of starting, the stator windings have at least two windings. One set is used to start at full-line voltage which, in effect, acts as a small motor. As the motor comes to speed, the second set of windings is parallel to the first and the motor accelerates smoothly since there is no open switching period to interrupt the current.

41. What is a delta-star starting switch?

Answer.   A delta-star starting switch uses six motor leads instead of three; both ends of the phase windings are brought out. The starting side is connected in star and the running side is hooked up in delta. The effect of starting in star is to reduce the starting current, since the voltage is $V/\sqrt{3}$. When the motor is brought up to speed, the switch is moved to the "run" position, giving a voltage of V per phase. It is assumed that the starting in rush is 5 times full load and that the torque is $1\frac{1}{2}$ times normal full-load torque when an across-the-line start is made. The delta-star method reduces starting current to 1.67 times full load and the starting torque is one-half normal.

42. An induction motor operates at 60 cycles and has a speed rating of 1750 r/min. What are the number of poles, slip, and rotor frequency?

Answer.   The nearest synchronous speed is 1800 r/min:

$$\text{No. of poles} = \frac{60 \times 120}{1800} = 4 \text{ from pole, speed, and frequency equation}$$

$$\text{Slip is } \frac{1800 - 1750}{1800} = 0.0277 = 2.8 \text{ percent}$$

$$\text{Rotor frequency} = \text{slip} \times \text{stator frequency} = 0.0277 \times 60 = 1.67 \text{ cycles per second}$$

43. What are two common insulation resistance tests?

Answer.   Using a megger to take spot readings of insulation resistance and a time-run absorption test with a motor-driven megger.

44. How may shorts between turns be found?

Answer. By taking impedance or resistance tests and comparing these readings to the values when the machine was new or after being rewound.

45. What is a thermal overload relay?

Answer. A thermal relay is a protective device that operates on the principle of the heat expansion of metals. A heating coil element is used to heat the relays. Since this is proportional to the current, an overload will actuate the relay. Time is involved, and an overload of short duration won't cause enough heat to actuate the relay. This is called the inverse time-limit effect.

46. A synchronous motor shows the following rating: 400 hp, 2200 V, 60 cycles, Y connected, three phase, 0.8 P.F., and efficiency of 0.93. What is the motor's kVA input rating?

Answer. It is:

$$\text{kVA} = \frac{400 \times 746}{0.8 \times 0.93 \times 1000} = 401 \text{ kVA}$$

47. What is a synchronizing device?

Answer. A synchronizing device operates when two ac circuits are within the desired limits of frequency, phase angle, and voltage to permit paralleling of the circuits, such as two generators.

48. What is a differential protective relay?

Answer. Differential protective relays are used to protect an alternator on a percentage or phase angle difference of two currents in a three-phase system. The relay thus picks up a fault if one occurs in one of the windings in a three-phase generator or motor.

49. What will prevent an alternator from developing full voltage?

Answer. The problem could be (1) a defective voltage regulator, (2) low speed, (3) overload.

50. Why is a field discharge resistor required on an ac generator?

Answer. A suitable resistor is required across the field switch or breaker just before the switch or breaker is opened in order to keep the voltage induced in the rotor by the sudden change in field current when a switch is opened to within allowable limits. The resistor should be regularly inspected to make sure it is not burned out or open circuited. An open field discharge resistor can subject the rotor winding to high voltage that can puncture the insulation, which can produce a failure to ground, or between turns.

CHAPTER 13

ELECTRIC POWER DISTRIBUTION, TRANSFORMERS, RECTIFIERS, AND LIGHTING

Facility services include operating and maintaining the electric energy that is distributed through a facility by means of transformers, cables or busway systems, switchboard equipment, and the many protective devices for the circuits that may be involved in a facility. Electric power is essential in all types of facilities for supplying services, such as refrigeration, air conditioning, and heating, in addition to process uses in industrial plants.

The major electrical equipment encountered by plant service operators is:

1. Purchased power substations and associated transformers and switching equipment or power that is generated in-house and distributed, which may include emergency standby equipment, such as a diesel-generator

2. Secondary distribution equipment such as feeders, transformers, switchgear, and protective devices

3. Lighting circuits and associated equipment

4. Electric and electronic control and instrumentation systems

5. Communications equipment such as telephones, paging, and intercommunications

6. Auxiliary systems that could include fire alarms, burglar surveillance and alarms, and guard station systems

7. Process electrical equipment such as batteries, rectifiers, elevators, electric tow trucks, conveyors and cranes, welding machines, and associated control equipment

8. Outdoor lighting and yard surveillance systems

NEC CODE REQUIREMENTS

The *NEC* has many requirements that involve facility distribution equipment (see Fig. 13.1). This code should be referred to when making any additions or repairing distribution equipment. Figure 13.1 lists the subject matter of the code articles that should be

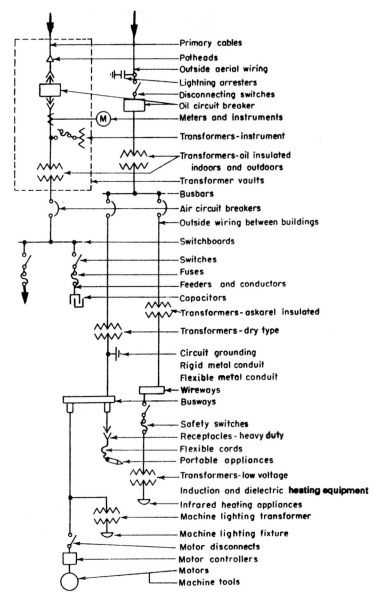

FIGURE 13.1 The NEC has many specific requirements for electrical distribution equipment that are important in any additions or repairs to be made in a facility. Check the code articles per above subject matter.

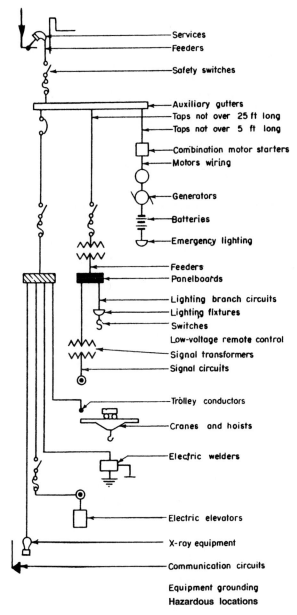

Services
Feeders
Safety switches
Auxiliary gutters
Taps not over 25 ft long
Taps not over 5 ft long
Combination motor starters
Motors wiring
Generators
Batteries
Emergency lighting
Feeders
Panelboards
Lighting branch circuits
Lighting fixtures
Switches
Low-voltage remote control
Signal transformers
Signal circuits
Trolley conductors
Cranes and hoists
Electric welders
Electric elevators
X-ray equipment
Communication circuits
Equipment grounding
Hazardous locations

FIGURE 13.1 (*Continued*)   The NEC has many specific requirements for electrical distribution equipment that are important in any
additions or repairs to be made in a facility. Check the code articles per
above subject matter.

checked for requirements. Because of the variety and extent of electrical distribution systems, only major items that could affect a facility service operator will be reviewed here and then only in a general manner.

A *one-line diagram* is extremely useful in laying out the distribution system that a facility has, and it can be very important when tracing power interruptions. Figure 13.2 shows a typical one-line electrical diagram for a manufacturing plant. It shows incoming voltages, transformer sizes and secondary voltages, the location of oil switches or circuit breakers, and the load that is being supplied in the facility.

CONTINUITY OF SERVICE

Plant service operators should draw a one-line electric diagram for their facility in the event none is available because this will help them understand the size and scope of the electrical distribution system. The type of electrical distribution system in a facility is influenced by the value that may have been placed on the dependability and flexibility that may be required in the facility operation. For example, some facilities may suffer severe losses if a substation suffers any lengthy outage. Continuity of service is thus essential, and a redundant distribution system is required. One-line diagrams are very useful in locating weak links in the distribution system. Prudent management decisions would dictate obtaining, for example, spare circuit breakers for those circuits that are vital to the facility's operation. The location of spare transformers, if not on the premises, is another vital concern if the consequential damage can be severe from the outage of a transformer. The term *emergency planning* is used extensively for those facilities that analyze their electrical systems and plan what to do if the service or part of it is interrupted. Some facilities, such as hospitals with operating rooms and nursing homes of certain sizes, are required by law to have emergency generators.

The power supply shown in Fig. 13.2 is classified as a *simple radial system* with one primary service connection and one transformer supplying the service. A fault in the feeder or distribution transformers and switchboard equipment will result in an area outage. Many facilities with single objects supplying area loads accept the risk of outages if repair or replacement organizations are nearby to respond to them. Where continuity of service is essential, most facilities seek and install a more flexible system, starting with two or more primary feeders of power to the plant. *Tie breakers,* when used to connect two power supplies may be operated upon or closed. Using them in the closed position at all times results in increased short-circuit currents on feeders, thus requiring larger feeder breakers. For this reason, many facilities normally leave tie breakers open and only close them to shift load from one source to the other.

Feeder distribution systems may be overhead or underground. Many older industrial plants have overhead lines supplying electric power to the different sections of the plant by means of (1) bare or weatherproofed cables installed on pole lines or towers, and (2) catenary suspended conduit, weatherproof wireways or busways carried between buildings.

Underground feeders may be (1) armored or synthetic-covered cable placed in a trench that is refilled, (2) underground raceways with manholes at intervals for repair access, and (3) exposed cables in tunnels or enclosed in raceways.

The advantage of overhead distribution is primarily cost. The disadvantages are that (1) weather conditions can knock out the service, (2) poles and wires may affect traffic, and (3) a downed wire may pose a danger to people. Underground cables have two classifications: (1) metal or lead covered and (2) nonmetallic.

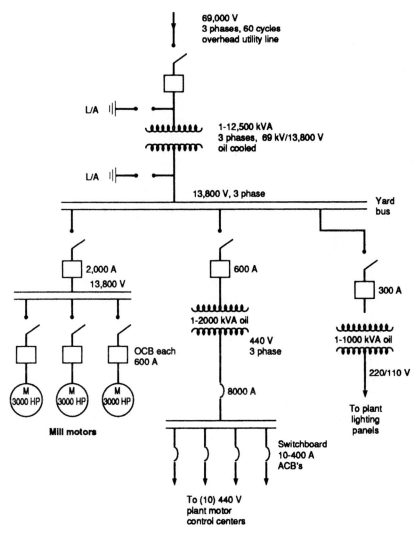

FIGURE 13.2 Simple radial-type one-line electrical diagram assists in understanding facility distribution systems.

Lead-covered cable prevents moisture, acids, and alkalies from reaching the insulation underneath, which would affect the dielectric strength of the cable. It is resistant to corrosion, and the lead sheath provides good grounding. Lead-sheathed cable should be inspected for vibration damage where it is subjected to overhead traffic. It requires potheading and wiped sleeve joints.

Nonmetallic covered cables are less expensive than metallic-covered cable. They are not as susceptible to mechanical damage and do not require wiped joints or potheading at 5 kV and under. One disadvantage of nonmetallic cables made of rubber is that they should not be exposed to oil or direct sunlight unless they are protected by a suitable synthetic covering.

Interior wiring systems should comply with the NEC requirements, and the reader should refer to this code for details on the many interior wiring systems that are permitted in distributing electric power within a premises.

In wet and damp locations, the wiring raceway systems, including all boxes and fittings, must be of watertight construction and be of corrosion-resistant material.

Hazardous locations are also covered by the NEC. Facilities with processes that create fumes that when mixed with air can be very explosive and dust-laden atmospheres that also may be dangerously flammable require electrical installations that will minimize the chance for explosions and fire. The *NEC* has classifications for hazardous locations, and it should be consulted for requirements.

It is important to provide extra maintenance to electrical equipment in hazardous areas in order to check on the tightness of enclosures, such as switch boxes and fittings, and to practice good housekeeping where dust or fibers may be present. Adequate ventilation should always be maintained so that no combustible mixtures may develop.

Tracking power consumption in a large facility can be performed by installing meters per department, service, or manufacturing section. This will assist in tracking power losses and inefficient use of power and will also help in assigning the appropriate percentage of power costs to the subdivision in the facility. The kilowatthour meter is usually used, and if supplemented by an ammeter and voltmeter, the power factor can also be determined for that particular department or division.

POWER FACTOR PENALTY

Utilities apply power factor penalties for a low power factor; therefore, there is an economic advantage in maintaining the power factor as high as possible. In addition, a low power factor inside the facility causes these additional costs or requirements:

1. A low power factor requires heavier cables and conductors to carry the higher amperage required per kilowatt. This may limit loading, especially if any additional load due to expansion is being considered.

2. The I^2R losses on the distribution system are higher per kilowatt carried.

3. Higher current per kilowatt will also cause higher volt drops (IR).

The voltage drop may affect the operations of larger motors in starting and in operation since more current will be needed at lower voltage to produce the same horsepower output.

The power factor can be improved by using synchronous motors on larger-horsepower drives and by strategically installing capacitors to provide leading reactive kilovoltamperes. Synchronous motors perform the double duty of driving a load and, if properly selected, improving the power factor of the facility, especially in those facilities with a large induction motor load. Figure 13.3 illustrates several methods of installing capacitors or synchronous motors for power factor correction.

Also used are *synchronous condensers,* which are similar to synchronous motors; however, they are manufactured primarily to operate without mechanical load. They

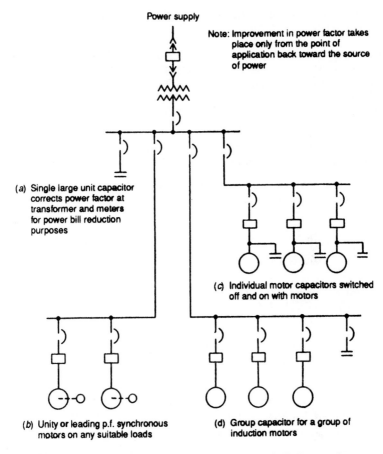

FIGURE 13.3   Various methods that can be used to improve a facility's power factor.

concentrate on correcting the power factor, or regulating voltage, by supplying the kilo-vars necessary to raise the power factor in a facility.

DEMAND FACTORS

Utilities use a demand charge in billing customers that is somewhat related to the facil-ity's ratio of the maximum use or demand of power to the total connected capability, expressed as a percentage. The utilities must supply equipment to meet the maximum demand in a facility, but there is a charge for this, usually based on the peak load in a month that lasts about $\frac{1}{2}$ h. Contracts can require that this demand, once established, must be paid for in the next 12 months. In other contracts, there is only the charge for the month incurred.

Operation of the electrical distribution system may have a bearing on the demand factor. For example:

1. Must all equipment run simultaneously and under what load? It is possible sometimes to alternate the operation of heavy equipment or operate at night and perform finishing operations during the day. This will limit the demand on the electric distribution system.
2. Can some equipment be started at different times so that the in-rush starting current can be limited, thus reducing demand?
3. Can load densities from department to department be reduced by varying the manufacturing steps?

It is necessary to carefully analyze how much electric power is required for the different facility loads and how the method of operation affects the demand factor.

LIGHTNING HAZARD

Damage from lightning is especially possible on electrical distribution equipment that is connected to overhead or aerial power lines. Protective equipment is installed to minimize the possible damage from lightning, and this equipment needs inspection and testing to make sure it is functional. Protection usually is based on the fact that surge voltages on lines are limited to maximum values so that rated flashovers over the insulation to ground are not exceeded. The conditions that lightning arresters must meet include the following:

1. Voltage across the arrester during a discharge should not exceed the value that is safe for the insulation that is to be protected.
2. The arrester should drain the stroke while not interrupting the electric circuit protected.
3. The arrester's characteristics should remain constant, and its normal life should compare with the equipment it is protecting.

Lightning protection is provided by:

1. An overhead fourth wire, or ground wire, that is grounded at appropriate intervals of length. It is important to check the ground resistance of any lightning arrester per the NEC requirements; it is usually a maximum of 5 Ω to ground. It should be checked during dry spells when ground resistance may be the highest.
2. Lightning arresters that provide a path to ground without lightning flashing over the insulation of line equipment. Line-type arresters are available that have gaps sized for sparkover to occur at certain voltages and thus lead the stroke to ground. It is necessary to have a sparkover voltage that is below the basic insulation level of the electric distribution system.

Two types of arresters are:

1. The *valve-type,* which employs a nonlinear resistance called a valve element in series with the spark gap. The valve element permits the flow of surge current at the instant the gap sparks over from a voltage surge.

2. *Expulsion-type* arresters use an arc-extinguishing chamber in series with the gaps to interrupt the current after the gaps are sparked over. Arresters are also rated by voltage classes per national standards.

GROUNDING

There are two types of grounding employed in electrical distribution systems:

1. *Equipment grounding* where the non-current-carrying metal parts or frames of electrical equipment and wiring enclosures are grounded so that a person touching a grounded electrical device as a result of internal electrical fault does not receive a shock as current tries to flow through the person's body to ground. The metallic parts that must be grounded include conduits, sheaths or armor of cables, frames or cases of switches, motor starters, controllers, fuse boxes, junction boxes, switchgear, transformers, generators, motors, and similar electrical equipment. Equipment grounding is a life-safety matter and is extensively covered by Article 250 of the NEC. Many people are electrocuted because of defective, or missing grounds on metallic parts of electrical appliances. It is certainly one of the functions of facility service operators to check to make sure proper frame grounding still exists on electrical equipment that they may maintain.

2. Electric circuit *system grounding* clears or immediately detects a fault before the fault develops into a double-line-to-ground fault, which would cause a facility short circuit to develop that could burn up other equipment due to the in-rush of current to feed the double fault. For example, an ungrounded delta system supplying a motor load may have many motors fail when a double ground develops on the feeder circuit. Arcing faults cause surge voltages that will destroy weak insulation on electrical equipment. System grounding conductors must provide a low-impedance path to facilitate the operation of protective devices when an accidental ground occurs. The NEC should be referred to on the specific requirements for system grounding.

System grounding and the setting of protective devices may require a coordinated effort with the utility since it may also affect their relays.

Ground detectors, such as lamps that light up when a ground on a circuit develops, have been used to warn of a grounded condition somewhere on the circuit, but they are often ignored and do not locate the fault.

The NEC requires secondary ac systems supplying interior wiring and systems to be so grounded that the maximum voltage to ground does not exceed 150 V.

A major concern on electric systems has been the 440 V Y-connected distribution systems that may have a low-amperage simmering-type ground fault that the normally sized short-circuit current-rated breaker may not interrupt due to the high amperage setting of the breaker. Article 230-95 of the NEC now has a *ground-fault protection* requirement that states,

> Ground-fault protection of equipment shall be provided for solidly grounded wye *electrical services* of more than 150 volts to ground, but not exceeding 600 volts phase-to-phase for

each service disconnecting means rated 1000 amperes or more. The ground-fault protection system shall operate to cause the disconnecting means to open all ungrounded conductors of the faulted circuit. The maximum setting of the ground-fault protection shall be 1200 amperes, and the maximum time delay shall be one second for ground-fault currents equal or greater than 3000 amperes.

There have been many switchboard failures from low-amperage simmering faults on 440-V Y-connected systems because the breakers were set too high for the low-amperage fault and thus only tripped when a major short circuit developed. This is the reason the NEC adopted the ground-fault protection of equipment article. It will be the ground fault relay that will open the breaker and not the set-tripping amperes. Study your system to see if it needs ground-fault protection.

The NEC also requires ground-fault interrupters for certain occupancies that pose an immediate danger, such as swimming pools. The NEC should be referred to for specific requirements.

Surge arresters are used to limit surge voltages by discharging or bypassing surge current, and they also prevent continued flow of follow current while remaining capable of repeating the surge arrester function. Surge arresters are applicable to circuits that have delicate electronic equipment, computers, and other sensitive electrical equipment that may be damaged by circuit voltage disturbances. Article 280 of the NEC lists some of the requirements in installing surge arresters, including grounding.

Short-circuit and protective device coordination studies are the modern method of determining the size and location of protective interrupting devices in the larger facilities that may be required for clearing a fault on an electric circuit. There are two main considerations in selecting the proper protective devices; (1) Does the device have the permissible overcurrent rating for the equipment it is protecting per the NEC requirements so as to prevent damaging effects to that equipment? The NEC provides setting ranges on the different electrical equipment that must be protected against overload and short circuits. (2) The protective devices must be capable of withstanding, without injury, the high momentary fault currents that may exist before the device opens the circuit. In other words, a breaker, for example, cannot be destroyed by the overcurrents that may develop in the circuit it is protecting.

FAULT CURRENT CALCULATIONS

It is possible to calculate the possible fault current in a circuit at a given voltage by determining the impedance of that circuit. It has been simplified to some extent but still requires good one-line electrical diagrams that list ratings, voltages, and impedances of the connected equipment. The method used involves the calculation of the maximum symmetrical short-circuit current, which is then multiplied by established factors to obtain the total short-circuit current to be interrupted and also the momentary peak current which the protective device must withstand. Therefore, modern breakers have nameplates to show their continuous current rating and their short-circuit rating.

Figure 13.4 illustrates a typical curve of a short-circuit current on a 60-cycle system. Note how its value diminishes with cycle time.

An example for calculating the approximate short-circuit current is given to show the method that is used.

The reactance or impedance of a generator or transformer is expressed in *percentage* values based upon the rated kilovoltamperes instead of in ohms. For example, 5 percent

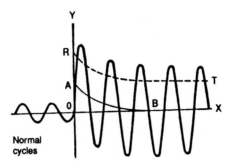

FIGURE 13.4 Typical jump in a 60-cycle current when a short circuit develops in a circuit. Note peak values are reached within one-half a cycle.

reactance in a circuit means that, when *rated current* is flowing, the voltage drop due to the reactance is 5 percent of the rated voltage. It can also be used to calculate *approximately* the current that a transformer may deliver to a short.

Example. A 5000-kVA transformer is connected to an unlimited utility power supply at 13,800 to 440 V secondary, three phase, 60 cycle with 7.5 percent impedance. What is the possible short-circuit current this transformer can deliver on the 440-V side if all other reactance on the 440-V side is not considered? First calculate *rated current* on the 440-V side:

$$I = \frac{5000 \times 1000}{1.732 \times 440} = 6561 \text{ A}$$

Possible short-circuit current is 6561/0.075 = 87,480 A.

A 1.25 factor is applied to the above short-circuit current for the *dc component* that exists in the first half cycle, thus making the possible short-circuit current 1.25 × 87,480 = 109,350 A approximately.

Motor components contribute to short-circuit currents as the stored energy at time of failure feeds the fault. For estimating purposes, the approximate symmetrical values of short-circuit current contributed by motors at an instant ½ cycle after the short happens is 3.6 times full load current rating of induction motors and 4.8 times rated full load current for synchronous motors. The literature indicates that to estimate the motor contribution to a short, the following can be used:

1. For a transformer supplying 208/120 V, the motor amperes to be added is 2.5 times the calculated *rated amps* of the transformer.

2. For a 240- to 480- or 600-V secondary transformer, the motor amperes to a short can be estimated as 5 times the calculated *rated amperes* of the transformer.

Example. What would be the total short-circuit amperes possible for the previous problem if the motor contribution is considered? The rated amperes for the 5000 kVA transformer on the 440-V side was 6561 A. The motor contribution would be 5 × 6561 = 28,255 A. The combined short-circuit current possible is thus 109,350 + 28,255 = 137,605 A.

The nearest breaker with at least 137,605-A *interrupting capacity* should be installed on the secondary side of the transformer. The *continuous rating* of the breaker should at least equal the rated amperes of the transformer, or the nearest standard size to the 6561 A.

Short-circuit studies are made on a circuit-by-circuit basis in order to determine the protection required in addition to NEC requirements. They are also made to coordinate the protection so that a fault is cleared in the circuit it occurs and not by the operation of the main supply breaker.

When a breaker trips due to a fault, it should be inspected to make sure contacts are not burned and all moving parts are still functional. *High-capacity fuses* have been developed to protect circuits for possible short-circuit currents. They have to be replaced if they open a circuit due to a fault because the element inside the fuse has melted from the short-circuit current.

TRANSFORMERS

Transformers are used extensively in electrical distribution systems, but because they transform electric energy from one voltage to another in a static manner, they are often neglected. Serious fires can result from internal electrical faults, and of course, the outage can affect facility services and production. They are important pieces in the electric distribution of power and, in some cases, in converting alternating current to direct current in a static manner and thus require the attention of facility service operators. Maintenance includes periodic checks on loading, inspecting connections and casings for tightness and corrosion, making sure that cooling flow is adequate, determining if indicating and protective devices are in working order, checking liquid cooling levels and the dielectric strength of the cooling liquid, performing periodic winding insulation checks or making certain they are performed, and reconditioning or repairing any deficiencies found during inspections or tests. The trend is to employ firms that specialize in transformer testing; however, facility service operators are still responsible for having the tests performed at proper intervals and to even perhaps interpret readings.

Principles

A brief review of some basics of *transformer principles* will assist or refresh the reader's knowledge of this important electrical machine.

Transformers operate on the principle that energy may be transferred by *induction* from one set of coils, called *the primary,* to another set of coils, called *the secondary,* by means of a varying magnetic flux, provided that the two windings are on a common magnetic circuit. An electromotive force (emf) is induced by a change in flux, such as occurs in alternating current. The magnetic circuit is formed by an iron core that links the windings together. Since the flux is the same for the two windings, relationships are established between windings that are based on numbers of turns in the primary and secondary windings (see Fig. 13.5). The emf is the same *per turn* in each winding, and this means that the *total* induced emf per winding is dependent on the number of turns in that winding. The relationship between primary and secondary voltage is then

$$\frac{E_1}{E_2} = \frac{N_1}{N_2}$$

with E_1 and E_2 = the primary and secondary induced emfs, and N_1 and N_2 = the number of turns in the primary and secondary windings.

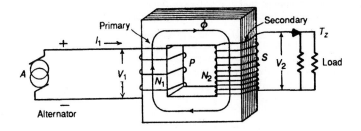

Single-phase relationships, unity power factor, $\dfrac{I_1}{I_2} = \dfrac{V_2}{V_1} = \dfrac{N_2}{N_1}$

FIGURE 13.5 By means of magnetically induced voltages and currents, energy can be transferred from one voltage to another in a transformer.

Another relationship is the fact that the ampere turns in the primary equal the ampere turns in the secondary, which results in this relationships:

$$N_1 I_1 = N_2 I_2$$

with I_1 and I_2 = the primary and secondary winding currents, amperes. By transposing,

$$\frac{I_1}{I_2} = \frac{N_2}{N_1}$$

The primary and secondary currents vary inversely from the respective number of turns. As can be noted by equating to the common-number-of-turn relationship, neglecting transformer losses and assuming unity power factor,

$$V_1 I_1 = V_2 I_2$$

$$\frac{I_1}{I_2} = \frac{V_2}{V_1} = \frac{N_2}{N_1}$$

The above relationships are useful in determining approximate values. Transformers have leakage reactance because all the flux produced by the primary does not go through the secondary; some flux completes its magnetic circuit by passing through the air rather than around the core. Transformers have high efficiencies with losses being 1 to 3 percent of the rating.

Example.   A three-phase transformer is rated 13,800 V primary and 2300 V secondary. An ammeter on the 2300-V side shows a load of 627.6 A. What would the amperage load be on the primary if losses were neglected and unity power factor is assumed? Use *the equation*

$$I_1 V_1 = I_2 V_2$$

therefore,

$$I_1 = \frac{I_2 V_2}{V_1} = \frac{627.6 \times 2300}{13,800} = 104.6 \text{ A}$$

What is the kilovoltampere load in above example? Either the primary or secondary current and respective voltages can be used. For the *primary:*

$$kVA = \frac{13,800 \times 104.6 \times 1.732}{1000} = 2500 \text{ kVA}$$

For the *secondary:*

$$kVA = \frac{2300 \times 627.6 \times 1.732}{1000} = 2500 \text{ kVA}$$

Step-up transformers have more turns on the secondary windings than on the primary windings and thus increase the voltage.

Example. A power plant has a 10,000-kVA 3-phase transformer that steps up the generated voltage from 4160 to 13,800 V. How many turns are required on the secondary if the primary has 2000 turns, neglecting losses and power factor and assuming Y-Y connection? Phase voltages must be used to calculate turns per phase:

$$V_1 = \frac{4160}{1.732} \qquad V_2 = \frac{13,800}{1.732} \qquad \text{or} \qquad V_1 = 2402 \text{ and } V_2 = 7968 \text{ V}$$

Use

$$\frac{V_1}{V_2} = \frac{N_1}{N_2} \qquad \text{or} \qquad \frac{2402}{7968} = \frac{2000}{N_2}$$

$$\text{then} \qquad N_2 = 6634 \text{ turns}$$

Step-down transformers decrease the primary voltage to the secondary voltage, and thus the secondary has less turns.

Transformer classification is by application as follows:

1. Power transformers 500 kVA and over are used in power plants and in substations to distribute generated power.

2. Distribution transformers step down voltages for residential and commercial use, usually 600 V or less and under 500-kVA size.

3. Instrument transformers are further classified as potential or current and are used to serve low-voltage meters and relays so that line voltages, or large currents in the case of current transformers, are avoided on these devices.

4. There are numerous specialty transformers for specific applications such as rectifiers, arc lamps, furnace transformers, and other special industrial and medical applications.

Core and Shell Construction

Transformers are also divided by their core- and/or shell-type arrangement (see Fig. 13.6). In the core type, the windings surround the iron core. The core is in the form of a hollow square made of special sheet-steel laminations. By having both a primary and a secondary on each leg as shown in Fig. 13.6a, the leakage flux is reduced to a small value, and good mutual inductance is obtained between the primary and secondary. Note also that the low-voltage winding is placed near the iron laminations, thus reducing the insulation needed between the iron and the windings. By placing the high-voltage wind-

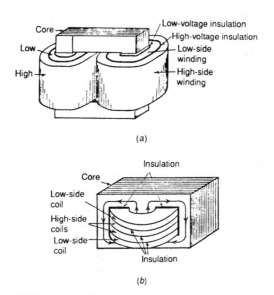

FIGURE 13.6 (*a*) Core-type transformer construction; (*b*) shell-type transformer construction.

ing outside and around the low-voltage winding, only one layer of high-voltage insulation is needed, that between the windings.

In the shell-type construction, Fig. 13.6*b,* the iron surrounds the windings, with the core having the form of a figure 8. The coils are made in the shape of pancakes with the primary and secondary stacked so that each primary is adjacent to a secondary, and this reduces the leakage flux. The low-voltage coils are placed next to the iron in order to reduce the insulation required.

In general, the core type has a lesser cross section of iron and, therefore, can have a greater number of turns. The core type is used for high voltage because there is more space for insulation. The shell type permits better bracing of the windings and thus is less easily moved by high emf's that may occur during a short circuit.

The improved methods of producing silicon steel have increased the permeability in the direction of grain orientation, and this has resulted in better assemblies of cores and windings to avoid flux transverse to the grain. For example, GE uses a spirakore assembly, Westinghouse a hypersil core, and others make a line material round-wound transformer.

Methods of Cooling

Transformers are also identified by the method used to cool and insulate them from the energy that is lost inside the windings. The nameplate may show different kilovoltampere ratings based on the method of cooling. For example, a natural-cooled transformer may have a rating of 30,000 kVA with temperature rise of 55°C and with the class letters OA/FA/FOA on its nameplate may have the following kilovoltampere ratings depending on how it is cooled:

1. The letters *OA* indicate it is self-cooled by casing fins that dissipate the heat by natural air circulating around the casing. Its rating for the class OA would be 30,000 kVA with a 55°C rise. The letter O indicates oil cooling.

2. The OA/FA designation indicates that it is self-cooled with oil but can also use forced-air cooling on the fins and casing with fans. The transformer now would be rated 40,000 kVA with a 55°C rise.

3. The FOA designation indicates forced-oil cooling and forced-air cooling that permit a 50,000-kVA rating with a 55°C rise in temperature are possible.

Remember that the dial-type temperature indicator supplied with transformers indicates an *operating* temperature, *not a temperature rise*. In order to obtain the temperature rise, subtract the ambient temperature from the indicated operating temperature. For example, if the transformer thermometer reads 65°C in an ambient temperature of 30°C, the temperature rise is 35°C.

The forced-air letter designation (FA) indicates fans are used to give additional cooling, while forced oil (FO) indicates pumps are used to promote oil or liquid cooling circulation inside the transformer.

Dry-type transformers may be cooled by air or may be gas insulated and cooled by inert high-dielectric gases such as Freon, sulfur hexafluoride, or perfluropropane. Figure 13.7 illustrates a dry-type air-cooled transformer. Dry-type transformers using forced air-cooling are also called air-blast dry type and have the letters AFA. Instrument transformers, except those used in high-voltage service, are of the dry-type construction. The great growth in the use of dry-type transformers for power and lighting loads is due primarily to the fact they can be installed indoors and do not need fireproof vaults if the size does not exceed 112½ kVA. If over this size, per the NEC the transformer must be installed in a transformer room of fire-resistant construction.

Oil-type transformers require fireproof vault installation per NEC requirements if they are installed indoors. Oil-type transformers have different casing arrangements to allow the oil to expand and contract with temperature changes, called *breathing*. Older units had gooseneck vents to the atmosphere that permitted gases to escape when the temperature rose and cool air to enter when the oil temperature dropped. Unless preventive measures were taken, this type of operation permitted moisture-laden air to be drawn into the transformer's oil, which eventually affected its dielectric strength. Oxygen causes the oil to oxidize forming sludge that blocks the internal cooling passages. In case of an electrical failure within the transformer, a fire erupts due to the presence of air. It is possible to construct a casing for smaller transformers that can withstand the pressure rise as the oil is heated. This is usually limited to units under 2000 kVA and 44,000 V. Free-breathing units may be equipped with a unidirectional dehydrating-type breather. Figure 13.8 shows some of the methods used to prevent oil contamination by outside air. The methods used include: (1) expansion tank with weatherproof breather (see Fig. 13.8a), (2) sealed tank (see Fig. 13.8b) with an inert gas such as nitrogen acting as a cushion for expansion and contraction of the oil with temperature changes, and (3) positive pressure automatic inert-gas cushion that keeps a constant pressure in the space above the oil, usually a nitrogen cushion, and which is connected to a nitrogen pressurized tank with a pressure-reducing valve supplying the cushion to maintain positive pressure over the oil at all times (see Fig. 13.8c).

Outdoor oil-filled transformers should be placed on a concrete base with a concrete dam enclosure high enough to catch all the oil from the transformer tank so that any spillage will not affect the water supply or spread a fire. Most insurance companies also require fireproof concrete walls between large transformers so that a fire will not affect the adjoining transformer(s).

NEC-classified *askarel transformers* were developed for their higher flammable temperatures than oil-filled transformers and at one time were treated like air-cooled units for indoor installation. These are now considered to contain polychlorinated biphenyls (PCBs) that are potentially harmful to human health if spills occur or if an

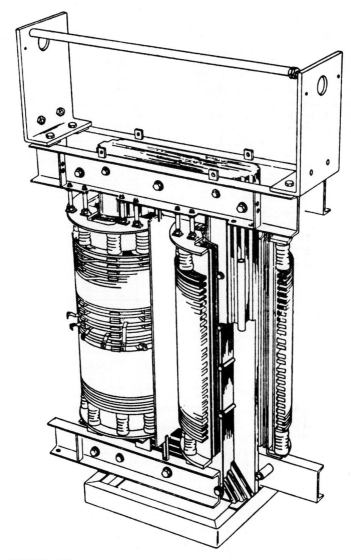

FIGURE 13.7 Interior view of a dry-type air-cooled transformer. (*Courtesy, Westinghouse Electric Corp.*)

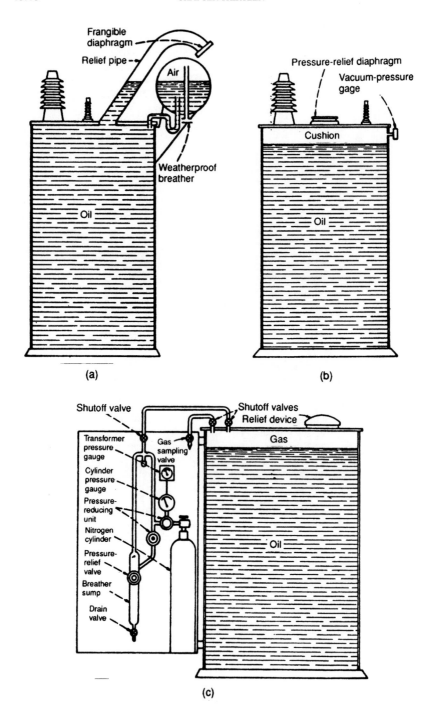

FIGURE 13.8   Three methods to provide for oil expansion and contraction. (*a*) Expansion tank breather; (*b*) sealed-tank with nitrogen cushion; (*c*) constant positive-pressure nitrogen cushion.

electrical short within the unit develops. Manufacture of PCBs was halted in the United States around 1979 due to federal government regulations. Existing units had to be tested and labeled PCB contaminated if they were above 50 ppm in the liquid medium. A program of draining, flushing, and refilling transformers with a cooling medium that lowers the PCB contamination to acceptable levels has been adopted, or the contaminated transformer must be completely replaced in order to meet federal requirements.

Oil Tests

Transformer oil testing is a standard maintenance test on facility transformers that are critical from a repair or replacement cost perspective or which will affect facility services or manufacturing operations. The frequency of testing depends on this criticality and also on the age of the transformer and its loading. The most common dielectric liquids are highly refined mineral oils, and these are in contact with the critical parts of the transformer. If the dielectric deteriorates chemically or physically, an outage can occur; therefore testing is performed to note any degradation from previous test results. The oil is analyzed after drawing a sample per ASTM guidelines, and the following tests are performed, usually by firms specializing in oil testing:

1. *Dielectric strength.* This is a spark-over test across two electrodes, spaced 0.1 in apart. The spark should not occur for any voltage across the spark below 25 kV. This test indicates the ability of an oil to withstand electrical stress without failure. Low dielectric strength usually indicates contamination with water and/or suspended solids from the insulation of the windings. Low readings are usually corrected by oil filtering or replacement. ASTM D887 provides a standard for this test.

2. *Neutralization number* per ASTM D974 of an oil is the measure of the acidic constituents in the oil and equals the milligrams of potassium hydroxide required to neutralize 1 g of oil. The test indicates oil oxidation. Oil starts to sludge above 0.2 acidity. New oil will read almost 0, while used oil may read between 0 to 0.2 and still be considered satisfactory. Above 0.2 the oil should be reclaimed by filtering, and if the reading goes above 0.5, oil replacement is recommended due to sludging.

3. *Interfacial-tension tests* are used to determine the strength of the surface film of the oil in contact with water. Its value is also affected by acidic compounds in the oil. New oil may have an interfacial tension up to 50 dyn/cm, while used oil may be as low as 15 dyn/cm, indicating that it may need reclaiming.

4. *Color* changes in the oil in service may show some type of contamination and are used as a diagnostic test that supplements the other tests described. The color of the oil is compared to ASTM color charts. Most testers look for the results of other tests if the color chart reads above 3.0.

5. *Gas chromatography* tests are being used to detect arcing damage, such as may occur in tap-changing switches. The test indicates the combustible gases present in the oil, such as hydrogen, methane, and acetylene. Permissible percentages are provided by the organizations specializing in gas chromatography. These tests are conducted annually on critical production units, such as electric arc-furnace transformers used in the steel industry.

The gas chromatography test is useful in detecting the small amounts of gas that can be produced in a sealed transformer because of brief overloads, internal corona, loose ground straps, or grounded core bolts. An analysis of the gas above the oil in larger transformers that have a gas-sampling accessory may disclose internal trouble before a serious failure occurs. The type and quantity of gases found assist in analyzing what the internal problem may be.

Tank leaks can occur from corrosion, gasket failures, weld cracks, and insulator bushing leaks that can admit water into a transformer and cause a ground arc failure. For this reason, it is important to inspect transformer tanks and repair leaks as found. To prevent corrosion, paint the tank surfaces with recommended paints to maintain the surfaces of the tank before serious corrosion results.

Electric Tests

In addition to oil tests, electric tests are made to detect weaknesses in the solid insulation of a transformer. Most transformer manufacturers and/or electrical testing companies perform these tests, as do some insurance companies that provide breakdown coverage. Here are some of the tests that facility service operators should be familiar with:

1. *Turn-to-turn tests* to detect internal shorts between winding turns. Resistance and impedance readings are taken and then compared to original specs.

2. *Megger insulation spot tests* to note if any ground or low value of insulation exists. Primary to ground, secondary to ground, and primary to secondary are the usual spot checks made.

3. *Dielectric absorption tests,* which are modified megger checks that are operated for 10 min with readings taken every 1 min and plotted. The polarization index is the ratio of the 10- to 1-min reading. Insulation resistance values of good insulation rise with time. Ratios have been established that determine if the insulation is poor, weak, or acceptable, with the latter being about 2.5.

4. *Power factor testing* is a method used to measure the power leakage through the insulation medium. Voltage, current, and power loss in watts are read from meters, and the power factor is calculated. The measurement is normally between the primary and secondary windings but may include primary and secondary to ground. Oil-filled transformers are judged satisfactory when the power factor so determined is 2 percent or less.

5. *AC Hi-pot tests* are made on new or rewound transformers and on used transformers to assure that they can withstand transient surges, switching arcs, and lightning strikes. For new transformers or those rewound, the test voltage is 2 times rated voltage plus 1000 applied primary to ground and secondary to ground at applicable rated voltages with the other windings grounded. For field testing used transformers, the test voltage recommended is 65 percent of that for a new transformer.

6. *Induced voltage tests* per ASA C57.1292 use high-frequency excess voltage across usually the low-voltage winding and then measure the induced voltage in the high-voltage winding. This test stresses the insulation—turn-to-turn, coil-to-coil, and phase-to-phase—and should be performed by the manufacturer since it is potentially destructive.

7. *Step-voltage tests* are made per winding, and at each step the leakage current is read. This is plotted, and if there is an abrupt jump in the leakage current, the test is stopped, and the voltage at which this occurs is then noted. If it is below the rated voltage, corrections are needed on the transformer. This test is usually conducted by manufacturers' representatives.

Protective Devices

The NEC requires *overcurrent protection* that is detailed in Article 450-2. Fuses or circuit breakers must have ratings or settings that are based on the transformer impedance and what the primary and secondary voltages are. For example, on the secondary side of a transformer rated 600 V or less, the fuse or circuit breaker must interrupt the circuit at 250 percent of secondary ampere rating. See the NEC requirements for proper overcurrent protection.

Lightning arresters are needed on transformers because transformer installations are subject to lightning surges. They should be installed as close to the transformer as possible and should have a maximum voltage drain rating that is within 125 percent of the normal line-to-ground voltage of the transformer.

Transformer tanks should be solidly grounded to eliminate the danger of electric shock. This also applies to wire fences that may surround a high-voltage installation.

Pressure relief is required on sealed units. A short inside the transformer tank releases gases and heat that cause rapid pressure rises that can rupture the transformer tank. *Gas detectors* and *pressure rise relays* are often installed to ring alarms or even trip the power supply to the transformer. Relief diaphragms provide the backup overpressure protection and are usually designed to rupture well below the allowable pressure that the transformer tank was designed for. If a pressure relief device is found to have discharged, it is a sign of internal arcing and should be further investigated.

On large, critical transformers, additional alarms are quite often installed: Liquid and hot-spot winding temperature alarms, air flow indicators, single-phase relays, and high-pressure and low-oil-level alarms are some that also may be installed.

Tap-changers are used to permit changes in voltage ratios. It is very important to determine if the tap changer is of the *load* or *no-load type*. The no-load tap changer must only be operated with the transformer deenergized because the tap changer contact points were not designed for load switching. Burned contacts and arcing will result if this is ignored. Some units have interlocks to prevent this, but the interlock should not be bypassed. The load-type tap changer permits manual or automatic tap changing while the transformer is in service.

Safety practices are essential when inspecting transformers. When the transformer is energized, it is important to determine if the tank, connected metallic parts, and wire fences are grounded per NEC requirements. It is essential not to touch or even come too close to conductors and other energized parts. Never assume a transformer is deenergized. Check the fuses or circuit breakers on the primary and secondary sides to make sure the switches are open or the fuses are pulled.

In general, only experienced personnel should work on transformers because of the inherent threat that exists with electric energy, and transformers may have extra-high-voltage ratings that need special care in inspection and maintenance work.

Three-phase transformers are extensively used because they are considerably lighter and occupy less space than three single-phase units of equal rating. However, if a three-phase transformer fails, the entire unit is out of service. With three single-phase units, it is quite often possible to reconnect the bank into an open delta system at 58 percent of the bank rating. Figure 13.9 illustrates a three-phase core-and-shell-type transformer.

Three-Phase Transformer Connections

The most common connections are; (1) Y-Y, (2) delta-delta, (3) delta-Y, and (4) V or open delta. Figure 13.10 illustrates these connections. The delta-Y connection is generally used to step up the voltage, while the Y-delta connection is used to step down the

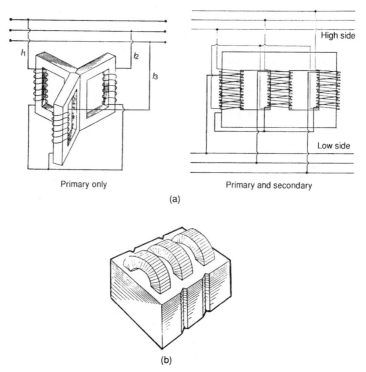

FIGURE 13.9 (*a*) Three-phase, core-type transformer connected ungrounded Y-Y; (*b*) three-phase shell-type transformer with coils and laminations.

voltage. The V or open delta connection provides one-third, or 58 percent, of the rated output that the full three-phase transformer supplies. Some utilities use the open delta on pole-line installation on the first installation when the load does not justify a full three-phase installation and then add a third transformer when the load increases.

The *Scott,* or T, connection makes it possible to transform three- to three-phase by means of two transformers, and it also permits converting three- to two-phase and two- to three-phase. Figure 13.11*a* illustrates the connections for three- to three-phase by using two transformers, one called the main and the other, the teaser transformer. By reconnecting two transformers as shown in Fig. 13.11*b,* a three- to two-phase three-wire system results.

Phasing transformers is an important consideration when connecting transformers, and the manufacturer's instructions should be followed because if the windings are not connected properly, a short circuit could develop. It is important to check the connections with ammeters and voltmeters before placing them in service.

In general, single-phase transformers that are to be operated in parallel must have the same voltage ratio and same impedance, and the polarity must be the same. Transformers to operate in three-phase must also have the same voltage ratio, about the same impedance (3.0 percent variation), and the same polarity.

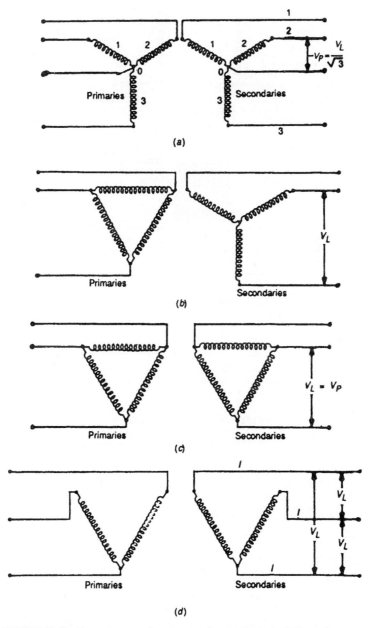

FIGURE 13.10 Three-phase transformer connection: (*a*) Four-wire Y-Y transformer connection; (*b*) delta-Y transformer connection; (*c*) delta-delta transformer connection; (*d*) V, or open, delta connection using two windings to obtain three-phases.

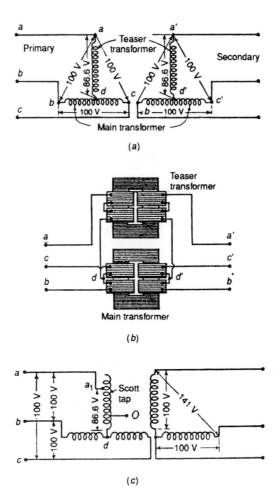

FIGURE 13.11 Scott, or T, connection of transformers. (*a*) Three phase to three phase; (*b*) three phase to two phase, four wire phase.

Transformer Nameplates

Modern transformers have nameplates to show the connections to be made for the voltages involved. All winding terminations are labeled—H_1, H_2, and H_3 to designate the high-voltage side—and low voltages are labeled corresponding—X_1, X_2, and X_3. *Polarity* is a term used to determine the instantaneous vector relationship between the primary and secondary voltages for single-phase transformers, and for a three-phase transformer, a vector diagram on the nameplate displays the angular displacement between the high- and low-voltage windings. A system of marking has been adopted to show additive or subtractive polarity. Figure 13.12 shows this system for single-phase

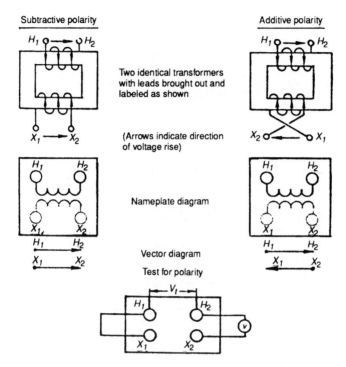

FIGURE 13.12   Subtractive and additive polarity for a single-phase transformer is shown on nameplate. (*Courtesy,* Power Magazine.)

transformers. ANSI standard C57.12.70 provides further details on combinations of different polarity transformers and their connections.

Instrument transformers are used to avoid connecting the meters directly into a circuit, thus exposing the instrument to either high current for a current transformer or high voltage for a potential transformer. Figure 13.13 shows typical instrument transformer connections. If a primary has a large current rating, the current transformer will have a straight conductor passing through the center of a hollow core as shown in Fig. 13.13*a*, while the secondary with several turns is wound around the laminated core. The secondaries of current transformers are usually rated at 5 A regardless of the primary current rating. For example, a 600-A current transformer has a ratio of 600/5 = 120.

Example.   A current transformer with a ratio of 5 to 2000 shows 3.5 A on its secondary. What is the current carried in the conductor it is measuring? The conductor current is 3.5 × 400 = 1400 A.

Potential transformers (see Fig. 13.13*b*) are similar to constant-potential transformers except that their power rating is small, usually from 40 to 500 W. For primary volt-

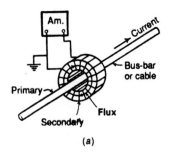

(a)

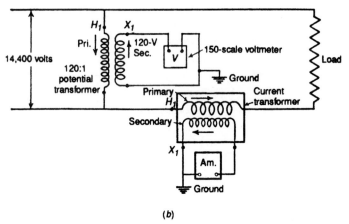

(b)

FIGURE 13.13 Instrument transformer connections. (a) Current transformer is connected around a conductor on the primary with the secondary reducing the amperes to be shown on a meter; (b) method of connecting current and potential transformers to a high-voltage circuit.

ages of 34,500 V and above, the secondaries of potential transformers are rated at 115 V, which supplies a voltmeter. Under 34,500 V, the secondaries are rated 120 V. Dry-type units are used below 5000 V; between 5000 and 13,800, they may be dry or oil immersed, and above 13,800 V, they are oil immersed. The secondary of a potential transformer must be grounded to avoid personal injury.

Example. A potential transformer is connected to a high-voltage line and is marked 14,400 to 120 V. A meter shows 19.2 V. What is the line voltage? The line voltage is 14,400/120 × 19.2 = 2304 V.

Transformer inspections should be recorded and stored in equipment files for future reference, especially for transformers that are critical to vital services or production. An outline of items to be inspected is usually the best procedure to follow in recording findings. One such outline is shown in Fig. 13.14. Each plant should draw up a list that will help detail the evaluation. In addition, if outside firms do the testing, they will issue a report of their own on their findings and conclusions.

SCOPE OF INSPECTION
(Dismantled, Operating, or Shutdown) ____
KIND (POWER, LIGHTING, REGULATOR) ____
(Oil/Air cooled) ____
RATING ____
 Manufacturer ____
 Kilovoltamperes ____
 Volts-Line (Transf. Primary) ____
  (Transf. Secondary) ____
 Phase and Frequency
  (Number, cycles per second) ____
 Reactance—Percent ____
 Year Built ____
NUMBER—Owner's ____
  —Manufacturer's Serial
LOCATION OF EQUIPMENT—Building ____
   —Room ____
   —Department ____
LOAD-TIME OF INSPECTION ____
 Kilovoltamperes ____
 Volts-Line ____
 Amperes-Line ____
 Power Factor—Percent ____
VENTILATION—Cooling Air ____
 Quantity—(Scant or Ample) ____
 Quality—(Injurious, Stagnant, Tolerable, Good) ____
TANKS AND FITTINGS
 Oil Leakage, Excessive Sweating, None ____
 Bushings and Terminals—Cracked, Chipped, Dirty, Normal ____
INSULATING OIL ____
 Quantity—(Scant or Ample) ____
 Quality—(Sludged, Hydrous, Dirty, Clean) ____
 Temperature—Degrees Cent. ____
 Dielectric Strength—Kilovolts ____
 (Electrodes, 1.0 in diam., 0.1 in apart) ____
 Date of Test ____
LINE SWITCH—Input Side ____
 Contact Surfaces—(Burnt, Blistered, Smooth) ____
 Temp.—Current Carrying Parts
  (Excessive, Hot, Normal) ____
 Switch Oil—Quantity—(Scant or Ample) ____
 Switch Oil—Quality—(Sludged, Hydrous, Dirty, Clean) ____
 Bushings and Terminals (Cracked, Chipped, Dirty, Normal) ____
PROTECTIVE EQUIPMENT
 Input Side (See NEC)
 Overload Relay Setting—(High, Low, Suitable) ____
 Fuses or Cutouts—Rating—(High, Low, Suitable) ____

FIGURE 13.14 An outline of items to be checked on a transformer will assist in recording conditions for evaluation and future reference.

GROUNDING/LIGHTNING PROTECTION
  Transformer Neutral—(Ungrounded, Defective, Adequate) ____
  Tank or Framework—(Ungrounded, Defective, Adequate)
  (Ground Resistance in ohms if measured) ____
FOUNDATION OR SUPPORT
  (Dangerous, Yielding or Rigid) ____
CONDITION OF CASE AND FRAME
  (Corroded, Cracked, Patched) ____
HEAT TRANSFER SURFACE (Clean, dirty)
  Oil circulating through heat exchange (yes, no) ____
  Water circulating through heat exchange (yes, no) ____
  Hydro applied to heat exchanger ____
  Shell   Spsi   Tube   Tpsi ____
  Date of Test ____
WINDINGS
  (Dismantled/Shut-down) ____
  Primary-Insulation—
    (Soft, Brittle, Broken, Normal) ____
  Secondary-Insulation—
    (Soft, Brittle, broken, Normal)
  Temperature—Degrees Cent. (if operating) ____
INSULATION RESISTANCE
  Prim. Winding to Ground—Megohms ____
  Sec. Winding to Ground—Megohms ____
  Bet'n Transf. Windings—Megohms ____
  Time between Shutdown and Test
    (Megger Test Voltage - 500 dc) ____
  Evidence of Moisture (yes, no) ____
ELECTRICAL CONNECTIONS INSIDE TANK
  Loose, Tight, Burnt ____
COMMENTS/RECOMMENDATIONS:

_____

_____

_____

_____

_____

_____

_____

_____

_____

_____

FIGURE 13.14 (*Continued*) An outline of items to be checked on a transformer will assist in recording conditions for evaluation and future reference.

POWER RECTIFIERS

AC power was converted to dc by means of ac motor and dc generator sets prior to the introduction of static devices called rectifiers. DC power is still extensively used where speed control and starting torques may be important. However, rectifier units for large power requirements are more economical than rotating machinery for converting ac to dc. Some advantages of the rectifier are (1) no magnetic field, therefore, no copper loss in supplying the magnetic force required for a dc generator, (2) in the rectifier, there are no core losses such as occur in the armature iron of a dc generator, (3) in the rectifier, there are no moving parts; therefore, mechanical losses are negligible, and (4) the cost of maintaining brush assemblies, commutators, bearings, and similar electric rotating machinery expenses does not exist on static rectifiers.

The definition of a *rectifier* by the American Standard Association (ASA), now American National Standards Institute (ANSI), is "a device which converts alternating current into unidirectional current by virtue of a characteristic permitting flow of current in only one direction." An *inverter* converts unidirectional current (dc) back into alternating current. Vacuum tubes have been used for rectification, but these were used for small power needs, such as communication equipment. They have been replaced by *semiconductor-type rectifiers.* These are devices that have the unique solid-state properties to block the flow of current in one direction while permitting flow in the other. Semiconductor devices are generally two different substances joined together so that the rectifying process takes place at the junction, or barrier, between them. At present, the silicon-controlled rectifier (SCR) is the dominant rectifier type for converting alternating current to direct current.

The principle of ac-to-dc conversion in a rectifier circuit is a matter of the current passing in only one direction and also the extent of wave rectification (see Fig. 13.15). The early rectifiers had simple circuits that produced *half-wave* rectification with no current flowing during the negative half of the sine wave. *Full-wave* rectification is shown in Fig. 13.15*b,* where four half-wave rectifiers are connected in the form of a Wheatstone bridge. The direction of the direct, or positive, current, is indicated by arrows on the diagram. Figure 13.15*c* shows a rectifier circuit using six rectifiers with transformers connected delta, six-phase, and double-wye, and the corresponding rectification diagram shows the rectification waves approaching a straight line.

Transformers for rectifier circuits are specially designed for the phase and voltage characteristics required for these circuits. One hazard that rectifier transformers must resist is arc back, where the flow of direct current is accidentally reversed until the protective devices operate. The secondary winding of the transformer may be subject to feedback from the dc system in addition to the normal alternating current that may flow due to a fault on the ac secondary side. Greater bracing of the windings is required to resist these currents. High-speed breakers are also required to quickly clear any fault currents.

Silicon Power Rectifier

Figure 13.16 shows a *pn* junction device that has high-ampere forward-carrying capability. The operation of most semiconductor devices is dependent upon the *pn* junction. This is the boundary formed between a *p* region and an *n* region in a monocrystalline piece of semiconductor material. The silicon rectifier is a unidirectional device that passes current only in one direction. Silicon rectifiers can operate at juncture temperatures as high as 175°C (347°F), while other solid-state rectifiers such as copper oxide,

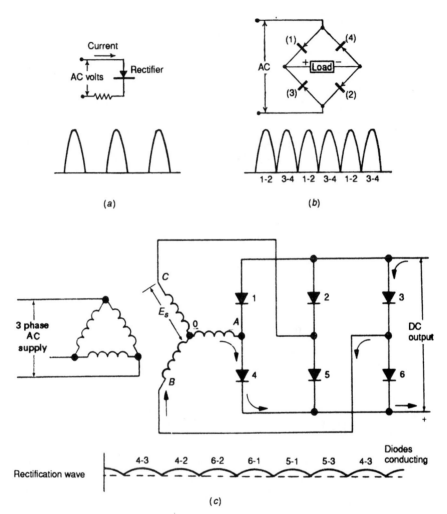

FIGURE 13.15 Rectification waves. (*a*) Simple rectifier circuit (top) produces half-wave rectification (bottom); (*b*) a Wheatstone bridge circuit (top) uses four rectifiers to produce full-wave rectification (bottom); (*c*) a delta, six-phase, double-Y transformer connection with six rectifiers produces an almost straight line rectification.

selenium, and germanium are limited to temperatures below 100°C (212°F). High-current devices such as a silicon rectifier require the use of large pieces of silicon that require special fabrication methods. The silicon wafer is alloyed directly to a material with similar coefficient of expansion, such as molybdenum or tungsten, and this assembly is then joined to a copper stud and counter electrode.

With proper circuit design, silicon rectifiers can have a wide range of power and voltage application. For example, by suitable series connection, they may be used for high-voltage applications.

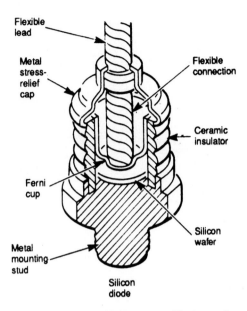

FIGURE 13.16 The silicon power rectifier is a semiconductor that is extensively used to convert alternating current to direct current.

Thyristors

A thyristor, an SCR, is used to obtain adjustable dc voltages. A *thyrite* is a special resistance material with negative resistance properties used in surge protectors. The thyrite has high resistance at normal voltages, but as the voltage rises, the resistance of the thyrite drops so that the voltage surge is limited, similar to a lightning arrestor.

Selenium Rectifier

Selenium rectifiers were used before the advent of the juncture rectifier, such as the germanium and silicon rectifiers. The selenium rectifier (see Fig. 13.17a) is also based on the unilateral conduction property of a thin layer of selenium that is placed between and in intimate contact with two metallic electrodes. The selenium is deposited as a very thin film of about 0.05 mm on one side of a carrier plate made of either iron or aluminum. Heat treatment is used to bake the selenium to the surface. Under controlled conditions, this produces the required crystalline structure. A low-melting-point alloy is sprayed on the selenium surface forming the counter electrode. Further chemical treatment produces a blocking film, or barrier layer, between the selenium coating and the counter electrode.

Mercury-Arc Rectifier

Mercury-arc rectifiers were extensively used in industries where large dc output was required. These rectifiers depend on the existence of an electron-emitting spot located

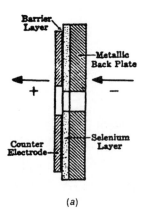

(a)

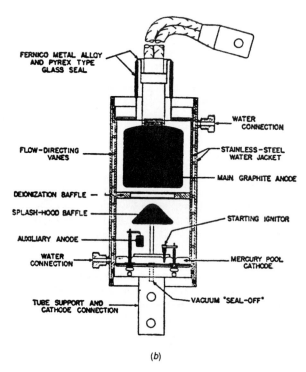

(b)

FIGURE 13.17 Two types of rectifiers. (a) Selenium rectifier stack; (b) mercury-arc ignition-type rectifier. (*Courtesy, General Electric Co.*)

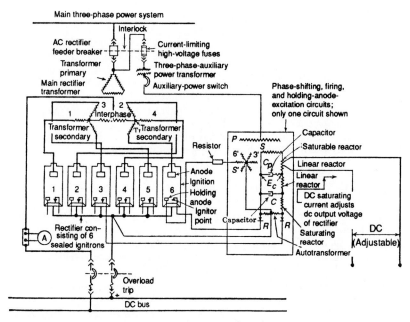

FIGURE 13.18   One-line diagram showing six sealed ignitrons connected to a three-phase power supply and delivering 250 Volts of dc power. (*Courtesy, General Electric Co.*)

in a pool of pure mercury, and they also have an evacuation tank and an anode to collect electrons. Figure 13.17*b* illustrates an *ignitron,* which is a single half-wave mercury-arc unit. The main anode is constructed of graphite, and the cathode is a mercury pool. The firing or starting ignitor is a pointed rod of high-resistance material, with the pointed end dipping into the mercury pool. When starting current is delivered to the ignitor, a spark occurs at the juncture with the mercury pool, and this ionizes the pool. This in turn causes an arc to strike from anode to cathode. The arc is extinguished during the next half-cycle, when the anode-cathode potential reverses, thus giving rectification of half-wave. Therefore, the ignitron is conducting only during the firing of the ignitor. Figure 13.18 shows an electrical connection diagram for a three-phase, 60-cycle power supply being converted to a 250-V dc power supply.

Plug-In Subassemblies

Plug-in subassembly construction is widely used in modern rectifier equipment. This permits rapid replacement of faulted parts by installing the spare subassembly. Inspection and maintenance consists mostly of making sure that the equipment is clean, tight, and dry. Testing of circuits is the job for a specialist, usually from the manufacturer's service organization. Because of the high voltages involved in some rectifier circuits, extreme care is required to make sure that all power is removed from the equipment before any servicing is attempted.

FACILITY LIGHTING

There are many possible light sources and a variety of fixtures and equipment arrangements to provide lighting to a facility. The type of installation depends on the occupancy and the lighting design. Plant service operators may be responsible for maintaining this equipment. Besides aiding production, good lighting conserves eyesight without undue eyestrain and fatigue to the occupants of the facility. Good lighting also promotes good workmanship and even affects attitudes about a job; therefore, it is important to maintain this equipment.

Plant service operators should be familiar with the following illumination terms and definitions developed by The Illuminating Engineering Society:

Absorption. A certain amount of the light which hits an object is absorbed by that object. The amount absorbed depends upon the reflection factor of the surface.

Brightness. The brightness of an object depends upon the intensity of the illumination on it and the percentage of light it reflects, emits, or transmits. Brightness is one of the fundamental factors in seeing. When brightness values are low, seeing becomes difficult and slow. Extremely high brightness or wide variation in brightness in the visual field can cause glare and eye fatigue.

Diffusion. Many surfaces, such as plaster and opal or prismatic glass, break up the incident light from a point source and distribute it more or less over an entire surface. Diffusion tends to enlarge the image of light source and reduces its brightness by breaking up its outline. Glare is, therefore, reduced, and seeing becomes easier.

Fluorescent. Fluorescent lamps are electric-discharge lamps in which the radiant energy from an electric discharge is transferred by suitable material (phosphors) into wave lengths that give higher luminosity.

Filament. Filament lamps are light sources that consist of glass bulbs containing a coiled filament electrically maintained at incandescence by heat energy created by the electric current.

Footcandles. A footcandle is the measure of illumination on a surface which is 1 ft distant from, and perpendicular to, a source of illumination of 1 candlepower. This is the term used to determine the illumination on a working surface.

Foot-lambert. A foot-lambert is a measure of brightness. Numerically, it is approximately equal to the product of the incident footcandles and the reflection factor of the surface, the degree of brilliancy of any part of a surface or medium of transmitting light.

Glare. Glare may be defined as brightness out of place. The degree of brilliancy of any part of a surface or light source, when viewed from a designated direction, determines the amount of glare. Glare may be direct or reflected from any surface and is injurious to the eye.

Illumination. Illumination may be termed the effect or result of light.

Incandescent. See *Filament.*

Light. For the purpose of illumination engineering, light is radiant energy evaluated according to its capacity to produce visual sensation.

Lumen. A lumen is the unit of light flux representing a quantity of light. One lumen is the total amount of light falling upon 1 ft^2 of surface, every part of which is 1 ft distant from a point source of 1 candlepower.

Luminaire.   Luminaire is a term broadly used in industry to designate any type of lighting fixture, complete with globe, reflector, refractor, and housing.

A luminaire designed to distribute 90 percent or more of its light output downward below the horizontal is classed as *direct.*

A luminaire designed to distribute from 90 to 100 percent of its light output upward above the horizontal is classed as *indirect.*

A luminaire designed to distribute from 60 to 90 percent of its light output downward below the horizontal and from 10 to 40 percent of its light output upward above the horizontal is classed as *semidirect.*

A luminaire designed to distribute from 40 to 60 percent of its light output upward above the horizontal and the balance downward is classed as *direct-indirect.*

A luminaire designed to distribute from 60 to 90 percent of its light output above the horizontal and 10 to 40 percent of its light output below the horizontal is classed as *semiindirect.*

Mat finish.   Mat finish is a dull finish opposite that of specular. Light hitting a mat finish has a minimum of reflected glare.

Mazda.   Mazda is a trade name formerly used by General Electric and Westinghouse for most lamp bulbs made by them, both incandescent and fluorescent. The term is taken from the Persian word meaning "God of Light."

Reflection.   Light is reflected from any given surface, depending on its finish. The ratio of the reflected light to the incident light is termed *reflection factor.* A bright, shiny piece of metal, for example, will reflect much more light than a dark, dull piece of fabric.

Refraction.   Light in passing from one medium to another passes through the second medium at an oblique angle and might be referred to as "bent" light.

Specular surface.   A specular surface is a bright, shiny surface, opposite that of mat finish. Light hitting a specular surface is reflected therefrom at an angle equal to the angle of incidence. The light reflected from a specular surface is usually very bright and has much glare.

Surface brightness.   Surface brightness is the degree of brilliancy or luminous intensity of any part of a surface when viewed from a designated direction.

Transmission.   Transmission describes light going through a transparent or translucent medium. The ratio of light transmitted to the incident light is termed *transmission factor.*

Watt.   A watt is a unit of electrical power in which all lamp bulbs are rated.

Lighting systems are generally classified as *direct, semidirect,* and *indirect.* Direct systems cost less but can produce reflected glare and shadows that can cause eye fatigue. The semidirect system eliminates shadows and glare but is not so efficient for use with dark ceilings and walls. The indirect system is the best for eliminating glare and shadows but is more expensive to install and maintain.

All lighting wiring should comply to the NEC interior and exterior wiring code. Plants with chemical fumes, gas exposures, or a dust-explosion hazard require lighting and wiring that is explosionproof or dusttight. Flameproof fittings or lighting units that are under a small inert-gas pressure are used.

The most important factor in lighting systems is see-ability and comfort of the occupants. Some common lighting complaints are shown in Table 13.1. As the table indi-

TABLE 13.1   Common Lighting Complaints or Problems

Symptom of complaint	Possible remedy
1. Insufficient illumination on the work for the seeing task involved. Remember that brightness relationships are more important to good seeing conditions than the highest footcandles for the least wattage.	1. Sometimes a greater illumination level may be the only agency to make one see enough. It takes about 4 times as long to perceive small detail under 2 ftc than it does under 100 ftc. For a 1:10 ftc ratio the time is about twice as long.
2. Direct glare caused by bright light in the range of vision. Glare results at any and all intensities where unsupported by proper levels of total field brightness to stabilize adaptation. The eye receives clear impressions of detail only in the absence of glaring light. Glare is caused by too much light in the worker's eyes compared to that on the work. Discomfort from a light source of fixed brightness increases with its visible area and, vice versa, the comfort increases with a reduction of source brightness.	2. *a.* Shield lamps properly. Avoid bare lamps in continuously occupied areas. *b.* Mount lighting fixtures as high as practical. The higher the mounting, the less the glare effect. If the mounting height is doubled, 3 or 4 times the brightness of the source may be tolerated. *c.* Increasing lighting intensities helps, but the footcandles must be increased 10 times to stand doubling the brightness of the light sources. *d.* Best solution is to have a low-brightness broad-distribution light source. Comfort from a light source of fixed brightness increases as its visible area becomes larger. *e.* Make the lighting installation as inconspicuous as possible. Glare effect is multiplied by the number of lighting units visible from any work position. What may be suitable for a small office may be very uncomfortable in a large room.
3. Direct daylight glare caused by improper placement of work with respect to sunlight.	3. *a.* Properly shade windows. *b.* Shift working surfaces in such a way that the natural light falls on the working plane and away from the worker's eyes.
4. Reflected glare caused by reflections from polished metals, glossy paper, or any material possessing a sheen.	4. *a.* Position light sources so as to keep reflections away from the eyes. *b.* Again, the best solution is a large-area low-brightness source.
5. Reflected daylight glare caused by natural light that bounces from the work plane to the eyes.	5. *a.* Shade the source. *b.* Provide a high enough artificial-lighting-intensity level to offset reflected daylight.

TABLE 13.1 Common Lighting Complaints or Problems (*Continued*)

Symptom of complaint	Possible remedy
6. Specular reflection from machine surfaces which creates direct or diffused images of the light source on the work, polished furniture, and other shining surroundings.	6. *a.* Reflections may be reduced by increasing the general level of illumination with low-brightness light source. *b.* May be corrected sometimes by changing the mounting height of the offending units. *c.* Avoid the use of bare lamps.
7. Dense shadows caused by a worker on a piece of equipment cutting off the light.	7. Large, closely spaced units or continuous rows shine around objects and erase shadows.
8. Range of brightness within the visual field is too great. While the eye focuses on objects directly before the eye, it does a partial job of seeing masses well to the side of the direct line of vision.	8. *a.* Avoid competition between the nerves in the eye by keeping the brightness range between the work and the surroundings within reasonable limits—5:1 if possible. *b.* Splash enough light on the inside walls to hold a ratio of at least 10:1 or better. May look a bit inefficient by calculation but worth much more psychologically.
9. More light at the eyes than on work caused by improper placements of light source.	9. *a.* Angle lighting may cure bright surface jobs. *b.* Vertical or horizontal local lighting may solve some inspection problems. *c.* Install local lighting where needed.
10. Unpleasant surroundings, dingy, gloomy atmosphere. You get fooled by uncomfortable illumination even if the lighting intensity is adequate. Lighting cannot be evaluated by simple seeing.	10. *a.* Where the ceilings serve as a secondary light source, such as in semi-indirect or indirect lighting, the ceiling should be finished a flat white. Avoid the use of glossy paints because the shiny surfaces cause specular reflections of light sources. *b.* Walls should be colored because they are usually the background against which the task is viewed. *c.* Floors should be light—the lighter the finish of the floor covering, the higher the efficiency of the lighting system. The light reflected back to the ceiling helps relieve brightness contrasts between the luminaries and the ceiling. *d.* Desk tops should be dull finish and light in color. *e.* Plate glass should never be permitted to be placed on any desk top.

TABLE 13.1   Common Lighting Complaints or Problems (*Continued*)

Symptom of complaint	Possible remedy
11. Improper maintenance. Lamps age and blacken and give out less light. Dirt and dust reduce reflecting and transmitting qualities of the fixtures and reflecting surfaces.	11. Five musts for proper maintenance are: *a.* Clean lamps and fixtures frequently. *b.* Repaint walls and ceiling regularly. *c.* Replace burned-out lamps promptly. *d.* Maintain proper circuit voltage. *e.* Establish a definite maintenance program. Very few plants spend enough on lighting maintenance. Keeping the lighting system in best condition more than pays for itself.

cates, a good *lighting maintenance program* is essential. This includes (1) responding to lighting complaints by workers promptly, and determining what the problem may be, (2) keeping walls and ceilings clean and painted with the right colors to avoid glare and shadows that can irritate the occupants, (3) replacing burned-out lamps promptly, (4) keeping lamps and fixtures clean for good lighting and also for avoiding fixture deterioration and electrical failures, (5) maintaining at least the following minimum footcandles of lighting: 5 ftc for very easy work, 10 fc for casual seeing work, 30 fc for ordinary work, 50 fc for difficult seeing work, and at times several hundred footcandles may be needed for very difficult tasks, (6) always replacing lighting fixtures so that they are suitable for the surroundings (i.e., explosionproof where so required, dusttight, and moisture resistant).

QUESTIONS AND ANSWERS

1. At full load, a single-phase transformer operating at 2300 V shows 10.9 A, and the kilowatt meter shows 22.5 kW. What is the power factor at this load?

Answer.   The equation to use is

$$\text{P.F.} = \frac{\text{kW}}{\text{kVA}} = \frac{22.5}{(2300/1000)10.9} = 89.7\%$$

2. A series circuit has 12-Ω resistance, 32 Ω inductive reactance, and 20-Ω capacitive reactance. What is the impedance of the circuit?

Answer.   The impedance is

$$Z = \sqrt{12^2 + (32 - 20)^2} = 16.97 \ \Omega$$

3. A single-phase transformer of 100 kVA is rated 2300/575 V and has 200 turns on the primary. What are the number of turns in the secondary and the rated current in primary and secondary?

Answer.   The turns are

$$N_2 = \frac{200 \times 575}{2300} = 50 \text{ turns}$$

The rated current is

$$\text{Primary current} = \frac{100 \times 1000}{2300} = 43.5 \text{ A}$$

$$\text{Secondary current} = \frac{100 \times 1000}{575} = 173.9 \text{ A}$$

4. What is a Class 2, Division 1 location per the NEC?

Answer. A Class 2, Division 1 location is a location where combustible dust is present during normal operating conditions, and the electrical equipment in such a location must be constructed to be dust-ignitionproof.

5. What are the two types of grounding that may be encountered in a facility?

Answer. Equipment grounding is used to ground all metallic noncurrent-carrying parts of electrical equipment in order to avoid personal injury if an internal electrical failure occurs on electric equipment. (2) System grounds are used to provide a low impedance path for a ground so that protective devices immediately detect the ground and clear the fault.

6. What determines the magnitude of a short-circuit current?

Answer. The magnitude of short-circuit currents is dependent on the voltage in the circuit and the system impedance up to the fault location. In plants that purchase all of their power, the utility is assumed to have an unlimited power capacity to feed a fault, and the impedance of the supply transformer is used to calculate the possible short-circuit currents that may occur in a facility.

7. How do *solid, high-,* and *low*-resistance system grounds differ?

Answer. Solid grounding is accomplished by having a point in the electrical system directly connected to the switchboard or substation ground. High-resistance grounds are used to limit the ground current to a low value and are obtained by placing resistance in the system neutral ground. The ground current in high-resistance grounding is usually limited to 0.1 percent of the three-phase fault current. Low-resistance grounding limits the ground fault current to a value far below the three-phase fault current but high enough to permit fast protective relays to function. *Reactance grounding* is also used in system grounding designs. Here reactance is introduced in the line going to ground.

8. What is a coordination study of protective devices?

Answer. Coordination studies are made to determine the required setting of protective devices so that the protective device nearest to the fault clears that circuit first without other protective devices away from the fault opening or being damaged by the fault.

9. What are two problems that require considerations in rectifier assemblies?

Answer. The two problems are (1) overload or short-circuit currents because of the semiconductor's small thermal capacity and (2) protection against failure of a semiconductor device in the circuitry. Protection for the first condition is usually supplied by a current balance coil and a fast-acting current-limiting fuse. For the second condition, it is assumed that a semiconductor has failed when the fuse in series blows; therefore, provisions are made to monitor the fuse condition by sensing the voltage across the fuse and then having a small indicating light connected across the fuse to show by glow when the fuse has blown.

10. How is the impedance of a transformer shown on the nameplate?

    Answer. The impedance is shown as a percentage value of the rated kilovoltamperes of the transformer, not an ohmic value.

11. What are four transformer oil tests?

    Answer. (1) Dielectric strength of the oil, (2) neutralization number to show acidity, (3) interfacial tension to determine the strength of the oil in resisting water contamination, and (4) color to show if the oil is contaminated.

12. What is the purpose of a power factor test on a transformer?

    Answer. Power factor testing measures the power leakage through the insulating medium and is used to evaluate the condition of the winding and oil insulation. Sludge, water, acidity, and oxidation of the windings or oil all lower the insulation medium's strength.

13. A three-phase Y-delta-connected transformer has a rating of 2000 kVA with a primary voltage of 13,800 V and a secondary of 4160 V. What is the rated phase amperes on the primary and secondary sides?

    Answer. In a Y connection, the line amperes equals the phase amperes:

    $$\text{Y-side amperes} = \frac{2000 \times 1000}{1.732 \times 13,800} = 83.7 \text{ A} = \text{phase amperes}$$

    The delta-side phase currents equals line current divided by the $\sqrt{3}$:

    $$\text{Delta-side line current} = \frac{2000 \times 1000}{1.73(4160)} = 277.6 \text{ A}$$

    $$\text{Delta-side phase current} = \frac{277.6}{1.732} = 160.3 \text{ A}$$

14. A transformer shows a permitted 55°C rise. If the maximum ambient temperature is considered by standards to be 40°C, what is the maximum temperature allowed when you read the oil temperature?

    Answer. The maximum observable temperature would be $40 + 55 = 95°C$. Higher temperatures are an indication of (1) insufficient cooling or (2) overloaded transformer.

15. What does the basic insulation level, or BIL, on the nameplate indicate about transformer insulation?

    Answer. Another name is basic impulse insulation level, or BIIL. They both indicate the crest value of the voltage wave that the transformer can withstand during a transient system overvoltage. This value is also used in selecting lightning arresters to protect the transformer. A 20 percent margin between the arrester's spark-over voltage and the transformer BIL is considered a minimum margin to be used, or to restate, lightning arresters are selected with a spark-over voltage rating *less* than 80 percent of the BIL.

16. What is the relationship between turns and voltage in a single-phase transformer?

    Answer. Transformer voltages are calculated by the ratio of turns between the primary and secondary windings. For example, suppose a primary winding of 1800 turns has 440 V and the secondary winding delivers 110 V. How many turns would be in the secondary winding? Use the formula

$$\frac{E_p}{E_s} = \frac{T_p}{T_s}$$

where E_p = voltage, primary winding
E_s = voltage, secondary winding
T_p = turns, primary coil
T_s = turns, secondary coil

$$\frac{440}{110} = \frac{1800}{T_s} \qquad T_s = \frac{1800 \times 110}{440} = 450 \text{ turns}$$

17. How is amperage calculated from the number of turns in the primary and secondary windings of a single-phase transformer?

Answer. This amperage is calculated by the formula $I_p T_p = I_s T_s$. Used inversely, the more turns, the less current. For example, a transformer has 1800 turns in the primary winding and 450 turns in the secondary winding. If the secondary has a current of 5 A, how many amperes are in the primary?

$$I_p T_p = I_s T_s$$

$$I_p = \frac{450 \times 5}{1800} = 1.25 \text{ A}$$

18. How are current transformers used?

Answer. The function of all transformers is to change voltage. But some transformers also change current values for metering and relaying purposes (see Fig. 13.13). These are called *current transformers* (ct). The relationship between the primary and secondary voltages and the currents is the same as that in voltage transformers. The secondary winding of a current transformer is generally designed for 5 A with the primary winding carrying the full-load current.

Don't open a ct secondary circuit while the primary circuit is carrying full-load current. Close the short-circuiting switch on the ct itself. Otherwise, the open-circuit secondary voltage in some ct's may build up to a value dangerous to both you and the insulation. This happens because no bucking flux opposes the mutual flux linking the primary and secondary circuits. The secondary voltage depends on the core flux, a function of primary current. With no opposition from the secondary winding, the mutual flux will soar when any sizable primary current flows.

19. How can you determine transformer polarity?

Answer. See Fig. 13.12. The polarity mark of a transformer refers to the direction of the induced voltage in the transformer leads, as brought outside the case to a terminal strip. Transformer-winding leads are marked to show polarity and to distinguish between the high- and low-voltage side. Primary and secondary windings are not identified as such because they depend on input and output connections. The high-voltage side is marked H; each lead and tap is designated as H_1, H_2, H_3, etc. The low voltage side is tagged X; each lead and tap is marked X_1, X_2, X_3, etc.

20. How would you test a transformer for polarity?

Answer. See Fig. 13.12. To test a transformer for polarity, use a temporary jumper and voltmeter as shown. If the voltmeter shows that the high and low voltages subtract, the connection is subtractive polarity. If high and low voltages add, the connection is additive polarity.

21. An oil-filled power transformer supplies a facility utility power and is rated 7500 kVA, 13,800 to 2300 V, three-phase with 7.2 percent impedance. What is the potential short-circuit current that this transformer can feed an incoming switchboard short circuit if all other reactances on the 2300-V side are not considered and unity power factor is assumed?

Answer. The percentage impedance is based upon the rated kilovoltampere; therefore, calculate the rated secondary current as follows:

$$\text{Secondary rated amperes} = \frac{7500 \times 1000}{1.73 \times 2300} = 1{,}882.7 \text{ A}$$

$$\text{Possible short-circuit amperes} = \frac{1882.7}{0.072} = 26{,}149 \text{ A}$$

22. For what type of service are the core and shell types of transformer construction applied?

Answer. The core type (see Fig. 13.6) has less iron, and this permits using more turns and also permit installing more insulation for high-voltage service. The shell type permits better bracing of the coils and is used in transformers that have to withstand shock loading. In both types, the low-voltage winding is placed near the iron in order to reduce the chance of voltage jump. It reduces the amount of insulation required.

23. How are filament lamps rated?

Answer. Tungsten-filament lamps are rated in volts and watts.

24. What are the NEC requirements for lighting fixtures installed in damp or wet locations?

Answer. These fixtures must be so constructed or installed that water cannot enter or accumulate in wireways, lampholders, or other electrical parts, and for wet location installations they must be marked, "Suitable for wet locations," and for damp locations, "Suitable for damp locations." Damp locations are those protected from weather but subject to dampness such as some basements, barns, some cold storage warehouses, partially weather-protected under canopies, roofed open porches, and similar locations.

25. What does the term footcandle mean?

Answer. A footcandle describes the strength of illumination and is the illumination received from a 1 candlepower source of illumination at a distance of 1 ft from that illumination.

26. What kind of reflection should one expect from a specular surface?

Answer. A specular surface is defined as a bright, shiny surface; therefore, the light reflected from such a surface is usually very bright and glaring.

27. What voltage variations are usually permitted for fluorescent lamps?

Answer. Fluorescent lamps generally should be operated at voltages within $\pm$ 10 percent of their designed operating points. Decreased life and uncertain starting may result from operation at lower voltages, and at high voltages, there is danger of overheating of the ballast or transformer as well as decreased life.

CHAPTER 14
REFRIGERATION AND AIR CONDITIONING

Facility services include the operation and maintenance of the many occupancies that must have refrigeration and air conditioning equipment for food and similar storage and must maintain proper temperatures for human comfort, which is broadly called air conditioning. Modern refrigeration systems are designed to produce desired temperature conditions in a given space, even though the surrounding temperature may be higher. Air conditioning systems were developed to not only provide comfortable temperatures but also to provide desired humidity and air quality for the workplace. Because of process needs, air conditioning is required, for example, in the electronic chip manufacturing process.

BASICS

Since refrigeration deals with transferring heat to make a space cooler, some review is needed on some basics of how a body or space is made cooler than the surroundings.

Heat is the internal energy of molecules in a substance. When heat is added to a body, it is stored in that body by increased molecular activity. The Btu is the amount of heat required to raise the temperature of 1 lb of water 1°F between the temperatures of 32 to 212°F.

The mechanical equivalent of heat of 1 Btu is 778 ft · lb of work. The Btu is used in all types of heat calculations.

Temperature is a measure of the relative hotness of a body, and the most common scales used are Fahrenheit and Celsius, both taken as between the melting point of ice and the boiling point of water at atmospheric pressure. For Fahrenheit, the scale is between 32 and 212°F, and for the Celsius scale it is 0 to 100°C. Conversion to Fahrenheit from Celsius is

$$°F = \frac{9}{5} \, C° + 32$$

and from Fahrenheit to Celsius it is

$$°C = \frac{5}{9} \, (°F - 32)$$

Absolute temperature is based on the fact that if cooling were continued, all the energy would be extracted, and molecular activity would cease. Calculations and experiments show this would occur at about 460° below zero on the Fahrenheit scale, and at about 273° below zero on the Celsius scale. The absolute temperature used in thermodynamic equations for the Fahrenheit scale is 460 + °F, and for the Celsius scale it is 273 + °C.

There are also two heats involving change of state that are important in refrigeration. The *heat of fusion* is the amount of heat required to change a solid to liquid. For example, to change ice to water when it reaches 32°F requires about 144 Btu/lb. *The heat of vaporization* is the amount of heat required to change a liquid into the vapor or gaseous state. Thus to change water at 212°F from liquid to steam at 212°F temperature and atmospheric pressure, 970.3 Btu/lb of water is required.

Specific heat is the amount of heat energy required to raise the temperature of 1 lb of that substance 1°F. For water, it is 1 Btu in the temperature range previously described. When dealing with gases, two specific heats are used: (1) that at *constant volume* and (2) that at *constant pressure*. For example, if air is heated in a closed container, no change in volume is possible, so it has a specific heat of 0.171 Btu/lb at constant volume. If the air is free to expand at a certain constant pressure, the specific heat at constant pressure is 0.24 Btu/lb/°F. *Sensible heat* is the addition or removal of heat from a substance that causes a temperature change but does not cause a change in state, such as solid to liquid.

A common use of specific heats in refrigeration is to calculate the amount of Btu's that must be removed from a product brought in for storage.

CONDITIONS FOR STORING PRODUCTS

Table 14.1 lists some common conditions that must be obtained in storing products in refrigerated spaces. There is a wide diversity of temperature, humidity, and storage time that practice dictates must be met. Some of the different purposes of refrigeration include control of (1) biochemical processes such as fruit ripening and spoiling, (2) chemical reactions such as in chip manufacturing or cold room operations, and (3) the appearance of meat, vegetables, fruit, and similar edible products. Thus storage is involved with loss of moisture, bacterial growth, and similar concerns in addition to temperature.

Cooling loads include:

1. Product cooling to required temperatures.
2. Maintaining proper moisture content.
3. Ventilation to keep fumes and odors at a minimum and also to remove respiration action products. For example, some fruits due to respiration action generate carbon dioxide, which uses up the oxygen in the refrigerated space. This requires ventilation, which increases the cooling load.
4. Equipment in cold rooms, such as motors, give off heat that must be considered.
5. People walking in and out, add to the cooling load through door losses.
6. Basic losses through insulation and walls, sun loads, and similar location loads add to the cooling required.

Example.   What are the Btu's that must be removed from a load of beef delivered to a cold box at 80°F that is to be frozen to a temperature of 25°F if the weight is 2400 lb? See Table 14.1. There are two specific heats to be considered—above and below freez-

TABLE 14.1 Common Conditions of Products Placed in Refrigerated Storage

	Range of storage, °F		Optimum Rel. Hum, %	Freezing point, °F	Composition, % water	Specific heat, btu/lb/°F		Latent heat of fusion	Maximum storage period
Products	Short-time storage	Ware-house storage				Above freezing	Below freezing		
					Fruits				
Apples	35–40	30–32	85	28.5	85	0.90	0.49	122	8 months
Bananas	55–56	55–56	80	26–30	75	0.90		112	10 days
Grapes	35–40	30–32	80	28	77	0.90	0.61	130	1–6 months
Lemons	55–60	50–55	80	28	89	0.94	0.50	124	90 days
Oranges	40–45	32–34	80	28	86	0.90	0.47	128	2 months
Peaches	35–40	31–33	80	29.5	88	0.92	0.48	122	30 days
Pears	35–40	30–32	85	28.5	84	0.91	0.49	131	1–7 months
Strawberries	35–40	31–33	80	30	90.5	0.92	0.48		10 days
					Vegetables				
Asparagus	40–45	32–34	90	30	94	0.91	0.49	136	30 days
Beans (string)	40–45	32–34	85	30	68.5	0.80	0.46	98.5	30 days
Beets	40–45	32–34	85	27	88.5	0.86	0.48	128	7–90 days
Cabbage	35–40	32–34	90	31	91.5	0.93	0.47	132	4 months
Carrots	35–40	32–34	90	29.5	88	0.86	0.45	126	2–4 months
Celery	35–40	31–33	90	30	94.5	0.91	0.46	136	2–4 months
Corn (green)	35–40	31–33	85	29	75.5	0.86	0.38	108	10 days
Corn (dried)	50–60	35–40	60		10.5	0.29	0.24	15	12 months
Lettuce	35–40	32–34	95	31	94.5	0.90	0.46	136	20 days
Potatoes	36–50	38–42	85	29	78.5	0.86	0.47	113	6 months
Tomatoes	50–55	50–55	80	30.5	94.5	0.92	0.46	132	10 days
Vegetables, mixed	40–45	35–40	85	30	90	0.90	0.45	130	10 days

TABLE 14.1 Common Conditions of Products Placed in Refrigerated Storage *(Continued)*

Products	Range of storage, °F Short-time storage	Range of storage, °F Ware-house storage	Optimum Rel. Hum., %	Freezing point, °F	Composition, % water	Specific heat, btu/lb/°F Above freezing	Specific heat, btu/lb/°F Below freezing	Latent heat of fusion	Maximum storage period
				Meat and Fish					
Bacon	40–45	28–30	80		20	0.50	0.30	29	15 days
Beef (fresh)	35–40	30–32	84	27	68	0.75	0.40	98	3 weeks
Fish (frozen)	15–20	5–10	80	28	70	0.76	0.41	101	6 months
Fish (iced)	34–38	30–32	85		70	0.76	0.41	101	15 days
Hams and loins	34–38	28–30	80	27	60	0.68	0.38	86.5	3 weeks
Lamb	34–38	28–30	85	29	58	0.67	0.30	83.5	2 weeks
Pork (fresh)	34–38	30–32	80	28	60	0.68	0.38	86.5	15 days
Pork (smoked)	40–45	28–30	80		57	0.60	0.32		15 days
Poultry (fresh)	28–30	28–30	84	27	74	0.79	0.37	106	10 days
Poultry (frozen)	15–20	0–5	85	27	74	0.79	0.37	106	10 months
Sausage (fresh)	35–40	21–27	80	26	65	0.89	0.56	93	15 days
Sausage (smoked)	40–45	32–40	75	25	60	0.86	0.56	86	6 months
Veal	34–38	28–30	84	29	63	0.71	0.39	91	15 days
				Miscellaneous					
Beer	35–40	34–38	85	28	92	1.0			6 months
Butter	45–40		80	75–70	15	0.64			10 days
Cheese (American)	40–45	32–34	80	17	55	0.64	0.36	79	15 months
Cheese (Swiss)	40–45	38–42	80	15	55	0.64	0.36	79	60 days
Chocolate coating	65–70	60–75	55	95–85	0.5	0.30	0.56	40	6 months
Eggs (crated)	40–45	30–31	85	27	73	0.76	0.40	100	12 months
Eggs (frozen)	15–20	0–5	60	27			0.41	100	18 months
Flowers (cut)	40	35	85	32					1 week
Ice cream	0–10	–20–0	85	27–0	60	0.78	0.45	96	2 weeks
Lard	45–50	32–34	80			0.52			6 months
Milk	35–40	35–40	70	31	87.5	0.93	0.49	124	5 days
Nuts (dried)	35–40	30–32	75		3–10	.21–.29	.19–.24	4.3–14	8–12 months
Tobacco and cigars	42–44	42–44	90	25					

ing. Also, the heat of fusion must be considered in freezing the 2400 lb. The heat to be removed is

$$H = \text{weight} \times \text{specific heat} \times \text{temperature change}$$

Sensible heat is

$$H = 2400 \times 0.75 \times (80 - 27) = 95,400 \text{ Btu} \qquad (27° \text{ is the freeze point})$$

Freezing is

$$H = 2400 \times 98 = 235,200 \text{ Btu} \qquad (98° \text{ is latent heat of fusion})$$

Cooling to 25°F is

$$H = 2400 \times 0.40 \times (27 - 25) = 1920 \text{ Btu}$$

Total amount of heat to be removed $= 95,400 + 235,000 + 1,920 = 332,520$ Btu

A *ton of refrigeration* is defined as the removal of heat at a *rate* of 200 Btu/min, or 12,000 Btu/h. This is not power but expresses a rate of heat interchange. It was derived from the ice-making process, where 2000 lbs, or 1 ton, of ice required $2000 \times 144 = 288,000$ Btu's to convert water into ice during a 24-h period. Thus by hour, this is $288,000/24 = 12,000$ Btu/h, or 200 Btu/min.

Boiling or evaporation temperatures vary with pressure as does the condensing temperature, called the saturation temperature. Refrigeration systems use liquids with boiling points that can be many degrees below 0°F, but by varying the pressure, evaporation temperatures can be changed.

Ice refrigeration was used for many years. Since ice melts at 32°F, it could be placed in a space warmer than 32°F, and by changing the state from solid to liquid, 144 Btu/lb of ice cooling could be obtained in the space the ice was placed in.

Dry ice provides refrigeration by the use of solid carbon dioxide. At atmospheric pressure, carbon dioxide is a solid and goes directly into the vapor state when absorbing heat. This process is called sublimation. Sublimation occurs at $-109.3°F$, and the heat removed is about 246 Btu/lb. It is widely used for cooling small packages, with no liquid resulting as the CO_2 changes from solid to gas.

MECHANICAL VAPOR REFRIGERATION

Most refrigeration equipment used is of the mechanical vaporization type, where a liquid is vaporized in picking up heat and is then liquefied again to repeat the process of picking up heat from the space to be cooled. Mechanical refrigeration makes it possible to control the pressure and temperatures of a refrigerant within desired points and also allows using the same refrigerant over and over again.

The most commonly used refrigerants are ammonia and the Freons. Freons have been blamed for causing ozone depletion of the earth's atmosphere but were originally developed because they met most of the desirable features for a refrigerant. The EPA has proposed phasing out the production of chlorofluorocarbons (CFCs) by January 1, 1996. Industry is being requested to switch to alternative refrigerants. Alternatives to CFCs are available for all commercial refrigeration and air conditioning equipment, and many parallel the characteristics of the Freons but do not affect the earth's ozone layer.

Figure 14.1 shows the saturation pressure-temperature relationships of many previously used refrigerants.

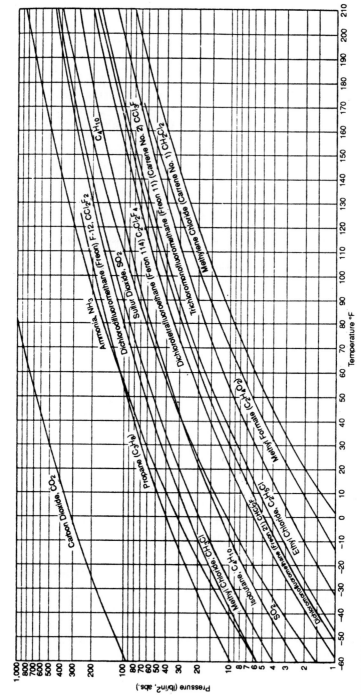

FIGURE 14.1 Past and currently used refrigerants and their saturation pressure-temperature characteristics.

MAIN COMPONENTS OF MECHANICAL REFRIGERATION

The main components of a mechanical compression-type refrigeration system are illustrated in Fig. 14.2, which shows brine being cooled and then being pumped to the space to be cooled. This is called an *indirect system* since the refrigerant is not cooling the space directly. In Fig. 14.2, the brine is cooled by the liquid refrigerant being supplied through an expansion valve to the brine heat exchanger. The refrigerant vaporizes as it picks up heat in this *evaporator*. The *compressor* pulls the evaporated refrigerant out of the evaporator and compresses the refrigerant vapor to a pressure sufficiently high to have a saturation temperature higher than the available condensing medium in the *condenser*. This could be water or air. In the *condenser,* the refrigerant vapor becomes a liquid again as it gives up its heat of vaporation *and of the work of compression by the compressor.* After that, the liquid is stored in a receiver to be used as needed again in the evaporator. Note that there is a high-pressure side from the compressor discharge to the condenser, receiver, and expansion valve, where the refrigerant is throttled, or dropped in pressure, to the evaporator pressure. In passing through the expansion valve, the liquid refrigerant cools itself at the expense of evaporating a portion of the liquid. The low pressure in the evaporator is determined by the desired temperature for the space to be cooled. The high pressure in the condenser is determined by the temperature of the available cooling medium to condense the refrigerant.

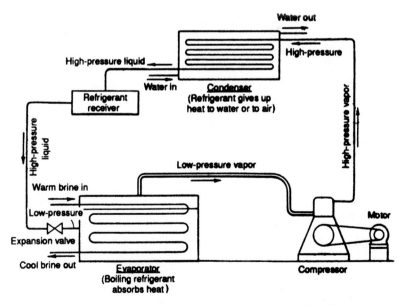

FIGURE 14.2   Components of a compression-type mechanical vapor refrigeration system.

Example.   An ammonia system uses water in a condenser. If a 10°F difference is allowed between the ammonia temperature and its condensing water temperature, what would be the probable discharge pressure if the condenser water temperature is 70°F? See Fig. 14.2. The ammonia temperature would be 70 + 10 = 80°F. The corresponding saturation pressure for ammonia is about 155 psi absolute, where absolute pressure equals gage pressure plus 14.7 psi.

Mechanical vapor refrigeration as shown in Fig. 14.2 has a theoretical *Carnot efficiency* that is expressed as follows:

$$\text{Efficiency} = \frac{Q_R - Q_C}{Q_R} = \frac{T_R - T_C}{T_R}$$

where Q_R = heat received in evaporator at temperature T_R
      Q_C = heat rejected in condenser at temperature T_C

The work done on the refrigerant is $W = Q_C - Q_R$. The heat added or subtracted is expressed in Btu's, while the temperatures are expressed in absolute degrees Fahrenheit.

In refrigeration, the *coefficient of performance* (C.P.) is used extensively, and this C.P. is the reciprocal of above equation, or

$$\text{C.P.} = \frac{Q_R}{Q_C - Q_R} = \frac{T_R}{T_C - T_R} = \frac{Q_R}{W}$$

The expression for C.P. is important in refrigeration because the ratio for C.P. as given above shows how much heat is removed per amount of work required to do it.

The *horsepower expanded per ton of refrigeration* is calculated from the fact that 1 ton of refrigeration = 200 Btu/min, and 1 hp is 33,000/778 = 42.4 Btu/min, then

$$\text{horsepower per ton of refrigeration} = \frac{200}{42.4 \text{ C.P.}} = \frac{4.717}{\text{C.P.}}$$

Example.   A refrigeration system has an evaporator temperature to be maintained of −20°F, and a condenser temperature of 80°F. What are (1) the Carnot C.P. and (2) the horsepower required per ton of refrigeration? The Carnot C.P. is

$$T_R = 460 - 20 = 440°\text{F absolute} \qquad T_C = 460 + 80 = 540°\text{F absolute}$$

$$\text{C.P.} = \frac{440}{540 - 440} = 4.4$$

The horsepower is

$$\text{hp per ton} = \frac{4.717}{4.4} = 1.07$$

REFRIGERANTS

Ammonia

Ammonia is used in many industrial applications, such as cold storage and meat packing plants. When it is water free, it is called anhydrous ammonia. Ammonia has a latent heat of vaporization of 500 Btu/lb; therefore, less has to be circulated than most refrigerants per ton of refrigeration. The condenser pressure seldom exceeds 200 psi. Ammonia corrodes copper, brass, and bronze; therefore, these materials should not be used in ammonia systems. Ammonia irritates the eyes and mucous membranes, and in quantities approaching 0.5 percent per volume, serious effects can occur if exposure is beyond a few minutes. Most ordinances require on-site gas masks for the operators because of this exposure. Ammonia can form *explosive mixtures* with air in concentrations of 16 to 25 percent by volume, but it is not easily ignited. Adequate ventilation is usually required for engine rooms to avoid this hazard. Ammonia *cannot* be used for air conditioning of the direct-expansion type because of the above hazards.

Freon 12

Freon 12 (CCl_2F_2), or dichlorodifluoromethane, has been extensively used for refrigeration and air conditioning equipment. The latent heat of vaporization is 50 to 80 Btu/lb. It is noncombustible and nontoxic, but in the presence of an open flame, it breaks down and forms toxic gases. Some other features of Freon 12 are:

1. Water and moisture must be removed completely from Freon 12 systems to avoid icing in the expansion valve. Various types of dehydrators are used to remove moisture, such as silica gel and activated alumina.

2. Mineral oils of selected grade are used for lubrication, and these must be free of water. Oil and Freon are mutually soluble in each other, and liquid carryover can result in oil being pulled from the crankcase on reciprocating compressors, causing valve damage and loss of oil pressure. *Crankcase heaters* are used to drive off any Freon that has migrated or condensed in the crankcase during idle periods.

Copper tubing with sweated joints is the preferred piping material. *Freon 22* is also nontoxic and has characteristics that are similar to those of Freon 12; however, it is suited for low-temperature service (-40 to $-100°F$). It has been extensively used as a replacement for the type of service for which ammonia was used.

Because of the changing use of the Freons due to federal regulations, facility service operators may have to learn the properties of new refrigerants that are being developed due to the hazard of ozone depletion in the atmosphere of some past used Freons, including Freon 11, Freon 21, Freon 113, and Freon 114; all are used on centrifugal compressors operating with vacuum on the suction side.

Carbon dioxide

Carbon dioxide is a very high-pressure refrigerant requiring extra heavy piping. The gas condenses at 1066 psia at 87.8°F. The latent heat of vaporization is 115/Btu/lb. The pressures are so high that the gas volume is small, and piston displacement on compressors per minute per ton of refrigeration is less than one-third of that required for ammonia. Liquid CO_2 cannot exist at pressures less than about 75 psi.

Pressure-Temperature Characteristics

Tables 14.2, 14.3, and 14.4 provide some thermodynamic properties of ammonia, Freon 12, and Freon 22. These tables are useful because they show the temperature-pressure relationships of the refrigerants and include data on the heat of vaporization. Service facility operators should obtain similar tables for the refrigerant they may be working with.

TABLE 14.2 Saturated Liquid and Vapor Table for Ammonia

Temp., °F t	Pres. Abs. lb/in² p	Sp. vol of liquid/ft³/lb V_f	Sp. vol. of vapor, ft³ V_g	Density of vapor, lb/ft³ $1/V_g$	Enthalpy Btu/lb above −40°F Liquid h_f	Vaporization h_{fg}	Vapor h_g	Entropy Liquid S_f	Vapor S_g
−40	10.41	0.02322	24.86	0.04022	0.0	597.6	597.6	0.0000	1.4242
−30	13.90	.02345	18.97	.05271	10.7	590.7	601.4	.0250	1.4001
−20	18.30	.02369	14.68	.06813	21.4	583.6	605.0	.0497	1.3774
−10	23.74	0.02393	11.50	0.08695	32.1	576.4	608.5	0.0738	1.3558
−9	24.35		11.23	.08904	33.2	575.6	608.8	.0762	.3537
−8	24.97		10.97	.09117	34.3	574.9	609.2	.0786	.3516
−7	25.61		10.71	.09334	35.4	574.1	609.5	.0809	.3495
−6	26.26		10.47	.09555	36.4	573.4	609.8	.0833	.3474
−5	26.92	0.02406	10.23	0.09780	37.5	572.6	610.1	0.0857	1.3454
−4	27.59		9.991	.1001	38.6	571.9	610.5	.0880	.3433
−3	28.28		9.763	.1024	39.7	571.1	610.8	.0904	.3413
−2	28.98		9.541	.1048	40.7	570.4	611.1	.0928	.3393
−1	29.69		9.326	.1072	41.8	569.6	611.4	.0951	.3372
0	30.42	0.02419	9.116	0.1097	42.9	568.9	611.8	0.0975	1.3352
2	31.92		8.714	.1148	45.1	567.3	612.4	.1022	.3312
4	33.47		8.333	.1200	47.2	565.8	613.0	.1069	.3273
5	34.27	0.02432	8.150	0.1227	48.3	565.0	613.3	0.1092	1.3253
6	35.09		7.971	.1254	49.4	564.2	613.6	.1115	.3234
8	36.77		7.629	.1311	51.6	562.7	614.3	.1162	.3195
10	38.51	0.02446	7.304	0.1369	53.8	561.1	614.9	0.1208	1.3157
11	39.40		7.148	.1399	54.9	560.3	615.2	.1231	.3137
12	40.31		6.996	.1429	56.0	559.5	615.5	.1254	.3118
13	41.24		6.847	.1460	57.1	558.7	615.8	.1277	.3099
14	42.18		6.703	.1492	58.2	557.9	616.1	.1300	.3081
15	43.14	0.02460	6.562	0.1524	59.2	557.1	616.3	0.1323	1.3062
16	44.12		6.425	.1556	60.3	556.3	616.6	.1346	.3043
17	45.12		6.291	.1590	61.4	555.5	616.9	.1369	.3025
18	46.13		6.161	.1623	62.5	554.7	617.2	.1392	.3006
19	47.16		6.034	.1657	63.6	553.9	617.5	.1415	.2988
20	48.21	0.02474	5.910	0.1692	64.7	553.1	617.8	0.1437	1.2969
21	49.28		5.789	.1728	65.8	552.2	618.0	.1460	.2951
22	50.36		5.671	.1763	66.9	551.4	618.3	.1483	.2933
23	51.47		5.556	.1800	68.0	550.6	618.6	.1505	.2915
24	52.59		5.443	.1837	69.1	549.8	618.9	.1528	.2897
25	53.73	0.02488	5.334	0.1875	70.2	548.9	619.1	0.1551	1.2879
26	54.90		5.227	.1913	71.3	548.1	619.4	.1573	.2861
27	56.08		5.123	.1952	72.4	547.3	619.7	.1596	.2843
28	57.28		5.021	.1992	73.5	546.4	619.9	.1618	.2825
29	58.50		4.922	.2032	74.6	545.6	620.2	.1641	.2808
30	59.74	0.02503	4.825	0.2073	75.7	544.8	620.5	0.1663	1.2790
31	61.00		4.730	.2114	76.8	543.9	620.7	.1686	.2773
32	62.29		4.637	.2156	77.9	543.1	621.0	.1708	.2755
33	63.59		4.547	.2199	79.0	542.2	621.2	.1730	.2738
34	64.91		4.459	.2243	80.1	541.4	621.5	.1753	.2721
35	66.26	0.02518	4.373	0.2287	81.2	540.5	621.7	0.1775	1.2704
36	67.63		4.289	.2332	82.3	539.7	622.0	.1797	.2686
37	69.02		4.207	.2377	83.4	538.8	622.2	.1819	.2669
38	70.43		4.126	.2423	84.6	537.9	622.5	.1841	.2652
39	71.87		4.048	.2470	85.7	537.0	622.7	.1863	.2635
40	73.32	0.02533	3.971	0.2518	86.8	536.2	623.0	0.1885	1.2618

TABLE 14.2 Saturated Liquid and Vapor Table for Ammonia *(Continued)*

Temp., °F t	Pres. Abs. lb/in² p	Sp. vol of liquid/ft³/lb V_f	Sp. vol. of vapor, ft³ V_g	Density of vapor, lb/ft³ $1/V_g$	Enthalpy Btu/lb above −40°F — Liquid h_f	Enthalpy — Vaporization h_{fg}	Enthalpy — Vapor h_g	Entropy — Liquid S_f	Entropy — Vapor S_g
41	74.80		3.897	.2566	87.9	535.3	623.2	.1908	.2602
42	76.31		3.823	.2616	89.0	534.4	623.4	.1930	.2585
43	77.83		3.752	.2665	90.1	533.6	623.7	.1952	.2568
44	79.38		3.682	.2716	91.2	532.7	623.9	.1974	.2552
45	80.96	0.02548	3.614	0.2767	92.3	531.8	624.1	0.1996	1.2535
46	82.55		3.547	.2819	93.5	530.9	624.4	.2018	.2519
47	84.18		3.481	.2872	94.6	530.0	624.6	.2040	.2502
48	85.82		3.418	.2926	95.7	529.1	624.8	.2062	.2486
49	87.49		3.355	.2981	96.8	528.2	625.0	.2083	.2469
50	89.19	0.02564	3.294	0.3036	97.9	527.3	625.2	0.2105	1.2453
51	90.91		3.234	.3092	99.1	526.4	625.5	.2127	.2437
52	92.66		3.176	.3149	100.2	525.5	625.7	.2149	.2421
53	94.43		3.119	.3207	101.3	524.6	625.9	.2171	.2405
54	96.23		3.063	.3265	102.4	523.7	626.1	.2192	.2389
55	98.06	0.02581	3.008	0.3325	103.5	522.8	626.3	0.2214	1.2373
56	99.91		2.954	.3385	104.7	521.8	626.5	.2236	.2357
57	101.8		2.902	.3446	105.8	520.9	626.7	.2257	.2341
58	103.7		2.851	.3508	106.9	520.0	626.9	.2279	.2325
59	105.6		2.800	.3571	108.1	519.0	627.1	.2301	.2310
60	107.6	0.02597	2.751	0.3635	109.2	518.1	627.3	0.2322	1.2294
61	109.6		2.703	.3700	110.3	517.2	627.5	.2344	.2278
62	111.6		2.656	.3765	111.5	516.2	627.7	.2365	.2262
63	113.6		2.610	.3832	112.6	515.3	627.9	.2387	.2247
64	115.7		2.565	.3899	113.7	514.3	628.0	.2408	.2231
65	117.8	0.02614	2.520	0.3968	114.8	513.4	628.2	0.2430	1.2216
66	120.0		2.477	.4037	116.0	512.4	628.4	.2451	.2201
67	122.1		2.435	.4108	117.1	511.5	628.6	.2473	.2186
68	124.3		2.393	.4179	118.3	510.5	628.8	.2494	.2170
69	126.5		2.352	.4251	119.4	509.5	628.9	.2515	.2155
70	128.8	0.02632	2.312	0.4325	120.5	508.6	629.1	0.2537	1.2140
71	131.1		2.273	.4399	121.7	507.6	629.3	.2558	.2125
72	133.4		2.235	.4474	122.8	506.6	629.4	.2579	.2110
73	135.7		2.197	.4551	124.0	505.6	629.6	.2601	.2095
74	138.1		2.161	.4628	125.1	504.7	629.8	.2622	.2080
75	140.5	0.02650	2.125	0.4707	126.2	503.7	629.9	0.2643	1.2065
76	143.0		2.089	.4786	127.4	502.7	630.1	.2664	.2050
77	145.4		2.055	.4867	128.5	501.7	630.2	.2685	.2035
78	147.9		2.021	.4949	129.7	500.7	630.4	.2706	.2020
79	150.5		1.988	.5031	130.8	499.7	630.5	.2728	.2006
80	153.0	0.02668	1.955	0.5115	132.0	498.7	630.7	0.2749	1.1991
81	155.6		1.923	.5200	133.1	497.7	630.8	.2769	.1976
82	158.3		1.892	.5287	134.3	496.7	631.0	.2791	.1962
83	161.0		1.861	.5374	135.4	495.7	631.1	.2812	.1947
84	163.7		1.831	.5462	136.6	494.7	631.3	.2833	.1933
85	166.4	0.02687	1.801	0.5552	137.8	493.6	631.4	0.2854	1.1918
86	169.2		1.772	.5643	138.9	492.6	631.5	.2875	.1904
87	172.0		1.744	.5735	140.1	491.6	631.7	.2895	.1889
88	174.8		1.716	.5828	141.2	490.6	631.8	.2917	.1875
89	177.7		1.688	.5923	142.4	489.5	631.9	.2937	.1860
90	180.6	0.02707	1.661	0.6019	143.5	488.5	632.0	0.2958	1.1846

TABLE 14.2   Saturated Liquid and Vapor Table for Ammonia   (*Continued*)

Temp., °F t	Pres. Abs. lb/in^2 p	Sp. vol of liquid/ft^3/lb V_f	Sp. vol. of vapor, ft^3 V_g	Density of vapor, lb/ft^3 $1/V_g$	Enthalpy Btu/lb above $-40°F$ Liquid h_f	Vaporization h_{fg}	Vapor h_g	Entropy Liquid S_f	Vapor S_g
91	183.6		1.635	.6116	144.7	487.4	632.1	.2979	.1832
92	186.6		1.609	.6214	145.8	486.4	632.2	.3000	.1818
93	189.6		1.584	.6314	147.0	485.3	632.3	.3021	.1804
94	192.7		1.559	.6415	148.2	484.3	632.5	.3041	.1789
95	195.8	0.02727	1.534	0.6517	149.4	483.2	632.6	0.3062	1.1775
96	198.9		1.510	.6620	150.5	482.1	632.6	.3083	.1761
97	202.1		1.487	.6725	151.7	481.1	632.8	.3104	.1747
98	205.3		1.464	.6832	152.9	480.0	632.9	.3125	.1733
99	208.6		1.441	.6939	154.0	478.9	632.9	.3145	.1719
100	211.9	0.02747	1.419	0.7048	155.2	477.8	633.0	0.3166	1.1705
102	218.6		1.375	.7270	157.6	475.6	633.2	.3207	.1677
104	225.4		1.334	.7498	159.9	473.5	633.4	.3248	.1649
105	228.9	0.02769	1.313	0.7615	161.1	472.3	633.4	0.3269	1.1635
106	232.5		1.293	.7732	162.3	471.2	633.5	.3289	.1621
108	239.7		1.254	.7972	164.6	469.0	633.6	.3330	.1593
110	247.0	0.02790	1.217	0.8219	167.0	466.7	633.7	0.3372	1.1566
115	266.2	.02813	1.128	.8862	173.0	460.9	633.9	.3474	1.1497
120	286.4	.02836	1.047	.9549	179.0	455.0	634.0	.3576	1.1427
125	307.8	.02860	0.973	1.028	185.1	448.9	634.0	.3679	1.1358

Vacuum refrigerants require a large volume of gas flow; therefore, centrifugal compressors are used for these types of refrigerants because reciprocating compressors would require enormous cylinder sizes to compress the large volume of flow required at low pressure.

Refrigerant leak detection varies with the refrigerant when looking for leaking tubes, valves, and piping joints as well as leaks on the different components of compressors. *Ammonia leaks* have strong odors, but fine leaks can be detected by the burning of sulfur candles that generate a cloud of white smoke in the presence of ammonia. *Freon leaks* can be detected with a halide torch. An alcohol flame burns in the presence of copper, and when air with Freon passes through the flame, the flame becomes greenish in color. These leak detectors are usually sold as a kit, and the instructions inside the kit should be followed. The use of suitable gas masks and clothing during any leak detection procedure is also recommended. Most codes require gas masks that are approved by the U.S. Bureau of Mines for the refrigerant in use. Class A systems with over 1000 lb of refrigerant require gas masks in the machinery room near the main exit and one or two more inside the plant at convenient points for quick use.

Health Hazard of Refrigerants

Ventilation required is also governed by the amount of refrigerants that are in the system. For Class A systems of over 1000 lb of refrigerants, an exhaust fan with a capacity of at least 2000 cfm is required. Also acceptable are windows that can be opened on

TABLE 14.3 Saturated Liquid and Vapor Table for Freon 12

Temp., °F t	Pres. Abs., lb/in² p	Sp. Vol. Liquid ft³/lb v_f	Sp. Vol. Vapor ft³/lb V_g	Density of vapor, lb/ft³ $1/V_g$	Enthalpy Liquid, Btu/lb h_f	Enthalpy Vaporization, Btu/lb h_{fg}	Enthalpy Vapor, Btu/lb h_g	Entropy Liquid S_f	Entropy Vapor S_g	Temp., °F t
−40	9.32	0.0106	3.911	0.2557	0	73.50	73.50	0	0.17517	−40
−30	12.02	.0107	3.088	.3238	2.03	72.67	74.70	0.00471	.17387	−30
−20	15.28	.0108	2.474	.4042	4.07	71.80	75.87	.00940	.17275	−20
−10	19.20	.0109	2.003	.4993	6.14	70.91	77.05	.01403	.17175	−10
0	23.87	.0110	1.637	.6109	8.25	69.96	78.21	.01869	.17091	0
2	24.89	.0110	1.574	.6352	8.67	69.77	78.44	.01961	.17075	2
4	25.96	.0111	1.514	.6606	9.10	69.57	78.67	.02052	.17060	4
5	26.51	.0111	1.485	.6735	9.32	69.47	78.79	.02097	.17052	5
6	27.05	.0111	1.457	.6864	9.53	69.37	78.90	.02143	.17045	6
8	28.18	.0111	1.403	.7129	9.96	69.17	79.13	.02235	.17030	8
10	29.35	0.0112	1.351	0.7402	10.39	68.97	79.36	0.02328	0.17015	10
12	30.56	.0112	1.301	.7687	10.82	68.77	79.59	.02419	.17001	12
14	31.80	.0112	1.253	.7981	11.26	68.56	79.82	.02510	.16987	14
16	33.08	.0112	1.207	.8288	11.70	68.35	80.05	.02601	.16974	16
18	34.40	.0113	1.163	.8598	12.12	68.15	80.27	.02692	.16961	18
20	35.75	0.0113	1.121	0.8921	12.55	67.94	80.49	0.02783	0.16949	20
22	37.15	.0113	1.081	.9251	13.00	67.72	80.72	.02873	.16938	22
24	38.58	.0113	1.043	.9588	13.44	67.51	80.95	.02963	.16926	24
26	40.07	.0114	1.007	.9930	13.88	67.29	81.17	.03053	.16913	26
28	41.59	.0114	0.973	1.028	14.32	67.07	81.39	.03143	.16900	28
30	43.16	0.0115	0.939	1.065	14.76	66.85	81.61	0.03233	0.16887	30
32	44.77	.0115	.908	1.102	15.21	66.62	81.83	.03323	.16876	32
34	46.42	.0115	.877	1.140	15.65	66.40	82.05	.03413	.16865	34
36	48.13	.0116	.848	1.180	16.10	66.17	82.27	.03502	.16854	36
38	49.88	.0116	.819	1.221	16.55	65.94	82.49	.03591	.16843	38

14.13

TABLE 14.3 Saturated Liquid and Vapor Table for Freon 12 (*Continued*)

Temp., °F t	Pres. Abs., lb/in² p	Sp. Vol. Liquid ft³/lb v_f	Sp. Vol. Vapor, ft³/lb v_g	Density of vapor, lb/ft³ $1/v_g$	Enthalpy Liquid, Btu/lb h_f	Enthalpy Vapor-ization, Btu/lb h_{fg}	Enthalpy Vapor, Btu/lb h_g	Entropy Liquid s_f	Entropy Vapor s_g	Temp., °F t
40	51.68	0.0116	0.792	1.263	17.00	65.71	82.71	0.03680	0.16833	40
42	53.51	.0116	.767	1.304	17.46	65.47	82.93	.03770	.16823	42
44	55.40	.0117	.742	1.349	17.91	65.24	83.15	.03859	.16813	44
46	57.35	.0117	.718	1.393	18.36	65.00	83.36	.03948	.16803	46
48	59.35	.0117	.695	1.438	18.82	64.74	83.57	.04037	.16794	48
50	61.39	0.0118	0.673	1.485	19.27	64.51	83.78	0.04126	0.16785	50
52	63.49	.0118	.652	1.534	19.72	64.27	83.99	.04215	.16776	52
54	65.63	.0118	.632	1.583	20.18	64.02	84.20	.04304	.16767	54
56	67.84	.0119	.612	1.633	20.64	63.77	84.41	.04392	.16758	56
58	70.10	.0119	.593	1.686	21.11	63.51	84.62	.04480	.16749	58
60	72.41	0.0119	0.575	1.740	21.57	63.25	84.82	0.04568	0.16741	60
62	74.77	.0120	.557	1.795	22.03	62.99	85.02	.04657	.16733	62
64	77.20	.0120	.540	1.851	22.49	62.73	85.22	.04745	.16725	64
68	82.24	.0121	.508	1.968	23.42	62.20	85.62	.04921	.16709	68
70	84.82	0.0121	0.493	2.028	23.90	61.92	85.82	0.05009	0.16701	70
72	87.50	.0121	.479	2.090	24.37	61.65	86.02	.05097	.16693	72
74	90.20	.0122	.464	2.153	24.84	61.38	86.22	.05185	.16685	74
76	93.00	.0122	.451	2.218	25.32	61.10	86.42	.05272	.16677	76
78	95.85	.0123	.438	2.284	25.80	60.81	86.61	.05359	.16669	78
80	98.76	0.0123	0.425	2.353	26.28	60.52	86.80	0.05446	0.16662	80
82	101.7	.0123	.413	2.423	26.76	60.23	86.99	.05534	.16655	82
86	107.9	.0124	.389	2.569	27.72	59.65	87.37	.05708	.16640	86
88	111.1	.0124	.378	2.645	28.21	59.35	87.56	.05795	.16632	88

90	114.3	0.0125	0.368	2.721	28.70	59.04	87.74	0.05882	0.16624	90
92	117.7	.0125	.357	2.799	29.19	58.73	87.92	.05969	.16616	92
94	121.0	.0126	.347	2.880	29.68	58.42	88.10	.06056	.16608	94
96	124.5	.0126	.338	2.963	30.18	58.10	88.28	.06143	.16600	96
98	128.0	.0126	.328	3.048	30.67	57.78	88.45	.06230	.16592	98
100	131.6	0.0127	0.319	3.135	31.16	57.46	88.62	0.06316	0.16584	100
102	135.3	.0127	.310	3.224	31.65	57.14	88.79	.06403	.16576	102
104	139.0	.0128	.302	3.316	32.15	56.80	88.95	.06490	.16568	104
110	150.7	.0129	.277	3.610	33.65	55.78	89.43	.06749	.16542	110
120	171.8	.0132	.240	4.167	36.16	53.99	90.15	.07180	.16495	120
130	194.9	.0134	.208	4.808	38.69	52.07	90.76	.07607	.16438	130
140	220.2	.0138	.180	5.571	41.24	50.00	91.24	.08024	.16363	140

14.15

TABLE 14.4 Saturated Liquid and Vapor Table for Freon 22

Temp. °F t	Pres. Abs., lb/in² p	Sp. Vol. Liquid, ft³/lb V_f	Sp. Vol. Vapor, ft³/lb V_g	Density of Vapor, lb/ft³ $1/V_g$	Enthalpy above −40°F Liquid, Btu/lb h_f	Enthalpy above −40°F Vaporization, Btu/lb h_{fg}	Enthalpy above −40°F Vapor, Btu/lb h_g	Entropy Liquid, S_f	Entropy Vapor, S_g	Temp., °F t
−155	0.19901	0.0102	188.13	0.00532	−29.05	115.85	86.80	−0.08075	0.29958	−155
−150	0.26049	.0103	146.06	0.00685	−27.77	115.15	87.38	−0.07670	0.29523	−150
−145	.33754	.0103	114.51	.00873	−26.50	114.46	87.96	−.07265	.29118	−145
−140	.43323	.0103	90.163	.01103	−25.23	113.78	88.55	−.06865	.28736	−140
−135	.55106	.0104	72.327	.01382	−23.97	113.10	89.13	−.06471	.28372	−135
−130	.69492	.0104	58.214	.01717	−22.71	112.43	89.72	−.06085	.28026	−130
−125	0.86922	0.0105	47.226	0.02117	−21.45	111.76	90.31	−0.05706	0.27695	−125
−120	1.0788	.0105	38.600	.02590	−20.20	111.10	90.90	−.05335	.27380	−120
−115	1.3291	.0106	31.773	.03147	−18.96	110.45	91.49	−.04970	.27082	−115
−110	1.6261	.0106	26.329	.03798	−17.71	109.80	92.09	−.04609	.26798	−110
−105	1.9760	.0106	21.960	.04553	−16.46	109.15	92.69	−.04254	.26527	−105
−100	2.3861	0.0107	18.426	0.05427	−15.21	108.50	93.29	−0.03903	0.26269	−100
−95	2.8649	.0107	15.544	.06433	−13.96	107.85	93.89	−.03557	.26023	−95
−90	3.4173	.0108	13.196	.07578	−12.71	107.20	94.49	−.03216	.25788	−90
−85	4.0554	.0108	11.256	.08884	−11.45	106.55	95.10	−.02881	.25563	−85
−80	4.7871	.0109	9.6497	.10363	−10.20	105.90	95.70	−.02551	.25347	−80
−75	5.6224	0.0109	8.3112	0.12032	− 8.94	105.24	96.30	−0.02224	0.25139	−75
−70	6.5711	.0110	7.1917	.13905	− 7.67	104.57	96.90	−.01899	.24941	−70
−65	7.6456	.0110	6.2488	.16003	− 6.41	103.91	97.50	−.01576	.24757	−65
−60	8.8562	.0111	5.4520	.18342	− 5.14	103.24	98.10	−.01256	.24580	−60
−55	10.224	.0112	4.7710	.20960	− 3.85	102.55	98.70	−.00939	.24407	−55
−50	11.744	0.0112	4.1948	0.23839	− 2.56	101.86	99.30	−0.00623	0.24245	−50
−45	13.428	.0113	3.7038	.26999	− 1.27	101.16	99.89	−.00311	.24088	−45
−40	15.309	.0114	3.2787	.30500	0.02	100.46	100.48	0.00000	.23942	−40
−35	17.391	.0114	2.9116	.34345	1.33	99.74	101.07	.00309	.23799	−35
−30	19.689	.0115	2.5936	.38557	2.64	99.01	101.65	.00616	.23663	−30

−25	22.217	0.0116	2.3170	.43159	3.97	98.26	102.23	0.00922	0.23531	−25
−20	24.992	.0116	2.0755	.48181	5.31	97.50	102.81	.01227	.23406	−20
−15	28.031	.0117	1.8639	.53650	6.66	96.72	103.38	.01531	.23285	−15
−10	31.344	.0118	1.6783	.59583	8.01	95.93	103.94	.01833	.23170	−10
− 5	34.951	.0118	1.5148	.66015	9.37	95.13	104.50	.02135	.23061	− 5
0	38.870	0.0119	1.3702	0.72980	10.74	94.30	105.04	0.02435	0.22953	0
5	43.118	.0120	1.2421	0.80510	12.13	93.45	105.58	.02735	.22849	5
10	47.66	.012081	1.1295	0.88532	12.89	93.02	105.91	.02884	.22692	10
20	58.00	.012250	0.93624	1.0683	15.68	91.27	106.95	.03468	.22498	20
30	69.97	.012430	0.78125	1.2800	18.55	89.40	107.95	.04054	.22314	30
40	83.72	0.012619	0.65591	1.5246	21.52	87.39	108.91	0.04642	0.22134	40
50	99.40	.012818	.55371	1.8060	24.55	85.25	109.80	.05235	.21964	50
60	117.2	.013029	.46951	2.1299	27.65	82.95	110.60	.05833	.21797	60
70	137.2	.013251	.40000	2.5000	30.81	80.50	111.31	.06436	.21636	70
80	159.7	.013487	.34174	2.9262	34.09	77.86	111.95	.07035	.21469	80
90	184.8	0.013735	0.29284	3.4148	37.43	75.06	112.49	0.07630	0.21281	90
100	212.6	.014015	.25169	3.9731	40.80	72.08	112.88	.08221	.21102	100
110	243.4	.014331	.21673	4.6140	44.17	68.94	113.11	.08810	.20913	110
120	277.3	.014685	.18709	5.3451	47.67	65.67	113.34	.09398	.20728	120

14.17

TABLE 14.5   Worker exposure and Limits of Concentration for Ammonia as Recommended by The Manufacturing Chemists Association

Gaseous, ppm	Concentration, %	Effects on unprotected worker	Exposure period
50	0.006	Least detectable odor	Permissible for 8-h working exposure
100	0.0125	No adverse effects for unprotected worker	
400	0.05	Causes irritation of throat	Ordinarily no serious results following
700	0.1	Causes irritation of eyes	infrequent short exposures (less than 1 h)
1720	0.22	Causes convulsive coughing	No exposure permissible—may be fatal after short exposure (less than ½ h)
5000–10,000	...	Causes respiratory spasm, strangulation, asphyxia	No exposure permissible (rapidly fatal)

opposite walls, each with a 25-ft² area. If only one side of the machinery room has windows, a window area of at least 60 ft² is required.

While *Freon-type* refrigerants are considered nontoxic, leaks with over 20 percent concentration with air cause unconsciousness due to lack of oxygen. Freons break down in the presence of hot surfaces over 1000°F and form toxic products, or poisonous gases.

Concentrations of 5 percent and over of *carbon dioxide* in air can also cause unconsciousness due to lack of oxygen. At about 9 percent concentration suffocation can result.

Ammonia at a temperature over 800°F decomposes into its constituent parts of nitrogen and hydrogen. Hydrogen can form explosive mixtures with air. At high temperatures, ammonia burns with a greenish-yellow flame.

The Manufacturing Chemists Association and the Compressed Gas Association have both published extensive recommendations and suggestions for safety and first aid in connection with ammonia. Following are some of the points covered: While ammonia is not a poisonous gas, it does severely irritate the mucous membranes of the eyes, nose, throat, and lungs. A very small concentration is easily detected by its sharp pungent odor. See Table 14.5 for exposure limits and concentrations. Liquid ammonia should never come in contact with the skin because it freezes tissue, subjecting it to caustic action. Symptoms of such action are similar to symptoms of a burn. Employees handling ammonia should understand the possible hazards and should be instructed and trained in the following: (1) Location of gas masks and other protective equipment. (2) Location of safety showers, bubbler drinking fountains, water hoses, exits, and first-aid equipment. An adequate water supply is extremely important because quick and thorough washing is essential when ammonia comes in contact with eyes or skin. (3) Proper use of gas masks and other protective equipment. (4) The urgency of immediately reporting any unusual odor of ammonia. (5) Proper conduct in case of an emergency. (6) Methods of properly handling ammonia containers and approved procedures for cleaning pipes and other equipment.

THERMODYNAMIC CONSIDERATIONS

Figure 14.2 shows the basic components of a compression-type refrigeration system. Each component performs a function that can be analyzed by using the properties of the refrigerant as shown in the saturated liquid and vapor tables for that refrigerant. The four components to be considered are the expansion valve, evaporator, compressor, and condenser.

Expansion Valve

Liquid refrigerant enters the expansion valve and drops in pressure. This causes some flashing into gas, which can accomplish no cooling but is a minor part of the cooling that does take place in the evaporator. The expansion valve equation for the transfer of energy is

$$h_{f1} = h_{fR} + x h_{fgR}$$

where h_{f1} = enthalpy of *liquid refrigerant* at the *temperature at which it enters the expansion valve,* Btu/lb
h_{fR} = enthalpy of *liquid* at *evaporator pressure*
h_{fgR} = latent heat of evaporation of refrigerant at *evaporator pressure*
x = quality, expressed as a decimal, of the refrigerant after passing through the expansion valve; this also expresses the weight in pounds of flash gas formed per pound of refrigerant.

Example. Freon 12 at 95 psi and at 70°F expands through an expansion valve to an evaporator at 34.4 psia and 18°F. What is the enthalpy h per pound of refrigerant before and after it passes through the expansion valve and the pounds of flash vapor formed in the expansion valve? See Table 14.3. The h_f at 70°F *liquid* is equal to 23.9 Btu/lb. Since no heat is added or rejected in the expansion valve, the enthalpy is the same before and after the expansion valve = 23.9 Btu/lb. To find the pounds of flash vapor, use the equation

$$h_{f1} = h_{fR} + x h_{fgR}$$

where h_{f1} = 23.9, and from Table 14.3 at 18°F, h_{fR} for *liquid* = 12.12, and h_{fgR} for *vaporization* at 18°F = 68.15. Then substituting to solve for x in the equation

$$23.9 = 12.12 + 68.15x$$

$$x = \frac{11.7}{68.15} = 0.172 \text{ lb of refrigerant vaporized}$$

per each pound of refrigerant passing through the expansion valve

Evaporator

In the evaporator, the refrigerant as a liquid from the expansion valve becomes vapor as it absorbs heat from the space being cooled. The heat absorbed is $Q_R = h_R - h_f$, where Q_R = Btu absorbed in the evaporator per pound of refrigerant, h_R = enthalpy of *vapor* leaving the evaporator, Btu per lb, and h_f = enthalpy of *liquid* refrigerant at temperature supplied to the expansion valve.

The pounds of refrigerant circulated per minute per ton of refrigeration are

$$W_R = \frac{200}{Q_R} = \frac{200}{h_R - h_f}$$

Example. A refrigeration system uses Freon 12. The dry vapor leaves the evaporator at 5°F and 26.51 psia. The liquid supply to the expansion valve is 70°F. What are the pounds of Freon 12 circulated per minute per ton of refrigeration? The h_R of vapor at

5°F from Table 14.3 is 78.79 Btu/lb. The h_f for *liquid* at 70°F from Table 14.3 is 23.9 Btu/lb. Substituting these values in the W_R equation,

$$W_R = \frac{200}{78.79 - 23.9} = 3.64 \text{ lb of refrigerant}$$

per minute per ton of refrigeration

Compressor

The function of the compressor is to pull the vapor out of the evaporator so that cooling can continue and then to compress the vapor to a pressure that will liquefy it again with the available condensing medium. The theoretical work performed by the compressor is equal to the enthalpy of the gas after compression less the enthalpy of the gas before the compression. From a thermodynamic perspective, the entropy is assumed to be constant. Therefore, the theoretical work of compression W_T is

$$W_T = (h_D - h_R) \text{ Btu per pound of refrigerant}$$

where h_D = enthalpy of vapor at discharge pressure
h_R = enthalpy of vapor entering the compressor

The theoretical *horsepower per ton* of refrigeration is calculated by

$$\text{hp} = \frac{\text{lb refrigerant}}{\text{min} \times \text{ton}} \times W_T \times \frac{1}{42.4}$$

Example. In the previous example for a Freon 12 system, the dry vapor leaves evaporator at 5°F and 26.51 psia. The gas is compressed to the condenser to a temperature of 86°F and 107.9 psia. What is the theoretical work of compression, and using the previously established pounds of refrigerant per minute per ton of refrigeration, what is the theoretical horsepower per ton of refrigeration?

h_R of vapor at 5°F = 78.79 Btu/lb   (suction condition from Table 14.3)

h_D of vapor at 86°F = 87.37 Btu/lb   (discharge condition)

Theoretical work of compression = 87.37 − 78.79 = 8.58 Btu/lb of refrigerant. Using 3.64 lb of refrigerant per minute per ton of refrigeration from the previous problem, the theoretical horsepower per ton of refrigeration is

$$\text{Theoretical hp} = \frac{3.64 \times 8.58}{42.4} = 0.74 \text{ hp/ton of refrigeration}$$

The actual horsepower will be determined by the volumetric efficiency of the compressor. The *volumetric, or charge, efficiency* of a compressor is defined as the amount of vapor handled in cubic feet per minute at suction pressure and temperature, divided by the piston displacement per minute. In addition to volumetric efficiency, mechanical and drive efficiency must also be considered.

Example. In the previous problem a theoretical horsepower of 0.74 hp/ton of refrigeration was established. What would the motor horsepower have to be for a 50-ton system with a compressor of 80 percent volumetric efficiency, 80 percent mechanical efficiency, and a motor drive with an 85 percent efficiency?

$$\text{Hp required} = \frac{0.74 \times 50}{0.8 \times 0.8 \times 0.85} = 68$$

Condenser

The vapor from the compressor, sometimes superheated above saturation temperature, is condensed by first removing the superheat and then the much larger latent heat of vaporization. Some subcooling below the condensation temperature may occur. The heat removed is equal to the enthalpy of the gas entering the condenser, less the enthalpy of the gas leaving the condenser, or

$$Q_C = (h_D - h_f)$$

where h_D = enthalpy of *vapor* entering the compressor, and h_F = enthalpy of *liquid* leaving the compressor.

Example.   Freon 12 enters a condenser at 86°F and 107.9 psia and leaves as a liquid at the same pressure. What is the heat removed in the condenser per pound of refrigerant?

$$h_D = 87.37 \text{ Btu/lb} \qquad \text{(from Table 14.3)}$$

$$h_f = 27.72 \text{ Btu/lb}$$

$$Q_C = 87.37 - 27.72 = 59.65 \text{ Btu/lb of refrigerant}$$

An energy balance of a compression system will reveal that the heat given up in the condenser almost equals the heat absorbed in the evaporator plus the heat equivalent of work in compressing the gas to a pressure so that it can be condensed by an available cooling medium in the condenser. There are some extraneous heat losses in piping and valves.

CLASSIFICATIONS OF REFRIGERATION SYSTEMS

Refrigeration systems may be classified as (1) *compression systems* using refrigerants such as ammonia, Freon, and carbon dioxide as a subheading (i.e., a Freon compression system), (2) *absorption systems* using a heat source to provide the driving force instead of a compressor, (3) *steam-jet compressor systems* using a steam jet to draw vapor from a flash chamber, (4) *multi-stage low-temperature* refrigeration in which compressors are cascaded to obtain low temperatures, and (5) *direct* and *indirect systems*. In a *direct system* or direct expansion evaporator, the boiling refrigerant in the evaporator coils cools the air or stored product by directly expanding in the place to be cooled. In the *indirect system,* water or brine is cooled in the evaporator, and this is delivered to the space or product to be cooled, hence the terms *chilled water system* and *brine system.*
The compression system using compressors has already been described.

Absorption Systems

An absorption system makes use of the ability of an absorber substance to absorb relatively large volumes of the vapor of another substance, usually a liquid (the refrigerant). The absorber material absorbs the liquid at cold temperature and becomes saturated. By

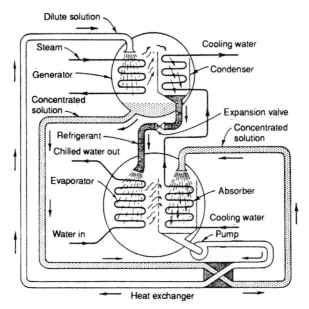

FIGURE 14.3 Lithium-bromide absorption refrigeration system is assembled as a machine package with automatic controls for fast installation.

pumping the saturated solution to a generator, the moisture is driven off, and the absorbent is returned to the absorber vessel.

Lithium Bromide Absorption System. (See Fig. 14.3.) In the lithium bromide system lithium bromide salt, which has great capacity for absorbing water vapor, is used as the absorbent, and pure water is used as the refrigerant. The basic components required are generator, condenser, evaporator, and absorber. The system makes use of the facts that, under a low pressure (high vacuum), water boils at a low temperature and a lithium bromide solution easily absorbs this water vapor.

In some installations the basic components of generator, condenser, evaporator, and absorber are separately housed. In others, they are installed into a single horizontal tank, or the "machine," as it is often called in the trade, placed in a vertical position. The refrigerant is usually lithium bromide.

The various units of an absorption system, and their purpose, are:

1. *Evaporator or cooler.* This cools the medium to be cooled by the evaporation of the refrigerant. The water being chilled (see Fig. 14.3) is circulated through a coil submerged in the water that is being continually evaporated in this unit. Evaporation is the result of a high level of vacuum over the surface of the water. This vacuum is maintained by the absorption of the water vapor in the adjacent absorber. This part of the cycle is compared with the steam-jet system, the steam jet creating a vacuum over the water in place of the absorber.

2. *Condenser.* This unit liquefies the water vapor that is being driven off the refrigerant-water solution in the generator. Cooling water for the condenser coil may be obtained from a cooling tower, an evaporative condenser, or even a cold lake or river, if one is handy.

3. *Absorber.*   Here the water vapor from the evaporator comes into contact with the refrigerant and is absorbed thereby. The mixture is termed the *weak solution,* naturally since the *strong solution* from the generator is diluted by the water.

Because heat is produced in the absorber by the reaction, cooling coils are provided in the absorber, the cold water for these coils being obtained from any convenient source, including cooling towers and evaporative condensers. Figure 14.3 shows the strong solution flowing into the absorber (from nozzles); other designs spread the strong solution over the cooling coils, by means of spray nozzles, and the water vapor from the evaporator is discharged into contact with the spray.

4. *Generator.*   This is the part of the system in which the heat of operation is applied. In Fig. 14.3 heat is applied through steam coils. But in some designs, the generator may be direct fired, using oil or gas as convenient or economical. In the generator, the water vapor that was originally created by being driven out of the solution in the generator, then was condensed in the condenser in order to be revaporized in the evaporator, was absorbed by the strong solution in the absorber, and now is back in the generator as part of the solution and again feels the heat of distillation and once more, in the form of vapor, sets out on a repetition of the cycle.

5. *Purger.*   Since the high vacuum necessary for effective operation would be affected adversely by noncondensable gases or air, a purge unit may be provided in the form of a vacuum pump or eductor.

6. *Controls and protective devices.*   These include a *low-temperature cutout,* which stops the machine if the evaporator temperature is too low, and liquid-level *float switches* that shut down operations if the liquid level in the evaporator becomes abnormal, either too high or too low. In addition, ammonia-water machines may be subject to a high-pressure switch, and *direct-fired* machines may be subject to a high-limit temperature switch.

The absorption system is best suited to applications where waste heat, or low-pressure steam, is available for boiling out the water from the lithium bromide solution in the generator. In this case, the absorption system tends to be less expensive.

Lithium bromide absorption systems use chilled water and are primarily used for air conditioning service. Freeze-up of the chilled water can occur if the lithium bromide concentration is out of the limit set by the manufacturer of the unit. A common mistake made by operators is to lower the chilled water cutout to obtain colder chilled water for higher heat transfer because the machine may be too small for the load. When mild weather arrives, the machine may reach this lower setting that an operator set for the hot weather, and if it is below 32°F, a chilled water freeze-up can result. Facility service operators should closely follow the manufacturer's instructions to avoid this type of problem.

Steam-Jet Compressor Refrigeration Systems

These are also called steam-jet ejectors. They use steam that is supplied to one or more nozzles as illustrated in Fig. 14.4 and then expanded. This causes the steam to leave the nozzles at high velocity and at the same time draws vapor from the flash chamber. The velocity energy of the mixture is reduced in a venturi-type nozzle called the booster compressor, and a resultant pressure increase occurs in the mixture. Design requires this final pressure to be sufficiently high for the steam to condense at the temperature available in the condenser.

In the booster condenser illustrated in Fig. 14.4, air and noncondensable gases that may be drawn into the system are removed in the air ejector. The chilled water from the

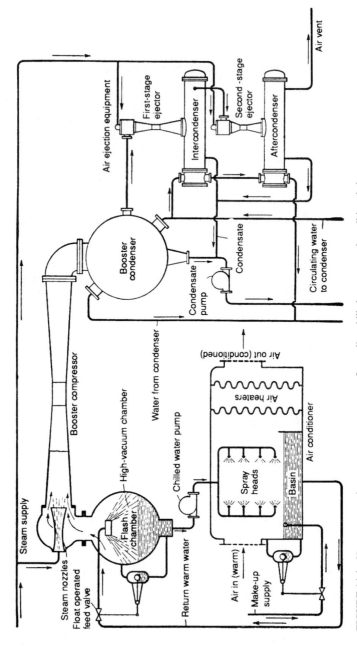

FIGURE 14.4 Vacuum steam-jet compressor arrangement for cooling chilled water to cool conditioned air.

14.24

flash chamber is pumped to the spaces to be cooled. This water, after absorbing the heat load, is returned to the flash chamber to be cooled again to repeat the cycle.

This system is feasible where steam costs are low; however, the condenser load, or water required, is 3 to 4 times greater than that for mechanical vapor compression systems. Practical chilled water temperatures cannot go below 40°F.

LOW-TEMPERATURE MULTISTAGE SYSTEMS

Commercial refrigeration that requires temperatures in the −40 to −160°F range require two- or three-stage compression. Another method used is to cascade the system by using two or more different refrigerants. In this system, a series of refrigerants having progressively lower boiling points are used. One refrigerant is the actual space cooler, while another is used to condense the refrigerant gas with the next lower boiling temperature. Figure 14.5 illustrates a two-stage cascade system. Starting with the *evaporator,* refrigerant gas is drawn into the *first compressor,* or low-pressure, first stage. This gas is compressed and passes through a water-cooled *intercooler.* It flows to the *interstage chamber* and is cooled to saturation temperature when it comes in contact with the liquid refrigerant vaporizing at interstage pressure. This vapor is drawn off by the second-stage compressor, which raises the head pressure so that it can be condensed in the condenser. The refrigerant is stored in a receiver. A cooling coil is used as the liquid refrigerant flows to the expansion valve (E) to supply the evaporator. The float valve (F) is used to control the refrigerant level in the interstage chamber that cools the liquid going to the evaporator and also cools the vapor going to the second-stage compressor. This compressor must handle the amount of flash and intercooling vapor that comes from the interstage chamber and thus handles more weight of refrigerant than does the first stage. The arrangement illustrated in Fig. 14.5 is also called a *booster system.*

Cryogenics

The term *cryogenic* refers to processes that use temperatures below −100°F. Special construction materials must be used since most carbon-steel materials become brittle at temperatures below −20°F and thus are subject to possible cracking failures.

Among the most economically and technically important physical reactions for the use of cryogenic temperatures are the cooling and liquefaction of gases and the distillation and fractional condensation of liquefied-gas mixtures to yield pure-component streams. Cooling and liquefaction are the basis of every cryogenic process. Figure 14.6 shows the liquefaction temperatures of some gases.

The first and still greatest application of the cryogenic field is the bulk production of liquefied gases. The second use is the application of very low temperatures to certain processes.

Present cryogenic plants are designed for the liquefaction and separation of ordinary atmospheric air into its components of oxygen, nitrogen, argon, neon, helium, krypton, and xenon. Hydrogen, ethylene, methane, and other gases are liquefied from petroleum gas streams.

Gases with low boiling points are used as a working media in cooling and liquefaction. For production of ultralow temperatures, helium is used, but below 1 K (−272°C or −443.6°F), liquid helium has a very low vapor pressure so that adiabatic demagnetization must be employed to reach temperatures close to absolute zero at present. Economically significant cryogenic processes do not occur at such extremes.

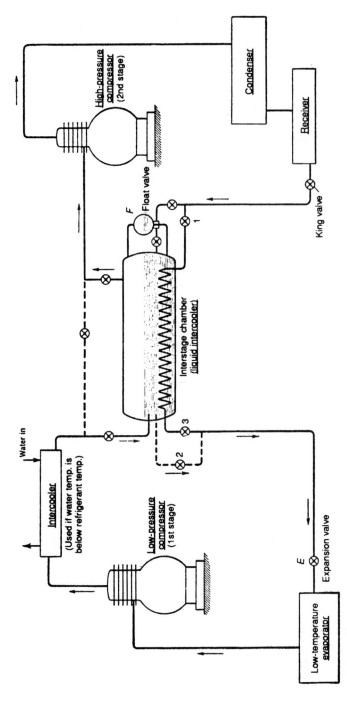

FIGURE 14.5 Two-compressor arrangement in a two-stage compression refrigeration system.

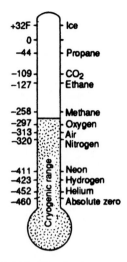

FIGURE 14.6 Liquefication temperatures of some gases.

Brine Systems

Brine is used for low-temperature refrigeration service and is an *indirect* cooling method. The usual brine is made from calcium chloride, and one of the functions of brine system operators is to maintain the proper concentration so that the brine does not freeze as it is cooled by a refrigerant such as ammonia in a submerged-type heat exchanger. Figure 14.7 shows the freezing temperatures, specific heat, and specific gravity of a pure calcium chloride brine. The brine should be tested monthly for specific gravity and pH and to note if any refrigerant is leaking into the brine.

The density of the brine should be such that the freezing point is from 10 to 15°F below the temperature to which the brine is to be cooled in the brine cooler or ice tank. Then there will be no danger of the brine freezing. The brine temperature upon leaving the cooler should also be about 15°F below the temperature to be held in the coldest rooms.

Example. With a room temperature of 24°F and a brine temperature of 9°F, the freezing point of brine is −5°F. To test brine for specific gravity, see Fig. 14.8. Pour a sample of brine (at 60°F) in a tall cylinder or salinometer jar. Float a hydrometer or salinometer in the liquid and read the scale where it coincides with the brine level (not top of meniscus). Insert a thermometer to determine the brine temperature; if it is above or below 60°F, the specific gravity reading must be corrected by a factor given in brine tables. If the specific gravity of the brine is not within the desired range, make the necessary adjustments.

Because iron corrodes more quickly in brine of a low pH and zinc is damaged at relatively high pH values, it is very important to maintain the pH level at the best intermediate point, which is about 7.5 to 8.5.

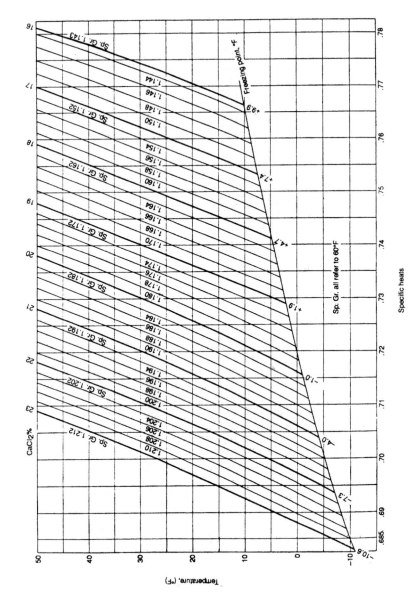

FIGURE 14.7 Characteristics of calcium chloride brine show freezing temperatures at different specific gravity or concentration of the brine solution.

14.28

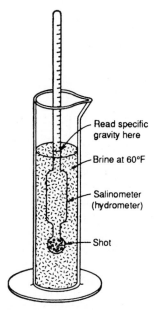

FIGURE 14.8 Brine should be tested at least once per month for specific gravity with a salinometer to detect any dilution that may result in the brine freezing in heat exchangers.

REFRIGERATION SYSTEM COMPONENTS

There are many variations in the components of a refrigeration system that needs review.

Evaporators

Evaporators are classified as *direct* and *indirect expansion, coil-type, bare* and *finned, shell and tube,* and *flooded* and *not flooded* (see Fig. 14.9). In a direct-expansion evaporator, the boiling refrigerant cools the air or substance to be cooled by heat transfer over the coils containing the refrigerant. In an indirect system, the refrigerant cools water, brine, or some other medium, and this is pumped to the space to be cooled as shown in Fig. 14.9*b*. In the flooded direct-expansion evaporator as shown in Fig. 14.10, the liquid refrigerant, controlled by a float-type feed valve, maintains a refrigerant level above the evaporator cooling coils by means of a surge drum (accumulator). The evaporated gas from the flooded coil is collected in a gas header and rises to the top of the surge drum, where the compressor suction line directs the gas back to the compressor.

Shell and tube evaporators, as the name implies, have a shell within which tubes are inserted for heat transfer. Shell and tube evaporators are usually employed in indirect systems, cooling either chilled water or brine. They may have several passes. Shell and tube evaporators may be operated flooded, which is common for large-tonnage units. In the flooded cooler, or evaporator, the refrigerant is in the shell, and the water or brine is

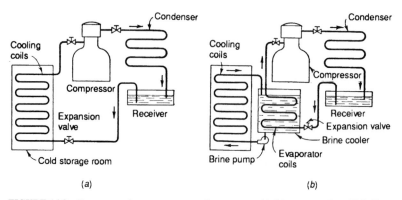

FIGURE 14.9   Two types of evaporator expansion systems. (*a*) Direct expansion; (*b*) indirect expansion using brine.

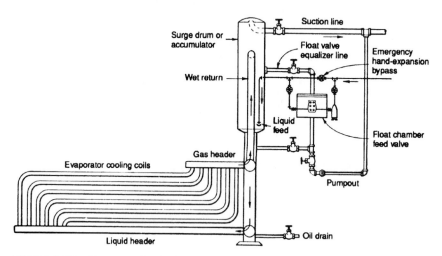

FIGURE 14.10   Direct-expansion flooded evaporator has surge drum and float-level drum control.

in the tubes. In the dry-type shell and tube heat exchanger, the refrigerant is in the tubes, and the water is pumped through the shell.

Shell and tube heat exchangers of the flooded type must have liquid level controls in order to prevent liquid carryover back into the compressor. It is common practice to use eliminator plates above the tubes on flooded units that separate out liquid particles and send the particles back into the evaporator.

A common hazard with a shell and tube evaporator is the danger of freezing the liquid being cooled inside the evaporator. Protection is now provided against this hazard by (1) having a low-temperature chilled water cutout, (2) having a low refrigerant temperature cutout as back-up, and (3) to prevent freeze-up in a dormant chiller, having a flow-switch to make sure there is circulation of the fluid before the refrigeration cycle starts.

Expansion Valves

Expansion valves are used to control the flow of the liquid refrigerant into the evaporator and, depending on the system, may be (1) manually operated, (2) one that uses capillary tubes on small domestic-type refrigerators, (3) thermostatic expansion valves that aim for a constant temperature, (4) the automatic diaphragm expansion or constant pressure type, or (5) the high- or low-side float type.

The *manual type* uses a needle valve to feed the liquid refrigerant to the evaporator and is quite often used in fast-freezer operations.

Although the *capillary tube* is not a valve, it does serve as an expansion valve in household units and in some small commercial systems. It is a coil or length of fine tubing that has a very small orifice, usually 0.03 to 0.10 in diameter. The high pressure is dissipated in forcing the liquid through this small orifice, and a predetermined amount of liquid at a reduced pressure is allowed to flow to the evaporator.

There are several types of thermostatic expansion valves (see Fig. 14.11a). All have a thermostatic bulb clamped to the exit from the evaporator or to the suction line to the compressor. The bulb has a refrigerant inside that reacts to the gas leaving the evaporator. If the refrigerant-leaving temperature from the evaporator is too high, the higher temperature causes a rise in the thermostatic bulb's pressure, which in turn actuates a bellows or diaphragm chamber, which eventually causes the refrigerant needle valve to open more so that more refrigerant is supplied to the evaporator. From 3 to 20° of superheat is incorporated in the device to give the desired rates of flow. Figure 14.11b shows the installation of an expansion valve in a direct-expansion system.

The *automatic diaphragm* expansion valve is shown in Fig. 14.12. This valve has a spring-loaded diaphragm which is acted on by evaporator pressure to control the flow to the evaporator with the goal of maintaining a constant suction pressure. It is also called a *constant-back-pressure* valve.

Automatic *high-side float valves* are similar to a steam trap. They are designed to deliver all the refrigerant coming from the condenser to the evaporator, thus requiring a refrigerant charge that stores the liquid largely in the evaporator and not in a receiver. This requires the evaporator to be designed to handle the liquid coming from the condenser without danger of sending liquid slugs over to the compressor.

Automatic low-side floats operate to maintain a definite level of liquid in a flooded-type evaporator. As the refrigerant in the evaporator vaporizes, the float drops from the assigned level, and liquid refrigerant flows into the evaporator to restore the assigned level.

Also used are *automatic electric-operated* valves that are connected to a thermostat, where a solenoid arrangement usually holds the valve open against a spring, and this spring closes the valve by the setting on the thermostat when the electric circuit is broken. This is basically an ON-OFF switch arrangement.

Refrigeration Controls

Controls have been developed for all types of automatic operation, surveillance, corrections, data printouts, alarms, and automatic shutdown as the electronic revolution has entered the control field and by the use of computer tracking systems. The reader should be familiar with some simple controls such as those illustrated in Fig. 14.11b for a simple direct-expansion system used to spot-cool a room. Many of the basic controls illustrated are still used today, but perhaps with electronic-controlled relays. Among the controls illustrated in Fig. 14.11b are:

1. High- and low-pressure cutout switches that stop the electric motor through the motor's magnetic starter.
2. Solenoid-operated condenser water valve that automatically shuts off the water flow to the condenser when the compressor is tripped or cut out from service by the motor being stopped through the magnetic starter.
3. Solenoid-liquid refrigerant valve, which is controlled by a thermostat to open when the cooled room needs refrigerant flow.
4. Thermostatic expansion valve, which is regulated by the bulb that is usually mounted into or on the compressor suction pipe.
5. A diffuser fan that goes ON and OFF with a signal from the room thermostat. Some installations have the diffuser fan operate continuously.

The advantage of having a solenoid-operated liquid valve in series with the expansion valve is that it closes when the compressor stops, thus stopping flow to the evaporator, and this prevents the compressor from being slugged with liquid refrigerant on start-up.

Chilled water systems have additional controls and alarms because of the hazard of freeze-up in the evaporator. Air conditioning systems require controls for humidity and air quality (fresh air makeup). Facility service operators should become thoroughly familiar with the controls that regulate the system under their supervision. This will help in diagnosing problems that may appear in operation.

Compressors

Smaller-capacity refrigerating systems use mostly *hermetically sealed* refrigeration compressors with the motor directly connected or on the same shaft as the crankshaft with refrigerant gas cooling the motor's windings. Reciprocating compressors are

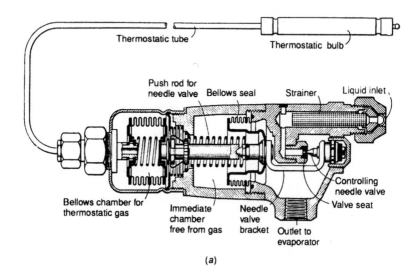

(a)

FIGURE 14.11 (a) Cross-sectional view of a thermostatic expansion valve.

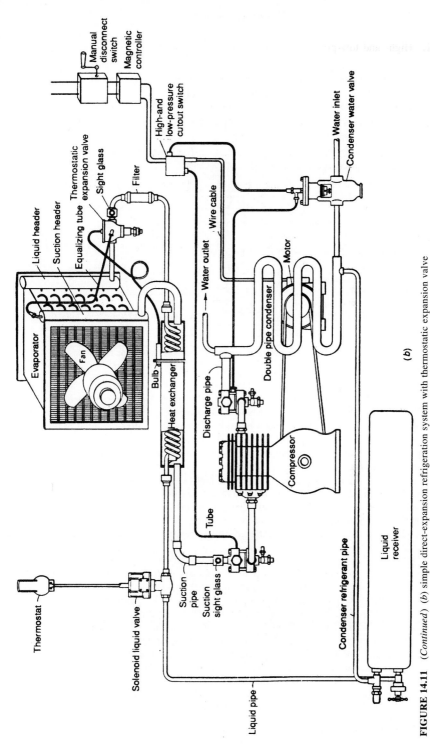

FIGURE 14.11 (*Continued*) (*b*) simple direct-expansion refrigeration system with thermostatic expansion valve and other controls.

14.33

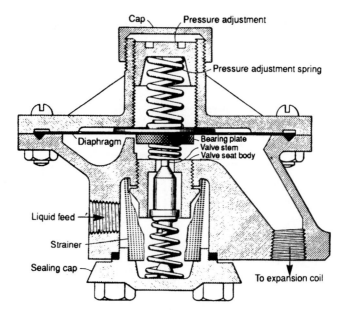

FIGURE 14.12   Cross-sectional view of constant-pressure type expansion valve.

arranged in straight horizontal or vertical, V or Y, formation. The radial units of the Chrysler Corp. are similar to reciprocating airplane engines. Hermetic compressors are enclosed in a factory-sealed gastight housing and cannot be disassembled in the field. Semihermetics (see Fig. 14.13a) also have the motor windings cooled by the refrigerant, but some field work is possible, such as when checking the crankcase after suitable pumpdown of the refrigerant.

Hermetic compressors and most semihermetics require special insulation on the windings since they are exposed to refrigerant gas flow over their surfaces. Normal insulation would be washed off or affected by the refrigerant. A burnout of windings in a hermetic or semihermetic is a more expensive repair since the refrigerant becomes polluted from the winding failure and generally has to be replaced. The compressor is also affected.

Figure 14.13b is a cross-sectional view of a two-cylinder, single-acting reciprocating refrigerant compressor. Poppet valves are shown, with the suction valves located in the top of the piston. This means the refrigerant passes through the crankcase and underside of the piston and is exposed to suction vapor pressure. Therefore, most automatic machines have crankcase heaters that prevent "Freon migration" or Freon condensing in the crankcase during idle periods. On start-up of the compressor, there is danger of the oil and Freon being sucked up on the suction stroke, thus damaging the suction valves as the liquid oil strikes the valves, even though the Freon may evaporate out of the oil. Always make sure the crankcase heaters are functional, especially in cooler weather when the Freon migration may occur.

Besides spring-loaded poppet valves, as are illustrated in Fig. 14.13b and operated by differential pressure on them, some compressors employ thin ribbons of steel that form flap-like closures, and these are called *feather valves*. Many compressors are built with *ring-plate valves*.

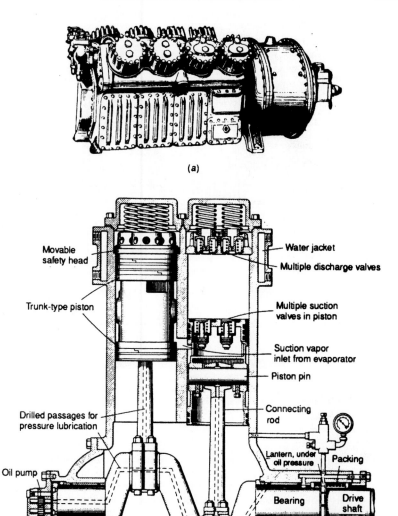

FIGURE 14.13   (*a*) Semihermetic Freon compressor with refrigerant-cooled electric motor; (*b*) vertical refrigeration compressor with suction valve on top of the piston.

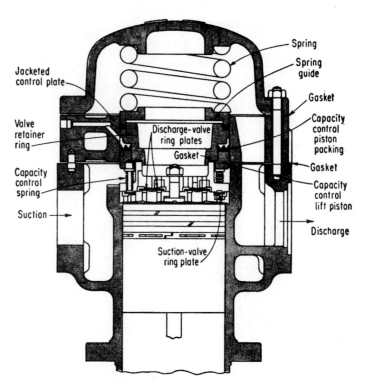

FIGURE 14.14 Safety heads are used to prevent cylinder head damage on the larger vertical reciprocating compressors.

Compressing the refrigerant vapor to a point where available condensing water or air can condense the refrigerant creates heat. Some of this heat is absorbed by the cylinder walls and if not removed can cause lubrication problems. Most ammonia and CO_2 compressors have water jackets to assist in removing this heat. Freon compressors usually depend on air-cooled fins since discharge temperatures are not as high as with ammonia and CO_2.

A safety head is illustrated in Fig. 14.14. In large vertical compressors the discharge valve is usually mounted in a separate plate. This plate is held down by heavy springs that bear against the cylinder head. If liquid refrigerant or oil (which are noncompressible) enters the cylinder, the safety head will lift against the springs. This prevents serious damage to the compressors.

Since reciprocating compressors are positive displacement types, it is important to have *overpressure protection* on the discharge side of the compressor *before* any closing valve. *Safety valves* should be installed and set at a pressure per the rating of the compressor or system with a capacity at least equal to the compressors capacity in cubic feet per minute.

Larger reciprocating compressors may have *bypass valves* that are opened in order to reduce the starting load (see Fig. 14.15). The starting bypass equalizes the pressure between the suction and discharge side of the compressor so that the driver has practically no load on it. After the compressor is up to operating conditions, the discharge valve is opened, and the bypass is closed.

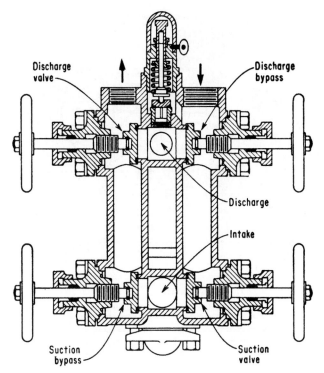

FIGURE 14.15 Crossover valves are used to equalize discharge and suction pressures on start-up of large compressors.

Unloading devices are used for starting and for load control. By means of tubing and actuator rings, a rod forces the suction valve off its seat, so the refrigerant gas merely surges back and forth through the open suction valve. After starting, the built-up oil pressure actuates the device through a bellows arrangement, and the suction valve returns to its normal functioning.

Centrifugal compressors are used to handle large volumes of gas flow and are usually used for large tonnage. They are similar in construction to centrifugal pumps, where the incoming gas enters the eye of the spinning impeller (Fig. 14.16a) and is thrown by centrifugal force to the periphery of the impeller. That causes the blades of the impeller to impart a high velocity to the gas and also to build up the pressure. From the impeller, the gas then flows into diffuser blades or into a volute, where some of the kinetic energy is converted into pressure.

Essentially high-speed units, centrifugal compressors may be directly driven by steam turbines, an electric motor (as in Fig. 14.16a), or an internal-combustion engine.

Most machines are multistage. Temperatures in multistage units may go as low as −100°F in the evaporator. These units are widely used for chilling water to about 45°F in air conditioning systems. Figure 14.16b shows a hermetic centrifugal compressor. Also, these compressors operate with adiabatic compression efficiencies of up to 80 percent.

The compressor in Fig. 14.16b is of the hermetic type, in which gas (refrigerant) flows through the electric motor windings (for cooling the motor) to the suction side of the compressor impellers. These machines may be driven at motor speed or, by means

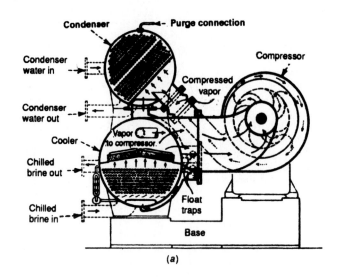

(a)

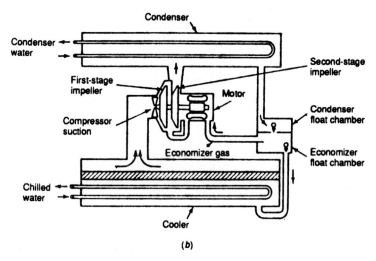

(b)

FIGURE 14.16 (*a*) Centrifugal compressor packaged with condenser and chiller (*Courtesy,* Power Magazine); (*b*) hermetic-type centrifugal compressor features refrigerant flowing through the motor windings with the two-stage compressor in the same casing.

of a speed-increasing gear between the motor and compressor, at a higher speed. Most machines have inlet guide vanes for capacity control.

Condensers

Condensers must remove the heat absorbed in the evaporator plus the equivalent Btu's of compressing the refrigerant gas to a pressure so that it can be liquefied by the available condensing medium. Therefore, condensers are manufactured in different arrangements, such as a (1) shell-and-tube heat exchanger, (2) shell-and-coil heat exchanger,

(3) double-pipe heat exchanger, (4) evaporative-condenser type, (5) atmospheric type, and (6) air-cooled type.

Two types of shell-and-tube condensers are illustrated in Fig. 14.17. A *horizontal shell-and-tube condenser* (see Fig. 14.17a) is equipped with enclosed water boxes and is mounted in a horizontal position. Baffles and gaskets in the enclosed water boxes cause the water to pass through the condenser several times. This allows the coolest water to enter at the bottom of the condenser and make several passes before leaving at the top. The result is a greater increase in water temperature with a corresponding decrease in water quantity. This type of condenser may be located at any point in a building. It should be installed so the heads may be easily removed for cleaning the tubes.

The *vertical shell-and-tube condenser* is illustrated in Fig. 14.17b. This type of condenser stands on end. Water is distributed over the entire head and enters each tube through a swirler to distribute the water evenly against the inside of the tube walls, where it flows downward by gravity. Some advantages are (1) large capacity installed in small floor space, (2) low pumping heads required, (3) good gas and liquid separating space, (4) simple purging connections, (5) cleaning of tubes possible without stopping water flow (allowing water with some dirt to be used), and (6) ability to carry overloads by increasing water volume without increase in pump friction head.

The disadvantages of vertical shell-and-tube condensers are (1) inside installations have a tendency to "steam" in cold weather, (2) the condenser must be mounted over an open water-collecting sump, and (3) relatively large quantities of water may have to be circulated because a single pass limits the water temperature rise.

The *double-pipe condenser* illustrated in Fig. 14.18a is used for high-pressure service. The water flows in the inner tube, while the gas to be condensed flows between the inner and outer tube. Water flows in the opposite direction to the refrigerant flow. Double-pipe condensers are extensively used in the petrochemical industry, where high pressure prevails. They were used in CO_2 refrigeration plants, which operate at high pressure. They require long lengths to provide the cooling surface, and this results in many joints that may leak.

The *shell-and-coil condensers* consist of a shell with a continuous coil inside for the water to flow through while it condenses the refrigerant in the shell. The waterside of the coil must be cleaned chemically, and in case of a coil leak, the entire coil must be replaced.

A simple *atmospheric condenser* is illustrated in Chap. 4, Fig. 4.6a. Figure 14.18b illustrates an *evaporative condenser* used on moderate-tonnage refrigeration systems. Hot vapor from the compressor is circulated through a bank of finned tubes. A circulating pump takes water from the basin and sprays it over the tubes containing the hot refrigerant vapor. A fan drives air over the lower portion of the cabinet and then around the wetted condenser tubes, which causes a portion of the water to evaporate into the air and thus pick up heat in changing from water to vapor. The recirculated spray water leaving the bottom of the tubes has a temperature at which the refrigerant condenses in the tubes. About 0.03 to 0.06 gpm of water per ton of refrigeration capacity is lost through evaporation into the air and carryout mist. A float in the bottom basin provides makeup water by keeping the basin at a predetermined level.

Air-cooled condensers use finned tubes through which the refrigerant gas flows, with fan-driven air blown over the tubes. They require less maintenance than water-cooled condensers and are used where water and sewer discharge rates may pose problems.

Water Savers

Water-saving devices are quite often decreed by local legislation. By using evaporative condensers, cooling towers (Chap. 4), and similar water conservation methods, only

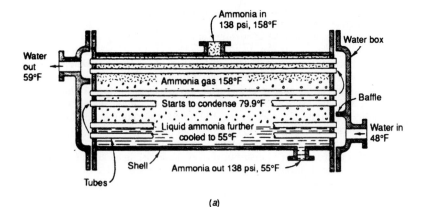

(a)

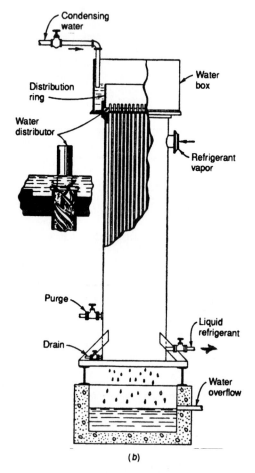

(b)

FIGURE 14.17 (a) Horizontal shell-and-tube condenser with three passes; (b) vertical shell-and-tube condenser with gravity water flow through the tubes.

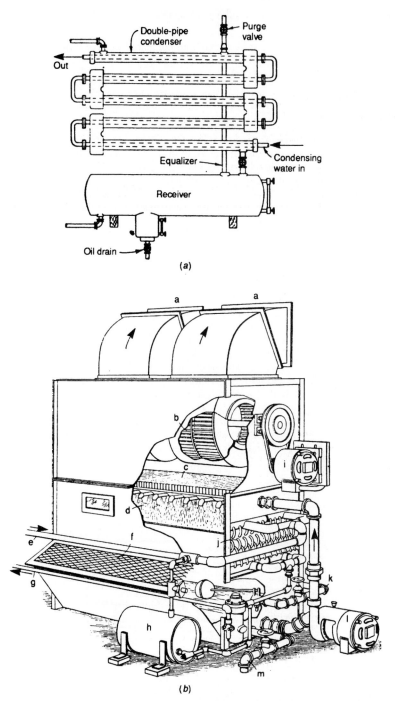

FIGURE 14.18 (*a*) Double-pipe condensers are used for high-pressure condensing service with water flowing through the inside pipe, and the refrigerant between the inner and outer pipe; (*b*) an evaporative condenser has refrigerant coils with water sprayed over the coils. Cooling is obtained by water evaporating by picking up heat.

14.41

about 4 to 8 percent makeup water is required compared to once-through condenser water flows. Chapter 4 also reviews the water treatment needed for recirculated water systems.

Defrosting

Defrosting is another operating procedure in refrigeration systems. Frost on evaporative coils in direct-expansion systems can act as an insulator, requiring longer time for the refrigerant to circulate to satisfy temperature requirements. It thus wastes power and adds to the wear and tear of machinery because it has to operate longer and harder. Frost on suspended ceiling evaporative coils can cause severe vibration on the circulating cold air fans in cold boxes. This vibration can break expansion coil joints, which may cause loss of the product in the cold box.

Three methods of defrosting in commercial services are (1) warm air, (2) use of hot water, and (3) use of hot refrigerant gas.

When the coil is defrosted with warm air, it is isolated temporarily from the space which it cools, and warm ambient (at room temperature) air is blown over the coil. The water defrost system floods the outside of the coil with water until the frost is melted.

Defrosting evaporator coils (see Fig. 14.19) with the hot-gas method permit defrosting the coils without raising the temperature of the compartment above 32°F. For this method a hot-gas line is connected from the discharge side of the compressor to a point just beyond the expansion valve in the evaporator coil. Stop valves at either end of the hot-gas line are provided for control. To operate: (1) close the liquid line stop valve ahead of the expansion valve, (2) close the compressor discharge valve, (3) open the valves in the hot-gas line, and start the compressor. *Caution:* Open the valves very slowly and not too wide. There is always danger of liquid from the evaporator slugging the compressor badly and causing damage.

Cold-storage brine plants may have a central defrosting system that heats the brine in a heat exchanger and circulates the hot brine through the pipes and coils to be defrosted. *Electric heaters* are also used to defrost evaporator coils.

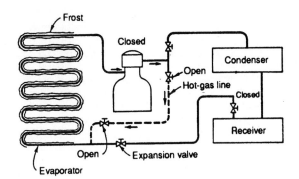

FIGURE 14.19 Hot-gas defrosting involves sending compressed gas from the compressor directly to the evaporator coil(s).

AIR CONDITIONING

The first "air conditioning" in movie houses consisted of air that was blown over cakes of ice to cool the air. From this humble beginning, air conditioning now consists of not only cooling the air but also regulating the humidity and air movement for ventilation, controlling the quality of the air, such as removing dust by filtering, controlling of odor, and making sure there is sufficient fresh air makeup to prevent bacteria buildup in the space being air conditioned.

The refrigeration systems used to cool air in air conditioning are still one of the major pieces of machinery in a system. Air conditioning also involves an understanding of air composition and how the humidity of this air can be controlled, either for comfort or for process needs. Air is a physical mixture of gases, water vapor, and dust, subdivided as follows:

1. Air by volume is 78.1 percent nitrogen, 21 percent oxygen, and 0.9 percent miscellaneous gases, primarily argon.

2. Air may also include foreign gases, such as carbon dioxide, sulfur dioxide, and ammonia, which may produce toxic or odorous effects.

3. Air contains water vapor, or moisture in a gaseous state, which is dependent on the air temperature with respect to full or partial saturation of the air with water vapor.

4. Air may also contain solid particles such as dust, smoke, and bacteria that may be suspended in the air.

Air Conditioning Terms

The purpose of air conditioning is to control the above items with suitable equipment. It is necessary to define some terms that are used in air conditioning.

Dew point is the saturation temperature corresponding to the actual partial pressure of water vapor (steam) in air. Dew-point temperatures cannot be read directly but can be calculated from readings of *dry-* and *wet-bulb* temperatures. Dry-bulb temperature is the temperature of the air as shown by an ordinary thermometer. *Wet-bulb temperature* is the temperature at which water, by evaporating into the surrounding air, can bring the air to saturation. When a thermometer with its bulb covered by a wick wetted with water is moved through the air that is unsaturated with water vapor, the water from the wick evaporates in proportion to the capacity of the air to absorb the evaporated moisture, and the temperature indicated by a wet-bulb thermometer drops below the dry-bulb temperature. The equilibrium that is reached by the wet-bulb thermometer in evaporating the water from the wick is the wet-bulb temperature. The readings from the two thermometers are used to calculate the exact humidity characteristics of the air. The sling psychrometer is illustrated in Fig. 14.20*a,* and an aspiration psychrometer is shown in Fig. 14.20*b*. Both instruments read dry- and wet-bulb temperatures. The sling unit is swirled manually until the wet-bulb reading stabilizes, or reaches equilibrium. In the aspiration type a small fan swirls the air over the wet-bulb thermometer.

The difference between the dry- and wet-bulb temperatures is the *depression,* and if the temperature of the water on the wick is above the dew-point temperature, evaporation takes place from this surface into the air.

Saturated air occurs when a given volume of space contains so much water vapor, expressed as grains per cubic foot, that the air cannot absorb any more, and condensa-

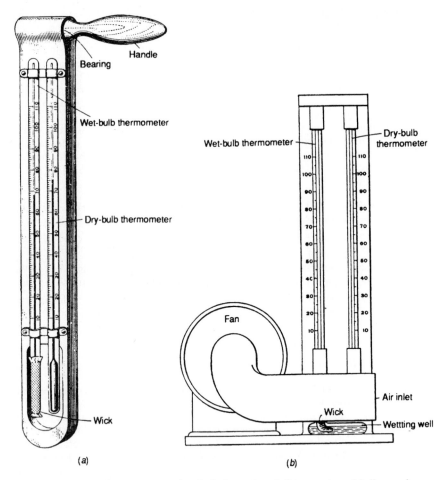

FIGURE 14.20 Psychrometers are used to obtain dry- and wet-bulb temperatures. (*a*) sling psychrometer; (*b*) aspiration psychrometer.

tion of the vapor in the air takes place. This is completely dependent on the air temperature (1 lb = 7000 gr).

Atmospheric air exists at a pressure that is made up of the sum of the partial pressure exerted by each gas in the atmosphere, including water vapor (steam at very low pressure). This follows Dalton's Law of Partial Pressure: "In a given mixture of gases or vapors, each gas or vapor exerts the same pressure it would exert if it occurred alone in the same space and at the same temperature as exists in the mixture."

Vapor pressure in air conditioning is the pressure exerted by the water-vapor component in the existing atmospheric air pressure, or barometric pressure, which is measured in inches of mercury.

Absolute humidity refers to the vapor density, or weight of water vapor, per unit volume, usually expressed as grains of water vapor per cubic foot of air. *Relative humidity* is the actual weight of water vapor per cubic foot divided by the maximum weight of water per cubic foot that could exist at the *same temperature,* or at 100 percent saturation.

Example. See Table 14.6. Air at 70°F can hold 0.0011507 lb of water vapor per cubic foot at saturated conditions. What would be the relative humidity if the air had 0.0004 lb of vapor per cubic foot? The degree of saturation, or relative humidity, is 0.0004/0.0011507, or 34.8 percent.

Useful Equations

Some useful equations in air conditioning involving humidity are presented before the subject of psychrometric charts is approached.

Relative humidity (R.h.) in terms of partial pressures and density is:

$$R.h. = \frac{p_s}{p_d} = \frac{d_s}{d_d}$$

where p_s = partial pressure of the water vapor in the air
p_d = pressure of saturated water vapor at the air temperature (dry bulb)
d_s = density of the water vapor in the air, lb/ft^3
d_d = density of saturated water vapor, lb/ft^3 at the air temperature (dry bulb)

Example. The temperature of a room is 70°F with a relative humidity of 30 percent. The barometric pressure is 29.2 in mercury. What is the (1) partial pressure of steam in the air, (2) weight of steam per cubic foot, and (3) weight of steam per each pound of dry air?

1. See Table 14.6. At 70°F, the pressure of saturated steam is 0.3628 psi. Use the relative humidity equation with the density relations, or

$$p_s = R.h. \times p_d = 0.30 \times 0.3628 = 0.1088 \text{ psi}$$

2. From Table 14.6,

$$d_d \text{ at } 70°F \text{ dry bulb} = 0.0011507 \text{ lb/ft}^3$$

Use the relative humidity equation but use densities, or

$$d_s = R.h. \times d_d = 0.30 \times 0.0011507 = 0.000345 \text{ lb/ ft}^3$$

vapor density = weight of steam per cubic foot

3. Per Dalton's Law, barometric pressure = partial air pressure p_a plus partial steam pressure p_s. Barometric pressure = 29.2 × 0.491; therefore, with p_s = 0.1088 psi per the first problem,

$$p_a = (29.2 \times 0.491) - 0.1088 = 14.23 \text{ psi}$$

To find the volume this air occupies at 70°F, use $PV = WRT$ where $R = 53.3$ for air, $W = 1$ lb of air; solve for V. It is necessary to use absolute terms per thermodynamic equation rules, or

$$144 \times 14.23 \times V = 1 \times 53.3 \times (460 + 70)$$

and solving for V, it equals 13.79 ft^3 occupied by 1 lb of dry air. Now from the answer to 2. above, the weight of steam per cubic foot of space is equal to 0.000345 lb/ft^3, or in 13.79 ft^3 of dry air, the weight of steam in this space is 13.79 × 0.000345 = 0.00476 lb of steam per each pound of dry air.

TABLE 14.6 Typical properties of saturated water vapor with air at different temperatures.

Temp. °F	Pressure of saturated vapor lb/in²	Weight of saturated vapor per ft³ lb	Weight of saturated vapor per lb of dry air		Volume ft³ Barometer, 29.92 in Hg		Enthalpy per lb		
			lb	gr	of 1 lb of dry air	of 1 lb of dry air + vapor to saturate it	Dry air, 0°F datum	Vapor, 32°F datum	Dry air with vapor to saturate it
50	0.1780	0.0005866	0.007626	53.38	12.84	12.99	12.00	1081.7	20.25
51	.1848	.0006078	.007921	55.45	12.86	13.02	12.23	1082.2	20.80
52	.1918	.0006296	.008226	57.58	12.89	13.06	12.47	1082.6	21.38
53	.1989	.0006516	.008534	59.74	12.91	13.09	12.71	1083.1	21.95
54	.2063	.0006746	.008856	61.99	12.94	13.12	12.95	1083.5	22.55
55	0.2140	.0006984	0.009192	64.34	12.96	13.15	13.19	1084.0	23.15
56	.2219	.0007228	.009536	66.75	12.99	13.19	13.43	1084.4	23.77
57	.2300	.0007477	.009890	69.23	13.01	13.22	13.67	1084.9	24.40
58	.2384	.0007735	.01026	71.82	13.04	13.25	13.91	1085.3	25.05
59	.2471	.0008003	.01064	74.48	13.06	13.29	14.15	1085.8	25.70
60	0.2561	0.0008278	0.01103	77.21	13.09	13.32	14.39	1086.2	26.37
61	.2654	.0008562	.01144	80.08	13.11	13.35	14.63	1086.7	27.06
62	.2749	.0008852	.01186	83.02	13.14	13.39	14.87	1087.1	27.76
63	.2848	.0009153	.01229	86.03	13.16	13.42	15.11	1087.6	28.48
64	.2949	.0009460	.01274	89.18	13.19	13.46	15.35	1088.0	29.21
65	0.3054	0.0009778	0.01320	92.40	13.21	13.49	15.59	1088.5	29.96
66	.3162	.0010105	.01368	95.76	13.24	13.53	15.83	1088.9	30.73
67	.3273	.0010440	.01417	99.19	13.26	13.57	16.07	1089.4	31.51
68	.3388	.0010816	.01468	102.8	13.29	13.60	16.31	1089.8	32.31
69	.3506	.0011140	.01520	106.4	13.31	13.64	16.55	1090.3	33.12

70	0.3628	0.0011507	110.2	13.34	13.68	16.79	1090.7	33.96
71	.3754	.0011884	114.2	13.37	13.71	17.03	1091.2	34.83
72	.3883	.0012269	118.2	13.40	13.75	17.27	1091.6	35.70
73	.4016	.0012667	122.4	13.42	13.79	17.51	1092.1	36.60
74	.4153	.0013075	126.6	13.44	13.83	17.75	1092.5	37.51
75	0.4295	0.0013497	131.1	13.47	13.87	17.99	1093.0	38.46
76	.4440	.0013927	135.7	13.49	13.91	18.23	1093.4	39.42
77	.4590	.0014371	140.4	13.52	13.95	18.47	1093.9	40.40
78	.4744	.0014825	145.3	13.54	13.99	18.71	1094.3	41.42
79	.4903	.0015295	150.3	13.57	14.03	18.95	1094.8	42.46
80	0.5067	0.0015777	155.5	13.59	14.08	19.19	1095.2	43.51
81	.5236	.0016273	160.9	13.62	14.12	19.43	1095.7	44.61
82	.5409	.0016781	166.4	13.64	14.16	19.67	1096.1	45.72
83	.5588	.0017304	172.1	13.67	14.21	19.91	1096.6	46.88
84	.5772	.0017841	178.0	13.69	15.26	20.15	1097.0	48.05
85	0.5960	0.0018389	184.0	13.72	14.30	20.39	1097.5	49.24
86	.6153	.0018950	190.3	13.74	14.34	20.63	1097.0	50.47
87	.6352	.0019531	196.7	13.77	14.39	20.87	1098.4	51.74
88	.6555	.0020116	203.3	13.79	14.44	21.11	1098.8	53.02
89	.6765	.0020725	210.1	13.82	14.48	21.35	1099.3	54.35
90	0.6980	0.0021344	217.1	13.84	14.53	21.59	1099.7	55.70
91	.7201	.0021982	224.4	13.87	14.58	21.83	1100.2	57.09
92	.7429	.0022634	231.8	13.89	14.63	22.07	1100.6	58.52
93	.7662	.0023304	239.5	13.92	14.69	22.32	1101.1	59.99
94	.7902	.0023992	247.5	13.94	14.73	22.56	1101.5	61.50
95	0.8149	0.0024697	255.6	13.97	14.79	22.80	1102.0	63.05
96	.8403	.0025425	264.0	13.99	14.84	23.04	1102.4	64.62
97	.8663	.0026164	272.7	14.02	14.90	23.28	1102.9	66.25
98	.8930	.0026925	281.7	14.04	14.95	23.52	1103.3	67.92
99	.9205	.0027700	290.9	14.07	15.01	23.76	1103.8	69.63

Source: American Society of Heating and Ventilating Engineers

The *Carrier equation* was developed to calculate the pressure of the water vapor in terms of dry- and wet-bulb pressures and temperatures:

$$P_s = p_w - \frac{(p_B - p_w)(t_d - t_w)}{2800 - 1.3t_w}$$

where p_s = pressure of the water vapor in the atmosphere
$\quad\;\; p_w$ = pressure of saturated water vapor at the wet-bulb temperature
$\quad\;\; p_B$ = barometric pressure, in of mercury, or lb/in^2
$\quad\;\; t_d$ = dry-bulb temperature, °F
$\quad\;\; t_w$ = wet-bulb temperature, °F

Example. Air has 70°F dry-bulb temperature and 64°F wet-bulb temperature with the barometer indicating 29.9 in of mercury. Find (1) the relative humidity of the air, (2) the water vapor density of the air, and (3) the dew-point temperature:

1. To find relative humidity, we must find p_s and p_d. Using the Carrier equation to find p_s, p_w = 0.2949 from Table 14.6 for a 64°F wet-bulb temperature and p_B = 29.9 × 0.491 = 14.68 lb/in^2. Now substituting in the Carrier equation,

$$p_s = 0.2949 - \frac{(14.68 - 0.2949) \times (70 - 64)}{2800 - 1.3 \times 64}$$

$$= 0.2631 \text{ psi partial pressure of the vapor in the air}$$

$$R.h. = \frac{p_s}{p_d} = \frac{0.2631}{0.3628} = 72.5 \text{ percent relative humidity}$$

Note that p_d = pressure of saturated vapor at 70°F, and the value is obtained from Table 14.6 at 0.3628 psi.

2. Use the relative humidity equation d_s/d_d and the density at 70°F from Table 14.6 is 0.0011507 lb/ft^3; therefore,

$$d_s = \text{R.h.} \times d_d = 0.725 \times 0.0011507 = 0.000834 \text{ lb/ft}^3 \text{ of vapor in the air}$$

3. The dew-point temperature can be found by using Table 14.6 with the saturation pressure equal to the pressure of the vapor in the air, developed above by the Carrier equation, equaling 0.2631 psi. The corresponding temperature from Table 14.6 is 60.8°F.

Psychrometry

To avoid many calculations, the relationship between dry- and wet-bulb temperatures, dew point, and relative humidity are put in psychrometric charts. Many charts are available from the manufacturers of air conditioning equipment. One such chart is shown in Fig. 14.21, while Fig. 14.22 shows a skeleton chart.

The skeleton chart is a Trane Psychrometric Chart. Lines of the constant dry-bulb temperature are almost vertical. Lines of the constant dew-point temperature are horizontal. Lines of the constant wet-bulb temperature slope downward to the right. Lines of the constant percentage humidity are curved. Before using the chart, any two of the above values must be known. The remaining two can then be found on the chart.

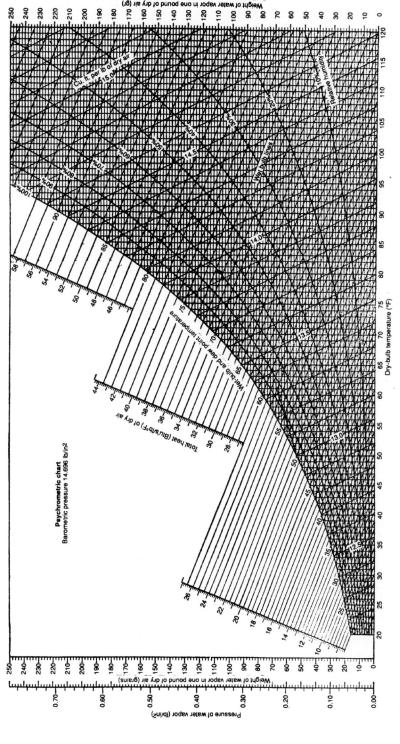

FIGURE 14.21 Psychrometric charts are used to calculate air conditioning problems on humidity control. *(Courtesy, General Electric, Co.)*

14.49

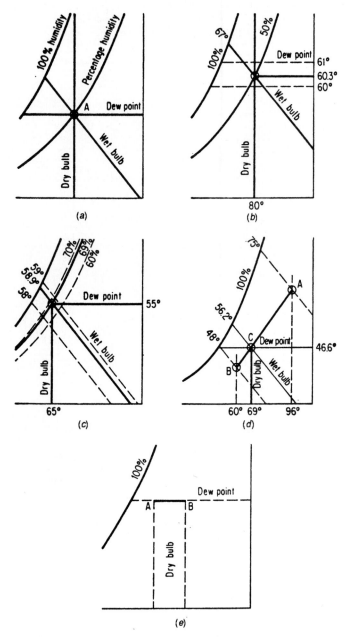

FIGURE 14.22   Excerpt, or skeleton, psychrometric chart for illustrating the use of psychrometric charts. (*Courtesy, The Trane Co.*)

Example.   To find humidity on a psychrometric chart proceed as follows: Given air having a dry-bulb temperature of 80°F and a wet-bulb temperature of 67°F, find the dew-point temperature and the percentage of humidity. Refer to the skeleton chart in Fig. 14.22a. Then try this problem on the psychrometric chart in Fig. 14.22b. You will find that the dew-point temperature of the air is 60.3°F and that the humidity is 50 percent.

Example.   To find the wet-bulb reading proceed as follows: Given air having a dry-bulb temperature of 65°F and a dew-point temperature of 55°F, find the wet-bulb temperature and the percentage of humidity. Refer to the skeleton chart in Fig. 14.22c. Then try this problem on the psychrometric chart. You will find that the wet-bulb temperature is 58.9°F and the humidity is 69 percent.

Example.   Find air mixtures on the Trane chart. The chart can be used to determine resultant wet-bulb and dew-point temperatures of a mixture of two streams of air at different conditions. Suppose that 25 percent of the mixture has a dry-bulb temperature of 96°F and a wet-bulb temperature of 75°F, as represented by point A in the sketch in Fig. 14.22d. The balance, or 75 percent, has a dry-bulb temperature of 60°F and a wet-bulb temperature of 48°F, as shown by point B. The resultant dry-bulb temperature of the mixture is figured as follows:

$$0.25 \times 96 = 24°F$$

$$0.75 \times 60 = 45°F$$

$$24 + 45 = 69°F, \text{ the resultant dry-bulb temperature of the air mixture}$$

Now, draw a straight line between points A and B. The intersection of the vertical line of 69°F dry bulb with the line from A to B gives point C. The wet-bulb and dew-point lines running through point C are the resultant wet-bulb and dew-point temperatures of the mixture. In this case the wet-bulb temperature of the mixture is 56.2°F and the dew-point temperature is 46.6°F.

Besides giving the properties of air, the Trane chart can be used to trace air-conditioning cycles. Processes involving heating or cooling and humidification or dehumidification can be clearly followed on the chart. In tracing any process, keep this fact in mind: the dew-point temperature is constant as long as there is no change in the moisture content of the air. Thus, heating of air without changing its moisture content takes place along a horizontal line of the constant dew point. The dew-point line is determined by the initial condition of the air. In Fig. 14.22e the air that is initially in the condition designated by point A is heated to point B along the horizontal line AB. Also, if the air is to be cooled without the moisture being condensed, the process is represented by a straight line drawn from point A and B in Fig. 14.22e.

Human Comfort

Human comfort air conditioning requires considering the following:

1. Temperature of occupancy needed, where refrigeration is an important consideration to cool the air to the desired temperatures.
2. Maintaining proper humidity. This involves humidifying or dehumidifying the air in conjunction with air cooling. For example, the air to be conditioned can be cooled

to the dew point to condense the moisture in the air and then heated to the desired temperature for the space under consideration and with a comfortable humidity.

3. Air ventilation and distribution by proper fan and duct sizing and arrangement. This also includes providing fresh air makeup.

4. Maintaining proper air quality by filtering the air of dust particles and minimizing odors.

Good centralized systems strive to achieve simultaneous control of all of the above factors, while packaged, direct-expansion units, as shown in Fig. 14.23, primarily cool and filtered air and, in the unit illustrated, also have provisions for cold weather heating. Relative humidity can affect human comfort. If the relative humidity is too low, evaporation from the skin may make a person feel too cool and will also cause the skin to dry out. If the relative humidity is too high, evaporation from the skin will be very low; therefore, comfort cooling derives mostly by having a low dry-bulb temperature, supplied by refrigeration in most cases.

Table 14.7 provides a guide on comfort cooling for human occupancy as recommended by the Air Conditioning Manufacturers' Association. The term *effective temperature* (ET) is an index temperature which expresses the composite effect of air temperature, relative humidity, and air motion over the human body. The numerical values shown in the table are made equal to the temperatures of calm saturated air which produce an equal sensation of warmth to that existent under the given air condition.

Cooling loads for a conditioned space have two parts: (1) keeping the dry-bulb temperature low enough for comfort, termed the *sensible heat load* and (2) maintaining desired humidity, termed the *latent heat load*. The *sensible heat load* is calculated by

$$Q_S = (h_C - h_A) \text{ Btu/lb}$$

$$= 0.24 \times (t_C - t_A) \text{ Btu/lb, where } 0.24 = \text{specific heat from Btu/lb}$$

where h_C = enthalpy of dry-bulb temperature of conditioned air
h_A = enthalpy of supplied air temperature
t_C = dry-bulb temperature of conditioned air
t_A = supplied air temperature

The *latent heat* is calculated by

$$Q_L = (h_B - h_C) \text{ Btu/lb}$$

where h_B = enthalpy of conditioned air in space
h_C = enthalpy of dry-bulb temperature of conditioned air

The total heat that can be taken up is $Q_T = (h_B - h_A)$ Btu/lb

The following example is provided only to illustrate how humidity calculations are made in air conditioning work, which in turn is the basis for sizing the equipment.

Example.   The maximum temperature allowed for a space is 76°F dry bulb and 66°F wet bulb with a *sensible heat load* of 300,000 Btu/h and requiring 900,000/gr of moisture per hour to be removed (7000 gr = 1 lb). The air is supplied at 64°F dry bulb. Calculate (1) pound of air that must be supplied, (2) the dew-point and wet-bulb tem-

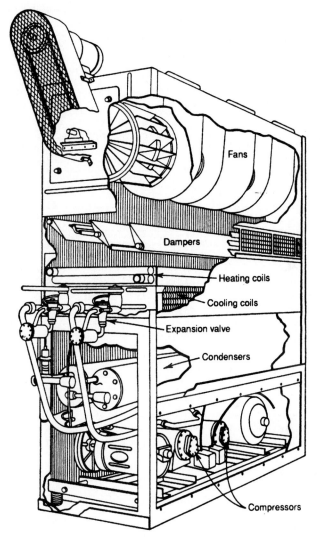

FIGURE 14.23 Factory-assembled direct-expansion air conditioning and heating unit.

peratures of the supply air, (3) the cubic feet per minute supplied, and (4) the latent heat load picked up from the space.

1. Each pound of air can absorb $Q_S = 0.24 \times (76 - 64) = 2.88$ Btu/lb. *Note:* Use dry-bulb temperatures on sensible heat calculations.

$$\frac{300,000}{2.88} = 104,200 \text{ lb/h of air must be supplied to the space}$$

TABLE 14.7 Recommended Comfort Cooling Conditions for Human Occupancy per Air Conditioning Manufacturers' Association Guidelines.

Outside design dry bulb, °F	Class A, occupancy over 40 min				Class B, occupancy under 40 min			
	Dry bulb, °F	Wet bulb, °F	Rel. Hum., %	Eff. Temp., E. T.	Dry bulb, °F	Wet bulb, °F	Rel. Hum, %	Eff. Temp., E. T.
	75	65	60	71	76	66	61	72
	76	64	53	71	77	65	54	72
80	77	63	47	71	78	64	47	72
	78	62	40	71	79	63	41	72
	79	61	35	71	80	62	36	72
	76	66	61	72	77	67	61	73
	77	65	54	72	78	66	54	73
85	78	64	47	72	79	65	48	73
	79	63	41	72	80	64	42	73
	80	62	36	72	81	63	36	73
	77	67	61	73	78	69	64	74
	78	66	54	73	79	68	58	74
90	79	65	48	73	80	67	52	74
	80	64	42	73	81	66	46	74
	81	63	36	73	82	65	40	74
					83	64	35	74
	78	69	64	74	79	70	65	75
	79	68	58	74	80	69	58	75
	80	67	52	74	81	68	52	75
95	81	66	46	74	82	67	47	75
	82	65	40	74	83	66	41	75
	83	64	35	74	84	65	36	75
	79	70	65	75	81	71	63	76
	80	69	58	75	82	70	56	76
	81	68	52	75	83	69	50	76
100	82	67	47	75	84	68	44	76
	83	66	41	75	85	67	38	76
	84	65	36	75				
	80	71	65	75.5	81	72	65	76.5
	81	70	58	75.5	82	71	59	76.5
	82	69	52	75.5	83	70	54	76.5
105	83	68	47	75.5	84	69	47	76.5
	84	67	42	75.5	85	68	41	76.5
	85	66	37	75.5	86	67	36	76.5

2. Each pound of air must absorb $900,000/104,200 = 8.64$ gr of moisture. It is necessary to go to the psychromatic chart in Fig. 14.21. It is given that air leaving the space cannot exceed 76°F *dry bulb* and 66°F *wet bulb.* For these two temperatures, from Fig. 14.21, the weight of water vapor in 1 lb of air $= 79.9$ gr; therefore, moisture to be absorbed $= 79.9 - 8.64 = 71.3$ gr. Using this value and a *dry-bulb temperature* of 64°F, from Fig. 14.21, the dew-point temperature $= 57.6$°F, and the *wet-bulb temperature* $= 60$°F.

3. Using 64°F *dry bulb* and 60°F *wet bulb,* follow the cubic feet per pound of dry air line in Fig. 14.21 and obtain 13.4 ft³/lb as the specific volume. For 104,200 lb/h,

$$\text{cfm} = \frac{104,200}{60} \times 13.4 = 23,300 \text{ cfm}$$

4. Using the psychromatic chart in Fig. 14.21, the enthalpy at 76°F and 79.9 gr is 30.78 Btu/lb, and the enthalpy at 76°F and 71.3 gr is 29.5 Btu/lb; therefore, for 104,200 lb,

$$Q_L = 104,200 \times (30.78 - 29.5) = 133,400 \text{ Btu/h of latent heat (moisture) to be}$$
$$\text{removed}$$

The *cooling load* calculation is beyond the scope of this book, but it involves applying heat-transfer equations as well as considering inside design temperatures and average outside temperatures for the area. The cooling load is used to determine the tonnage required for the space to be cooled, including humidity control.

Makeup air is an important consideration in air conditioning because all occupancies generate carbon dioxide and perhaps unhealthy or disagreeable gases and fumes are generated from paints, plastics, rugs, furniture shellacs, and similar indoor conditions (see Fig. 14.24). The growth of recirculated air to save energy has brought forth the term *building syndromes* that implies the inside air is becoming contaminated and that there is not enough outside fresh air being introduced. Ventilation standards have been published, but these are under review because the modern office building, for example, has more data processing, fax machines, and computer-type equipment than when many of the standards were established. In general, makeup air required may vary from 15 to 40 cfm outside air per person, depending on the type of occupancy. It is also recommended that local health standards be checked for the regulations for makeup air. Conditioned air temperatures to a space have a thumb-rule guide for summer air conditioning to be 5 to 20° below the room temperature, again depending on how many persons are in the room.

Ducts to distribute the conditioned air must be sized by determining the maximum air velocities which can be used without causing excessive and annoying noises or causing excessive friction losses. The best duct system is as direct as possible and without sharp bends.

Ducts in central systems are connected to chambers where the conditioning of the air takes place (see Fig. 14.25). These conditioning chambers are strategically located near the load, with chilled water piped to the chambers. The *chiller* is located by the refrigerant compressors. As Fig. 14.25 illustrates, this chamber has heating and cooling coils to prepare air for summer and winter occupancy. Filters are used to remove dust particles. The air is washed and cooled, and the sprays are also used to control the humidity. Reheaters are used to bring the humidity to the proper level.

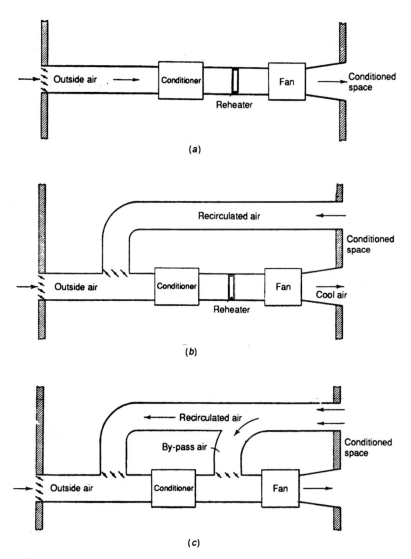

FIGURE 14.24 Three types of conditioned air flow. (*a*) Conditioned air flow, no recirculation, 100% outside air; (*b*) conditioned air flow with recirculated air to save energy; (*c*) recirculated air is also used to reheat air for humidity control by using bypass air.

Odors are removed in some installations by the use of activated carbon that is supplied in canisters or in panels. Dust particles are removed by the use of (1) air washers, (2) dry-type filters, (3) viscous-type filters that cause the dust to "stick" to the exposed surfaces, and (4) some form of electrostatic filters.

On direct-expansion unit air conditioners, dirty filters are a frequent cause of low

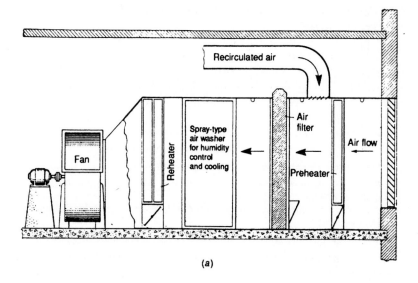

(a)

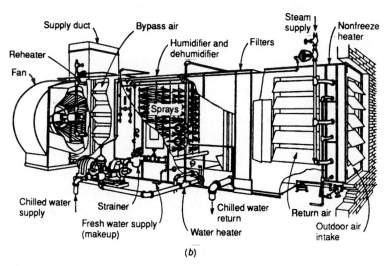

(b)

FIGURE 14.25   (a) Conditioning chambers are used to prepare the air for delivery to the space to be cooled and heated with the desirable humidity; (b) conditioning chambers are used in central systems with the refrigeration equipment in "engine rooms," and chilled water piped to the conditioning chambers.

suction pressure and frosting on the suction line. Dirty filters can hamper air flow over the expansion coils; thus the refrigerant does not completely vaporize in the evaporator coils but returns to the compressor as saturated vapor. This may cause liquid slugging of the compressor, which is usually a reciprocating unit. Many refrigeration service people install a *liquid separator* in the suction line to avoid this type of occurrence.

REFRIGERATION AND AIR CONDITIONING TERMS AND DEFINITIONS

Absolute humidity.   The weight of water vapor in a unit volume, usually expressed as grains per cubic foot (7000 gr equals 1 lb).

Air bound.   Air trapped in piping, equipment, etc., such as a steam radiator, which prevents maximum heat transfer; or air trapped in the suction side of a pump which causes loss of suction.

Ambient temperature.   The temperature of air in a space (e.g., room temperature).

Anhydrous.   Free of water, especially water of crystallization.

Backpressure.   Another term for suction pressure.

Brine.   In refrigeration systems, any liquid that is cooled by the refrigerant and pumped through the cooling coils to pick up heat. It does not undergo any change in state, but only in temperature. Brine is used in indirect systems; refrigerant is used in direct systems.

Calorie.   The quantity of heat required to raise the temperature of 1 g of water 1°C.

Chilled water.   A cooling medium that removes heat from the area to be cooled and gives up the heat in the chiller.

Chiller.   A heat exchanger in which low-pressure refrigerant boils or vaporizes, thus absorbing the heat that was removed from the refrigerated area by the cooling medium (water).

Chiller load.   An indication of the number of tons of refrigerant being produced.

Coefficient of performance.   The ratio of refrigerating effect to work of compression. A high coefficient of performance means high efficiency. The theoretical coefficients range from about 2.5 to more than 5.

Cooling medium.   A fluid used for picking up heat which is circulated to the heat exchanger, where heat is removed; examples are chilled water and brine.

Degree-day.   For any given day, the number of heating degree-days is the difference, in degrees, between the average (mean) temperature for that day and 65°F. For example, if the mean temperature for a day is 40°F, the number of degree-days for that day equals 65 − 40 = 25 degree-days. Thus, when the mean temperature is less than 65°F, there are as many degree-days as there are Fahrenheit degrees difference between the mean temperature for the day and 65°F.

Dehumidify.   To reduce the quantity of water vapor within a space.

Dehydrate.   To remove water from any form of matter.

Dew point.   The temperature at which the water vapor in the air begins to condense, or the temperature at which the relative humidity of air becomes 100 percent.

Enthalpy.   The total heat or heat content of a substance, expressed in Btu per pound.

Flooded refrigeration system.   A type of system where only part of the circulated refrigerant is evaporated, with the remainder being separated from the vapor and then recirculated.

Freeze-up.   Ice formation on a refrigeration system at the expansion device, making the device inoperative.

Head pressure.   Pressure at the discharge of a compressor or in the condenser. It is also known as *high-side* pressure.

High side.   The portion of a refrigeration system that is under discharge or condenser pressure. It extends from the compressor discharge to the expansion valve(s) inlet.

Horsepower per ton.   Mechanical input in horsepower, divided by tons of refrigerating effect produced. If the coefficient of performance is known, the horsepower per ton can be figured directly; divide 12,000 Btu/h by 2545 Btu/hp · hr and the coefficient of performance.

Latent heat of fusion.   The heat added or extracted when a substance changes from the solid to the liquid state or from the liquid to the solid state. For example, when ice melts in a refrigerator, 144 Btu/lb must be added to produce the melting, or when ice freezes in an ice tank, 144 Btu/lb must be extracted; 144 Btu/lb is the latent heat of fusion.

Liquid line.   Refrigerant piping through which liquid refrigerant flows from the condenser to the expansion valves.

Low side.   The portion of a refrigeration system in which the refrigerant is at low pressure. It extends from the expansion valve(s) outlet to the suction inlet of the compressor.

Mechanical equivalent of heat.   One Btu equals 778.2 ft · lb of mechanical energy.

Plenum chamber.   An air compartment maintained under pressure with connections to one or more disturbing ducts.

Pump-down.   The operation by which the refrigerant in a charge system is pumped in liquid form into the condenser/receiver.

Refrigerant handled.   The amount of refrigerant circulated. Dividing 200 Btu/min by the refrigerating effect, in Btu/lb of refrigerant, gives the number of pounds of refrigerant circulated each minute.

Refrigerating effect.   The amount of heat absorbed in the evaporator, which is the same as the amount of heat removed from the space to be cooled. It is measured by subtracting the heat content of 1 lb of refrigerant as it enters the expansion valve from the heat content of the same pound of refrigerant as it enters the compressor.

Standard-ton conditions.   An evaporating temperature of 5°F, a condensing temperature of 86°F, liquid before the expansion valve at 77°F, and a suction gas temperature of 14°F produce standard-ton conditions. Refrigerating machines are often rated under standard-ton conditions.

Suction pressure.   Pressure in the compressor suction or at the outlet of the evaporator. It is also known as *low-side* pressure.

Volatile.   Easily evaporated. This is a necessary property of all compression refrigerants.

Wind-chill factor.   Temperature effect on exposed flesh at certain wind speeds and temperatures. If the weather temperature is 10° above zero, for example, and the wind is blowing at 20 mph, the wind-chill factor is 25° below zero.

Work of compression.   The amount of heat added to the refrigerant in the compressor. It is measured by subtracting the heat content of 1 lb of refrigerant at the compressor suction from the heat content of the same pound of refrigerant at the compressor discharge. Multiplying the work of compression, in Btu/lb by the number of pounds of refrigerant handled in an hour, and dividing by 2545 Btu/h · h, gives the theoretical power requirements.

TROUBLESHOOTING

The following table "Troubleshooter's Guide to Refrigeration and Air Conditioning" can be used to help diagnose and correct various problems with equipment.

Troubleshooter's Guide to Refrigeration and Air Conditioning Problems

Symptoms	Probable cause	Recommended action
	Compressor "short-cycles"	
1. Normal operation except too frequent stopping and starting	1. Intermittent contact in electrical control circuit	1. Repair or replace faulty electrical control circuit
2. Normal operation except too frequent stopping and starting	2. Low-pressure controller differential set too close	2. Reset differential in accordance with proper job conditions
3. Valve may hiss when closed. Also temperature change in refrigerant line through valve	3. Leaky liquid line solenoid valve	3. Repair or replace

4. Reduced airflow
  a. Dirty air filters
  b. Broken fan belt
  c. Fan belt tension improperly adjusted
	4. Dirty or iced evaporator	4. Clean or defrost evaporator. Check filters and fan drive

5. Excessively high discharge pressure	5. Faulty condensing	5. Check for water failure or evaporative condenser trouble
6. High discharge pressure	6. Overcharge of refrigerant or noncondensable gas	6. Remove excess refrigerant or purge noncondensable gas
7. Normal operation except too frequent stopping and starting on low-pressure control switch	7. Lack of refrigerant	7. Repair refrigerant leak and recharge
8. High discharge pressure	8. Water-regulating valve inoperative or restricted by dirt, or water temperature too high	8. Clean or repair water valve
9. High discharge pressure	9. Water piping restricted or supply water pressure too low	9. Determine cause and correct
10. Suction pressure too low and frosting of strainer	10. Restricted liquid line strainer	10. Clean strainer
11. Motor starts and stops rapidly	11. Faulty motor	11. Repair or replace faulty motor
12. Compressor cuts off on high-pressure cutout	12. Fouled shell-and-tube condenser	12. Clean condenser tubes

		13. Determine cause and correct
13. Compressor cuts off on high pressure cutout a. No water b. Spray nozzles clogged c. Water pump not operating d. Coil surface dirty e. Air inlet or outlet obstructed f. Fan not operating	13. Faulty operation of evaporative condenser	a. Fill with water b. Clean spray nozzles c. Repair faulty pump d. Clean coil e. Remove obstruction f. Repair

Compressor runs continously

1. High temperature in conditioned area	1. Excessive load	1. Check for excessive fresh air or infiltration. Check for inadequate insulation of space
2. Low temperature in conditioned area	2. Thermostat controlling of too low a temperature	2. Reset or repair faulty thermostat
3. Low temperature in conditioned space	3. "Welded" contacts on electrical controls in motor starter circuit	3. Repair or replace faulty control
4. Bubbles in sightglass	4. Lack of refrigerant	4. Repair leak and charge
5. High discharge pressure	5. Overcharge of refrigerant	5. Purge or remove excess
6. Compressor noisy or operating at abnormally low discharge pressure or abnormally high suction pressure	6. Leaky valves in compressor	6. Overhaul compressor
7. Air-conditioned space too cold	7. Solenoid stop valve stuck open or held open by manual lift stem	7. Repair valve or restore to automatic operation

Compressor loses oil

1. Oil level too low	1. Insufficient oil charge	1. Add sufficient amount of proper compressor oil
2. Oil level gradually drops	2. Clogged strainers or valves	2. Clean or repair and replace
3. Excessively cold suction	3. Loose expansion valve or remote bulb	3. Provide good contact between remote bulb and suction line
4. Excessively cold suction. Noisy compressor operation	4. Liquid flooding back to compressor	4. Readjust superheat setting or check remote bulb contact

Troubleshooter's Guide to Refrigeration and Air Conditioning Problems (*Continued*)

Symptoms	Probable cause	Recommended action action
	Compressor loses oil (*Cont.*)	
5. Too frequent starting and stopping of compressor	5. Short cycling	5. Defrost; check pressure cutout
6. Oil around compressor base and low crankcase oil level	6. Crankcase fittings leak oil	6. Repair oil leak and add proper refrigerant oil
	Compressor is noisy	
1. Coupling bolts loose	1. Loose compressor drive coupling	1. Tighten coupling and check alignment
2. Compressor cuts out on oil failure control	2. Lack of oil	2. Add oil
3. Squeak or squeal when compressor runs	3. Dry or scored seal	3. Check oil level
4. Compressor knocks	4. Internal parts of compressor broken	4. Overhaul compressor
5. Abnormally cold suction line. Compressor knocks	5. Liquid "flood back"	5. Check and adjust superheat. Valve may be too large or remote bulb loose on suction line. Air entering evaporator too cold for complete evaporation of liquid
6. Water valve chatters or hammers	6. Dirty water regulating valve, too high water pressure or intermittent water pressure	6. Clean water regulating valve. Install air chamber ahead of valve
7. Abnormally cold suction line. Compressor knocks	7. Expansion valve stuck in open position	7. Repair or replace
8. Compressor or motor jumps on base	8. Compressor or motor loose on base	8. Tighten motor or compressor hold-down bolts
	System short of capacity	
1. Expansion valve hisses	1. Flash gas in liquid line	1. Add refrigerant
2. Temperature change in refrigerant line through strainer or solenoid stop valve	2. Clogged strainer or solenoid stop valve	2. Clean or replace
3. Reduced airflow	3. Ice or dirt on evaporator	3. Clean coil or defrost
4. Short-cycling or continuous running	4. Expansion valve stuck or obstructed	4. Repair or replace expansion valve
5. Superheat too high	5. Excess pressure drop in evaporator	5. Check superheat and reset thermostatic expansion valve

14.62

Symptom	Cause	Remedy
6. Short-cycling or continuous running	6. Improper superheat adjustment	6. Adjust expansion valve. Check superheat and reset thermostatic expansion valve
7. Short-cycling or continuous running	7. Expansion valve improperly sized	7. Replace with correct valve
Discharge pressure too high		
1. Excessively warm water leaving condenser	1. Too little or too warm condenser water	1. Provide adequate cool water, adjust water-regulating valve
2. Excessively cool water leaving condenser	2. Fouled tubes in shell-and-tube condenser	2. Clean tubes
3. Low air or spray water volume. Scaled surface	3. Improper operation of evaporative condenser	3. Correct air or water flow. Clean coil surface
4. Exceptionally hot condenser and excessive discharge pressure	4. Air or noncondensable gas in system	4. Purge
5. Exceptionally hot condenser and excessive discharge pressure	5. Overcharge of refrigerant	5. Remove excess or purge
Discharge pressure too low		
1. Excessively cold water leaving condenser	1. Too much condenser water	1. Adjust water-regulating valve
2. Bubbles in sightglass	2. Lack of refrigerant	2. Repair leak and charge
3. Suction pressure rises faster after pressure shutdown than 5 lb/min	3. Broken or leaky compressor discharge valves	3. Remove head, examine valves, replace faulty ones
4. Low discharge pressure and high suction pressure	4. Leaky relief bypass valve	4. Inspect valve to determine if replacement is necessary
Suction pressure too high		
1. Compressor runs continuously	1. Excessive load on evaporator	1. Check for excessive fresh air or infiltration, poor insulation of spaces
2. Abnormally cold suction line. Liquid flooding to compressor	2. Overfeeding of expansion valve	2. Regulate superheat setting expansion valve, see if remote bulb is okay on suction line
3. Abnormally cold suction line. Liquid flooding to compressor	3. Expansion valve stuck open	3. Repair or replace valve
4. Abnormally cold suction line. Liquid flooding to compressor	4. Expansion valve too large	4. Check valve rating, replace if necessary
5. Noisy compressor	5. Broken suction valves in compressor	5. Remove head, examine valves, repair faulty ones

Troubleshooter's Guide to Refrigeration and Air Conditioning Problems (*Continued*)

Symptoms	Probable cause	Recommended action action
	Suction pressure too low	
1. Bubbles in sightglass	1. Lack of refrigerant	1. Repair leak, then charge system
2. Compressor short-cycles	2. Light load on evaporator	2. Not enough refrigerant
3. Temp. change in refrigerant line through strainer or solenoid stop valve	3. Clogged liquid-line strainer	3. Clean strainer
4. No flow of refrigerant through valve	4. Expansion-valve power assembly has lost charge	4. Replace expansion valve power assembly
5. Loss of capacity	5. Obstructed expansion valve	5. Clean valve or replace if necessary
6. Conditioned space too cold	6. Contacts on control thermostat stuck on closed position	6. Repair thermostat or replace if necessary
7. Compressor short-cycles	7. Compressor capacity control range set too low	7. Reset compressor capacity control range
8. Lack of capacity	8. Expansion valve too small	8. Check valve rating table for correct sizing and replace if necessary
9. Too high superheat	9. Too much pressure drop through evaporator	9. Check for plugged external equalizer

QUESTIONS AND ANSWERS

1. What is heat of vaporization?

Answer. The heat of vaporization is the amount of heat in Btu's that are required to change a liquid into the vapor or gaseous state.

2. What are latent heat of fusion, vaporization, and condensation?

Answer. The latent heat of fusion is the amount of heat needed to change a substance from the solid to the liquid state. It is usually expressed in Btu per pound. The latent heat of vaporization is the amount of heat required to change a substance from the liquid to the vapor state. It is also expressed in Btu per pound. The latent heat of condensation is the amount of heat that must be removed from a vapor to change it to a liquid (condense it) and is the same as the latent heat of vaporization. For example, 970 Btu must be added to 1 lb of water to change it to 1 lb of steam. If the pound of steam is to be returned to a liquid, or condensed, 970 Btu must be removed.

3. What is a ton of refrigeration?

Answer. A ton of refrigeration is equivalent to melting 1 ton of ice in 24 h. The latent heat of fusion (freezing) of the ice is 144 Btu/lb (i.e., the amount of heat that is removed from 32°F water to turn it into ice at 32°F). This takes place without any change in temperature. If 144 Btu must be taken out of every pound of water to freeze it into ice, ice will absorb 144 Btu when melting to water. Then 1 ton of ice must absorb 144 × 2000 (lb) or 288,000 Btu to melt. If it takes 24 h to melt completely, 288,000/24 = 12,000 Btu/h or 12,000/60 = 200 Btu/min.

4. What is a refrigerant?

Answer. A refrigerant is the substance used for heat transfer in a refrigerating system. It picks up heat by evaporating at a low temperature and pressure and then gives up this heat by condensing at a higher temperature and pressure.

5. What are the main components of a mechanical-vapor refrigeration system?

Answer. See Fig. 14.2. The main components are (1) compressor, (2) condenser, (3) receiver, (4) expansion valve, and (4) evaporator.

6. A compression refrigeration system has an evaporator temperature of −15°F and an 85°F condensing temperature. What is the coefficient of performance (CP) and the horsepower required per ton of refrigeration?

Answer. It is:

$$CP = \frac{T_R}{T_C - T_R}$$

where $T_R = 460 - 15 = 445°F$ absolute, and $T_C = 460 + 80 = 540°F$ absolute. Then,

$$CP = \frac{445}{540 - 445} = 4.68$$

$$hp/ton = \frac{200}{CP \times 42.4} = 1.008$$

7. What is meant by standard-ton conditions?

Answer. An evaporating temperature of 5°F, a condensing temperature of 86°F, liquid before expansion valve at 77°F, and suction-gas temperature of 14°F produce standard-ton conditions. Refrigerating machines are often rated under standard-ton conditions.

8. What is meant by anhydrous?

Answer. Anhydrous means free from water, especially water of crystallization.

9. Why is it necessary to keep moisture from mixing with refrigerants?

Answer. Anhydrous ammonia has no effect on metals, but the addition of small quantities of water causes ammonia to attack copper and its alloys. Moisture in ammonia systems can also cause sludging of lubricating oils. Moisture in systems using halocarbon refrigerants is much more critical. Very small quantities of water are enough to cause ice to form at the expansion valve and plug the system. A more serious effect is that water in contact with these refrigerants will create acids that attack all metal parts of the system. Great care should be taken to keep moisture out of the system, and adequate dryers should be kept in the system to remove any moisture that may get in.

10. What is absolute zero?

Answer. All bodies are known to be made up of a large number of small particles known as molecules. Molecules are in constant motion, vibrating to and fro. The faster they move, the hotter the body. On the Fahrenheit scale of 459.8°F below zero (−460°F) is known as absolute zero. On the Celsius scale absolute zero is 273.2°C below zero. At this point there is absolutely no vibration of molecules; there is therefore no heat.

11. A Freon 12 direct-expansion system has an evaporator with the dry gas leaving the evaporator at 10°F and 29.35 psia. The liquid supply to the expansion valve is at a temperature of 65°F. What is the heat per pound of refrigerant absorbed in the evaporator and what are the pounds of refrigerant circulated per minute per ton of refrigeration?

Answer. See Table 14.3. The enthalpy of the vapor at 10°F is 79.36 Btu/lb. The enthalpy of the liquid at 65°F is 22.61 Btu/lb. The heat absorbed is $Q_R = 79.36 - 22.61 = 56.75$ Btu/lb. The pounds of refrigerant circulated per ton of refrigeration are $W_R = 200/56.75 = 3.524$ lb of refrigerant per minute per ton of refrigeration.

12. At what temperature is a refrigeration process considered to be in cryogenic service?

Answer. Cryogenic temperatures are considered to start at temperatures below −100°F, such as are used to liquefy the constituents of air and petroleum-type gases, including natural gas.

13. What is the purpose of a purge unit on a centrifugal machine, and how does it work?

Answer. Since centrifugal machines operate under vacuum on the suction side in most applications, there is always the chance of pulling in noncondensable air and vapor. The purge unit draws vapor and noncondensable gases from a cool portion of the condenser, and these are compressed by a separate small reciprocating compressor. The vapor is then condensed in the purge unit with the liquid refrigerant returned to the main system and the noncondensables vented to the atmosphere.

14. What four types of condensers are used in refrigeration service?

Answer.   Four types are: (1) shell-and-tube, (2) double-pipe, (3) atmospheric type, and (4) evaporative condenser.

15. How are evaporators classified?

Answer.   Classification is as follows: direct and indirect expansion, bare and finned tube, shell and tube, flooded and not flooded, and coil type.

16. On what kind of refrigeration system would you expect to find capillary tubes as expansion valves?

Answer.   On small domestic-type refrigerators.

17. What is a dry-type shell-and-tube evaporator heat exchanger?

Answer.   See Fig. 14.26. In this type of evaporator, the refrigerant is in the tubes, and the water is in the shell.

18. What is a bleedoff line on an evaporative condenser, and what purpose does it serve?

Answer.   A bleedoff line is a small line leading from the pump discharge directly to the sewer. Since an evaporative condenser evaporates large quantities of water, the remaining water becomes increasingly concentrated in salts and solids. If these are allowed to build up, the problem of scale formation is increased. The bleedoff line assures that enough water will be pumped to the sewer to prevent the concentration from becoming too great. This idea is similar to a continuous blowoff line used on boilers.

19. What is a solenoid valve, and where can it be used?

Answer.   A solenoid valve (sometimes called magnetic valve) is a shutoff valve that is actuated by an electromagnetic coil so designed that when the coil is energized, the magnetic field attracts an armature or plunger up into the core of the coil.

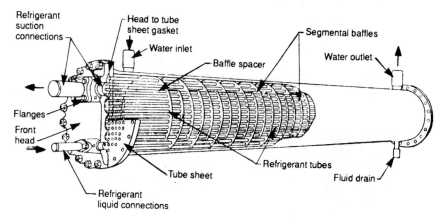

FIGURE 14.26   Dry-type shell-and-tube evaporator has refrigerant in the tubes and chilled water in the shell.

By means of a valve stem or pin attached to the plunger, the valve is opened. Deenergizing the coil destroys the magnetic field, and the plunger falls of its own weight, closing the valve.

20. Why is the discharge pressure gage so important in a refrigeration system?

Answer. A discharge pressure gage shows the pressure carried on the condenser, on the liquid receiver, and on the piping from the compressor through the condenser and receiver to the expansion valves. This gage gives warning when the pressure reaches a dangerous point. Reasons for this can be lack of cooling water, water that is too warm, or excessive amounts of noncondensable gases in the system. It also shows when condensing surfaces are scaled or dirty.

21. What precaution would you take when starting an ammonia reciprocating compressor with no liquid separator?

Answer. When a plant has been closed down for a time, suction pressures may increase up to 50 psi in some ammonia systems, and liquid may accumulate at the low points in the suction line. Therefore, you must open the suction valves on the compressor slowly until the line is pumped down to normal working pressures.

22. What causes surging in centrifugal compressors?

Answer. Surging in many centrifugals is normal and nothing to worry about, except at low loads. At 10 to 20 percent of full load surging can cause the compressor to overheat, raising the bearing temperatures. So don't operate continuously under these conditions without getting advice from the manufacturer.

23. What causes a compressor to short-cycle?

Answer. (1) Intermittent contact in the electrical control circuit, (2) a low-pressure controller differential that is set too close, (3) a high-pressure controller differential that is set too close, (4) a leaky solenoid valve in the liquid line, (5) a dirty or iced evaporator, (6) a faulty condenser, (7) an overcharge of refrigerant or noncondensable gas, (8) lack of refrigerant, (9) a water-regulating valve that is inoperative or restricted by dirt, (10) water temperature that is too high, (11) restricted water piping, (12) supply water pressure that is too low, (13) a restricted strainer in the liquid line, (14) a faulty motor, (15) a fouled shell-and-tube condenser, or (16) faulty operation of the evaporative condenser.

24. Where would you look for trouble in an absorption refrigeration system?

Answer. The main sources of trouble in absorption units are leaks, crystallization, and scale formation where water is used to cool the condenser and absorber. The ammonia-water machines do not have crystallization problems, and the air-cooled machines do not have water scale problems. Crystallization occurs when there is an abnormal operating condition, such as that caused by failure of an operating control.

25. What is crystallization in an absorption system, and how would you cure it?

Answer. Crystallization means that the solution at some point in the system becomes too concentrated with lithium salt and will not flow. It does not occur where the operating controls are working normally. When it does happen, the condition may easily be corrected by: (1) applying heat externally, (2) adding distilled water, or (3) bypassing diluted water-salt solution internally to the section which contains the highly concentrated solution.

26. What is the difference between flooded and dry-expansion shell-and-tube brine or water coolers?

Answer. The flooded cooler has the refrigerant in the shell area surrounding the tubes; the dry-expansion cooler has the refrigerant in the tubes, and the fluid to be cooled is pumped through the shell.

27. What causes a compressor to run continuously?

Answer. (1) An excessive load, (2) thermostat controlling at too low a temperature, (3) "welded" contacts on the electrical control in the motor-starter circuit, (4) lack of refrigerant, (5) an overcharge of refrigerant, (6) a leaky valve in the compressor, or (7) a solenoid stop valve that is stuck open or held open by a manual lift stem.

28. What causes noise in reciprocating compressors?

Answer. (1) A loose compressor-drive coupling, (2) lack of oil, (3) a dry or scored seal, (4) broken or worn internal parts, (5) liquid floodback, (6) a dirty water-regulating valve, (7) water pressure that is too high or intermittent, (8) an expansion valve that is stuck in the open position, or (9) a compressor or motor that has worked loose from the base.

29. What are five important steps in starting an ammonia system?

Answer. (1) Turn on the cooling water, (2) open the compressor discharge valve, (3) start the compressor, (4) open the compressor suction valve slightly until evaporator pressure is pumped down to about 20 psi, then open it wide, and (5) open the liquid valve and adjust the expansion valve to give the desired suction pressure.

30. What are some important steps in shutting down an ammonia system?

Answer. (1) Close the liquid valve, (2) close the suction valve, (3) shut down the compressor, (4) close the compressor discharge valve, and (5) shut off the cooling water to the equipment.

31. How are condenser and chiller tubes tested for thickness reduction and cracks?

Answer. Pressure tests used to be applied, and "sniffers" used such as halide torches for detecting leaks. This method shows a tube failure has occurred by leaks being present. Therefore, *eddy-current inspections,* which can reveal defects, such as thinning around intermediate tube supports on long heat exchangers due to vibration erosion are now made of tubes, and the suspected tubes can be replaced *before* leakage occurs.

32. What are two safety devices on reciprocating compressors that prevent damage due to Freon migration into the crankcase?

Answer. Freon in the crankcase oil can pull the oil with it when the compressor is started. To prevent this, crankcase heaters are used to keep the oil temperature above the condensing temperature of the Freon. The low oil-pressure cutout is a backup safety in the event oil pressure to the compressor is too low.

33. How do halocarbon refrigerants affect lube oil?

Answer. All halocarbon refrigerants thin lube oil greatly. This means that a heavier oil must be used with them. Thinning of oil by halocarbon refrigerants also reduces the pour point and cloud point of the oil. Refrigerants such as sulfur dioxide, ammonia, and carbon dioxide do not have a thinning effect on oil.

34. What characteristics must lube oil have for an ammonia system?

Answer. An SAE 10 or 20 oil is usually best, the higher grade being preferred for splash-lubricated machines to reduce the risk of foaming. Heavier oil is for pressure lubrication to avoid too much oil being thrown into the cylinders. Heavier oil is also for machines with separate cylinder lubrication. Some large, slow-running machines need SAE 30 for the cylinder and for the bearings, if supplied by drop or other intermittent oiler. Vaporization is not a serious problem because temperature at the end of compression is low in most installations—usually under 300°F or even under 200°F.

35. How are extremely low temperatures obtained?

Answer. See Fig. 14.27. Extremely low temperatures can be obtained in several ways. One method is the cascade system. Here, a series of refrigerants having progressively lower boiling points are used. One refrigerant serves as the coolant to condense the refrigerant gas with the next lower boiling temperature. Cascading is usually used to obtain temperatures of below −135°F.

The cascade system in Fig. 14.27 uses ammonia, ethylene, methane, and nitrogen as refrigerants. These all have progressively lower boiling points. The evaporator of one refrigerant is used to condense the discharge from the next-lower-temperature compressor. The total system horsepower input per unit of nitrogen produced is lower than that in other systems.

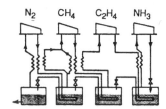

FIGURE 14.27 Several compressors and condensers using different refrigerants are used to obtain very low temperatures in the cascade system.

36. What is a heat pump?

Answer. A heat pump is a refrigeration compressor which provides both cooling and heating by using the heat of compression that is rejected at the condenser (see Fig. 14.28). During the cooling cycle, the heat pump operates as a straight compression-refrigeration cycle. Whenever heat or cooling is needed, a changeover switch of summer or winter operation is provided, and this causes the function of the condenser and evaporator-heat exchanger to be automatically reversed by a changeover valve. By rerouting flows, the air-cooled condenser becomes an evaporator and picks up heat from the outside air.

Extracting heat from outside air at 0°F, for example, requires a pressure drop in the refrigerant. Some newer units include a second compressor in series to boost pressure for heating. In these systems, a centrifugal compressor may handle the base load, with reciprocating units supplying the second compression state.

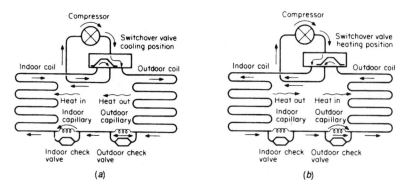

FIGURE 14.28 Heat pumps reverse refrigerant flows from the compressor. In summer, the outdoor coil is the condenser, while in winter, the indoor coil is the condenser, thus, giving up the heat of compression that is used to heat the premises. (*a*) cooling; (*b*) heating.

Cooling Towers

37. How are cooling towers classified?

Answer. Cooling towers are usually classified by the way their housing is arranged and by their draft-producing methods. They are broadly classified as (1) natural draft, (2) induced draft, and (3) forced draft. Natural-draft towers may be further classified as spray towers and deck towers.

38. What mechanical maintenance does a cooling tower need?

Answer. Set up a preventive maintenance schedule for your cooling tower. Grease and oil the bearings and check the motor insulation yearly. Refill the speed reducers every 3000 h. Look for oil leaks around the fan driveshaft caused by worn oil seals. Keep the fan blades and hubs painted to avoid corrosion. Tighten the blade clamps and hub bolts and rebalance the fans when necessary. Check the clearances, lubrication, and air gaps periodically.

39. What causes scale formation?

Answer. The most common type of scale found in heat-exchange equipment is calcium carbonate, which is formed by the breakdown of calcium bicarbonate in the cooling water at high temperatures. Magnesium compounds and calcium sulfate are rarely the cause of scale in cooling systems. Scaling is affected by temperature, rate of heat transfer, calcium, magnesium, silica, sulfate, alkalinity concentrations, and the pH.

40. What causes the formation of slime and algae?

Answer. Algae growths are composed of millions of tiny plant cells, which multiply and produce large masses of plant material in a short time. Slime growths are a gelatinous mass of microorganisms, which cling tightly to secluded surfaces in the system, trapping organic and inorganic matter and debris along with scale-forming materials. Any appreciable buildup seriously interferes with heat-transfer efficiency.

41. How are accumulated slime and algae removed?

Answer. The best way to get rid of these growths is by mechanical cleaning. In the case of algae the algicide may only loosen the growth enough to set it free in the system to plug lines and damage pumps. After the system is completely cleaned, start out on the right foot with a good chemical treatment program to prevent future buildup of these growths as recommended by a water treatment company.

Air Conditioning

42. What is complete air conditioning?

Answer.   Complete air conditioning means heating, cooling, humidifying and dehumidifying, circulating, and cleaning the air.

43. What factors affect the conditioning load?

Answer.   A group of complex factors affect the conditioning load. They are (1) heat transmission, (2) solar radiation or sun effect, (3) people, (4) light and power equipment, (5) ventilation air or infiltration, (6) product load, and (7) miscellaneous factors.

44. How are air temperature and humidity related?

Answer.   Air temperature and humidity are related through the basic properties of steam. For any given air-water temperature, each cubic foot of air can hold a specific weight of water vapor; this is the saturation-temperature dew point.

45. What is saturation temperature?

Answer.   For any given pressure there is one temperature—the saturation temperature—at which steam starts to vaporize or condense. Let us say that 1 ft^3 of moist air at 70°F contains 0.0004 lb of moisture. But the same air can hold 0.0011 lb of steam. The reason why it is less is that the steam is superheated. In this case, 70 − 40 = 30°F of superheat.

46. What is relative humidity?

Answer.   Because the cubic foot of air in the above problem holds less moisture than it is capable of holding, we have relative humidity. Here, we have 0.0004 lb of moisture when we could have 0.0011 lb. So the ratio is 4 to 11, or 36 percent relative humidity. This is based on volume, which is cubic feet of moist air.

47. A room has a temperature of 72°F with a relative humidity of 50 percent. The barometric pressure is 29.5 in of mercury. What is (*a*) a partial pressure of steam in the air, (*b*) the weight of steam per cubic foot in the air, and (*c*) the weight of steam per each pound of dry air.

Answer.   See Table 14.6. (*a*) At 72°F, the pressure of saturated vapor is 0.3883 psi. Relative humidity = p_S/p_D with p_D = 0.3883 and relative humidity is given as 0.50; the p_S = 0.50 × 0.3883 = 0.194 psi. (*b*) Density of air at 72°F from Table 14.6 is 0.0012269 lb/ft^3 and density of steam = 0.5 × 0.0012269 = 0.000613 lb/ft^3 vapor in the air. (*c*) Barometric pressure is total pressure of air, which consists of partial air pressure p_a plus partial vapor pressure p_s. Then, P_a = (29.5 × 0.491) − 0.194 = 14.29 psi. To find the volume the air occupies, use $PV = WRT$, with p = 14.29, W = 1 lb, R = 53.3 for air. Then, 144 × 14.29 × V = 1 × 53.3 × (460 + 72), and solving for V, V = 13.78 ft^3 occupied by 1 lb of air. The weight of steam per pound of air is 13.78 × 0.000613 from part (*b*) = 0.00845-lb steam per pound of dry air.

48. What is a dew point?

Answer.   A dew point is the saturation temperature corresponding to the actual partial pressure of water vapor in the air.

49. What is a psychrometric chart?

Answer. This is a chart showing the graphical values of air-vapor mixtures including wet- and dry-bulb temperatures, pressure of water vapor, and relative humidity.

50. What two heats must be considered in human comfort air conditioning?

Answer. Sensible and latent heat.

51. What is latent heat?

Answer. Heat must be added to vaporize water into steam, but heat must be removed to condense steam back into water. In each case, the same amount of heat is removed or added. This heat is called latent heat. Since the amount of latent heat varies with the dew point, it is used to change a liquid to vapor, or a vapor to liquid, without changing the temperature.

52. What is sensible heat?

Answer. Any heat needed to change the temperature of a liquid or vapor is sensible heat. The amount of sensible heat depends on the temperature and specific heat of the substance.

53. What is wet-bulb temperature?

Answer. To take the wet-bulb temperature, you need a wet-bulb thermometer and a psychrometric chart. The thermometer has a bulb that is covered with a wetted silk gauze and placed in the airstream. Some of the water in the gauze will evaporate. Because vaporizing takes heat from the remaining water, the water temperature will drop. The amount of this temperature drop depends on the dryness and temperature of the air. The thermometer reads wet-bulb temperature and tells us the amount of moisture in the air. Then, by checking a psychrometric chart, we learn the other things we need to know.

54. What winter maintenance is needed for heat-rejection equipment?

Answer. If cooling towers and condensers are subject to freezing, remove all water to protect tubes, water boxes, and piping. Remove all sludge and accumulated dirt from the cooling tower; then paint it. Inspect spray nozzles and strainers in the waterline to the condenser; clean if needed. Check blade roots of fans for cracks and gears for wear, and also check the condition of the oil in the gear case. Swelling and shifting of the cooling tower may have misaligned the motor and gear case. Inspect the condenser tubes for effectiveness of water treatment. Coated heat-transfer surfaces greatly increase power costs.

Safety Practices

55. What are the refrigeration codes?

Answer. Refrigerating equipment should be designed, installed, and repaired according to the ASA Code B9.1 (Safety Code for Mechanical Refrigeration, American Standards Association); also according to the code and regulations of the National Board of Fire Underwriters (NBFU), Standards for the Unit Refrigeration Systems and Standards for Air-Conditioning and Commercial Refrigeration Equipment; and according to all the local municipal and state codes.

These codes cover refrigeration plant location with respect to flammability and toxicity, type of occupancy of the building, foundations, electrical equipment and wiring, machine rooms, ventilation, pipes and fittings, strength and test pressures, safety devices, gages, discharge means, markings, adding or withdrawing of refrigerants, testing for leaks, gas masks, maintenance, and periodic inspection. Codes also provide tables showing permissible pressures, quantities of various types of refrigerants allowed, sizes of air ducts and openings, line sizes, and sizes of relief devices.

If there is no state or local code, or if the local codes do not cover specific installations, follow the B9.1 code. Consult it when the equipment is installed and tested. The suppliers should be asked to furnish evidence of compliance with the requirements of the local codes.

Electrical equipment, wiring, and motors should comply with the national code (NBFU) and ASA. All persons responsible for installation, testing, operation, maintenance, and repair of refrigerating equipment should know the provisions of all codes which apply. After major repairs the system should be tested in compliance with the codes as if it were a new installation.

56. What are the major hazards encountered with refrigeration?

Answer. The main hazards are (1) explosions, (2) fire, and (3) toxic effects from the gases used. The explosion hazard tends to increase as more refrigerant is used. If gas escapes, it may damage goods in cold storage. If gas is toxic, it may cause serious personal injury. If it is flammable, it may form an explosive concentration.

57. What are some of the main causes of explosions in the crankcase of a refrigeration compressor?

Answer. Two main causes of explosions in the crankcase of a compressor are shots of liquid impurities in the gas and leakage past the piston rings and packing. This leakage allows oil vapors to mix in the crankcase with the refrigerating gas vapors at high temperatures, creating a flammable mixture.

58. What safety feature should a gage glass installed in a refrigeration system have?

Answer. A gage glass in a refrigeration system should have a device which will close automatically if the glass breaks. The glass also should have a guard to protect it from external blows.

59. What precaution would you take before opening a system for repairs?

Answer. No part of a system should be opened for repairs until all pressure has been relieved from that section and all refrigerant gas pumped out of it. The section to be repaired should be pumped down to a vacuum of 10 to 15 in if possible, or to at least atmospheric pressure, and held for a minimum of 10 min. To assure that all refrigerant has been removed, pump down at least three times. Pour water on the piping as a further test. If frost appears, there may still be too much refrigerant inside to open the section safely.

60. Are fire extinguishers needed to protect refrigeration systems?

Answer. Yes. Fire extinguishers of approved types should be kept near the equipment and doors or at other accessible locations in the machine room. The fire extinguishers should be plainly visible and well marked. Operators should know their location and be instructed in their use. Vaporizing liquid and carbon dioxide extinguishers are suitable for electrical and refrigeration fires. The carbon dioxide type is preferable because it is not as toxic as the vaporizing liquid on a hot fire. For emergencies it is good practice to have two or more doors to all refrigeration machine rooms.

61. Can you use a gas mask of the canister type for all types of refrigerants?

Answer. No. The type of canister should be approved by the U.S. Bureau of Mines for the refrigerant being used in the system. Gas masks of the canister type do not give suitable protection against either carbon dioxide gas or oxygen deficiency. An oxygen-supplied gas mask is the only type which protects in such cases.

62. Is it dangerous to smoke while drawing oil from an oil trap?

Answer. Yes. Oil vapor and ammonia gas explode from an open light. In one plant when a gasket blew out of a compressor head the operator pulled the switch because the plant was filled with ammonia fumes. The switch arced, and an explosion wrecked the plant. Freon decomposes in a fire to form extremely toxic gases.

63. Are refrigerants toxic?

Answer. In a strict sense all refrigerants except air can cause suffocation by oxygen deficiency. The term *toxic* is usually applied to refrigerants which are actually injurious to human beings. Some cause death or serious injury even though mixed in small doses with air. In these cases enough oxygen cannot be obtained from the air-refrigerant vapor mixture. A mixture of toxic vapors and air may attack the membranes of the lungs and be carried to other parts of the body. So, know your refrigerants and what to do in case of an accident.

64. How would you enter a room that had a bad ammonia leak?

Answer. Before entering a room where there is an ammonia leak, put on canister-type gas mask approved for ammonia use by the U.S. Bureau of Mines. Then spray the room with water because it absorbs ammonia quickly and washes it to the floor. Use rubber gloves to protect your hands. Don't use an all-purpose gas mask if the air contains more than 3 percent ammonia. You will know when that point is reached because ammonia will start to penetrate the mask's filter.

65. What should every operator know about the dangers of carbon dioxide?

Answer. When carbon dioxide leaks, it mixes with air. There is nothing to worry about if the leak is small. But when air is mixed with more than 5 percent carbon dioxide, you get headaches and feel drowsy; larger doses will cause unconsciousness. When the air contains about 9 percent carbon dioxide, you suffocate. So, if the leak is large, use an oxygen-breathing apparatus or an air-line mask when fixing it. Don't use an all-purpose gas mask because it relies on the atmospheric condition in the room. Human beings suffocate with less than 7 percent oxygen in the air. The flame of a safety lamp goes out when there is 16 percent oxygen in the air. If this flame won't burn, don't use an all-purpose gas mask.

66. Are Freon-type refrigerants (halocarbons) dangerous?

Answer. Halocarbons form a nonpoisonous and nonirritating gas; but a large leak (over 20 percent concentration with air) causes unconsciousness. The refrigerant itself won't burn, but in the presence of a flame or hot surfaces above 1000°F it decomposes into toxic products that are extremely irritating and poisonous. So take all precautions when welding, brazing, etc. Unlike ammonia, a halocarbon refrigerant does not form explosive mixtures when it breaks down into its constituent parts.

67. How many helmets and gas masks are required in a refrigerating machinery room? Where should they be located?

Answer.   Class A systems with over 1000 lb of refrigerant should have two masks in the machinery room near the main exit and one or two masks at convenient points throughout the plant. All codes require masks approved by the U.S. Bureau of Mines for the refrigerant in use.

68. What are the general requirements for ventilation in refrigerating machinery rooms?

Answer.   Requirements vary with the amount of refrigerant in the system, the type of refrigerant, and whether the ventilation is natural circulation or forced-draft. Thus, a Class A system with over 1000 lb of refrigerant needs (1) an exhaust fan with a capacity of at least 2000 cfm, or (2) a window area to outside air of 25 ft^2 if on opposite walls, or (3) a window area of 60 ft^2 if windows open on one side of machinery room. See your local code.

69. What material is forbidden by OSHA in ducts, plenums, and ceilings where there is human occupation?

Answer.   Asbestos; this material can cause lung cancer.

70. What are fire dampers and why are they important?

Answer.   Fire dampers are automatic dampers and are usually rectangular "flaps" or else louvers. They act when heat, fumes, or smoke is sensed in the system, as in the ducts, plenums, or intakes and prevent the spread of smoke, fumes, and hot air due to fire. For example, internal air being recirculated may carry fumes or fire into other areas. Fresh-air intakes may carry fire and smoke from an adjacent building fire. Many modern materials and building contents are either made of or coated with plastics, which release heavy smoke and extremely lethal gases during combustion. These plastic materials are confined in a sealed air-conditioning building, and it is also necessary to consider including pressurized fireproof stairwells, areas of refuge, and video (TV) as well as voice communication systems as part of the system. The blower systems should automatically stop when smoke, fumes, or heat is sensed in ductwork.

71. Are there other health hazards to an air-conditioning system?

Answer.   Yes, harmful bacteria sometimes build up in parts of the system, especially in cooling towers. Bacterial problems are ordinarily part of water treatment, and both the manufacturer and the water treatment service will have instructions. Local emergencies will be the concern of health officials.

72. What precautions should be taken when charging or withdrawing gas?

Answer.   Be sure to ventilate the refrigeration machine room thoroughly. Start the ventilation system or fans and keep them running until the operation is complete. If there is no ventilating equipment, open all doors and windows.

CHAPTER 15
GENERAL MAINTENANCE AND ADMINISTRATION

HOUSEKEEPING

Keeping service equipment rooms clean and free of oil spills, rag accumulations, and trash is an often neglected but important part of plant maintenance. Good housekeeping habits promote good operation and maintenance practices because they promote correcting minor problems, such as oil leaks around bearings and seals and water leaks around valves and pumps, before the problem becomes one of major repairs. The best housekeeping practice is to perform it daily and whenever cleanup is necessary, such as after a repair or inspection of service equipment.

Painting is another good housekeeping practice that will make equipment rooms reflect the attitude of plant operation and maintenance practices. Some industrial and commercial environments are severe, and if protective coatings are not applied, rapid corrosion may result on machine elements as well as on structural building components. Coatings resist hostile, corrosive environments and also maintain a visual and psychological attractiveness in the workplace. Surface preparation before a protective coating is applied is an important consideration, as is the type of coating that should be applied, for example, to room walls versus machinery frames. Most paint manufacturers and suppliers have excellent instructions on how to prepare surfaces and the type of coating to use.

CORROSION

Corrosion in its many forms can attack steel structures, boilers, and pressure vessels and machinery elements. The main reason that inspections must be made is to detect the weakening effects of corrosion on metals. Identification of the many different forms of corrosion can assist in determining the corrective actions to take to reduce the degradation caused by corrosion. Some forms are:

1. General corrosion occurs uniformly over the entire surface of metal parts; its presence indicates some form of attack by acids or hydroxides.

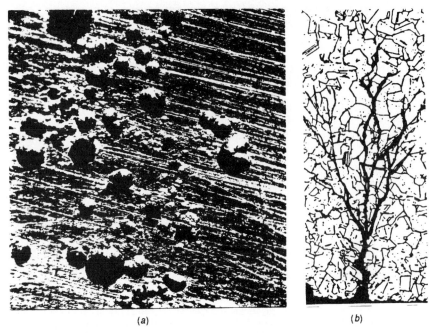

(a)   (b)

FIGURE 15.1 Corrosion has many forms. (*a*) Pitting corrosion (*Courtesy, Nalco Chemical Co.*); (*b*) stress-corrosion cracks progress along grain boundaries.

2. Pitting corrosion (see Fig. 15.1*a*) results in the formation of small holes in an otherwise relatively unattacked surface. Pits act as stress concentration points that can magnify normal stresses, allowing cracking to start in the pit and travel from pit hole to pit hole, sometimes with disastrous results. Pits are also places for acids to settle and attack the metal.

3. Crevice corrosion is similar to pitting because any crevice acts as a concentration cell in which corrodents can settle out and attack the metal. Crevice corrosion can occur on bolts and gasketed joints.

4. Fretting corrosion occurs where metals slide over each other to cause metal-to-metal rubbing. This removes the protective oxide coating. The freshly exposed metal surface is then attacked by the corrodents in the fluid. Fretting corrosion is quite often noted in heat exchangers with long tubes. Intermediate supports may not have the tubes rolled in, resulting in tube vibration and, consequently, tube metal rubbing against the intermediate support.

5. Corrosion fatigue is a combination of corrosion and the application of a repetitive stress, such as on a steam turbine blade. The repetitive stress is magnified by stress concentration as a result of pit corrosion, for example.

6. Intergranular corrosion is a localized attack on the material's grain boundaries by a corrodent fluid or gas. The grain boundaries look rough to the naked eye, with loose grains evident at times. Good heat treatment of heat-affected zones (HAZs) in welding, for example, reduce the chance for intergranular corrosion as does good alloy material selection.

7. Stress-corrosion cracking is a progressive type of failure that causes cracks to occur well below a material's yield point. It occurs on highly stressed parts that operate in a medium that can chemically attack the grain structure of the material at relatively low concentration levels. Stress-corrosion cracking is generally intergranular in low-carbon steel and transgranular in austenitic stainless steels. Figure 15.1b shows a stress-corrosion crack.

It is now common practice in maintenance work to inspect for corrosion and cracks with instruments, such as thickness testers and flaw detectors, in addition to making visual inspections.

Methods to Minimize Corrosion's Effects

Broad strategies in combating corrosion include:

1. Designing to avoid crevices, sharp corners, bends, corrosion cells, and similar areas that have been shown to be prime candidates for corrosion.
2. Designing, operating, and maintaining the system to limit corrodent concentrations. This is possible in some operations, such as water treatment to control water chemistry.
3. Selecting materials that are more corrosion-resistant, if economically justified.
4. Applying a corrosion-resistant lining, such as stainless steel or glass.
5. Establishing a periodic inspection program to track deterioration and establishing minimum requirements for the pressure and temperature of the process so that repairs or replacements are instituted when these minimum requirements are reached, thus avoiding serious accidents and process interruptions. Periodic inspections include nondestructive testing (NDT) inspections where appropriate.
6. Using monitors to keep track of chemistry changes in a process stream and to assist in analyzing the concentration of possible corrodents. For example, in the power-generation field, on-line monitoring of contaminants in steam and condensate is being used. Items being measured include pH, O_2, conductivity, cation conductivity, and sodium and chloride concentrations. This is combined with liquid ion chromatography showing the parts per billion of sodium, chloride, ammonium, potassium, fluoride, and sulfate ions. The tests are performed on grab samples from selected loops of the steam-condensate flow. This permits identification of impurities and their sources so that corrective actions can be taken before more damage can result from a corrodent's presence in the process loop.

NONDESTRUCTIVE TESTING AND EXAMINATION

The terms *nondestructive testing* and *nondestructive examination* denote that a test is made to find flaws or defects but that no damage is done to the material being tested. A *visual examination* is a nondestructive test. Equipment in certain types of service require periodic inspections to determine if the harmful effects of overtemperature, overstress, creep, fatigue, and similar long-term actions on materials have caused defects that require repair in order to avoid sudden breakdowns while in service. Nondestructive tests are also used to predict the future life of equipment.

Selection of NDT in Maintenance Inspection

Nondestructive testing is an important tool in maintenance inspection and field repair and is not confined only to thickness testing. The NDT method to be selected is dictated by several factors: the type of defect to be located, the geometry of the component, access (whether on one or both sides of the component), and the type of metal to be tested. Other considerations may exist, such as whether testing must be performed under on-line conditions or during a shutdown.

Selection of the best NDT method to use is dictated not only by the type of defect but also by other factors as well. Magnetic-particle testing is used for defects on or close to the surface but not for nonmagnetic materials such as stainless steel. Subsurface defects can be found using the ultrasonic method or radiography.

Ultrasonic testing has the advantage that testing can be performed even if only one side is accessible.

Types of NDT

The most commonly used types of NDT are (1) visual inspections, including the use of borescopes and fiberscopes on sections of a structure that are difficult to see because of blocked lines of vision; (2) radiography; (3) magnetic-particle tests; (4) dye-penetrant tests; (5) ultrasonic tests; (6) eddy-current tests; and (7) acoustic emission tests.

Visual inspections. All inspections start with a visual examination for leaks, wear, cracks, bulges, loosening, and similar signs of trouble.

A borescope or fiberscope can be used in hard-to-see equipment areas, such as in tubing in a heat exchanger. The borescope is a slender, hand-held tube consisting of a series of lenses and internal light sources, which permit viewing an area of a part that normally cannot be seen. The advent of fiber optics has led to the development of the fiberscope, which consists of bundles of lights and image-transmitting fibers within a long flexible sheath having an eyepiece lens at one end. Cameras can be attached to the eyepiece to obtain a permanent record of what is seen.

Radiography. Radiographic inspection uses x-ray and gamma rays obtained from isotopes, such as cobalt 60 and iridium 192, for which the resultant radiant energy can be safely controlled.

Magnetic-particle testing. Magnetic-particle testing is used to detect surface faults by means of setting up a magnetic field or magnetic lines of force between two electrodes. Powdered magnetic material is sprinkled over the work to be tested. The magnetic field will affect the magnetic powder, and these particles will align themselves in a fault.

Magnetic-particle testing equipment is available in fairly lightweight (35–90 lb, or 16–41 kg) power source units that can be taken to the test site. This equipment has not been routinely used in industry by plant personnel but has been used extensively by NDT contract service personnel. As with ultrasonic inspection, the magnetic-particle method requires that the operator be well trained. This is especially important for reliable interpretation of the results.

Yoke-type magnetizing units (see Fig. 15.2) are extensively used in field inspections. The yoke generates a magnetic field, and if a defect is present, a leakage field is created around the defect. The closer the defect is to the surface, the sharper will be the indications because the field will become more distorted. For deeper cracks, the indications get wider, a fact that can be used as a guide in judging the depth of the defect.

Liquid-penetrant inspection. Liquid-, or dye-penetrant inspection is used somewhat like magnetic-particle testing, except that it is used primarily on nonmagnetic material; it can, however, be used on magnetic material. The dye penetrant contains a visible dye, usually red. Indications of defects appear as red lines or dots against a white developer background. It is primarily a surface-defect indicator.

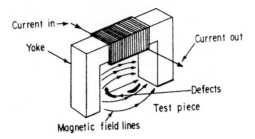

FIGURE 15.2 Portable magnetic yoke for making magnetic-particle inspections.

Ultrasonic testing. Ultrasonic testing makes use of high-frequency sound waves in the range of 0.5 to 10.0 MHz for the inspection of material for flaws and also for thickness testing. The basic principle used in an ultrasonic system is the transformation of an electrical impulse into mechanical vibrations and then the transformation of the mechanical vibrations into electric pulses that can be measured or displayed on a CRT screen. The transfer of mechanical energy to electric energy is performed by means of a transducer, which is capable of transforming one form of energy to another (see Fig. 15.3).

Ultrasonic tests are grouped into two basic categories: *pulse echo testing* and *resonance testing*. The pulse-echo method involves transmitting a short burst of high-frequency sound (well above human hearing) through the piece being tested and then detecting the echoes that are received from either a construction detail, such as a shoulder or hole, or a defect in the material with a separation or void in the material sufficiently large that sound cannot be transferred across the interface. In operation, a pulse-echo unit will produce, through an electronic pulser, a short burst of high-frequency electric signal. This is transmitted to the transducer, which is forced to vibrate, usually at its resonant frequency. The probe must be coupled to the test piece with oil, water, or some other liquid or grease. The sound wave train then travels through the test piece until some form of discontinuity or boundary (or the back side of the test piece) is

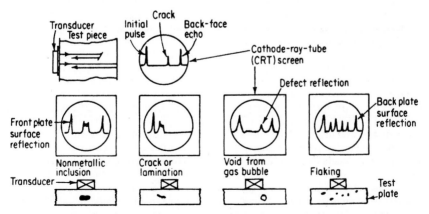

FIGURE 15.3 High-frequency sound waves are used in ultrasonic inspections with a CRT screen used to show deflections if a back surface or defect redirects the sound wave back to the transducer. (*Courtesy,* Power Magazine.)

encountered. This discontinuity in the medium causes the sound wave to be reflected to the receiver transducer. The vibrational energy of the sound wave sets the transducer in motion to produce an electric impulse, which is fed into an amplifier. The output of the amplifier is displayed on a CRT, which shows the signals on a linear time baseline. If a linear amplifier is used, the amplitude of the returned echoes can be used as a measure of the area producing the reflected signal.

Resonance testing makes use of a tunable, continuous-wave system. This method is usually employed for measuring small or thin walls to 2 or 3 in. The resonance of a crystal is tuned to the piece under test. In practice, a loud pip is heard and can also be seen because the electronic circuit is also in resonance electrically. By proper calibration, direct thickness readings can be made.

Eddy-current testing.   The underlying principle of eddy-current testing is the measurement of impedance of electron flow in the part being tested. Weak electric currents or eddy currents are induced by a probe containing inducing and sensing coils. Any changes in the geometry of a test part, such as a pit, crack, or thinning, will affect the flow of the eddy currents. This change in the flow of the eddy currents will be detected by the sensing coil and displayed on a strip chart, a CRT display, or both.

Eddy-current testing is an appropriate NDT for loss-prevention work on heat exchanger tubes. Regular eddy-current testing of tubes in heat exchangers will help detect some of the common problems found in large heat exchangers, such as feedwater heaters, condensers, chillers in air conditioning units, and numerous process heat exchangers, such as those used in the petrochemical industry. Common causes of tube failures that may be detected include:

1. Tube wear at intermediate supports, usually due to tube vibration. Gas and liquid flows across the tubes can cause this vibration if near the natural frequency of the tubes.
2. Corrosion from acid formation when noncondensable gases are trapped in the exchanger. For example, CO_2 becomes carbonic acid.
3. Fatigue cracks from overrolling, expansion and contraction of the tubes with load, and similar repetitive stresses that may be imposed on the tubes.
4. Stress-corrosion cracking. Nonferrous tubes may be affected by ammonia and mercury.
5. Bulged tubes in air conditioning chillers, caused by control failures that permit the water in the tubes to freeze.

Acoustic emission testing.   Acoustic emission is another developing NDT method that could be very effective in the future in the inspection of large pressure vessels, such as those in the petrochemical industry. This method of nondestructive testing is being developed to monitor large pressure vessels such as digesters and nuclear reactors for crack growth. Acoustic emission is based on the principle that a growing defect releases bursts of energy or stress waves that can be detected by sensitive and suitably designed transducers. If a transducer is strategically installed in known highly stressed areas of the pressure vessel, it can convert the minute sound emitted when a material "gives" into electric signals. The signals can be recorded on a computer for immediate analysis or future use. When transducers are installed in a triangular mode, the source of any abnormal sound can be determined by quick trigonometric calculations, and thus the defect can be located for analysis and repair.

MAINTENANCE

Periodic tests such as NDT, vibration reading and data logging, and electrical insulation resistance testing are all used to determine what the condition of equipment may be and what may be required to restore it to near its original design condition. It tries to avoid the unexpected breakdown.

Breakdown maintenance, if practiced, can be expensive. The unexpected shutdown can cause process lines to produce no income. Labor for repair is unplanned and thus may involve premium overtime pay. Parts may not be available from normal sources and thus may be more expensive in the rush to restore production. It is important for a plant to have an effective organization to avoid the pitfalls of breakdown maintenance so that proper maintenance is carried out in a planned manner and within budget requirements.

Planned or Preventive Maintenance

There are many terms used to define maintenance work. *Planned or preventive maintenance* (PM) is maintenance carried out on a scheduled basis in order to catch wear and tear of components before a failure occurs. A modification of a regularly planned and scheduled maintenance program is to combine this scheduled maintenance, based on review of past inspections, testing, and wear rates, with a program of diagnostic readings.

Predictive Maintenance

Predictive maintenance is another term for a controlled method of instituting corrective action as a result of periodic measurements that point out problems that may be developing on the equipment. An effective predictive maintenance program can reduce costly, unexpected downtimes by recognizing deteriorating conditions through the use of monitoring measurements. One result of a monitoring program may be that overhauls are not necessary because the variables subject to periodic measurement prove to be constant and within acceptable limits. Measurements can be made with installed sensors and recorded for continuous monitoring. Supplementary measurements can be made with portable equipment to verify sensor readings or to obtain even more accuracy.

Performance Monitoring

Performance monitoring is a real adjunct of maintenance and thus requires close coordination and cooperation between operating people, testing people, and the maintenance department. It is common to establish benchmarks of equipment performance when the equipment is new or after a major overhaul. Performance is then subsequently monitored in order to track efficiency of performance. When the efficiency of performance drops to a calculated level that justifies the expense of shutting down for cleaning, repairs, or other maintenance work, operating and maintenance personnel work with the test group in an effort to restore performance efficiency.

Loss in performance output is now recognized as a major reason for preventive maintenance, in addition to the traditional purpose of preventing breakdowns. In a highly competitive world, efficiency of operation is becoming a critical item from a profit-and-loss point of view. For this reason, while outside maintenance might prove beneficial for major planned shutdowns, the daily routine of maintaining equipment for good performance and reliability is still an essential part of good plant management programs.

VIBRATION MONITORING

Major rotating machinery in a facility, such as critical turbines, pumps, fans, and blowers as well as electrical motors and generators can manifest problems by increases in vibration and, therefore, should be subject to either continuous monitoring or periodic testing with portable vibration equipment. Vibration monitoring assists a predictive maintenance program by recording conditions during operation and then scheduling overhauls based on conditions observed instead of inspecting or repairing machines at specific time intervals.

Definitions

Vibration is a periodic motion that moves in the opposite direction from the position of equilibrium of a machine part and then back.

Periodic motion repeats itself at definite intervals of time or periods of vibration.

Free vibration is the harmonic motion of a body after being displaced from its position of rest.

Forced vibration is the periodic vibration of a body after being acted upon by an external periodic force.

Natural frequency is the number of cycles per unit of time that a freely vibrating body will undergo once it is set in motion from rest.

Forced frequency is the number of cycles of vibration that a body undergoes when an external force acts upon it.

Resonance is the vibration frequency a body experiences when an external force acting on the body has the same frequency as the natural frequency of the body, resulting in increased amplitude of vibration.

Critical speed is the speed at which the speed of motion has the same frequency, or near it, as the natural frequency of the body set in motion.

Amplitude is the maximum distance a body vibrates from a neutral or mean position, usually expressed in mils of an inch.

Damping is the introduction of friction into a vibrating body in order to change its natural frequency.

Vibration Instruments

Vibration instruments today are based on the piezoelectric principle, in which a crystal is energized or moved by a pickup that may ride the shaft or may be of the proximity type. The crystal or transducer, when moved by the pickup, generates a voltage which is proportional to acceleration and which, by means of integrated electronic circuits built into the vibration meter, permits direct reading of rms of displacement, velocity, and the acceleration of the vibration. The output of such meters can be used for alarming and tripping a machine and for recording the data inside control rooms or on computer printouts.

Vibration analyzers are used for measuring several different occurring frequencies or harmonics. The analyzer contains tuning circuits which sweep through the entire frequency range and then separate out each vibration frequency that may be present so that it can be measured separately for machine analysis purposes.

Continuous monitoring is applied only to critical machinery because the instrumentation required is quite costly. Continuous monitoring has the advantage of immediately detecting a dangerous condition. It also helps to detect trends of vibration so that corrective actions can be taken before serious damage results. Vibration is an indicator of many machine faults. Misalignment, worn bearings, oil whip, and being out of balance, perhaps due to internal failures, are a few of the physical events that manifest themselves by increased vibration levels. By careful study, experienced vibration specialists can analyze the cause of the vibration and prevent further machine damage.

MAINTENANCE OBJECTIVES

A maintenance program's success is measured by a facility's capability to produce quality service when required without interruptions, and this depends on how effectively the facility services are maintained and operated. Economic constraints, or an effort to reduce operating costs, have also affected facility operation and maintenance; however, this should be viewed as an attempt to continue to produce good operating results with fewer resources.

The major maintenance and facility operation objective is still the same—to provide the essential services and avoid equipment downtime that could affect these services. What are continuously changing are the methods to achieve these objectives. Innovations that facility operators will have to cope with in the future include remote monitoring and diagnostics to note if any corrective actions are needed from the data being monitored. *Computerized maintenance* programs are already being used in industrial plants as is a program that keeps track of inventory of spare parts and where to obtain them as quickly as possible. Remote monitoring and diagnostics may be through a central control room or through outside service organizations that assist the maintenance and operation function by using high-level technical specialists. The service organizations, however, are still dependent on plant facility operators detecting the problem because there is a limit to "Christmas-tree" gadget installations for remote monitoring as well as economic constraints. It is just too expensive.

Service facility operators have always worked closely with manufacturer's representatives to solve a system or machine problem. Most manufacturers provide technical assistance after working with plant operators to analyze the possible reasons for the problem(s). They still depend on facility operators to detect conditions or to schedule maintenance outages under their guidelines. Thus, remote diagnostics and remote data logging do not eliminate the need for good facility operating personnel.

BENCHMARK READINGS

Benchmark readings are important in equipment evaluation. All facility service equipment and systems permit establishing benchmark condition readings when the equipment is new or has been reconditioned. From these benchmark readings, changes in the conditions can be noted in future inspections to determine if repairs may be needed.

ADMINISTRATION

Maintenance Organizations

In large process plants there are several organizational formats for maintenance programs. The organization may be of the area, central, or modified area-central type, using either contract or owner maintenance personnel, depending on many factors peculiar to the specific plant. The following points should prove helpful in organizing the maintenance function, especially for large industrial facilities.

1. It is essential in complex plants that continuous training be applied; mechanics and supervisors must have the continuing experience of working on and learning about specialized equipment. This precludes the type of central organization in which different mechanics are sent in to perform maintenance on the equipment each time it is shut down.

2. Some plants have had good experience using a minimum number of assigned personnel for maintaining specialized equipment and supplementing these with additional people as the work and work load requires.

3. The organization should include technical personnel organized so they can focus their efforts on the technical and problem-solving aspects of maintenance.

As plants become larger and more complex, the use of first-class engineering talent in maintenance becomes essential. Engineers can find both challenge and reward in a well-organized department oriented toward the goals of optimum maintenance; engineers can be effectively used as maintenance supervisors, planners, technical supervisors, maintenance specialists, or equipment problem solvers. In-depth training in maintenance methods, equipment design, and equipment operation is essential for engineers in large, complex plants. This training may be obtained in many ways, some of which are cooperation with equipment manufacturers by means of factory training programs or visits and discussions with equipment designers; study of equipment, via manufacturers' drawings and literature; study of equipment during construction and maintenance; on-the-job training; and visits to other plants.

Equipment Inventory

One of the first requirements in a preventive or predictive maintenance program is to identify the equipment that requires periodic surveillance and inspection. The frequency of this surveillance and inspection should be determined on the basis of hazard, importance to production or service, expected rate of deterioration, and extent of sensor and control installation for the equipment or system. Determining this frequency is most conveniently accomplished by establishing an equipment card system. The card should show:

1. Name of equipment and manufacturer, date of manufacture, and date of installation.

2. ASME, National, Board or serial numbers with other data attached for the equipment. These should show AWP, horsepower, kilowatt, and other relevant data.

3. In-plant use of the equipment and its location in the plant.

4. Prior inspection findings or past test results on output performance.

5. Past repairs made and the reasons for these repairs. Included in the repairs should be a description of materials, and if welding was applied, welding procedures used,

preweld and postweld heat treatment that may have been applied, and the NDT used to prove the soundness of the repair.

6. A note showing the potential effect on the service or process if the equipment has to be taken out of service.

7. Any bypasses to hooked-up spares. The same applies if a spare is available but needs to be hooked up in order to restore service. If there are no spares, the name of the nearest supplier should appear on the card with phone numbers and contacts shown.

8. Blueprints of the equipment as shown by the manufacturer, along with blueprints of associated piping and control hookups to the vessel or machinery.

9. Any other relevant data on contents, precautions to be used before starting inspections or repairs, and similar inspection problems, such as test equipment to be used.

Work Order and Checklist System

Some companies use a work order system to manage and control the expenses of maintenance work. Checklists on what to do to make the equipment available for inspection and maintenance are also a management tool to ensure that an organized effort is being applied in preventive maintenance. Management must exercise control of maintenance because it can become a significant part of operating costs. There is an increased emphasis on life extension of equipment, and the long run reliability and availability of equipment without problems are promoted by planned preventive or predictive maintenance. It has been found that a modified preventive or predictive maintenance program combined with interpretation of performance efficiency and diagnostic readings can reduce maintenance costs in the long run and thus increase output at a lower cost.

Computerized Maintenance Programs

Computerized maintenance programs are becoming more prevalent. With these, management can compare maintenance work achieved or proposed with expected expenditures much more quickly and accurately. Poor equipment performance requires more maintenance, and one result of monitoring this performance on the computer will be that poor performing equipment will not be purchased in the future.

With advances in computer technology and the subsequent reduction in system costs, computers are now being used in more preventive maintenance management systems. Multiple CRT terminals allow the operations of the various departments to be integrated more easily. When setting up any computer maintenance program, it is necessary to include a history of each piece of equipment, details of maintenance to be performed, frequency of maintenance, automatic issuing of work orders, and inventory control of spare parts. Obviously, there are many other functions that can be added, but the cost benefits may not always be there.

Part of all programs is the control and recording of labor costs and the materials used in carrying out the work. Work orders, which can be issued automatically by the computer on a weekly basis, are used to control this function. In addition, work orders should be manually issued to cover work due on unscheduled breakdowns or additional work revealed during normal maintenance and inspection. This allows a complete history of the equipment to be maintained. All costs associated with labor and materials used should be entered on the work order for entry into the computer at a later time. This allows historical data to be maintained on all equipment for use in preparing department budgets and for analysis of equipment problems. Inventory control of parts and materials should be tied into the work order procedure so that stock supply can be optimized.

No plant should be without some type of maintenance management system, which, if designed and operated correctly, will more than pay for itself within a couple of years. When management, operations, and maintenance all work as one cooperative group, facilities will function to their maximum efficiency and unscheduled outages will be minimized.

Frequency of Inspection Programs

Inspection programs are usually patterned on plant needs and the physical, chemical, and economic factors that can influence output or plant reliability. From a broad safety perspective, an inspection program can consist of the following activities:

1. Establish benchmark conditions when equipment is new or has had an extensive overhaul. Equipment record cards and construction data will assist in this initial evaluation.

2. Set up an operating inspection program to note if the equipment is operating within acceptable parameters and within the pressure and temperature ratings. This would include a check on the alarms and shutdown and pressure-relieving devices that provide protection against abnormal conditions. Log your readings and observations.

3. Review with operation and test personnel the efficiency of performance and whether the equipment may require some maintenance work to restore output efficiency. A review of pressures, temperatures, flows, and similar indicators of normal and abnormal performance will assist in determining if this may be required.

4. Review past incidents of failure and abnormal conditions, such as crack discovery, in order to develop future schedules of inspection for fatigue cracks after a given number of cycles. This will require coordination with the operating personnel and careful record keeping.

5. Perform similar reviews on such items as water treatment, concentrations of mix, etc., where past analysis of failures indicates a need for controlling these items in order to avoid, for example, stress-corrosion cracks.

6. When calculations of wear rate and records indicate that corrosion allowances may have been used up, conduct more frequent internal inspections to track this thinning action.

7. Determine the condition of connected valves during operating inspections in order to note if leakage is occurring through wearable items such as packing and valve seats.

8. Follow manufacturers' instructions; most manufacturers have explicit instructions on what inspections and testing may be required on their equipment.

9. Cooperate with the jurisdictional inspector when mandated inspection is due.

Electrical equipment needs to be inspected and tested for moisture, cleanliness, connection tightness, and deterioration caused by wear and tear from aging of insulation systems, insulating oil deterioration, and similar effects of lengthy operation. Electrical equipment has characteristics that permit detecting abnormal conditions by testing methods of various types. These may include insulation resistance tests, oil dielectric tests, circuit breaker tripping tests, and relay coordination. Abnormal temperature in operation can be due to overload, single-phasing, and/or inadequate cooling. Many mechanical inspection methods also apply to electrical equipment, such as NDT, vibration analysis, and signs of looseness of assembled parts, such as wedges or windings.

The requirements for safe and efficient operation of service equipment include establishing sound frequency of testing and maintenance programs. Qualified trained personnel should be knowledgeable about all equipment under their control so that they can recognize any dangerous or abnormal developing conditions. The ability to know when outside expertise may be required is also important. Logging will assist in noting changes from the normal. Logs can be designed for the facility's different types of equipment such as for boilers, air conditioning, generating equipment, diesels, etc. A well-trained staff in a facility will assure continuous, safe, and economical operation of the equipment comprising facility services.

EMERGENCY DISASTER PLANS

It is important for facility service operators and managers to have a planned emergency action plan, even if it only involves the steps to be taken by the facility staff for such common problems as electrical supply failure; water interruption; loss of heat, refrigeration, or air conditioning; or vital process equipment breakdowns. Emergencies also can arise from floods, hurricanes, tornadoes, earthquakes, and similar natural disasters.

The best procedure for determining the steps to be taken in planning for an emergency disaster is to incorporate the following in the plan:

1. *Hazard analysis.* The hazard analysis should identify the critical services or process equipment that could be affected and that would have a significant impact on repair costs, time to repair, and the loss of service or ability to operate.
2. The analysis should assist in identifying the *possible failure modes* to the structure, equipment, surroundings, services, and process.
3. *Written procedures* should be prepared to deal with the failure mode that may be experienced as a result of a natural or other major disaster, such as fire and explosion. It is important to provide the supervisors and operators with planned emergency procedures to be followed if a disaster should occur. Emergency procedures should incorporate
   (*a*) Escape route directions.
   (*b*) Type of alerts to be used.
   (*c*) Who to notify of the emergency. This should include name, address, and phone numbers.
   (*d*) Emergency management assignments and operator assignments should detail who is responsible for what within the facility.
   (*e*) Detailed instructions on medical and rescue operations to be followed.
   (*f*) Procedures for notifying outside services such as police, fire department, highway department, electric and gas utility suppliers, and emergency medical assistance—doctors and ambulance services.
   (*g*) Repair organization lists to help in restoring the facility or its equipment, such as repairers of boilers, turbines, piping, electrical motors, and similar facility equipment.
4. Facility service operators will normally be the *first to respond* to emergencies by closing valves, isolating breaks in piping, and/or opening electrical switches to isolate electrical faults. Operators must therefore become familiar with all pipes and valves that carry steam, water, oil, gas, or chemicals so that valves can be quickly used to isolate a pipe break and thus minimize damage. Piping diagrams of the facility will assist an operator in isolating a fault in an emergency. The steps required to secure a boiler, turbine, pump, transformer, etc., are important for facility operators who may have to respond to an emergency.

5. A similar knowledge of the electrical distribution system is necessary in order to quickly remove the electric energy from an electrical fire, flood, or other similar hazard. One-line electrical diagrams will assist in understanding the electrical distribution system of the facility.
6. Loss of electric power requires planning for:
   (*a*) Occupants' safety without electric power, especially if the premises are multi-storied.
   (*b*) Food and product spoilage and actions to take to minimize the effect of electric power failure.
   (*c*) Interruption of production—a facility plan for this possibility is an important consideration.
   (*d*) Loss of electric power can affect computer operations, water supply, heat and similar services; therefore, many side-effects are possible, and these must be considered by the facility operators and managers.

It is important to have emergency disaster plans for the facility since they may save lives, minimize the effect of injuries from a disaster, avoid more serious consequential damage, and possibly expedite restoring the facility to a productive operation again.

QUESTIONS AND ANSWERS

1. Why is it necessary to periodically defrost evaporator coils used in refrigerated storage rooms?

   Answer. It is necessary to periodically defrost the evaporator coils because the frosting acts like an insulator to prevent heat transfer and causes lower suction pressure with possible liquid return to the compressor, and the compressor experiences higher head pressure, which requires more power input. This also causes longer running time, increasing the wear and tear on the compressor and system.

2. How may the fuel feed to each cylinder of a diesel engine be checked in order to make sure the cylinders are evenly loaded?

   Answer. The most common practice is to periodically check the exhaust temperatures of each cylinder with a pyrometer. For example, some engines have permanent pyrometer hookups; therefore, readings should be taken of the exhaust temperatures per cylinder and logged on an hourly basis. Fuel oil flow to each cylinder should be adjusted so that the cylinder-exhaust temperatures are within 10°F of each other.

3. What is the purpose of crankshaft deflection gage readings on reciprocating diesel engines or compressors?

   Answer. This check is made between webs of each crank by means of dial indicators. It is a check for crank alignment that could be affected by worn bearings or misalignment. The readings in mils can be converted to imposed stress that may be caused by the misalignment; this depends on the stroke and distance between the webs of each crank.

4. What is the Charpy V-notch test and what is its purpose?

   Answer. The Charpy V-notch test is used to measure the ability of a material to resist impact loading, or brittleness. Brittle material fractures and does not "give" as does ductile material. The Charpy V-notch test consists of a pendulum-type apparatus, usually notched, that is set in motion to strike the material and the foot pounds of energy required to cause a fracture are correlated to a scale or standard so that the material can be rated as being brittle or not. Low-carbon and low-alloy steel frequently exhibit a transition zone from ductile behavior to brittle failure over a small temperature range, called the *nil-ductility temperature.*

5. What is creep, and what equipment is exposed to this hazard?

Answer. Creep is defined as the slow deformation of material with time due to operation at high temperatures even with no increase in stress. The material slowly stretches until it is no longer ductile, and unexpected brittle failure of the material can result because of long-term creep. Equipment exposed to high temperatures may include steam boilers and associated steam piping, steam turbines, and gas turbines. Periodic inspection programs for creep are an important part of loss prevention on this type of equipment.

6. What is the purpose of an electrical conductivity test of boiler water?

Answer. This test is made to measure the extent to which dissolved substances are concentrated in the boiler water in order to prevent carryover of dissolved solids into steam-using machinery, such as turbines, where the dissolved solids condense on steam turbine blades.

7. How does vanadium in a fuel oil affect boiler tubes or gas turbine hot gas-paths?

Answer. Vanadium in fuel oils has a tendency to form a fluid plastic-type ash, starting around 1100°F. The ash deposits that form on metal parts of boilers or gas turbines that are exposed to products of combustion act as an insulator, which reduces heat transfer and may cause boiler tubes to overheat locally where the plastic-type coating is heaviest. Vanadium also causes high-temperature corrosion by combining with iron while in the molten state. Vanadium deposits are hard to remove. Specifications are now available from boiler and gas turbine manufacturers on levels of vanadium their equipment can tolerate, and these should be used to specify fuel oil purchase requirements and for tests for the presence of harmful vanadium.

8. On what basis are internal inspections scheduled on gas turbines that are used in cogeneration plants?

Answer. Gas turbines have sections that are exposed to high temperatures, and these sections, such as combustors, inlet guide passages, and gas turbine blades, require more frequent inspections because of the exposure to high temperatures that quite often produce small "fire cracks" on metal surfaces. Normally it is recommended that this hot gas path be inspected after 5000 h of operation; however, this is reduced by the number of starts and stops that the turbine may have experienced. Frequent starts and stops, such as in peaking service, introduce the possibility of thermal cycling with fatigue cracks forming. The compressor section is normally inspected after 25,000 h of operation, provided the blades are periodically cleaned and past inspections showed no adverse conditions.

9. What are permissible water temperatures during hydrostatic tests of boilers or pressure vessels?

Answer. Water temperatures for hydrostatic tests should not be below 60°F in order to avoid brittle failures on some steel due to transition-temperature problems; this minimum test temperature will also prevent water vapor in the air from condensing on the boiler or pressure vessel's exterior surfaces and thus make it difficult to recognize leaks from welds and similar areas of the boiler or pressure vessel. A higher permissible water temperature, per ASME rules, is a maximum of 120°F so that parts can be safely touched and also to avoid evaporation at leak spots.

Boiler or pressure vessels installed outside in the open should not be hydrostatically tested during inclement weather that may conceal leak points.

10. What is the purpose of a repair order in maintenance work?

Answer.   The repair order system is a means of properly recording the work that has to be done in terms of material, equipment, and labor and documenting that the maintenance work is necessary. It is an excellent tool for a supervisor to use to analyze the requirements and schedule when the work should be performed. The estimate of cost can then be compared to the actual cost after the work is performed, and this information can be used for future reference and budget preparations.

11. What are the four essential features of a preventive maintenance program?

Answer.   The four features, or the scope of preventive maintenance, are: (1) conduct routine maintenance of equipment based on visual inspections and operating conditions of the equipment, (2) standardize procedures for conducting routine oiling; scale removal or cleaning; routine repairs, such as leak corrections; and establishing frequencies for these tasks; (3) maintaining a spare part inventory for wearable items so that quick repairs can be made, and (4) establishing overhaul and testing schedules with durations over 1 day for all major equipment in the facility.

12. What are some factors to consider in establishing a frequency schedule for major overhauls of equipment?

Answer.   There are many items to consider, such as (1) age of equipment, (2) past condition reports and what was found, (3) manufacturer's recommendation, (4) importance to plant operations, (5) performance with respect to efficiency and operating conditions (vibration, temperatures, pressures, etc.), (6) frequency of ON-OFF operation, load swings, and operator log readings.

13. What is sensitization of austenitic stainless steel?

Answer.   Austenitic stainless steels may be susceptible to intergranular corrosion if any heating or cooling is performed between 800 and 1650°F. This causes the chromium in the stainless steel along the grain boundaries to combine with the carbon in the steel to form chromium carbides, thereby depleting the grain boundary of chromium, which in turn makes the stainless steel less resistant to attack from a corrodent in the vessel or piping along the depleted grain boundaries. This process of chromium depletion along grain boundaries is classified as *sensitizing* the stainless steel.

Sensitization is avoided by maintaining a low carbon level in the stainless steel or by adding a strong carbide former such as titanium or columbium to the stainless steel. Low carbon is considered to be a level below 0.03 percent, and these steels are referred to as the *L* grades, such as 304L. Types 321 and 347 have the carbide formers.

14. What are some causes of black smoke coming from stacks of fuel burning equipment?

Answer.   Black smoke coming from stacks may be due to (1) improper air-fuel ratios in the combustion process, (2) fuel viscosity of oil-burning equipment not maintained correctly, (3) improper adjustment of air registers, (4) dirty oil fuel atomizer, (5) flame flutter caused by incorrectly positioned atomizer, and (6) carbon formation on the throat tile.

15. What are considered emergency repairs?

Answer. Any condition that can immediately affect the safety of the property or its occupants is considered a repair that must be performed immediately.

16. What is the purpose of thermographic inspections of electrical equipment?

Answer. Thermographic inspections are conducted on electrical equipment in order to detect conditions of high temperature, or "hot spots," that may be a sign of overload, single-phase operation on three-phase systems, loose connections, cracked and deteriorated insulation, and similar distress conditions that if not corrected could result in major electrical damage.

17. What is a *hot work* permit?

Answer. A hot work permit indicates that maintenance work is to be performed, such as welding, that could be a source for igniting combustibles in the work area, and therefore, precautions in the repair are required, including establishing a fire watch and having the means to quickly extinguish any fire. It also requires covering with flameproof mats any combustibles that are near the site of the work, including equipment below grates that could be exposed to overhead sparks from the repair procedure.

18. What effect does cycling service have on a steam turbine?

Answer. Cycling service creates temperature swings on the turbine that can produce thermal-induced repetitive stresses on shells, rotors, and steam chests, and this may produce fatigue cracks in transition sections or in sharp corners of the steam turbine.

19. What is the usual source of water induction on steam turbo-generators with extraction lines?

Answer. The most prevalent source of water induction is from leaking tubes of a feedwater heater that is supplied steam through the extraction lines and which have nonexistent or defective nonreturn valves in the extraction line.

20. What are some abnormal operating conditions that may develop on an ac generator?

Answer. Abnormal conditions that must be watched for include (1) dirty and weakened insulation on the stator or rotor, (2) stator overheating from loss of cooling medium, (3) loss of excitation, (4) phase-current unbalance, (5) reverse current flow, motorizing the generator, (6) improper voltage, which could be high or low, (7) loss of proper or rated frequency, and (8) single-phase operation of a three-phase generator.

21. What are some causes of brush sparking on dc motors and generators?

Answer. Brush sparking on dc motors and generators is a sign that corrective maintenance may be required. Among the reasons for excessive sparking are (1) excessive carbon dust buildup around brushes and holder causing conductive leakage, (2) motor or generator is operating overloaded, (3) brushes cannot move due to dirt accumulation or carbon dust buildup, which causes poor contact between the brushes and the commutator, (4) brushes are not fitted properly to the contour of the commutator, resulting in poor brush-to-commutator contact, (5) wrong grade of brushes were installed, (6) commutator mica is too high, (7) commutator bars are too high, flat, or rough, causing poor contact between the brushes and the commutator, and (8) grounded armature or field pole or short circuits within the field or armature.

22. What are some important safety controls on a chilled water system that uses a reciprocating Freon refrigeration compressor?

Answer. The following should be checked for proper operation at least annually: (1) high and low pressure Freon cutouts, (2) crankcase heater, (3) low chilled water temperature cutout, (4) low oil pressure cutout, (5) chilled water lack of flow switch, and (6) moisture indicator.

23. Why shouldn't globe valves be used for boiler blowdown application?

Answer. Globe valves have a pocket under the seat which can clog up with scale, mud, and sediment, and this prevents all the sediments from being blown out of the boiler and may even partially block the flow.

24. What factors must be considered in treating cooling tower water?

Answer. The cooling tower and its circulating water may be affected by (1) corrosion, (2) deposits and scale, (3) biological or algae growth, and (4) chemical attack on tower components, such as wood.

25. What is an oxygen-breathing apparatus?

Answer. An oxygen-breathing apparatus is used in places where an ordinary gas mask cannot be used because the oxygen content of the air is dangerously low or where an area has dangerous concentrations of poisonous or asphyxiating gases. The facepiece of an oxygen-breathing apparatus is supplied with oxygen from cylinders that are connected to it.

26. What is outside, or contract, maintenance?

Answer. Outside, or contract, maintenance is used when the facility's maintenance staff is inadequate to perform the testing and maintenance that may be required on a piece of equipment or system. However, it is still the responsibility of facility management to contract for this outside work so that the equipment or system is functional and operated efficiently and safely. Facility operators can assist outside maintenance staffs by pointing out problem areas or conditions that require corrections. This will make their jobs easier in operating the equipment in a sound manner.

Most mechanical and electrical machinery manufacturers provide service and maintenance personnel for their equipment. Some have agreements to have local, approved service organizations provide this service. It is more economical in the long run to have testing and maintenance performed by trained and qualified personnel on expensive facility service equipment such as steam and gas turbines, large refrigeration and air conditioning equipment, large motors, generators, and transformers. Manufacturers also can provide spare parts and design tolerance information to service and maintenance personnel. This information assists in determining if a part is worn excessively and should be replaced.

APPENDIX A
COMMONLY USED CONSTANTS

1 mil = 0.001 inch (in)
2000 pounds (lb) = 1 short ton (t)
2240 pounds = 1 long ton
5280 feet (ft) = 1 statute mile (mi)
6080 feet (ft) = 1 nautical mile (nmi)
Cross-sectional area in circular mils (mil) = diameter of round wire in mils squared (mil^2)
Atmosphere (standard) = 29.92 inches of mercury
Atmosphere (standard) = 14.7 pounds per square inch (lb/in^2)
1 horsepower (hp) = 746 watts (W)
1 horsepower (hp) = 33,000 foot-pounds (ft · lb) of work per minute
1 British thermal unit (Btu) = 778 foot-pounds (ft · lb)
1 cubic foot (ft^3) = 7.48 gallons (gal)
1 gallon (gal) = 231 cubic inches (in^3)
1 cubic foot (ft^3) of fresh water = 62.5 pounds (lb)
1 cubic foot (ft^3) of salt water = 64 pounds (lb)
1 foot (ft) of head water = 0.434 pound per square inch (lb/in^2)
1 inch (in) of head of mercury = 0.491 pound per square inch (lb/in^2)
1 gallon (gal) of fresh water = 8.33 pounds (lb)
1 barrel (bbl) (oil) = 42 gallons (gal)
1 long ton (t) of fresh water = 36 cubic feet (ft^3)
1 long ton (t) of salt water = 35 cubic feet (ft^3)
1 ounce (oz) (avoirdupois) = 437.5 grains (gr)
1 therm-hour = 100,000 Btu per hour (Btu/h)
1 brake horsepower (bhp) = 2544 Btu per hour (Btu/h)
1 brake horsepower = 2544/100,000 = 0.02544 therm-hour
1 therm-hour = 100,000/2544 = 39.3082 brake horsepower
1 therm-hour = 100,000/33,475 = 2.9873 boiler horsepower
Example: How many therm-hours in a 100-hp engine? 100 × 0.02544 = 2.544 therm-hours
1 boiler hp = 33,475 Btu per hour (Btu/h)
      = 34.5 pound (lb) steam per hour at 212°F
      = 139 square feet (ft^2) equivalent direct radiation (EDR)
1 EDR = 240 Btu per hour (Btu/h)
1 kW = 3413 Btu per hour (Btu/h)
3413 British thermal units (Btu) = 1 kilowatthour (kWh)
1000 watts (W) = 1 kilowatt (kW)
1.341 horsepower (hp) = 1 kilowatt (kW)
2545 Btu = 1 horsepower-hour (hp · h)
0.746 kilowatt = 1 horsepower (hp)
1 micron (micrometer) (μm) = one millionth of a meter (m) (unit of length)

APPENDIX B
QUANTITY CONVERSION TO ANOTHER QUANTITY

Quantity	Multiply quantity by	To obtain
	Lengths	
centimeters	0.03281	feet
centimeters	0.3937	inches
fathom	6.0	feet
feet	30.48	centimeters
feet	0.3048	meters
inches	2.540	centimeters
meters	100.0	centimeters
meters	3.281	feet
meters	39.37	inches
meters	1.094	yards
mils	0.001	inches
mils	0.00254	centimeters
yards	91.44	centimeters
yards	0.9144	meters
square centimeters	0.155	square inches
square centimeters	0.001076	square feet
square inches	6.452	square centimeters
square inches	645.2	square millimeters
square inches	1.273×10^6	circular mils
	Volumes	
barrels (U.S. liquid)	31.5	gallons
barrels (oil)	42.0	gallons (oil)
cubic centimeters	0.06102	cubic inches
cubic centimeters	2.642×10^{-4}	gallons (U.S. liquid)
cubic centimeters	1.057×10^{-3}	quarts (U.S. liquid)
cubic feet	28.32	liters
cubic feet	29.92	quarts (U.S. liquid)
cubic feet	1728	cubic inches
cubic feet	28.320	cubic centimeters
cubic inches	4.329×10^{-3}	gallons (U.S. liquid)
cubic inches	0.01732	quarts
cubic inches	0.01639	liters

Quantity	Multiply quantity by	To obtain
	Volumes (*Cont.*)	
gallons	3785	cubic centimeters
gallons	0.1337	cubic feet
gallons	231	cubic inches
gallons	3.785×10^{-3}	cubic meters
gallons	3.785	liters
liters	1000.0	cubic centimeters
liters	61.02	cubic inches
liters	0.2642	gallons (U.S. liquid)
liters	1.057	quarts (U.S. liquid)
pints	28.87	cubic inches
pints	0.125	gallons
pints	0.4732	liters
quarts (liquid)	57.75	cubic inches
quarts	946.4	cubic centimeters
quarts	0.25	gallons
quarts	0.9643	liters
	Weights	
grams	2.205×10^{-3}	pounds
grams	0.001	kilograms
kilograms	2.205	pounds
kilograms	1000	grams
ounces	28.35	grams
ounces	0.0625	pounds
pounds	16	ounces
pounds	453.5924	grams
	Force	
newtons	0.22481	pounds
pounds	4.44822	newtons
	Pressure	
atmospheres	29.92	inches of mercury
atmospheres	1.0333	kilograms per square centimeter
atmospheres	14.7	psi
bars (metric unit)	0.9869	atmospheres
bars (metric unit)	14.50	psi
bars (metric unit)	1.020	kilograms per square centimeter
bars (metric unit)	10	megapascals (MPa)
feet of water	0.4335	psi
feet of water	0.03048	kilograms per square centimeter
feet of water	0.8826	inches of mercury
inches of mercury	0.03342	atmospheres
inches of mercury	0.03453	kilograms per square centimeter
inches of mercury	0.4912	psi

ELECTRICAL TERMS AND HANDY FORMULAS

ELECTRICAL TERMS

Ampere = unit of current, rate of flow of electricity; I symbol is used.
Volt = unit of electromotive force to cause current flow; E or V is used.
Ohm = unit of resistance to current flow. In alternating currents, resistance is called impedance.
  For direct currents, $I = V/R$ (Ohm's law).
Megohm = 1,000,000 ohms.
Watt = unit of true power; For direct current it equals $V \times I$; For alternating current it equals
  $V \times I \times$ power factor for single-phase circuit.
Voltamperes = unit of apparent power in ac circuits.
Power factor = ratio of true power divided by apparent power.

ELECTRICAL FORMULAS

The following table lists several handy formulas for determining desired units when quantities are expressed in terms of other units. Formulas are given for direct current and for single-phase, two-phase, four-wire, and three-phase alternating current.

To Determine	Direct Current	Alternating current		
		Single-phase	Two-phase, four-wire	Three-phase
Amperes when horsepower is given	$\dfrac{\text{hp} \times 746}{V \times \text{Eff.}}$	$\dfrac{\text{hp} \times 746}{V \times \text{Eff.} \times \text{P.F.}}$	$\dfrac{\text{hp} \times 746}{2 \times V \times \text{Eff.} \times \text{P.F.}}$	$\dfrac{\text{hp} \times 746}{1.73 \times V \times \text{Eff.} \times \text{P.F.}}$
Amperes when kilowatts are given	$\dfrac{\text{kW} \times 1000}{V}$	$\dfrac{\text{kW} \times 1000}{V \times \text{P.F.}}$	$\dfrac{\text{kW} \times 1000}{2V \times \text{P.F.}}$	$\dfrac{\text{kW} \times 1000}{1.73 \times V \times \text{P.F.}}$
Amperes when kilovoltamperes are given		$\dfrac{\text{kVa} \times 1000}{V}$	$\dfrac{\text{kVa} \times 1000}{2V}$	$\dfrac{\text{kVa} \times 1000}{1.73 \times V}$
Kilowatts	$\dfrac{I \times V}{1000}$	$\dfrac{I \times V \times \text{P.F.}}{1000}$	$\dfrac{2 \times I \times V \times \text{P.F.}}{1000}$	$\dfrac{1.73 \times I \times V \times \text{P.F.}}{1000}$
Kilovoltamperes		$\dfrac{I \times V}{1000}$	$\dfrac{2 \times I \times V}{1000}$	$\dfrac{1.73 \times I \times V}{1000}$
Horsepower output	$\dfrac{I \times V \times \text{Eff.}}{746}$	$\dfrac{I \times V \times \text{Eff.} \times \text{P.F.}}{746}$	$\dfrac{2 \times I \times V \times \text{Eff.} \times \text{P.F.}}{746}$	$\dfrac{1.73 \times I \times V \times \text{Eff.} \times \text{P.F.}}{746}$

APPENDIX D
JURISDICTIONS REQUIRING OPERATOR LICENSE

Check with jurisdiction on class of license, age, and experience requirements as well as applicable equipment size and occupancy. For most the minimum age is 21.

State and cities*	Low-pressure boiler	High-pressure boiler	Steam engine, steam turbine	Refrigeration
American Requirements				
Alabama (No state license law)				
Mobile:	X	X		
Board of Engineers Examiners,				
City Hall, Mobile, AL 36602				
Alaska				
Dept. of Labor,	X	X		
650 West International Airport Rd.,				
Anchorage, AK 99502				
Arkansas				
Boiler Inspection Div.,	X	X		
Capitol Hill Bldg.,				
Little Rock, AR 72203				
California (No state license law)				
Los Angeles:	X	X	X	
Elevator & Pressure Vessel Div.,				
City Hall, Los Angeles, CA 90012				
San Jose:				
Bureau of Fire Prevention,				
476 Park Ave., San Jose, CA 95110	X	X		
Colorado (No state license law)				
Denver:	X	X	X	X
Bldg. Inspection Dept.,				
1445 Cleveland Pl.,				
Denver, CO 80202				
Pueblo:	X	X	X	
City Engineer,				
Division of Inspection,				
211 East D St.,				
Pueblo, CO 80103				

State and cities*	Low-pressure boiler	High-pressure boiler	Steam engine, steam turbine	Refrigeration
American Requirements (*Cont.*)				
Connecticut (No state license law)				
Bridgeport:	X	X	X	
Power Engineers Board of Examiners,				
City Hall, Bridgeport, CT 06115				
New Haven:				X
Board of Examiners of Engineers,				
City Hall, Church Street,				
New Haven, CT 06510				
Delaware (No state license law)				
Wilmington:		X	X	
Board of Examining Engineers,				
Public Bldg., Wilmington, DE 19801				
District of Columbia				
Occupational & Professional	X	X		
Licensing Division,				
614 H St., N.W., Washington, 20001				
Florida (No state license law)				
Tampa:	X	X	X	X
Boiler Bureau,				
301 N Florida Ave.,				
Tampa, FL 33602				
Illinois (No state license law)				
Chicago:	X	X	X	
Boiler & Pressure Vessel Inspection,				
121 North LaSalle St.,				
Chicago, IL 60602				
Decatur:		X	X	
Board of Examiners, Steam Engineers,				
Decatur, IL 62521				
East St. Louis:		X		
Boiler & Elevator Inspector,				
City Hall, East St. Louis, IL 66201				
Elgin:		X		
City Hall, Elgin, IL 60120				
Evanston:		X	X	
Building Dept., Municipal Bldg.,				
Evanston IL 60201				
Peoria:		X		
City Boiler Inspector,				
City Hall, Peoria, IL 61602				
Indiana (No state license law)				
Hammond:	X	X	X	
Board of Examiners,				
City Hall, Hammond, IN 46320				
Terre Haute:	X	X	X	
Board of Examiners, City Hall,				
Terre Haute, IN 47801				

State and cities*	Low-pressure boiler	High-pressure boiler	Steam engine, steam turbine	Refrigeration
American Requirements (*Cont.*)				
Iowa (No state license law)				
Des Moines:	X	X	X	X
Dept. Building Inspectors, City Hall, Des Moines, IA 50307				
Sioux City:	X	X	X	X
6th and Douglas Sts., Sioux City, IA 51102				
Kentucky (No state license law)				
Covington:	X	X		
Examiner of Engineers, City Hall, Covington, KY 41011				
Louisiana (No state license law)				
New Orleans:	X	X	X	X
Mechanical Inspection Section, City Hall, New Orleans, LA 70112				
Maine				
Boiler Rules & Regulations,	X	X		
State Office Bldg. Annex, Augusta, ME 04330				
Maryland				
Department of Licensing & Regulations,			X	X
203 E. Baltimore St., Baltimore, MD 21202				
Massachusetts				
Board of Boiler Rules,	X	X	X	
1010 Commonwealth Ave., Boston, MA 02215				
Michigan (No state license law)				
Dearborn:	X	X		
Bldg. & Safety Division, 4500 Maple, Dearborn, MI 48126				
Detroit:	X	X	X	
Safety Engineering, Examination Div., City-County Bldg., Detroit, MI 48226				
Grand Rapids:		X		
Inspection Services, City-County Bldg., Grand Rapids, MI 49502				
Saginaw:	X	X		
Board of Examiners, Stationary Engineers, City Hall, Saginaw, MI 48602				
Minnesota				
Division of Boiler Inspection,	X	X	X	
444 Lafayette Rd., St. Paul, MN 55101				

State and cities*	Low-pressure boiler	High-pressure boiler	Steam engine, steam turbine	Refrigeration
American Requirements (*Cont.*)				
Missouri (No state license law)				
Kansas City: Codes Administration Division, City Hall, Kansas City, MO 64106	X	X	X	
St. Joseph: Dept. of Public Works, City Hall, St. Joseph, MO 64501	X	X	X	X
St. Louis: Dept. of Public Safety, City Hall, St. Louis, MO 63103		X	X	X
Montana				
Dept. of Labor & Industry, 815 Front St., Helena, MT 59601	X	X	X	
Nebraska (No state license law)				
Lincoln: Board of Examiners, City-County Bldg., Lincoln, NE 68508		X		
Omaha: Dept. of Public Safety, City Hall, Omaha, NE 68102	X	X	X	X
New Jersey				
Mechanical Inspection Bureau, Trenton, NJ 08625	X	X	X	X
New York (No state license law)				
Buffalo: Division Fuel Devices, City Hall, Buffalo, NY 14202		X	X	X
Geneva: Board of Examining Engineers, City Hall, Geneva, NY 14456				X
Mt. Vernon: Boiler Inspector, City Hall, Mt. Vernon, NY 10500		X		X
New York City: Bureau of Examinations, Civil Service Comm., Municipal Bldg., New York, NY 10013		X	X	X
Niagara Falls: Board of Examiners, Stationary Engineers, City Hall, Niagara Falls, NY 24302		X		
Rochester: City Public Safety Bldg., Rochester, NY 14614		X	X	

State and cities*	Low-pressure boiler	High-pressure boiler	Steam engine, steam turbine	Refrigeration
American Requirements (*Cont.*)				
Tonawanda: Smoke Abatement Office, City Hall, Tonawanda, NY 14150		X	X	
White Plains: Fire Dept., Municipal Bldg., White Plains, NY 10601		X	X	
Yonkers: Board of Examiners, City Hall, Yonkers, NY 10700	X	X		
Ohio				
Division of Steam Engineers, 2323 West Fifth Ave., Columbus, OH 43216	X	X	X	
Oklahoma (No state license law)				
Oklahoma City: Boiler Inspector, City Hall, Oklahoma City, OK 73102		X		
Tulsa: Board of Examiners, 200 Civic Center, Tulsa, OK 74103	X	X	X	X
Pennsylvania (No state license law)				
Erie: Bureau of Licenses, Municipal Bldg., Erie, PA 16501		X	X	
Philadelphia: Bureau of Licenses & Inspections, Municipal Services Bldg., Philadelphia, PA 19107		X	X	X
Pittsburgh: Bureau Bldg. & Inspections, 100 Grant St., Pittsburgh, PA 15219	X	X		
Rhode Island (No state license law)				
Providence: Dept. Public Service, City Hall, Providence, RI 02903		X	X	X
Woonsocket: Licensing Steam Engineer, City Hall, Woonsocket, RI 02895		X		
Tennessee (No state license law)				
Memphis: Safety Engineer Division, 125 N. Main St., Memphis, TN 38103		X	X	X

State and cities*	Low-pressure boiler	High-pressure boiler	Steam engine, steam turbine	Refrigeration
American Requirements (*Cont.*)				
Texas (No state license law)				
Houston:	X	X		
Public Works Department Bldg.,				
Inspection Division,				
Air Condition & Boiler Section,				
Houston, TX 77001				
Utah (No state license law)				
Salt Lake City:	X	X		
Power and Heating Div.,				
City & County Bldg.,				
Salt Lake City, UT 84111				
Washington (No state license law)				
Seattle:	X	X	X	
Dept. of Bldgs.,				
503 Seattle Municipal Bldg.,				
Seattle, WA 98104				
Spokane:	X	X	X	
Dept. of Buildings,				
City Hall, Spokane, WA 99201				
Tacoma:	X	X		
Boiler Inspector,				
930 Tacoma Ave. South,				
Tacoma, WA 98402				
Wisconsin (No state license law)				
Kenosha:		X	X	
Board of Examiners,				
City Hall, Kenosha, WI 53140				
Milwaukee:	X	X	X	
Dept. Bldg. Inspection & Safety				
Engineering, Municipal Bldg.,				
Milwaukee, WI 53202				
Racine:		X	X	
Stationary Engineer Examiner,				
City Hall, Racine, WI 53203				
Canadian jurisdictional operator license requirements				
Alberta				
Labour, General Safety	X	X	X	
Services Division,				
10339-124 Street,				
Edmonton, T5N 3W1				
British Columbia				
Safety Engineering	X	X	X	
Services Division,				
501 West 12th Ave.,				
Vancouver V5Z IM6				

State and cities*	Low-pressure boiler	High-pressure boiler	Steam engine, steam turbine	Refrigeration
Canadian Requirements (*Cont.*)				
Manitoba				
Mechanical & Engineering Div., 611 Norquay Bldg., Winnipeg, R3C 0P8	X	X	X	X
New Brunswick				
Dept. of Labour, P.O. Box 6000, Fredericton, E3B 5HI	X	X	X	
Newfoundland and Labrador				
Engineering & Technical Services Div., Confederation Bldg., St. John's A1C 5T7	X	X	X	X
Northwest Territories				
Yellowknife, X0E, 1HQ	X	X	X	X
Nova Scotia				
Board of Examiners, P.O. Box 697, Halifax, B3J 2T8	X	X	X	X
Ontario				
Operating Engineers Branch, 400 University Ave., Toronto, M7A 2J9	X	X	X	X
Prince Edward Island				
Boiler Inspection Branch, Box 2000, Charlottetown, C1A 7N8	X	X	X	
Quebec				
Edifice La Laurentienne, 425 St-Amable, 3ième étage Québec, G1R 4Z1	X	X	X	X
Saskatchewan				
Boiler & Pressure Vessel Unit, 1150 Rose St., Regina, S4P 2YR	X	X	X	X
Yukon Territory				
Box 2703, Whitehorse, Y1A 2C6	X	X	X	

*If a state or city is not listed, no operator license requirements apply.

U.S. Merchant Marine Operator Licenses

U.S. Coast Guard	Class rating or license

Apply to:
Merchant Marine Licensing,
U. S. Coast Guard,
any U.S. port
	1. QMED (Qualified Member of Engine Department)
	  a. Fireman
	  b. Oiler
	  c. Water tender
	  d. Machinist
	  e. Refrigerating engineer
	  f. Electrician
	  g. Deck engineer
	  h. Junior engineer
	  i. Boilermaker
	  j. Deck and engine mechanic
	  k. Pumpman
	  l. Engineman

APPENDIX E
BIBLIOGRAPHY

ASME Boiler and Pressure Vessel Codes, Sections I through IX, American Society of Mechanical Engineers, New York.

Elonka, S. M.: *Standard Plant Operator's Manual,* McGraw-Hill, New York, 1980.

————, and A. L. Kohan: *Standard Heating and Power Boiler Plant Questions and Answers,* McGraw-Hill, New York, 1984.

————, and Q. W. Minich, *Standard Refrigeration and Air Conditioning Questions and Answers,* McGraw-Hill, New York, 1983.

Fink, Donald G. (ed.), *Standard Handbook for Electrical Engineers,* McGraw-Hill, New York, 1993.

Fundamentals of Welding, American Welding Society, Miami, Fla.

Kohan, A. L.: *Pressure Vessel Systems,* McGraw-Hill, New York, 1987.

————and Harry M. Spring, *Boiler Operator's Guide,* McGraw-Hill, New York, 1991.

National Board Inspection Code, National Board of Boiler and Pressure Vessel Inspectors, Columbus, Ohio.

National Fire Protection Codes, National Fire Protection Association, Quincy, Mass.

Power Piping Code, ANSI B31.1, American National Standards Institute, New York.

Recommended Practices for NDT Personnel Qualifications and Certification, American Society for Nondestructive Testing, Evanston, Ill.

State, County, and City Synopsis of Boiler and Pressure Vessel Laws on Design, Installation, and Reinspection Requirements, Uniform Boiler and Pressure Vessel Laws Society, Louisville, Kentucky.

INDEX

ABOUT THE AUTHOR

Anthony L. Kohan is a consultant with more than 35 years of experience as a power plant technician, tester, and insurance company inspector and manager. He was formerly the manager of the Boiler and Machinery Technical Specialists for the Royal Insurance Companies. Mr. Kohan is also the author of *Boiler Operator's Guide*, Third Edition, *Standard Heating and Power Boiler Plant Questions & Answers*, Second Edition, *Pressure Vessel Systems*, and *Standard Boiler Operators' Questions and Answers*, all available from McGraw-Hill.